This book comes with access to more content online.

Quiz yourself, track your progress,
and improve your grade!

Register your book or ebook at
www.dummies.com/go/getaccess.

Select your product, and then follow the prompts
to validate your purchase.

You'll receive an email with your PIN and instructions.

Algebra II

ALL-IN-ONE

by Mary Jane Sterling

Algebra II All-in-One For Dummies®

Published by: **John Wiley & Sons, Inc.,** 111 River Street, Hoboken, NJ 07030-5774, www.wiley.com

Published simultaneously in Canada

For general information on our other products and services, please contact our Customer Care Department within the U.S. at 877-762-2974, outside the U.S. at 317-572-3993, or fax 317-572-4002. For technical support, please visit https://hub.wiley.com/community/support/dummies.

Wiley publishes in a variety of print and electronic formats and by print-on-demand. Some material included with standard print versions of this book may not be included in e-books or in print-on-demand. If this book refers to media such as a CD or DVD that is not included in the version you purchased, you may download this material at http://booksupport.wiley.com. For more information about Wiley products, visit www.wiley.com.

Library of Congress Control Number: 2022940643

ISBN 978-1-119-89626-5 (pbk); ISBN 978-1-119-89627-2 (ebk); ISBN 978-1-119-89628-9 (ebk)

SKY10035398_072622

Contents at a Glance

Table of Contents

PART 4: MAKING LISTS AND CHECKING FOR IMAGINARY NUMBERS

Introduction

Here you are: contemplating the study of Algebra II. You just couldn't get enough of Algebra I? Good for you! You're moving along and preparing yourself for more of the fun and mystery and challenge of higher mathematics.

Algebra is really the basis of most courses that you take in high school and college. You can't do anything in calculus without a good algebra background. And there's a lot of algebra in geometry. You even need algebra in computer science! Algebra was created, modified, and continues to be tweaked so that ideas and procedures can be shared by everyone. With all people speaking the same "language," there are fewer misinterpretations.

What you find in this book is a glimpse into the way I teach: uncovering mysteries, working in historical perspectives, providing information, and introducing the topic of Algebra II with good-natured humor. Over the years, I've tried many approaches to teaching algebra, and I hope that with this book I'm helping you cope with and incorporate other teaching methods.

About This Book

Because you're interested in this book, you probably fall into one of four categories:

- You're fresh off Algebra I and feel eager to start on this new venture.
- You've been away from algebra for a while, but math has always been your main interest, so you don't want to start too far back.
- You're a parent of a student embarking on or having some trouble with an Algebra II class and you want to help.
- You're just naturally curious about science and mathematics and you want to get to the good stuff that's in Algebra II.

Whichever category you represent (and I may have missed one or two), you'll find what you need in this book. You can find some advanced algebraic topics, but I also cover the necessary basics, too. You can also find plenty of connections — the ways different algebraic topics connect with each other and the ways that algebra connects with other areas of mathematics.

After all, many other math areas drive Algebra II. Algebra is the passport to studying calculus, trigonometry, number theory, geometry, all sorts of good mathematics, and much of science. Algebra is basic, and the algebra you find here will help you grow your skills and knowledge so you can do well in math courses and possibly pursue other math topics.

Each new topic provides:

- Example problems with answers and solutions
- Practice problems with answers and solutions

Each chapter provides:

- An end-of-chapter quiz with problems representing the topics covered
- Solutions to those quiz questions

Online quizzes are also available for even more practice and confidence-building.

Foolish Assumptions

You are reading this book to learn more about algebra, so I'm assuming that you already have some of the other basic math skills: familiarity with fractions and their operations, comfort with handling decimals and the operations involved, some experience with integers (signed whole numbers) and how they operate, and some graphing knowledge — how to place points on a graphing plane. If you don't have as much knowledge as you'd like of some items mentioned, you may want to refer to some resources such as *Algebra I All In One For Dummies*, *Basic Math & Pre-Algebra For Dummies*, or *Pre-Algebra Essentials For Dummies* (John Wiley & Sons, Inc.).

My second assumption is that you're as excited about mathematics as I am. Oh, okay, you don't have to be that excited. But you're interested and eager and anxious to increase your mathematical abilities. That's the main thing you need.

Read on. Work through the book at your own pace and in the order that works for you.

Icons Used in This Book

In this book, I use these five icons to signal what's most important along the way:

Each example is an algebra question based on the discussion and explanation, followed by a step-by-step solution. Work through these examples, and then refer to them to help you solve the practice test problems at the end of the chapter.

This icon points out important information that you need to focus on. Make sure that you understand this information fully before moving on. You can skim through these icons when reading a chapter to make sure that you remember the highlights.

TIP

Tips are hints that can help speed you along when answering a question. See whether you find them useful when working on practice problems.

WARNING

This icon flags common mistakes that students make if they're not careful. Take note and proceed with caution!

YOUR TURN

When you see this icon, it's time to put on your thinking cap and work out a few practice problems on your own. The answers and detailed solutions are available so you can feel confident about your progress.

Beyond the Book

In addition to what you're reading right now, this book comes with a Cheat Sheet that provides quick access to some formulas, rules, and processes that are frequently used. To get this Cheat Sheet, simply go to www.dummies.com and type **Algebra II All in One For Dummies Cheat Sheet** in the Search box.

You'll also have access to online quizzes related to each chapter. These quizzes provide a whole new set of problems for practice and confidence-building. To access the quizzes, follow these simple steps:

1. **Register your book or ebook at Dummies.com to get your PIN.** Go to www.dummies.com/go/getaccess.
2. **Select your product from the drop-down list on that page.**
3. **Follow the prompts to validate your product, and then check your email for a confirmation message that includes your PIN and instructions for logging in.**

If you do not receive this email within two hours, please check your spam folder before contacting us through our Technical Support website at http://support.wiley.com or by phone at 877-762-2974.

Now you're ready to go! You can come back to the practice material as often as you want — simply log on with the username and password you created during your initial login. No need to enter the access code a second time.

Your registration is good for one year from the day you activate your PIN.

Where to Go from Here

This book is organized so that you can safely move from whichever chapter you choose to start with and in whatever order you like. You can strengthen skills you feel less confident in or work on those that need some attention.

If you haven't worked on any algebra recently, I'd recommend that you start out with Chapter 1 and some other chapters in the first unit. It's important to know the vocabulary and basic notation so you understand what is being presented in later chapters.

I'm so pleased that you're willing, able, and ready to begin an investigation of Algebra II. If you're so pumped up that you want to tackle the material cover to cover, great! But you don't have to read the material from page 1 to page 2 and so on. You can go straight to the topic or topics you want or need, and refer to earlier material if necessary. You can also jump ahead if so inclined. I include clear cross-references in chapters that point you to the chapter or section where you can find a particular topic — especially if it's something you need for the material you're looking at or if it extends or furthers the discussion at hand.

You can use the table of contents at the beginning of the book and the index in the back to navigate your way to the topic that you need to brush up on. Regardless of your motivation or what technique you use to jump into the book, you won't get lost because you can go in any direction from there.

Enjoy!

1

Getting to First Base with the Basics

In This Unit. . .

IN THIS CHAPTER

» Recognizing and using the rules of algebra

» Adding products to your list

» Raising your exponential power and taming radicals

» Looking at special products and factoring

Chapter 1
Beginning at the Beginning of Algebra

Algebra is a branch of mathematics that people study before they move on to other areas or branches of mathematics and science. For example, you use the processes and mechanics of algebra in calculus to complete the study of change; you use algebra in probability and statistics to study averages and expectations; and you use algebra in chemistry to solve equations involving chemicals.

Any study of science or mathematics involves rules and patterns. You approach the subject with the rules and patterns you already know, and you build on them with further study. And any discussion of algebra presumes that you're using the correct notation and terminology.

Going into a bit more detail, the basics of algebra include rules for dealing with equations, rules for using and combining terms with exponents, patterns to use when factoring expressions, and a general order for combining all of the above. In this chapter, I present these basics so you can further your study of algebra and feel confident in your algebraic ability. Refer to these rules whenever needed as you investigate the many advanced topics in algebra.

Following the Order of Operations and Other Properties

Mathematicians developed the rules and properties you use in algebra so that every student, researcher, curious scholar, and bored geek working on the same problem would get the same answer — regardless of the time or place. You don't want the rules changing on you every day; you want consistency and security, which you get from the strong algebra rules and properties presented in this section.

Ordering your operations

When mathematicians switched from words to symbols to describe mathematical processes, their goal was to make dealing with problems as simple as possible; however, at the same time, they wanted everyone to know what was meant by an expression and for everyone to get the same answer to the same problem. Along with the special notation came a special set of rules on how to handle more than one operation in an expression. For instance, if you do the problem $4+3^2-5\cdot 6+\sqrt{23-7}+\frac{14}{2}$, you have to decide when to add, subtract, multiply, divide, take the root, and deal with the exponent.

REMEMBER

The *order of operations* dictates that you follow this sequence:

1. **Perform all operations inside/on grouping symbols (parentheses, brackets, braces, fraction lines, and so on).**
2. **Raise to powers or find roots.**
3. **Multiply and divide, moving from left to right.**
4. **Add and subtract, moving from left to right.**

EXAMPLE

Q. Simplify: $4+3^2-5\cdot 6+\sqrt{23-7}+\frac{14}{2}$.

A. $\mathbf{4+9-30+4+7=-6}$**.** Follow the order of operations. The radical acts like a grouping symbol, so you subtract what's in the radical first: $4+3^2-5\cdot 6+\sqrt{16}+\frac{14}{2}$.

Raise the power and find the root: $4+9-5\cdot 6+4+\frac{14}{2}$.

Multiply and divide, working from left to right: $4+9-30+4+7$.

Add and subtract, moving from left to right: $4+9-30+4+7=-6$.

Q. Simplify: $\frac{-3 \pm \sqrt{(-3)^2 - 4(2)(-2)}}{2 \cdot 2}$.

A. $\frac{1}{2}$ or -2.

Follow the order of operations: The fraction line serves as a grouping symbol, so you simplify the numerator and denominator separately before dividing. Also, the radical is a grouping symbol, so the value under the radical is determined before finding the square root. Under the radical, first square the −3, then multiply, and finally subtract. You can then take the root.

$$\frac{-3 \pm \sqrt{(-3)^2 - 4(2)(-2)}}{2 \cdot 2} = \frac{-3 \pm \sqrt{9-(-16)}}{2 \cdot 2} = \frac{-3 \pm \sqrt{25}}{2 \cdot 2} = \frac{-3 \pm 5}{2 \cdot 2}$$

Now multiply the two numbers in the denominator: $\frac{-3 \pm 5}{2 \cdot 2} = \frac{-3 \pm 5}{4}$.

Apply the two operations, and perform those operations before dividing: $\frac{-3+5}{4} = \frac{2}{4} = \frac{1}{2}$ **or** $\frac{-3-5}{4} = \frac{-8}{4} = -2$.

YOUR TURN

1 Simplify: 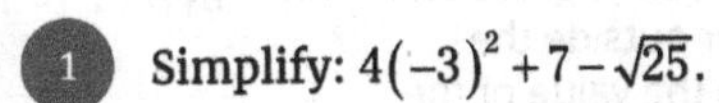 $4(-3)^2 + 7 - \sqrt{25}$.

2 Simplify: 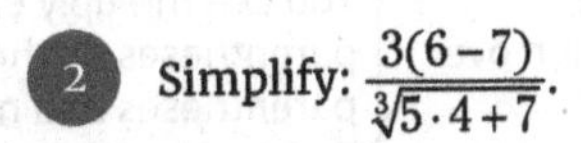 $\frac{3(6-7)}{\sqrt[3]{5 \cdot 4 + 7}}$.

Examining various properties

Mathematics has a wonderful structure that you can depend on in every situation. Most importantly, it uses various properties, such as associative, commutative, distributive, and so on. These properties allow you to adjust mathematical statements and expressions without changing their values. You make them more usable without changing the outcome.

Property	Description	Math Statement
Commutative (of addition)	You can change the order of the values in an addition statement without changing the final result.	$a+b=b+a$
Commutative (of multiplication)	You can change the order of the values in a multiplication statement without changing the final result.	$a \cdot b=b \cdot a$
Associative (of addition)	You can change the grouping of operations in an addition statement without changing the final result.	$a+(b+c)=(a+b)+c$
Associative (of multiplication)	You can change the grouping of operations in a multiplication statement without changing the final result.	$a(b \cdot c)=(a \cdot b)c$
Distributive (multiplication over addition)	You can multiply each term in an expression within parentheses by the multiplier outside the parentheses and not change the value of the expression. It takes one operation, multiplication, and spreads it out over terms that you add to one another.	$a(b+c)=a \cdot b+a \cdot c$
Distributive (multiplication over subtraction)	You can multiply each term in an expression within parentheses by the multiplier outside the parentheses and not change the value of the expression. It takes one operation, multiplication, and spreads it out over terms that you subtract from one another.	$a(b-c)=a \cdot b-a \cdot c$
Identity (of addition)	The additive identity is zero. Adding zero to a number doesn't change that number; it keeps its identity.	$a+0=0+a=a$
Identity (of multiplication)	The multiplicative identity is 1. Multiplying a number by 1 doesn't change that number; it keeps its identity.	$a \cdot 1=1 \cdot a=a$
Multiplication property of zero	The only way the product of two or more values can be zero is for at least one of the values to actually be zero.	$a \cdot b \cdot c \cdot d \cdot e \cdot f=0$ $\rightarrow a,b,c,d,e$ or $f=0$
Additive inverse	A number and its additive inverse add up to zero.	$a+(-a)=0$
Multiplicative inverse	A number and its multiplicative inverse have a product of 1. The only exception is that zero does not have a multiplicative inverse.	$a \cdot \left(\frac{1}{a}\right)=1, a \neq 0$

EXAMPLE

Q. Use the commutative and associative properties of addition to simplify the expression: $(16a+11)+5a$.

A. $\mathbf{11+21a}$. First, reverse the order of the terms in the parentheses using the commutative property: $(11+16a)+5a$.

Next, reassociate the terms: $11+(16a+5a)$.

Finally, add the two terms in the parentheses: $11+21a$.

Q. Use the distributive property to simplify the expression: $12\left(\frac{1}{2}+\frac{2}{3}-\frac{3}{4}\right)$.

A. **5.**

$$12\left(\frac{1}{2}+\frac{2}{3}-\frac{3}{4}\right)=12\cdot\frac{1}{2}+12\cdot\frac{2}{3}-12\cdot\frac{3}{4}={}^{6}\cancel{12}\cdot\frac{1}{\cancel{2}_1}+{}^{4}\cancel{12}\cdot\frac{2}{\cancel{3}_1}-{}^{3}\cancel{12}\cdot\frac{3}{\cancel{4}_1}=6+8-9=5$$

Q. Use the multiplicative identity to find a common denominator and add the terms: $\frac{3}{8}+\frac{5}{6}$.

A. $\mathbf{1\frac{5}{24}}$. Multiply the first fraction by 1 in the form of the fraction $\frac{3}{3}$, and the second fraction by $\frac{4}{4}$:

$$\frac{3}{8}+\frac{5}{6}=\frac{3}{8}\cdot\frac{3}{3}+\frac{5}{6}\cdot\frac{4}{4}=\frac{9}{24}+\frac{20}{24}=\frac{29}{24}=1\frac{5}{24}$$

Q. Use an additive inverse to solve the equation $3x+y=112$ for y.

A. $\mathbf{y=112-3x}$. Add $-3x$ to each side of the equation:

$$3x-3x+y=112-3x \rightarrow y=112-3x$$

Q. Apply the multiplication property of zero to solve the equation: $x(x+5)(x-4)=0$.

A. **−5, 4, 0.** At least one of the factors has to equal 0. So, if $x=0$, then the product is 0. Or, if $x=-5$, then the second factor is 0 and the product is 0. Likewise if $x=4$, then the third factor is 0 and the product is 0.

Q. Use the additive inverse to add and subtract a number that makes the expression x^2-10x factorable as the square of a binomial.

A. $\mathbf{(x-5)^2-25}$. Adding and subtracting 25 will do the trick:

$$x^2-10x+25-25=\left(x^2-10x+25\right)-25=(x-5)^2-25$$

YOUR TURN

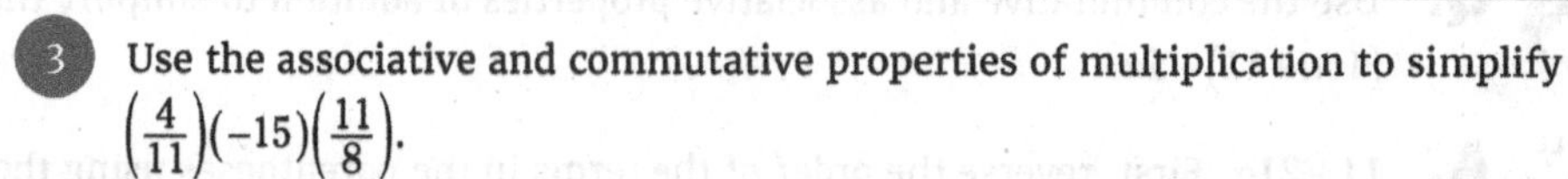

3 **Use the associative and commutative properties of multiplication to simplify $\left(\frac{4}{11}\right)(-15)\left(\frac{11}{8}\right)$.**

4 **Use the distributive property to simplify $10x\left(x-\frac{3}{5}\right)$.**

5 **Use a multiplicative inverse to solve the equation $-4x=100$ for the value of x.**

THE BIRTH OF NEGATIVE NUMBERS

In the early days of algebra, negative numbers weren't an accepted entity. Mathematicians had a hard time explaining exactly what the numbers illustrated; it was too tough to come up with concrete examples. One of the first mathematicians to accept negative numbers was Fibonacci, an Italian mathematician. When he was working on a financial problem, he saw that he needed what amounted to a negative number to finish the problem. He described it as a loss and proclaimed, "I have shown this to be insoluble unless it is conceded that the man had a debt."

Specializing in Products and "FOIL"

Multiplying numbers and variables is a pretty standard procedure in algebra, and you can take advantage of some of the rules, such as distributing and associating, to solve problems. Another timesaver is to use special situations. You don't want to slug through a messy multiplication process when there's an easier and more efficient way (and one that doesn't encourage little errors).

You've already seen the distributive property at work in the previous section. Distributing means to multiply each term in the grouping symbol by the designated factor. But what if there is more than one term to multiply by, or if there are more than two factors? This is where some nice rules step in to assist you.

First-Outer-Inner-Last

When multiplying binomials, a very nice method to use is called FOIL (First, Outer, Inner, Last). This describes the order in which you perform multiplication of the terms in the binomials before simplifying for the final answer.

When multiplying the two binomials $(a+b)(c+d)$, ac is the product of the first terms in each binomial; ad is the product of the two outer terms; bc is the product of the two inner terms; and bd is the product of the two last terms.

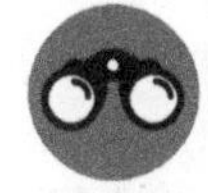

EXAMPLE

Q. Find the product of $(x-7)(x+11)$.

A. $\mathbf{x^2+4x-77}$. Using FOIL: F $=x^2$, O = $11x$, I = $-7x$, L = -77. Writing them together, you get $x^2+11x-7x-77$, which simplifies to $x^2+4x-77$.

Q. Find the product of $(2y+5)(3y+2)$.

A. $\mathbf{6y^2+19y+10}$. Using FOIL: F = $6y^2$, O = $4y$, I = $15y$, L = 10. Writing them together, you get $6y^2+4y+15y+10$, which simplifies to $6y^2+19y+10$.

YOUR TURN

6 Find the product: $(x+5)(x-8)$.

7 Find the product: $(2z-7)(3z+7)$.

8 Find the product: $(x+4y)(3x-5y)$.

Raising binomials to powers

A *binomial* is an algebraic expression that has two terms. The terms can both be variables, or they can be a variable and a number. You usually don't see two numbers, because then you could add them and you wouldn't have a binomial anymore. A common task is to raise a binomial to a power. For example:

$$(a+2)^2 = a^2+4a+4 \qquad (x+y)^3 = x^3+3x^2y+3xy^2+y^3 \qquad (z+1)^4 = z^4+4z^3+6z^2+4z+1$$

A big help in determining these products (powers) is Pascal's Triangle. This triangle gives you the coefficients when raising the binomial $(x+y)$ to a power. The first row gives the coefficient of $(x+y)^0$; the second row gives the coefficient of $(x+y)^1$; the third row gives the coefficient of $(x+y)^3$; and so on.

1
1 1
1 2 1
1 3 3 1
1 4 6 4 1
1 5 10 10 5 1
1 6 15 20 15 6 1

Q. Use Pascal's Triangle to expand $(x+y)^4$.

EXAMPLE

A. $x^4+4x^3y+6x^2y^2+4xy^3+y^4$. Go to the row with the second number 4. Write down the coefficients, leaving spaces for the variables: 1 4 6 4 1

Beginning with the first coefficient, multiply it by x^4, and then decrease the exponent by 1 as you go down the line of coefficients, ending with x^0 (or 1):

$1x^4 \quad 4x^3 \quad 6x^2 \quad 4x^1 \quad 1x^0$.

Next, beginning at the right end, multiply each term by a power of y, starting with y^4 and decreasing as you move to the left:

$1x^4y^0 \quad 4x^3y^1 \quad 6x^2y^2 \quad 4x^1y^3 \quad 1\cdot x^0y^4$.

Finally, simplify the terms and write the expression as the sum of the terms.

$x^4+4x^3y+6x^2y^2+4xy^3+y^4$

Q. Use Pascal's Triangle to expand $(x-3)^5$.

A. $x^5-15x^4+90x^3-270x^2+405x-243$. Go to the row with the second number 5. Write down the coefficients, leaving spaces for the variables: 1 5 10 10 5 1

Beginning with the first coefficient, multiply it by x^5, and then decrease the exponent by 1 as you go down the line of coefficients, ending with x^0 (or 1).

$1x^5 \quad 5x^4 \quad 10x^3 \quad 10x^2 \quad 5x^1 \quad 1x^0$

Next, beginning at the right end, multiply each term by a power of –3, starting with $(-3)^5$ and decreasing as you move to the left.

$1x^5(-3)^0 \quad 5x^4(-3)^1 \quad 10x^3(-3)^2 \quad 10x^2(-3)^3 \quad 5x^1(-3)^4 \quad 1x^0(-3)^5$

Finally, simplify the terms and write the expression as the sum of the terms.

$x^5-15x^4+90x^3-270x^2+405x-243$

YOUR TURN

9 Expand: $(z-1)^6$.

10 Expand: $(2z+3)^3$.

Expounding on Exponential Rules

Several hundred years ago, mathematicians introduced powers of variables and numbers called exponents. Exponents weren't immediately popular, however. Scholars around the world had to be convinced of their value; eventually, the quick, slick notation of exponents won over, and you benefit from their use today. Instead of writing *xxxxxxxx*, you use the exponent 8 by writing x^8. This form is easier to read and much quicker.

The expression a^n is an exponential expression with a base of a and an exponent of n. The n tells you how many times you multiply the a by itself.

You use radicals to show roots. When you see $\sqrt{16}$, you know that you're looking for the number that multiplies itself to give you 16. The answer? Four, of course. If you put a small superscript in front of the radical, you denote a cube root, a fourth root, and so on. For instance, $\sqrt[4]{81}=3$, because the number 3 multiplied by itself four times is 81. You can also replace radicals with fractional exponents — terms that make them easier to combine. Instead of using the radical with the root 4, you can write the expression as $81^{1/4}=3$. This use of exponents is very systematic and workable — thanks to the mathematicians who came before us.

Multiplying and dividing exponents

When two numbers or variables have the same base, you can multiply or divide those numbers or variables by adding or subtracting their exponents.

- $a^m \cdot a^n = a^{m+n}$: When multiplying numbers with the same base, you add the exponents.
- $\frac{a^m}{a^n} = a^{m-n}$: When dividing numbers with the same base, you subtract the exponents (numerator – denominator). And, in this case, $a \neq 0$.

REMEMBER

You must be sure that the bases of the expressions are the same. You can multiply 3^2 and 3^4, but you can't use the rule of exponents when multiplying 3^2 and 4^3.

EXAMPLE

Q. Multiply $x^4 \cdot x^5$.

A. x^9. Add the exponents: $x^{4+5}=x^9$.

Q. Divide $\frac{x^8}{x^5}$.

A. x^3. Subtract the exponents: $\frac{x^8}{x^5}=x^{8-5}=x^3$.

Getting to the roots of exponents

Radical expressions — such as square roots, cube roots, fourth roots, and so on — appear with a radical to show the root. Another way you can write these values is by using fractional exponents. You'll have an easier time combining variables with the same base if they have fractional exponents in place of radical forms.

- $\sqrt[n]{x} = x^{1/n}$: The root goes in the denominator of the fractional exponent.
- $\sqrt[n]{x^m} = x^{m/n}$: The root goes in the denominator of the fractional exponent, and the power goes in the numerator.

So, you can say $\sqrt{x} = x^{1/2}$, $\sqrt[3]{x} = x^{1/3}$, $\sqrt[4]{x} = x^{1/4}$, and so on, along with $\sqrt[5]{x^3} = x^{3/5}$.

EXAMPLE

Q. Simplify $\sqrt{x} \cdot \sqrt[4]{x}$.

A. $\sqrt[4]{x^3}$. First, change the radicals to an exponential form. Then add the exponents.

$$\sqrt{x} \cdot \sqrt[4]{x} = x^{1/2} \cdot x^{1/4} = x^{1/2 + 1/4} = x^{3/4} = \sqrt[4]{x^3}$$

Q. Simplify the radical expression $\dfrac{\sqrt[4]{x}\sqrt[6]{x^{11}}}{\sqrt[2]{x^3}}$.

A. $\sqrt[12]{x^7}$. Change the radicals to exponents and apply the rules for multiplication and division of values with the same base:

$$\frac{\sqrt[4]{x}\sqrt[6]{x^{11}}}{\sqrt[2]{x^3}} = \frac{x^{1/4} \cdot x^{11/6}}{x^{3/2}} = \frac{x^{1/4+11/6}}{x^{3/2}} = \frac{x^{3/12+22/12}}{x^{18/12}} = \frac{x^{25/12}}{x^{18/12}} = x^{25/12-18/12} = x^{7/12} = \sqrt[12]{x^7}$$

Raising or lowering the roof with exponents

You can raise numbers or variables with exponents to higher powers or reduce them to lower powers by taking roots. When raising a power to a power, you multiply the exponents. When taking the root of a power, you divide the exponents.

- $(a^m)^n = a^{m \cdot n}$: Raise a power to a power by multiplying the exponents.
- $\sqrt[m]{a^n} = \left(a^n\right)^{1/m} = a^{n/m}$: Reduce the power when taking a root by dividing the exponents.

The second rule may look familiar — it's one of the rules that govern changing from radicals to fractional exponents.

EXAMPLE

Q. Simplify $\left(x^4\right)^2$.

A. $\left(x^4\right)^2 = x^8$

Q. Apply the two rules to simplify $\sqrt[3]{\left(x^4\right)^6 \cdot x^9}$.

A. x^{11}. $\sqrt[3]{\left(x^4\right)^6 \cdot x^9} = \sqrt[3]{x^{24} \cdot x^9} = \sqrt[3]{x^{33}} = x^{33/3} = x^{11}$

Making nice with negative exponents

You use negative exponents to indicate that a number or variable belongs in the denominator of the term:

$$\frac{1}{a} = a^{-1} \text{ and } \frac{1}{a^n} = a^{-n}$$

Writing variables with negative exponents allows you to combine those variables with other factors that share the same base.

EXAMPLE

Q. Simplify $\left(x^{-5}\right)^2$.

A. $\frac{1}{x^{10}}$. Raise the power and then apply the negative exponent rule: $\left(x^{-5}\right)^2 = x^{-10} = \frac{1}{x^{10}}$.

Q. Simplify $\frac{1}{x^4} \cdot x^7 \cdot \frac{3}{x}$.

A. $3x^2$. Rewrite the fractions by using negative exponents, and then simplify by using the rules for multiplying factors with the same base:

$$\frac{1}{x^4} \cdot x^7 \cdot \frac{3}{x} = x^{-4} \cdot x^7 \cdot 3x^{-1} = 3x^{-4+7+(-1)} = 3x^2$$

YOUR TURN

Simplify the following expressions.

11 $\frac{8y^7}{2y^4}$

12 $\frac{18x^2y^3}{6x^4y^2}$

13 $x^6\left(x^2\right)^2$

14 $\sqrt[3]{4x^9y^6z^3}$

Taking On Special Operators

The operations you first learned were addition, subtraction, multiplication, and division. These are all called *binary operations*, because it takes two numbers to perform the operation. Then you learned about powers and roots, which only require one number in order to be performed. Some special operations that are important in mathematics are absolute value, greatest integer, and factorial.

Working with absolute value

The *absolute value* function takes a number and reports back its distance from 0. Basically, this means that it leaves all positive numbers positive and changes negative numbers to positive numbers. The formal rule is stated as follows:

$$|n| = \begin{cases} n, n \geq 0 \\ -n, n < 0 \end{cases}$$

EXAMPLE

Q. $|-7| =$

A. **7.** Change the negative number to a positive number. This gives you the distance from −7 to 0 on the number line: $|-7| = 7$.

Q. $|6^2 - 11| =$

A. **25.** First, perform the operations in the absolute value. Then find that absolute value.
$|6^2 - 11| = |36 - 11| = |25| = 25$

Doing great with the greatest integer function

The *greatest integer function* reports which integer the number being assessed is greater than or equal to. It is used frequently on fractions and decimals. Another way of looking at the function is to say, "On the number line, what integer is directly to the left of the number?" This is if the number isn't already an integer.

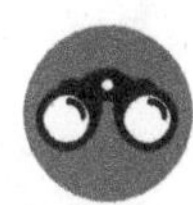

EXAMPLE

Q. $[5.8] =$

A. $[\mathbf{5.8}] = \mathbf{5}$**.** The number 5.8 is larger than 5. The integer 5 is to the left of 5.8 on the number line.

Q. $\left[-3\frac{7}{8}\right] =$

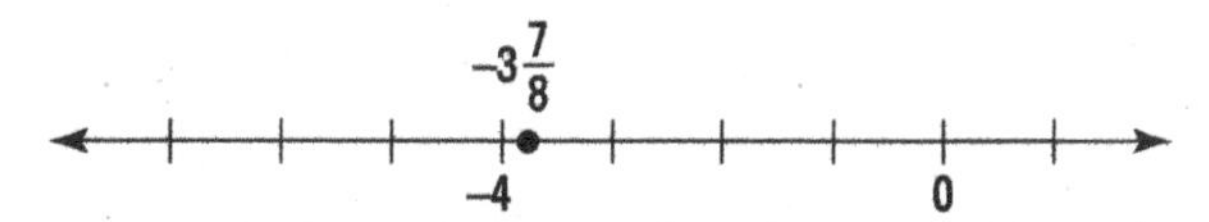

A. $\left[-3\frac{7}{8}\right] = -4$**.** Refer to the number line. The integer immediately to the left of $-3\frac{7}{8}$ is −4.

Getting the facts with factorial

The factorial function is indicated with an exclamation mark. This function tells you to take the number being operated upon and multiply it by every positive integer smaller than that number.

$$n! = n\cdot(n-1)\cdot(n-2)\cdot(n-3)\ldots3\cdot2\cdot1$$

And there's one special rule for 0!: it has been declared that $0! = 1$.

EXAMPLE

Q. $5! =$

A. **120.** $5! = 5\cdot4\cdot3\cdot2\cdot1 = 120$

Q. $\frac{8!}{6!} =$

A. **56.** $\frac{8!}{6!} = \frac{8\cdot7\cdot6!}{6!} = \frac{8\cdot7\cdot\cancel{6!}}{\cancel{6!}} = \frac{56}{1} = 56$

YOUR TURN

Evaluate each of the following.

15 $|-3.17| =$

16 $[15.9] =$

$\frac{4!}{0!} =$

Simplifying Radical Expressions

Making a radical expression simpler makes it more usable. When a number in a radical is simplified, it's easier to estimate the value of the expression (if the number isn't already a power of the root). Simplifying algebraic expressions in radicals makes them easier to factor or reduce when using the expression. Basically, you try to find some power of the root as a factor of the expression in the radical.

EXAMPLE

Q. Simplify $\sqrt{18}$.

A. $3\sqrt{2}$. First, factor the 18. Choose the factorization that gives you a perfect square as one of the choices: $\sqrt{18} = \sqrt{9 \cdot 2} = \sqrt{9} \cdot \sqrt{2} = 3\sqrt{2}$.

Q. Simplify $\sqrt{24x^3y^4}$.

A. $2xy^2\sqrt{6x}$. Write each factor under the radical as a product of a perfect square and another factor: $\sqrt{24x^3y^4} = \sqrt{4 \cdot 6 \cdot x^2 \cdot x \cdot y^4}$. Next, write a product of the roots of the perfect square factors times the remaining factors: $\sqrt{4 \cdot 6 \cdot x^2 \cdot x \cdot y^4} = \sqrt{4x^2y^4}\sqrt{6x}$.

Finally, simplify: $\sqrt{4x^2y^4}\sqrt{6x} = 2xy^2\sqrt{6x}$.

YOUR TURN

Simplify each of the following radical expressions.

18 $\sqrt{60x^{11}}$

19 $\sqrt{400y}$

20 $\dfrac{\sqrt{80x^3y^7}}{\sqrt{40xy}}$

Factoring in Some Factoring Techniques

When you factor an algebraic expression, you rewrite the sums and differences of the terms as a product. For instance, you write the three terms $x^2 - x - 42$ in factored form as $(x-7)(x+6)$. The expression changes from three terms to one big, multiplied-together term. You can factor two terms, three terms, four terms, and so on for many different purposes. Factorization comes in handy when you set the factored forms equal to zero to solve an equation. Factored numerators and denominators in fractions also make it possible to reduce the fractions.

You can think of factoring as the opposite of distributing. You have good reasons to distribute or multiply through by a value; for example, the process allows you to combine like terms and simplify expressions. Factoring out a common factor also has its purposes for solving equations and combining fractions. The different formats are equivalent; they just have different uses.

Factoring two terms

When an algebraic expression has two terms, you have four different choices for its factorization — if you can factor the expression at all. If you try the following four methods and none of them work, you can stop your attempt; you just can't factor the expression.

$ax + ay = a(x+y)$	Greatest common factor
$x^2 - a^2 = (x-a)(x+a)$	Difference of two perfect squares
$x^3 - a^3 = (x-a)(x^2+ax+a^2)$	Difference of two perfect cubes
$x^3 + a^3 = (x+a)(x^2-ax+a^2)$	Sum of two perfect cubes

In general, you check for a greatest common factor before attempting any of the other methods. By taking out the common factor, you often make the numbers smaller and more manageable, which helps you see clearly whether any other factoring is necessary.

EXAMPLE

Q. Factor $12x^2 - 15x^3$.

A. $\mathbf{3x^2(4-5x)}$. The greatest common factor of the two terms is $3x^2$: $12x^2 - 15x^3 = 3x^2(4-5x)$.

Q. Factor $100 - y^2$.

A. $\mathbf{(10-y)(10+y)}$. This is the difference of two perfect squares: $100 - y^2 = (10-y)(10+y)$.

Q. Factor $6x^4 - 6x$.

A. $\mathbf{6x(x-1)(x^2+x+1)}$. First, factor out the common factor, 6x, and then use the pattern for the difference of two perfect cubes:

$$6x^4 - 6x = 6x(x^3-1) = 6x(x-1)(x^2+x+1).$$

Q. Factor $z^4 - 81$.

A. $(z-3)(z+3)(z^2+9)$. **The expression $z^4 - 81$ is the difference of two perfect squares. When you factor it, you get $z^4 - 81 = (z^2 - 9)(z^2 + 9)$. Notice that the first factor is also the difference of two squares — you can factor again. The second term, however, is the sum of squares — you can't factor it. With perfect cubes, you can factor both differences and sums, but not with the squares. So, the factorization of $z^4 - 81$ is $(z-3)(z+3)(z^2+9)$.**

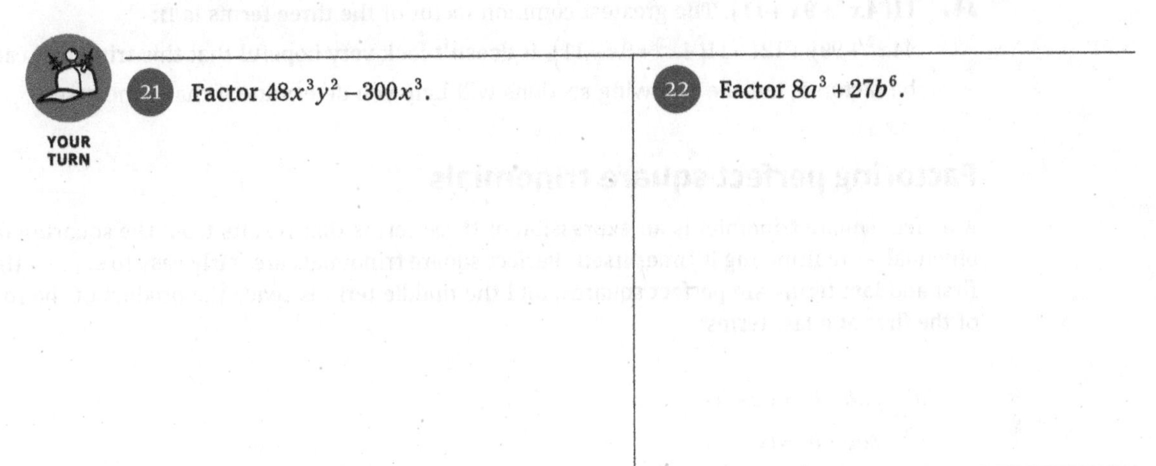

YOUR TURN

21 Factor $48x^3y^2 - 300x^3$.

22 Factor $8a^3 + 27b^6$.

Taking on three terms

A quadratic trinomial is a three-term polynomial with a term raised to the second power. When you see something like $x^2 + x + 1$, you often run through the possibilities of factoring it into the product of two binomials. In this case, you can just stop. These trinomials that crop up when factoring cubes just don't cooperate.

When a quadratic expression has three terms, making it a trinomial, you have two different ways to factor it. One method is factoring out a greatest common factor, and the other is finding two binomials whose product is identical to those three terms.

$ax + ay + az = a(x + y + z)$	Greatest common factor
$x^{2n} + (a+b)x^n + ab = (x^n + a)(x^n + b)$	Two binomials

You can often spot the greatest common factor with ease; you see a multiple of some number or variable in each term. With the product of two binomials, you either find the product or become satisfied that it doesn't exist.

Trinomials that factor into the product of two binomials have related powers on the variables in two of the terms. The relationship between the powers is that one is twice the other. When factoring a trinomial into the product of two binomials, you first look to see if you have a special product: a perfect square trinomial. If you don't, you can proceed to unFOIL. The acronym FOIL helps you multiply two binomials (First, Outer, Inner, Last); unFOIL helps you factor the product of those binomials. You find tips and techniques for factoring these products in the following sections.

EXAMPLE

Q. Factor $6x^3 - 15x^2y + 24xy^2$.

A. $\mathbf{3x(2x^2 - 5xy + 8y^2)}$. The greatest common factor of the three terms is $3x$: $6x^3 - 15x^2y + 24xy^2 = 3x(2x^2 - 5xy + 8y^2)$.

Q. Factor $44x^2 + 99y + 121$.

A. $\mathbf{11(4x^2 + 9x + 11)}$. The greatest common factor of the three terms is 11: $44x^2 + 99y + 121 = 11(4x^2 + 9x + 11)$. It doesn't look very hopeful that this trinomial can be factored, but the following sections will help you determine if that's the case.

Factoring perfect square trinomials

A perfect square trinomial is an expression of three terms that results from the squaring of a binomial — multiplying it times itself. Perfect square trinomials are fairly easy to spot — their first and last terms are perfect squares, and the middle term is twice the product of the roots of the first and last terms:

$$a^2 + 2ab + b^2 = (a+b)^2$$
$$a^2 - 2ab + b^2 = (a-b)^2$$

EXAMPLE

Q. Factor $x^2 - 20x + 100$.

A. $\mathbf{(x-10)^2}$. The first and third terms are both perfect squares. The middle term, $20x$, is twice the product of the root of x^2 and the root of 100; therefore, the factorization is $(x-10)^2$.

Q. Factor $25y^2 + 30y + 9$.

A. $\mathbf{(5y+3)^2}$. You can see that the first and last terms are perfect squares. The root of $25y^2$ is $5y$, and the root of 9 is 3. The middle term, $30y$, is twice the product of $5y$ and 3, so you have a perfect square trinomial that factors into $(5y+3)^2$.

Resorting to unFOIL

When you factor a trinomial that results from multiplying two binomials, you have to play detective and piece together the parts of the puzzle. Look at the following generalized product of binomials and the pattern that appears:

$$(ax+b)(cx+d) = acx^2 + adx + bcx + bd = acx^2 + (ad+bc)x + bd$$

So, where does FOIL come in? You need to FOIL before you can unFOIL, don't ya think? Just go back to the earlier section, "Specializing in Products and FOIL."

Now, think of every quadratic trinomial as being of the form $acx^2 + (ad+bc)x + bd$. The coefficient of the x^2 term, ac, is the product of the coefficients of the two x terms in the parentheses; the last term, bd, is the product of the two second terms in the parentheses; and the coefficient of the middle term is the sum of the outer and inner products. To factor these trinomials into the product of two binomials, you can use the opposite of FOIL, which I call *unFOIL.*

Here are the basic steps you take to unFOIL the trinomial $acx^2+(ad+bc)x+bd$:

1. **Determine all the ways you can multiply two numbers to get ac, the coefficient of the squared term.**
2. **Determine all the ways you can multiply two numbers to get bd, the constant term.**
3. **If the last term is positive, find the combination of factors from Steps 1 and 2 whose sum is that middle term; if the last term is negative, you want the combination of factors to be a difference.**
4. **Arrange your choices as binomials so that the factors line up correctly.**
5. **Insert the + and – signs to finish off the factoring and make the sign of the middle term come out right.**

EXAMPLE

Q. Factor $x^2+9x+20$.

A. $(x+4)(x+5)$. Find two terms whose product is 20 and whose sum is 9. The coefficient of the squared term is 1, so you don't have to take any other factors into consideration. You can produce the number 20 with $1\cdot 20$, $2\cdot 10$, or $4\cdot 5$. The last pair is your choice, because $4+5=9$. Arranging the factors and x's into two binomials, you get $x^2+9x+20=(x+4)(x+5)$.

Q. Factor $4y^2+5y-6$.

A. $(y+2)(4y-3)$. This is a little more complicated, because the coefficient of the y^2 term isn't 1. And this time you need a difference rather than a sum. So look at the possible factors of both that first term and the last term. For the factor 4, you can use $1\cdot 4$ or $2\cdot 2$. For the 6, you can use $1\cdot 6$ or $2\cdot 3$. You need to try out some combinations of the factors of each so that the difference between the products is 5. After some playing around, you see that if you use the $1\cdot 4$ and the $2\cdot 3$, you can create a product of 8 with the 4 and 2 and a product of 3 with the 1 and 3. Voila! You have a difference of 5. Start by setting up the binomials without signs: $(y\quad 2)(4y\quad 3)$. Then, to create $+5y$, you assign the + to the first binomial: $(y+2)(4y-3)=4y^2-3y+8y-6=4y^2+5y-6$.

Employing the "Box Method"

A very handy technique to use when factoring trinomials is called the *Box Method*. It helps you arrange the terms and their factors in such a way that the terms in the factorization just pop into view! Let me show you this method in the factorization of $3x^2+5x-12$:

1. **Create a 2-by-2 box.**

2. **Enter the first and last terms of the trinomial in the upper-left and lower-right areas.**

$3x^2$	
	-12

3. **Find the product of the two terms you entered.**

$$\left(3x^2\right)(-12) = -36x^2$$

4. **Find two factors of this product whose sum results in the middle term, 5x.**

$$(9x)(-4x) = -36x^2$$
$$9x + (-4x) = 5x$$

5. **Place the two factors in the open areas.**

$3x^2$	$9x$
$-4x$	-12

6. **Find the GCF of each row and column.**

$3x^2$	$9x$	$3x$
$-4x$	-12	-4
x	3	

7. **The terms along the right side and across the bottom are the terms in the factorization.**

$$(3x-4)(x+3) = 3x^2 + 5x - 12$$

YOUR TURN

23 Factor $x^2 - 8x + 15$.

24 Factor $12x^2 - 7x - 10$.

25 Factor $6x^2 - x - 35$.

26 Factor $x^4 + 4x^2 - 12$.

Factoring four or more terms by grouping

When four or more terms come together to form an expression, you have bigger challenges in the factoring. If you can't find a factor common to all the terms at the same time, another option is grouping. To group, you take the terms two at a time and look for common factors in each of the pairs on an individual basis. After factoring, you see if the new groupings have a common factor.

EXAMPLE

Q. Factor $x^3 - 4x^2 + 3x - 12$.

A. $\mathbf{(x-4)(x^3+3)}$. The four terms $x^3 - 4x^2 + 3x - 12$ don't have any common factors. However, the first two terms have a common factor of x^2, and the last two terms have a common factor of 3: $x^3 - 4x^2 + 3x - 12 = x^2(x-4) + 3(x-4)$. Now there are two terms, each with the common factor $(x-4)$. Factor that out of the two terms: $x^2(x-4) + 3(x-4) = (x-4)(x^3+3)$.

Q. Factor $xy^2 - 2y^2 - 5xy + 10y - 6x + 12$.

A. $\mathbf{(x-2)(y-6)(y+1)}$. The six terms don't have a common factor, but, taking them two at a time, you can pull out the factors y^2, $-5y$, and -6. Factoring by grouping, you get the following: $xy^2 - 2y^2 - 5xy + 10y - 6x + 12 = y^2(x-2) - 5y(x-2) - 6(x-2)$. The three new terms have a common factor of $(x-2)$, so the factorization becomes $(x-2)(y^2 - 5y - 6)$. The trinomial that you create also factors (see the previous section): $(x-2)(y^2 - 5y - 6) = (x-2)(y-6)(y+1)$. Factored, and ready to go!

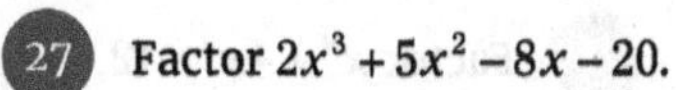

27 **Factor $2x^3 + 5x^2 - 8x - 20$.**

28 **Factor $x^3y - x^3 - 5x^2y + 5x^2 - 33xy + 33x - 27y + 27$.**

Practice Questions Answers and Explanations

1. **38.** First, evaluate the two powers. Then multiply. And, finally, add and subtract, moving left to right: $4(-3)^2+7-\sqrt{25}=4(9)+7-5=36+7-5=38$.

2. **-1.** Simplifying the numerator, first subtract in the parentheses, and then multiply by 3. In the denominator, first multiply and add under the radical, and then take the cube root. Finally, simplify the resulting fraction:

$$\frac{3(6-7)}{\sqrt[3]{5\cdot 4+7}}=\frac{3(-1)}{\sqrt[3]{5\cdot 4+7}}=\frac{-3}{\sqrt[3]{5\cdot 4+7}}=\frac{-3}{\sqrt[3]{20+7}}=\frac{-3}{\sqrt[3]{27}}=\frac{-3}{3}=-1$$

3. $\frac{-15}{2}$. Reverse the order of the last two factors. Then reduce the fractions and multiply.

$$\left(\frac{4}{11}\right)(-15)\left(\frac{11}{8}\right)=\left(\frac{4}{11}\right)\left(\frac{11}{8}\right)(-15)=\left(\frac{\not{4}}{\not{11}}\cdot\frac{\not{11}}{\not{8}_2}\right)(-15)=\left(\frac{1}{2}\right)(-15)=\frac{-15}{2}$$

4. $\mathbf{10x^2-6x}$. Multiply each term in the parentheses by $10x$.

$$10x\left(x-\frac{3}{5}\right)=10x\cdot x-10x\cdot\frac{3}{5}=10x^2-\not{10}^2x\cdot\frac{3}{\not{5}}=10x^2-6x$$

5. **-25.** Multiply each side of the equation by $-\frac{1}{4}$.

$$-\frac{1}{4}\cdot(-4)x=-\frac{1}{4}\cdot 100\rightarrow-\frac{1}{\not{4}}\cdot(-\not{4})x=-\frac{1}{\not{4}}\cdot\not{100}^{25}\rightarrow x=-25$$

6. $\mathbf{x^2-3x-40}$. Using FOIL: $(x+5)(x-8)=x^2-8x+5x-40=x^2-3x-40$.

7. $\mathbf{6z^2-7z-49}$. Using FOIL: $(2z-7)(3z+7)=6z^2+14z-21z-49=6z^2-7z-49$.

8. $\mathbf{3x^2+7xy-20y^2}$. Using FOIL: $(x+4y)(3x-5y)=3x^2-5xy+12xy-20y^2=3x^2+7xy-20y^2$.

9. $\mathbf{z^6-6z^5+15z^4-20z^3+15z^2-6z+1}$. Find the coefficients in Pascal's Triangle. Then insert decreasing powers of z moving from left to right and decreasing powers of -1 moving from right to left. Simplify the resulting terms.

$$1 \quad 6 \quad 15 \quad 20 \quad 15 \quad 6 \quad 1$$

$$1z^6 \quad 6z^5 \quad 15z^4 \quad 20z^3 \quad 15z^2 \quad 6z^1 \quad 1z^0$$

$$1z^6(-1)^0 \quad 6z^5(-1)^1 \quad 15z^4(-1)^2 \quad 20z^3(-1)^3 \quad 15z^2(-1)^4 \quad 6z^1(-1)^5 \quad 1z^0(-1)^6$$

$$z^6-6z^5+15z^4-20z^3+15z^2-6z+1$$

10. $\mathbf{8z^3+36z^2+54z+27}$. Find the coefficients in Pascal's Triangle. Then insert decreasing powers of 2z moving from left to right, and decreasing powers of 3 moving from right to left. Simplify the resulting terms.

$$1 \quad 3 \quad 3 \quad 1$$

$$1(2z)^3 \quad 3(2z)^2 \quad 3(2z)^1 \quad 1(2z)^0$$

$$1(2z)^3\cdot 3^0 \quad 3(2z)^2\cdot 3^1 \quad 3(2z)^1\cdot 3^2 \quad 1(2z)^0\cdot 3^3$$

$$8z^3+36z^2+54z+27$$

(11) $\mathbf{4y^3}$. Reduce the powers of y by subtracting: $\frac{8y^7}{2y^4} = \frac{\cancel{8}^4 y^{7-4}}{\cancel{2}} = 4y^3$.

(12) $\mathbf{\frac{3y}{x^2}}$. Reduce the powers of x and y by subtracting: $\frac{18x^2y^3}{6x^4y^2} = \frac{\cancel{18}^3 x^{2-4}y^{3-2}}{\cancel{6}} = 3x^{-2}y^1$. Rewrite the power of x as a positive number in the denominator: $3x^{-2}y^1 = \frac{3y}{x^2}$.

(13) $\mathbf{x^{10}}$. First, raise the power to a power by multiplying the exponents. Then find the product of the two factors by adding the exponents: $x^6\left(x^2\right)^2 = x^6\left(x^4\right) = x^{10}$.

(14) $\mathbf{4^{1/3}x^3y^2z}$. Change the cube root to a power of $\frac{1}{3}$. Then raise the variable factors to that power by multiplying the exponents: $\sqrt[3]{4x^9y^6z^3} = \left(4x^9y^6z^3\right)^{1/3} = (4)^{1/3}\left(x^9\right)^{1/3}\left(y^6\right)^{1/3}\left(z^3\right)^{1/3} = 4^{1/3}x^3y^2z^1$.

(15) **3.17.** Change the number to a positive value.

(16) **15.** The largest integer smaller than 15.9 is 15.

(17) **24.** Zero factorial is equal to 1: $\frac{4!}{0!} = \frac{4 \cdot 3 \cdot 2 \cdot 1}{1} = 24$.

(18) $\mathbf{2x^5\sqrt{15x}}$. Factor the number and variable, separating perfect square factors from the others. Then simplify, finding the square roots: $\sqrt{60x^{11}} = \sqrt{4 \cdot 15 \cdot x^{10} \cdot x} = \sqrt{4x^{10}} \cdot \sqrt{15x} = 2x^5\sqrt{15x}$.

(19) $\mathbf{20\sqrt{y}}$. Write the number and variable separately under radicals, and simplify: $\sqrt{400y} = \sqrt{400} \cdot \sqrt{y} = 20\sqrt{y}$.

(20) $\mathbf{xy^3\sqrt{2}}$. Write the fraction under one radical. Reduce the fraction and simplify the result: $\frac{\sqrt{80x^3y^7}}{\sqrt{40xy}} = \sqrt{\frac{80x^3y^7}{40xy}} = \sqrt{\frac{\cancel{80}^2 x^{3-1}y^{7-1}}{\cancel{40}}} = \sqrt{2x^2y^6} = xy^3\sqrt{2}$

(21) $\mathbf{12x^3(2y-5)(2y+5)}$. First, factor out the common factor $12x^3$: $48x^3y^2 - 300x^3 = 12x^3\left(4y^2 - 25\right)$. Now factor the binomial as the difference of perfect squares: $12x^3\left(4y^2 - 25\right) = 12x^3(2y-5)(2y+5)$.

(22) $\mathbf{(2a+3b^2)(4a^2-6ab^2+9b^4)}$. Use the pattern for the sum of perfect cubes: $8a^3 + 27b^6 = \left(2a + 3b^2\right)\left(4a^2 - 6ab^2 + 9b^4\right)$.

(23) $\mathbf{(x-3)(x-5)}$. Two factors of 15 are 3 and 5, and their sum is 8. They are both negative.
$x^2 - 8x + 15 = (x \quad 3)(x \quad 5) = (x-3)(x-5)$

(24) $\mathbf{(3x+2)(4x-5)}$. Using the Box Method, the two factors of $-120x^2$ that give you a difference of $-7x$ are $-15x$ and $8x$.

$12x^2$	$-15x$	$3x$
$8x$	-10	2
$4x$	-5	

(25) $\mathbf{(3x+7)(2x-5)}$. Two factors of 35 are 5 and 7. Two factors of 6 are 3 and 2. You want a difference of $-1x$: $6x^2 - x - 35 = (3x \quad 7)(2x \quad 5) = (3x+7)(2x-5)$.

(26) $\mathbf{(x^2+6)(x^2-2)}$. **Treat the trinomial like the quadratic trinomial $y^2+4y-12$:**
$x^4+4x^2-12=(x^2+6)(x^2-2)$.

(27) $\mathbf{(2x+5)(x-2)(x+2)}$. **First, factor by grouping: $2x^3+5x^2-8x-20=x^2(2x+5)-4(2x+5)=$ $(2x+5)(x^2-4)$. And now factor the difference of squares: $(2x+5)(x^2-4)=(2x+5)$ $(x-2)(x+2)$.**

(28) $\mathbf{(y-1)(x^3-5x^2-33x-27)}$. **Factor by grouping the four pairs of terms.**

$$\begin{aligned} x^3y-x^3-5x^2y+5x^2-33xy+33x-27y+27 &= x^3(y-1)-5x^2(y-1)-33x(y-1)-27(y-1) \\ &= (y-1)(x^3-5x^2-33x-27) \end{aligned}$$

The expression in the parentheses can be factored even further. See Chapter 3, for more information.

Whaddya Know? Chapter 1 Quiz

Quiz time! Complete each problem to test your knowledge on the various topics covered in this chapter. You can then find the solutions and explanations in the next section.

1. Simplify, applying the Order of Operations: $\frac{2(3-4\cdot 2)}{5} =$
2. Find the product: $(3x+10)(5x-11)$
3. Simplify: $\sqrt{200x^5y^6}$
4. Simplify, applying the Order of Operations: $-3(-2)^2+4(-2)^1+5(-2)^0 =$
5. Evaluate: $\frac{9!}{7!} =$
6. Simplify: $\left(\frac{3}{4}\right)^3 =$
7. Factor: $13x^2y-26xy^2$
8. Expand: $(x-2)^4$
9. Simplify: $\left(-\frac{1}{2}\right)^{-4} =$
10. Factor: $64-x^3$
11. Simplify: $25^{1/2} =$
12. Factor: $x^3+6x^2-2x-12$
13. Simplify: $64^{2/3} =$
14. Find the product: $(2x-z)(4x+5z)$
15. Simplify: $\frac{-27a^2b^3c}{-3ab^5c}$
16. Factor: $12x^2-31x+7$
17. Simplify: $\left(3^{-3}\right)^{-1} =$
18. Evaluate: $|7-11| =$
19. Simplify: $|6(2)-5|-|4+3(-5)| =$
20. Evaluate: $\left[-2\frac{1}{8}\right] =$
21. Factor: $64y^3-y^5$

Answers to Chapter 1 Quiz

1. **−2.** First, perform the multiplication in the parentheses and then subtract the result from 3: $\frac{2(3-4\cdot 2)}{5}=\frac{2(3-8)}{5}=\frac{2(-5)}{5}$. You can either reduce the fraction by dividing by 5 or multiply the two numbers in the numerator and then divide by 5. The result is −2.

2. $\mathbf{15x^2+17x-110}$. Using FOIL: $(3x+10)(5x-11)=15x^2-33x+50x-110=15x^2+17x-110$.

3. $\mathbf{10x^2y^3\sqrt{2x}}$. Factor the number and variables to create perfect square factors. Then write a product of two radicals and simplify:

   $\sqrt{200x^5y^6}=\sqrt{100\cdot 2\cdot x^4\cdot x\cdot y^6}=\sqrt{100x^4y^6}\cdot\sqrt{2x}=10x^2y^3\sqrt{2x}$.

4. **−15.** First, raise 2 to the indicated powers: $-3(-2)^2+4(-2)^1+5(-2)^0=-3(4)+4(-2)+5(1)$. Next, perform the multiplications: $=-12-8+5$. Now subtract and add, moving from left to right: $=-20+5=-15$.

5. **72.** Reduce the fraction: $\frac{9!}{7!}=\frac{9\cdot 8\cdot 7!}{7!}=\frac{9\cdot 8\cdot \cancel{7!}}{\cancel{7!}}=72$

6. $\mathbf{\frac{27}{64}}$. Raise both the numerator and denominator to the third power: $\left(\frac{3}{4}\right)^3=\frac{3^3}{4^3}=\frac{27}{64}$

7. $\mathbf{13xy(x-2y)}$. Factor out the greatest common factor, 13*xy*: $13x^2y-26xy^2=13xy(x-2y)$.

8. $\mathbf{x^4-8x^3+24x^2-32x+16}$. Use Pascal's Triangle and the descending powers of the two terms. Simplify the terms.

$$
\begin{array}{ccccc}
1 & 4 & 6 & 4 & 1 \\
1x^4 & 4x^3 & 6x^2 & 4x^1 & 1x^0 \\
1x^4\cdot(-2)^0 & 4x^3\cdot(-2)^1 & 6x^2\cdot(-2)^2 & 4x^1\cdot(-2)^3 & 1x^0\cdot(-2)^4
\end{array}
$$

$$x^4-8x^3+24x^2-32x+16$$

9. **16.** Change the negative exponent to a positive exponent when you write the reciprocal of the fraction: $\left(-\frac{1}{2}\right)^{-4}=\left(-\frac{2}{1}\right)^4=(-2)^4$. When raising the −2 to the fourth power, the result is a positive number: $(-2)^4=16$.

10. $\mathbf{(4-x)(16+4x+x^2)}$. Use the pattern for the difference of cubes: $64-x^3=(4-x)(16+4x+x^2)$.

11. **5.** An exponent of $1/2$ is equivalent to the square root of the number: $25^{1/2}=\sqrt{25}=5$.

12. $\mathbf{(x+6)(x^2-2)}$. Factor by grouping: $x^3+6x^2-2x-12=x^2(x+6)-2(x+6)=(x+6)(x^2-2)$.

13. **16.** The exponent of $2/3$ says to first find the cube root of the number and then square the result.

    $64^{2/3}=\left(\sqrt[3]{64}\right)^2=(4)^2=16$

14 $\mathbf{8x^2 + 6xz - 5z^2}$. Use FOIL to find the product:

$(2x - z)(4x + 5z) = 8x^2 + 10xz - 4xz - 5z^2 = 8x^2 + 6xz - 5z^2$

15 $\frac{\mathbf{9a}}{\mathbf{b^2}}$. Divide the numbers. Use division on the variables by subtracting the exponents: $\frac{-27a^2b^3c}{-3ab^5c} = 9a^{2-1}b^{3-5}c^{1-1} = 9a^1b^{-2}c^0$. Now rewrite the result, placing the variable with the negative exponent in the denominator and simplifying the other exponents: $9a^1b^{-2}c^0 = \frac{9a}{b^2}$

16 $\mathbf{(4x-1)(3x-7)}$. Using the Box Method, the two factors of $84x^2$ that give you a sum of $-31x$ are $-28x$ and $-3x$.

$12x^2$	$-28x$	$4x$
$-3x$	7	1
$3x$	7	

Both binomials will have a difference: $12x^2 - 31x + 7 = (4x-1)(3x-7)$.

17 **27.** When you raise a power to a power, you multiply the exponents: $\left(3^{-3}\right)^{-1} = 3^3 = 27$.

18 **4.** First, perform the subtraction, and then apply the absolute value to the result: $|7-11| = |-4| = 4$.

19 **−4.** First, perform the multiplications in the absolute value terms. Then perform the subtraction and addition and apply absolute value to the results:

$|6(2)-5| - |4+3(-5)| = |12-5| - |4-15| = |7| - |-11| = 7 - 11$. Finally, perform the last subtraction: $7 - 11 = -4$.

20 **−3.** This is the Greatest Integer Function. Put $-2\frac{1}{8}$ on a number line, and you see that the greatest integer (the integer immediately to the left) is −3.

21 $\mathbf{y^3(8-y)(8+y)}$. First, factor out the greatest common factor. Then factor the difference of squares: $64y^3 - y^5 = y^3\left(64 - y^2\right) = y^3(8-y)(8+y)$.

IN THIS CHAPTER

» Isolating values of x in linear equations

» Dealing with fractional terms

» Assessing absolute value in equations and inequalities

Chapter 2 Taking on Linear Equations and Inequalities

The term *linear* has the word *line* buried in it, and the obvious connection is that you can graph many linear equations as lines. But linear expressions can come in many types of packages, not just equations of lines. Add an interesting operation or two, put several first-degree terms together, throw in a funny connective, and you can construct all sorts of creative mathematical challenges. In this chapter, you find out how to deal with linear equations in many shapes and sizes, what to do with the answers in linear inequalities, and how to rewrite linear absolute value equations and inequalities so that you can solve them quickly, efficiently, and correctly.

Variables on the Side: Solving Linear Equations

Linear equations feature variables that reach only the first degree, meaning that the highest power of any variable you solve for is one. The general form of a linear equation with one variable is

$$ax + b = c$$

In this equation, the one variable is the x. The a, b, and c are coefficients and constants. (If you go to Chapter 14, you find linear equations with two or three variables.) But no matter how many variables you see, the common theme in linear equations is that each variable has only one solution or value that works in the equation.

The graph of the single solution of a linear equation, if you really want to graph it, is one point on the number line — the answer to the equation. When you up the ante to two variables in a linear equation, the graph of all the solutions (there are infinitely many) is a straight line. Any point on the line is a solution. Three variable solutions means you have a plane — a flat surface.

TIP

Generally, algebra uses the letters at the end of the alphabet for variables; the letters at the beginning of the alphabet are reserved for coefficients and constants.

Tackling basic linear equations

To solve a linear equation, you isolate the variable on one side of the equation. You do so by adding the same number to both sides — or you can subtract, multiply, or divide the same number on both sides. You want to maintain the balance; this preserves the original solution of the original version of the equation.

EXAMPLE

Q. Solve the equation $4x-7=21$ for x.

A. $x=7$. Add 7 to each side of the equation to isolate the variable term, and then divide each side by 4 to leave the variable on its own:

$$4x-7+7=21+7 \to 4x=28 \to \frac{4x}{4}=\frac{28}{4} \to x=7$$

It's always a good idea to check your answer. You do this by replacing the x with the answer in the original equation. Always check in the equation you started with.

Substituting 7 for x:

$4(7)-7=21 \to 28-7=21 \to 21=21$

When you get a true statement, you're confident of your solution.

Q. Solve for y in the equation $16y-5=7y+13$.

A. $y=2$. You want to isolate the variable terms on one side of the equation. Subtract $7y$ from each side and add 5 to each side.

$$\begin{array}{rcrcrcr} 16y & - & 5 & = & 7y & + & 13 \\ -7y & + & 5 & & -7y & + & 5 \\ \hline 9y & & & = & & & 18 \end{array}$$

Now divide each side by 9, and you find that $y=2$. Check your answer in the original equation.

$$\frac{9y}{9}=\frac{18}{9} \qquad 16(2)-5=7(2)+13$$
$$y=2 \qquad 32-5=14+13$$
$$27=27$$

Got it!

Clearing out the grouping symbols

When a linear equation has grouping symbols such as parentheses, brackets, or braces, you deal with any distributing across and simplifying within the grouping symbols before you isolate the variable.

EXAMPLE

Q. Solve the equation $5x-[3(x+2)-4(5-2x)+6]=20$ for x.

A. $x=-2$. You first distribute the 3 and –4 inside the brackets: $5x-[3x+6-20+8x+6]=20$. Then you combine the terms that you are able to and distribute the negative sign (–) in front of the brackets; it's like multiplying through by –1:

$$5x-[11x-8]=20 \rightarrow 5x-11x+8=20$$

Combine the x-terms, and solve for x by subtracting 8 from each side and then dividing each side by –6:

$$-6x+8=20 \rightarrow -6x=12 \rightarrow \frac{-6x}{-6}=\frac{12}{-6} \rightarrow x=-2$$

Checking the answer:

$$5(-2)-[3(-2+2)-4(5-2(-2))+6]=-10-[3(0)-4(5+4)+6]=20 \quad \text{Yes!}$$
$$-10-[-36+6]=-10-[-30]=-10+30=20$$

When distributing a number or negative sign over terms within a grouping symbol, make sure that you multiply every term by that value or sign. If you don't multiply each and every term, the new expression won't be equivalent to the original.

YOUR TURN

1 Solve for x: $10x+4-2x=5x-23$.

2 Solve for y: $2(3y-4)-5(2y+1)=y+2$.

Making Fractional Terms More Manageable

The problem with fractions, like cats, is that they aren't particularly easy to deal with. They always insist on having their own way, in the form of common denominators, before you can add or subtract. (Or, with cats, they get hissy.) And division? Don't get me started!

Seriously, though, the best way to deal with linear equations that involve variables tangled up with fractions is to get rid of the fractions. Your game plan is to multiply both sides of the equation by the least common denominator of all the fractions in the equation.

EXAMPLE

Q. Solve for x: $\frac{x+2}{5}+\frac{4x+2}{7}=\frac{9-x}{2}$.

A. **x = 3.** Multiply each term in the equation by 70, which is the least common denominator (also known as the least common multiple) for fractions with the denominators 5, 7, and 2:

$$^{14}\cancel{70}\left(\frac{x+2}{\cancel{5}_1}\right)+{}^{10}\cancel{70}\left(\frac{4x+2}{\cancel{7}_1}\right)={}^{35}\cancel{70}\left(\frac{9-x}{\cancel{2}_1}\right)$$

Now you distribute the reduced numbers over each set of parentheses, combine the like terms, and solve for x:

$$14(x+2)+10(4x+2)=35(9-x)\rightarrow 14x+28+40x+20=315-35x\rightarrow$$
$$54x+48=315-35x\rightarrow 89x=267\rightarrow x=3$$

And, of course, you want to check your answer. Insert $x=3$ into $\frac{x+2}{5}+\frac{4x+2}{7}=\frac{9-x}{2}$ and you get:

$$\frac{3+2}{5}+\frac{4(3)+2}{7}=\frac{9-3}{2}\rightarrow\frac{5}{5}+\frac{14}{7}=\frac{6}{2}\rightarrow 1+2=3 \text{ Checks!}$$

Q. Solve for x: $\frac{3(x+7)-5}{4}=2x+9$.

A. **x = −4.** Multiply each side of the equation by 4. Then distribute, isolate the x terms, and, finally, divide each side by −5.

$$\frac{\cancel{4}}{1}\cdot\frac{3(x+7)-5}{\cancel{4}}=4(2x+9)\rightarrow 3(x+7)-5=8x+36\rightarrow 3x+21-5=8x+36\rightarrow$$
$$3x+16=8x+36\rightarrow -5x=20\rightarrow x=-4$$

Checking the answer:

$$\frac{3(-4+7)-5}{4}=2(-4)+9$$
$$\frac{3(3)-5}{4}=-8+9 \quad \text{Checks!}$$
$$\frac{9-5}{4}=\frac{4}{4}=1$$

YOUR TURN

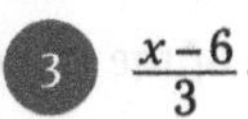

3 $\frac{x-6}{3}+\frac{x+10}{7}=2x-2$

4 $\frac{x}{3}-\frac{x}{6}+\frac{x}{2}=\frac{3x+2}{5}$

Solving for a Variable

When you see only one variable in an equation, you have a pretty clear idea what you're solving for. When you have an equation like $4x+2=11$ or $5(3z-11)+4z=15(8+z)$, you identify the one variable and start solving for it.

Life isn't always as easy as one-variable equations, however. Being able to solve an equation for some variable when it contains more than one unknown can be helpful in many situations. If you're repeating a task over and over — such as trying different widths of gardens or diameters of pools to find the best size — you can solve for one of the variables in the equation in terms of the others.

The equation $A=\frac{1}{2}h(b_1+b_2)$, for example, is the formula you use to find the area of a trapezoid. The letter A represents area, h stands for height (the distance between the two parallel bases), and the two b's are the two parallel sides called the bases of the trapezoid.

If you want to construct a trapezoid that has a set area, you need to figure out what dimensions give you that area. You'll find it easier to do the many computations if you solve for one of the components of the formula first — for h, b_1, or b_2.

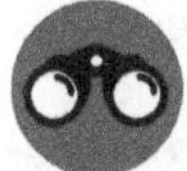

EXAMPLE

Q. Solve for h in $A=\frac{1}{2}h(b_1+b_2)$ in terms of the rest of the unknowns or letters.

A. $\frac{2A}{b_1+b_2}=h$. Multiply each side by two, which clears out the fraction, and then divide by the entire expression in the parentheses:

$$A=\frac{1}{2}h(b_1+b_2)\rightarrow 2A=\cancel{2}\cdot\frac{1}{\cancel{2}}h(b_1+b_2)\rightarrow\frac{2A}{(b_1+b_2)}=\frac{h\cancel{(b_1+b_2)}}{\cancel{(b_1+b_2)}}\rightarrow\frac{2A}{b_1+b_2}=h$$

Q. Solve for b_2 in $A = \frac{1}{2}h(b_1 + b_2)$, the measure of the second base of the trapezoid.

A. $\frac{2A}{h} - b_1 = b_2$ or $\frac{2A - b_1 h}{h} = b_2$. **Multiply each side of the equation by two, and then divide each side by** h**.**

$$A = \frac{1}{2}h(b_1 + b_2) \rightarrow 2A = \cancel{2} \cdot \frac{1}{\cancel{2}}h(b_1 + b_2) \rightarrow \frac{2A}{h} = \frac{\cancel{h}(b_1 + b_2)}{\cancel{h}} = b_1 + b_2$$

Next, subtract b_1 from each side of the equation.

$$\frac{2A}{h} - b_1 = b_2 \quad \text{or} \quad \frac{2A - b_1 h}{h} = b_2$$

When you rewrite a formula aimed at solving for a particular unknown, you can put the formula into a graphing calculator or spreadsheet to do some investigating into how changes in the individual values change the variable that you solve for (see the sidebar, "Paying off your mortgage with algebra").

YOUR TURN

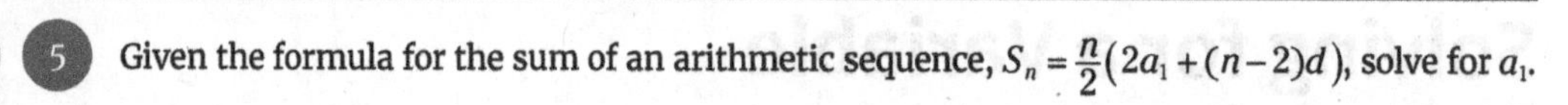

5 Given the formula for the sum of an arithmetic sequence, $S_n = \frac{n}{2}(2a_1 + (n-2)d)$, solve for a_1.

6 Given the formula for converting degrees Fahrenheit to degrees Celsius, $C° = \frac{5}{9}(F° - 32)$, solve for $F°$.

PAYING OFF YOUR MORTGAGE WITH ALGEBRA

A few years ago, one of my mathematically challenged friends asked me if I could help her figure out what would happen to her house payments if she paid $100 more each month on her mortgage. She knew that she'd pay off her house faster, and she'd pay less in interest. But how long would it take and how much would she save? I created a spreadsheet and used the formula for an amortized loan (mortgage). I made different columns showing the principal balance that remained (solved for P) and the amount of the payment going toward interest (solved for the difference), and I extended the spreadsheet down for the number of months of the loan. We put the different payment amounts into the original formula to see how they changed the total number of payments and the total amount paid. She was amazed. I was even amazed! She's paying off her mortgage much sooner than expected!

Making Linear Inequalities More Equitable

Equations — statements with equal signs — are one type of relationship or comparison between things; they say that terms, expressions, or other entities are exactly the same. An inequality is a bit less precise. Algebraic inequalities show relationships between two numbers, between a number and an expression, or between two expressions. In other words, you use inequalities for comparisons.

Inequalities in algebra are less than (<), greater than (>), less than or equal to (≤), and greater than or equal to (≥). A linear equation has only one solution, but a linear inequality has an infinite number of solutions. When you write $x \le 7$, for example, you can replace x with 6, 5, 4, −3, −100, and so on, including all the fractions that fall between the integers that work in the inequality.

REMEMBER

Here are the rules for operating on inequalities (you can replace the < symbol with any of the inequality symbols, and the rule will still hold):

- If $a < b$, then $a + c < b + c$ (adding any number c).
- If $a < b$, then $a - c < b - c$ (subtracting any number c).
- If $a < b$ and $c > 0$, then $a \cdot c < b \cdot c$ (multiplying by any *positive* number c).
- If $a < b$ and $c < 0$, then $a \cdot c > b \cdot c$ (multiplying by any *negative* number c).
- If $a < b$ and $c > 0$, then $\frac{a}{c} < \frac{b}{c}$ (dividing by any *positive* number c).
- If $a < b$ and $c < 0$, then $\frac{a}{c} > \frac{b}{c}$ (dividing by any *negative* number c).
- If $\frac{a}{c} < \frac{b}{d}$, then $\frac{c}{a} > \frac{d}{b}$ (reciprocating fractions).

Notice that the direction of the inequality changes only when multiplying or dividing by a negative number or when reciprocating (flipping) fractions. You must not multiply or divide each side of an inequality by zero. If you do so, you create an incorrect statement. You can't divide each side by 0, because you can never divide anything by 0 — no such number with 0 in the denominator exists.

Solving linear inequalities

To solve a basic linear inequality, you first move all the variable terms to one side of the inequality and the numbers to the other. After you simplify the inequality down to a variable and a number, you can find out what values of the variable will make the inequality into a true statement.

EXAMPLE

Q. Solve for x: $3x+4>11-4x$.

A. **$x>1$. Add 4x to each side and subtract 4 from each side. The inequality sign stays the same because no multiplication or division by negative numbers is involved. Now you have $7x>7$. Dividing each side by 7 also leaves the sense (direction of the inequality) untouched because 7 is a positive number. Your final solution is $x>1$. The answer says that any number larger than 1 can replace the x's in the original inequality and make the inequality into a true statement.**

Q. Solve for x: $4(x-3)-2\geq 3(2x+1)+7$.

A. **$x\leq -12$. The inequality has grouping symbols that you have to deal with. Distribute the 4 and 3 through their respective multipliers to make the inequality into $4x-12-2\geq 6x+3+7$. Simplify the terms on each side to get $4x-14\geq 6x+10$. Now you put your inequality skills to work. Subtract 6x from each side and add 14 to each side; the inequality becomes $-2x\geq 24$. When you divide each side by −2, you have to reverse the sense; you get the answer $x\leq -12$. Only numbers smaller than −12, or −12 itself, work in the original inequality.**

To check this out (with just one number), I randomly chose the number −20. Substituting this for x:

$$\begin{aligned}4(-20-3)-2&\geq 3(2(-20)+1)+7\\4(-23)-2&\geq 3(-39)+7\\-92-2&\geq -117+7\\-94&\geq -110\end{aligned}$$

Checking inequality solutions like this isn't really conclusive, because there are too many possible solutions. The main reason I do a "spot check" like this is to be sure my inequality symbol is facing in the correct direction.

YOUR TURN

7 Solve for x: $4-3x\geq 2x-1$.

8 Solve for x: $9x-3(2x+1)<5x-7$.

Introducing interval notation

You can alleviate the awkwardness of writing answers with inequality notation by using another format called interval notation. You use interval notation extensively in calculus, where you're constantly looking at different intervals involving the same function. Much of higher mathematics uses interval notation, although I really suspect that book publishers pushed its use because it's quicker and neater than inequality notation. Interval notation uses parentheses, brackets, commas, and the infinity symbol to bring clarity to the murky inequality waters.

Here are the rules for using interval notation:

- » You order any numbers used in the notation with the smaller number to the left of the larger number.
- » You indicate "or equal to" by using a bracket.
- » If the solution doesn't include the end number, you use a parenthesis.
- » When the interval doesn't end (it goes up to positive infinity or down to negative infinity), use $+\infty$ or $-\infty$, whichever is appropriate, and a parenthesis.

Q. Write $x < 3$ using interval notation.

A. $(-\infty, 3)$. The numbers come up from negative infinity and stop right before the number 3. Use the parenthesis to show that 3 is not included. The notation is $(-\infty, 3)$.

Q. Write $x \geq -2$ using interval notation.

A. $[-2, \infty)$. The numbers begin at 2 (including the 2) and continue on to positive infinity. The notation is $[-2, \infty)$.

Q. Write $4 \leq x < 9$ using interval notation.

A. $[4, 9)$. The numbers start at 4 (including the 4) and stop right before the number 9. Use the bracket to include the 4 and the parenthesis to show that 9 is not included. The notation is $[4, 9)$.

Q. Write $-3 < x < 7$ using interval notation.

A. $(-3, 7)$. The numbers start right after the number −3 and move up until right before 7. Use the parentheses to show that neither number is included. The notation is $(-3, 7)$. Taken out of context, how do you know if $(-3, 7)$ represents the interval containing all the numbers between −3 and 7 or if it represents the point $(-3, 7)$ on the coordinate plane? You can't tell. You consider the context. A problem containing such notation has to give you some sort of hint as to what it's trying to tell you.

YOUR TURN

9 **Solve for x and write your solution in interval notation: $4x-11 \le 2x+13$**

10 **Solve for x and write your solution in interval notation: $\frac{x+3}{4} > 3-2x$**

Compounding the Situation with Compound Statements

A compound inequality is an inequality with more than one comparison or inequality symbol — for instance, $-2 < x \le 5$. To solve compound inequalities for the value of the variables, you use the same inequality rules that you used in the previous section, and you expand the rules to apply to each part of the equation (intervals separated by inequality symbols).

EXAMPLE

Q. **Solve for x and write your solution in interval notation: $-8 \le 3x-5 < 10$**

A. $[-1,5)$. First, add 5 to each of the three sections. Then divide each section by 3.

$$\begin{array}{rcrcrcr} -8 & \le & 3x & - & 5 & < & 10 \\ +5 & & & + & 5 & & +5 \\ \hline -3 & \le & 3x & & & < & 15 \end{array} \rightarrow \begin{array}{ccccc} \frac{-3}{3} & \le & \frac{3x}{3} & < & \frac{15}{3} \\ -1 & \le & x & < & 5 \end{array}$$

In interval notation, this solution is written as $[-1,5)$.

Q. Solve for x and write your solution in interval notation: $-1 < 5 - 2x \le 7$

A. **[−1,3).** Subtract 5 from each section and then divide each section by −2. Of course, dividing by a negative means that you turn the senses around:

$$\begin{array}{rcrcrcr} -1 & < & 5 & - & 2x & \le & 7 \\ -5 & & -5 & & & & -5 \\ \hline -6 & < & & - & 2x & \le & 2 \end{array} \quad \rightarrow \quad \begin{array}{ccccc} \frac{-6}{-2} & \le & \frac{-2x}{-2} & < & \frac{2}{-2} \\ 3 & > & x & \ge & -1 \end{array}$$

You write the answer, $3 > x \ge -1$, which is backward compared to the order of the numbers on the number line; the number −1 is smaller than 3. To flip the inequality and its numbers in the opposite direction, you reverse the inequality symbols, too: $-1 \le x < 3$. In interval notation, you write the answer as $[-1, 3)$.

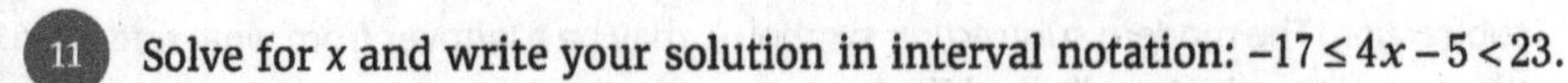

YOUR TURN

11 Solve for x and write your solution in interval notation: $-17 \le 4x - 5 < 23$.

12 Solve for x and write your solution in interval notation: $3 \le 6 - x \le 7$.

ANCIENT SYMBOLS FOR TIMELESS OPERATIONS

Many ancient cultures used their own symbols for mathematical operations, and the cultures that followed altered or modernized the symbols for their own use. You can see one of the first symbols used for addition in the following figure, located on the far left — a version of the Italian capital P for the word *piu*, meaning plus. Niccolò Tartaglia, a self-taught 16th-century Italian mathematician, regularly used this symbol for addition. The modern plus symbol, +, is probably a shortened form of the Latin word *et*, meaning and.

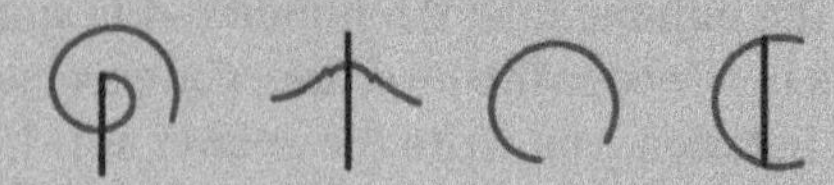

The second figure from the left was used in ancient Greek times by mathematician Diophantus for subtraction. The modern subtraction symbol, –, may be a leftover from what traders in medieval times used in order to indicate differences in product weights.

Gottfried Leibniz, a child prodigy from the 17th century who taught himself Latin, preferred the third symbol from the left for multiplication. One modern multiplication symbol, ×, is based on St. Andrew's Cross, but Leibniz used the open circle because he thought that the modern symbol looked too much like the unknown *x*.

The symbol on the far right is a somewhat backward D, used in the 18th century by French mathematician Gallimard for division. The modern division symbol, ÷, may come from a fraction line with dots added above and below.

Dealing with Linear Absolute Value Equations and Inequalities

When you perform an absolute value operation, you're not performing surgery at bargain-basement prices; you're taking a number inserted between the absolute value bars, $|a|$, and recording the distance of that number from zero on the number line. For instance $|3| = 3$, because 3 is three units away from zero. On the other hand, $|-4| = 4$, because –4 is four units away from zero.

The absolute value of a is defined as $|a| = \begin{cases} a \text{ if } a \geq 0 \\ -a \text{ if } a < 0 \end{cases}$. You read the definition as follows: "The absolute value of a is equal to a, itself, if a is positive or zero; the absolute value of a is equal to the opposite of a if a is negative."

Being absolutely fair with equations

A linear absolute value equation is an equation that takes the form $|ax + b| = c$. Taking the equation at face value, you don't know if you should change what's in between the bars to its opposite, because you don't know if the expression is positive or negative. The sign of the expression inside the absolute value bars all depends on the size and sign of the variable x. To solve an absolute value equation in this linear form, you have to consider both possibilities: $ax+b$ may be positive or it may be negative. To solve for the variable x in $|ax+b|=c$, you solve both $ax+b=c$ and $ax+b=-c$.

One restriction that you should be aware of when applying the rule for changing from absolute value to individual linear equations is that the absolute value term has to be alone on one side of the equation.

EXAMPLE

Q. Solve the absolute value equation $|4x+5|=13$.

A. 2 and $-\frac{9}{2}$. Write the two linear equations and solve each for x by subtracting 5 and dividing by 4.

If $4x+5=13$, then $4x=8$ and $x=2$.

If $4x+5=-13$, then $4x=-18$ and $x=\frac{-18}{4}=-\frac{9}{2}$.

You have two solutions: 2 and $-\frac{9}{2}$. Both solutions work when you replace the x in the original equation with their values.

Q. Solve the absolute value equation $3|4-3x|+7=25$.

A. $-\frac{2}{3}$ and $\frac{10}{3}$. Before writing the two linear equations, first subtract 7 from each side of the equation and then divide each side by 3. This leaves the absolute value alone on one side of the equation. $3|4-3x|=18$ becomes $|4-3x|=6$. Now you can write the two linear equations and solve them for x:

If $4-3x=6$, then $-3x=2$ and $x=-\frac{2}{3}$.

If $4-3x=-6$, then $-3x=-10$ and $x=\frac{10}{3}$.

YOUR TURN

 Solve for x: $|3x-7|=10$.

 Solve for x: $2|2x+3|-7=11$.

Seeing through absolute value inequality

An absolute value inequality contains both an absolute value, $|a|$, and an inequality: $<$, $>$, $\leq$, or $\geq$.

To solve an absolute value inequality, you have to change from absolute value inequality form to just plain inequality form. The way to handle the change from absolute value notation to

inequality notation depends on which direction the inequality points with respect to the absolute value term. The methods, depending on the direction, are quite different:

- To solve for x in $|ax+b|<c$ you solve $-c<ax+b<c$.
- To solve for x in $|ax+b|>c$, you solve both $ax+b>c$ and $ax+b<-c$.

The first change sandwiches the $ax+b$ between c and its opposite. The second change considers values greater than c (toward positive infinity) and smaller than $-c$ (toward negative infinity).

EXAMPLE

Q. Solve $|2x-1|\le 5$.

A. **[−2, 3]. You apply the first rule of solving absolute value inequalities to the inequality, because of the less-than direction of the inequality. You rewrite the inequality, using the rule for changing the format: $-5\le 2x-1\le 5$. Next, you add 1 to each section to isolate the variable; you get the inequality $-4\le 2x\le 6$. Divide each section by 2 to get $-2\le x\le 3$. You can write the solution in interval notation as [−2, 3].**

Q. Solve $2|3x+5|-7<11$.

A. $\left(-\frac{14}{3},\frac{4}{3}\right)$**. Be sure that the absolute value inequality is in the correct format before you apply the rule. The absolute value portion should be alone on its side of the inequality sign. With $2|3x+5|-7<11$, you need to add 7 to each side and divide each side by 2 before changing the form. When you perform these operations, the inequality becomes $|3x+5|<9$. Then, when you apply the first rule, the problem becomes $-9<3x+5<9$. Subtracting 5 from each interval gives you $-14<3x<4$. Then, dividing each interval by 3, you have $-\frac{14}{3}<x<\frac{4}{3}$. In interval notation, the answer is written $\left(-\frac{14}{3},\frac{4}{3}\right)$.**

Q. Solve $|7-2x|>11$.

A. **(−∞, −2) or (9, ∞). An absolute value inequality with a greater-than sign has solutions that go infinitely high to the right and infinitely low to the left on the number line. To solve for the values that work, you rewrite $|7-2x|>11$ using the rule for greater-than inequalities; you get two completely separate inequalities to solve: $7-2x>11$ and the inequality $7-2x<-11$. Notice that when the sign of the value 11 changes from positive to negative, the inequality symbol switches direction.**

When solving the two inequalities, be sure to remember to switch the sign when you divide by −2:

Solving $7-2x>11$, you first subtract 7 from each side to get $-2x>4$. Dividing by −2, you have $x<-2$.

Solving $7-2x<-11$, you subtract 7 to get $-2x<-18$. Dividing by −2, you have $x>9$. The solution of the absolute value inequality is $x<-2$ or $x>9$. In interval notation, you write the solution as (−∞, −2) or (9, ∞).

Q. Solve $2|3x-7|+8<6$.

A. **No solution.** Start by subtracting 8 from each side and then dividing each side by 2. The dividing value is positive, so you don't reverse the sense. After performing the initial steps, you use the rule where you change from an absolute value inequality to an inequality with the variable term sandwiched between inequalities. But watch it! This is a "problem child."

Subtracting 8, you get $2|3x-7|<-2$, and dividing by 2, you get $|3x-7|<-1$. Do you see the problem yet? This says that the absolute value of some number is less than –1. That just can't be. This problem has no solution. But let's say that you are in a hurry and don't notice this yet.

Under the format $-c<ax+b<c$, the inequality looks curious. Do you sandwich the variable term between –1 and 1 or between 1 and –1 (the first number on the left, and the second number on the right)? It turns out that neither works. First of all, you can throw out the option of writing $1<3x-7<-1$. Nothing is bigger than 1 and smaller than –1 at the same time. The other version seems, at first, to have possibilities, so you try to solve $-1<3x-7<1$ by adding 7 to each interval, giving you $6<3x<8$. Dividing each interval by 3, you have $2<x<\frac{8}{3}$.

The solution says that x is a number between 2 and $2\frac{2}{3}$. And this is where checking your answers can really pay off. If you check the solution by trying a number — say, 2.1 — in the original inequality, you get the following:

$$2|3(2.1)-7|+8<6 \rightarrow 2|6.3-7|+8<6 \rightarrow 2|-.7|+8<6 \rightarrow$$
$$2(.7)+8<6 \rightarrow 1.4+8<6 \rightarrow 9.4<6$$

Because 9.4 isn't less than 6, you know the number 2.1 doesn't work. You won't find any number that works. So, you can't find an answer to this problem. It's always better if you notice that impossibility early on — but you can still be saved here.

YOUR TURN

15 Solve $|8-3x|\le 20$.

16 Solve $3|4x+1|-1>74$.

Practice Questions Answers and Explanations

1 **-9.** **Combine the two variable terms on the left to get 8*x*. Then subtract 5*x* and 4 from each side.**

$$\begin{array}{rcrcrcr} 8x & + & 4 & = & 5x & - & 23 \\ -5x & - & 4 & & -5x & - & 4 \\ \hline 3x & & & = & & & -27 \end{array}$$

Dividing each side of the equation by 3, you have $x = -9$.

2 **-3.** **First, distribute the values over the parentheses and combine like terms.**

$$\begin{aligned} 2(3y-4)-5(2y+1) &= y+2 \\ 6y-8 \quad -10y-5 &= y+2 \\ -4y-13 &= y+2 \end{aligned}$$

Subtract *y* from each side and add 13 to each side.

$$\begin{array}{rcrcrcr} -4y & - & 13 & = & y & + & 2 \\ -y & + & 13 & & -y & + & 13 \\ \hline -5y & & & = & & & 15 \end{array}$$

Dividing each side by –5, you have $y = -3$.

3 $\frac{15}{16}$**.** **Multiply both sides of the equation by 21. Then distribute over the parentheses.**

$$\begin{aligned} \cancel{21}^{7} \cdot \frac{x-6}{\cancel{3}} + \cancel{21}^{3} \cdot \frac{x+10}{\cancel{7}} &= 21(2x-2) \\ 7(x-6)+3(x+10) &= 21(2x-2) \\ 7x-42+3x+30 &= 42x-42 \end{aligned}$$

Combine the like terms on the left.

$$10x - 12 = 42x - 42$$

Subtract 10*x* from each side and add 42 to each side.

$$\begin{array}{rcrcrcr} 10x & - & 12 & = & 42x & - & 42 \\ -10x & + & 42 & & -10x & + & 42 \\ \hline & & 30 & = & 32x & & \end{array}$$

Now, divide each side of the equation by 32 and reduce the fraction to get: $\frac{30}{32} = \frac{\cancel{32}x}{\cancel{32}}$, $x = \frac{15}{16}$

4 **6.** **Multiply each term by 30. Combine like terms on the left and distribute on the right.**

$$\begin{aligned} \cancel{30}^{10} \cdot \frac{x}{\cancel{3}} - \cancel{30}^{5} \cdot \frac{x}{\cancel{6}} + \cancel{30}^{15} \cdot \frac{x}{\cancel{2}} &= \cancel{30}^{6} \cdot \frac{3x+2}{\cancel{5}} \\ 10x - 5x + 15x &= 6(3x+2) \\ 20x &= 18x + 12 \end{aligned}$$

Subtract $18x$ from each side and then divide each side by 2.

$$20x = 18x + 12 \rightarrow 2x = 12 \rightarrow x = 6$$

5) $\mathbf{a_1 = \frac{2S_n - n(n-2)d}{2n}}$. First, multiply each side by 2, and then divide each side by n.

$$2 \cdot S_n = \cancel{2} \cdot \frac{n}{\cancel{2}}(2a_1 + (n-2)d)$$

$$2S_n = n(2a_1 + (n-2)d)$$

$$\frac{2S_n}{n} = \frac{\cancel{n}(2a_1 + (n-2)d)}{\cancel{n}}$$

$$\frac{2S_n}{n} = 2a_1 + (n-2)d$$

Subtract $(n-2)d$ from each side, and then divide by 2.

$$\frac{2S_n}{n} = 2a_1 + (n-2)d \rightarrow \frac{2S_n}{n} - (n-2)d = 2a_1 \rightarrow \frac{\frac{2S_n}{n} - (n-2)d}{2} = \frac{\cancel{2}a_1}{\cancel{2}} \rightarrow \frac{\frac{2S_n}{n} - (n-2)d}{2} = a_1$$

Simplify the fraction on the left by writing the second term in the numerator with a denominator of n and combining the terms.

$$\frac{\frac{2S_n}{n} - \frac{n(n-2)d}{n}}{2} \rightarrow \frac{\frac{2S_n - n(n-2)d}{n}}{2} \rightarrow \frac{2S_n - n(n-2)d}{2n} = a_1$$

The numerator can be further simplified, but this is probably the handiest version to have available.

6) $\mathbf{F° = \frac{9}{5}C° + 32}$. First, multiply both sides by $\frac{9}{5}$, and then add 32 to both sides.

$$\frac{9}{5}C° = \frac{\cancel{9}}{\cancel{5}} \cdot \frac{\cancel{5}}{\cancel{9}}(F° - 32) \rightarrow \frac{9}{5}C° = F° - 32 \rightarrow \frac{9}{5}C° + 32 = F°$$

7) $\mathbf{x \leq 1}$. Subtract $2x$ and 4 from both sides.

$$\begin{array}{rcrcrcr} 4 & - & 3x & \geq & 2x & - & 1 \\ -4 & - & 2x & & -2x & - & 4 \\ \hline & - & 5x & \geq & & - & 5 \end{array}$$

Now divide both sides of the inequality by -5; because this is a negative divisor, reverse the direction of the inequality.

$$-5x \geq -5 \rightarrow \frac{-5x}{-5} \leq \frac{-5}{-5} \rightarrow x \leq 1$$

8) $\mathbf{x > 2}$. First, distribute on the left side and combine like terms.

$$9x - 6x - 3 < 5x - 7 \rightarrow 3x - 3 < 5x - 7$$

Subtract $3x$ from each side and add 7 to each side.

$$\begin{array}{rrrrrrr} 3x & - & 3 & < & 5x & - & 7 \\ -3x & + & 7 & & -3x & + & 7 \\ \hline & & 4 & < & 2x & & \end{array}$$

Now divide each side by 2. This results in a backward statement, so, to put it in the same order as the number line, reverse the terms and reverse the inequality.

$$\frac{4}{2} < \frac{\cancel{2}x}{\cancel{2}} \to 2 < x \to x > 2$$

Another way to solve this problem is to subtract $5x$ from each side and add 3. This results in a negative coefficient on the x term, and dividing by a negative number reverses the inequality. Same result!!!

(9) $(-\infty, 12]$. Subtract $2x$ from each side and add 11 to each side.

$$\begin{array}{rrrrrrr} 4x & - & 11 & \le & 2x & + & 13 \\ -2x & + & 11 & & -2x & + & 11 \\ \hline 2x & & & \le & & & 24 \end{array}$$

Divide each side by 2.

$\frac{\cancel{2}x}{\cancel{2}} \le \frac{24}{2} \to x \le 12$. In interval notation, this is written $(-\infty, 12]$.

(10) $(1, \infty)$. Multiply both sides by 4 and distribute on the right.

$$\cancel{4} \cdot \frac{x+3}{\cancel{4}} > 4(3-2x)$$
$$x + 3 > 12 - 8x$$

Now add $8x$ to both sides and subtract 3 from both sides.

$$\begin{array}{rrrrrrr} x & + & 3 & > & 12 & - & 8x \\ 8x & - & 3 & & -3 & + & 8x \\ \hline 9x & & & > & 9 & & \end{array}$$

Divide both sides by 9: $\frac{\cancel{9}x}{\cancel{9}} > \frac{9}{9} \to x > 1$. In interval notation, the solution is $(1, \infty)$.

(11) $[-3, 7)$. Add 5 to each section. Then divide each section by 4.

$$\begin{array}{rrrrrrr} -17 & \le & 4x & - & 5 & < & 23 \\ 5 & & & + & 5 & & 5 \\ \hline -12 & \le & 4x & & & < & 28 \end{array}$$

$\frac{-12}{4} \le \frac{4x}{4} < \frac{28}{4} \to -3 \le x < 7$. In interval notation, the solution is written $[-3, 7)$.

(12) $[-1,3]$. Subtract 6 from each section. Then multiply through by –1 and reverse the inequality signs. Solve for x and write your solution in interval notation: $3 \le 6 - x \le 7$.

$$\begin{array}{rcccccc} 3 & \le & 6 & - & x & \le & 7 \\ -6 & & -6 & & & & -6 \\ \hline -3 & \le & & - & x & \le & 1 \end{array}$$

$-3 \le -x \le 1 \rightarrow 3 \ge x \ge -1$. The numbers are in the opposite order from the number line, so reverse the numbers and the directions of the inequalities: $-1 \le x \le 3$. This solution is written $[-1,3]$ in interval notation.

(13) $\mathbf{\frac{17}{3}}$ **and** $\mathbf{-1}$**.** Solve the two linear equations $3x - 7 = 10$ and $3x - 7 = -10$.

When $3x - 7 = 10$, $3x = 17$ and $x = \frac{17}{3}$.

When $3x - 7 = -10$, $3x = -3$ and $x = -1$.

(14) **3 and –6.** First, add 7 to each side of the equation, and then divide each side by 2.

$2|2x+3| - 7 = 11 \rightarrow 2|2x+3| = 18 \rightarrow |2x+3| = 9$

Now solve the two linear equations $2x + 3 = 9$ and $2x + 3 = -9$.

When $2x + 3 = 9$, $2x = 6$ and $x = 3$.

When $2x + 3 = -9$, $2x = -12$ and $x = -6$.

(15) $\mathbf{-4 \le x \le \frac{28}{3}}$**.** First, rewrite the absolute value inequality as $-20 \le 8 - 3x \le 20$. Subtract 8 from each section, and then divide through by –3. This division requires that you reverse the inequalities. Solve $|8 - 3x| \le 20$.

$$\begin{array}{rcccccc} -20 & \le & 8 & - & 3x & \le & 20 \\ -8 & & -8 & & & & -8 \\ \hline -28 & \le & & - & 3x & \le & 12 \end{array}$$

$$28 \le -3x \le 12 \rightarrow \frac{-28}{-3} \ge \frac{-3x}{-3} \ge \frac{12}{-3} \rightarrow \frac{28}{3} \ge x \ge -4$$

To put the statement in the order of the number line, reverse the numbers and the inequalities. In interval notation, the solution is written $\left[-4, \frac{28}{3}\right]$.

(16) $\mathbf{x > 6}$ **or** $\mathbf{x < -\frac{13}{2}}$**.** First, add 1 to each side of the inequality; then divide each side by 3.

$3|4x+1| - 1 > 74 \rightarrow 3|4x+1| > 75 \rightarrow |4x+1| > 25$

Now write the absolute value inequality as two inequalities: $4x + 1 > 25$ or $4x + 1 < -25$.

When $4x + 1 > 25 \rightarrow 4x > 24 \rightarrow x > 6$.

And when $4x + 1 < -25 \rightarrow 4x < -26 \rightarrow x < \frac{-26}{4} \rightarrow x < -\frac{13}{2}$.

When writing $x > 6$ or $x < -\frac{13}{2}$ in interval notation, you get $(6, \infty)$ or $\left(-\infty, -\frac{13}{2}\right)$.

Whaddya Know? Chapter 2 Quiz

Quiz time! Complete each problem to test your knowledge on the various topics covered in this chapter. You can then find the solutions and explanations in the next section.

1. **Solve for h in $S = 2\pi r(r + h)$.**
2. **Solve for x in $4x + 11 \leq 2(x - 3) + 1$. Write your answer in interval notation.**
3. **Solve for x: $3x - 4 = 17$.**
4. **Solve for x: $|5x - 2| > 13$. Write your answer in interval notation.**
5. **Solve for y: $5y + 11 = 9y - 13$.**
6. **Given the formula for the sum of an arithmetic sequence, $S_n = \frac{n}{2}(2a_1 + (n - 2)d)$, solve for d.**
7. **Solve for x: $\frac{2x}{3} - \frac{3x}{4} = 1$.**
8. **Solve for x: $2|5x - 2| + 3 \leq 19$. Write your answer in interval notation.**
9. **Solve for y: $\frac{8}{y} + \frac{4}{3y} = 112$.**
10. **Solve for x: $3x - 4y = 12$.**
11. **Solve for x: $|x + 5| = 6$.**
12. **Solve for y: $3|2y - 1| = 9$.**
13. **Solve for y: $|8 - 5y| = 7$.**
14. **Solve for z: $-10 < 3z - 4 \leq 14$. Write your answer in interval notation.**
15. **Solve for z: $4(z + 1) - 2(3 - z) = 7z + 5$.**

Answers to Chapter 2 Quiz

1. $\boldsymbol{h=\frac{S-2\pi r^2}{2\pi r}}$. First, divide each side of the equation by $2\pi r$. Then subtract r from each side: $\frac{S}{2\pi r}=\frac{2\pi r(r+h)}{2\pi r}\rightarrow\frac{S}{2\pi r}=r+h\rightarrow\frac{S}{2\pi r}-r=h$. This can be written as a single term by writing the r with the common denominator: $\frac{S}{2\pi r}-r=h\rightarrow\frac{S}{2\pi r}-\frac{r(2\pi r)}{2\pi r}=h\rightarrow\frac{S-2\pi r^2}{2\pi r}=h$.

2. $\boldsymbol{(-\infty,-8]}$. Distribute on the right, and combine like terms: $4x+11\leq 2x-6+1\rightarrow 4x+11\leq 2x-5$. Now subtract $2x$ and 11 from each side:

$$\begin{array}{rcrcrcr} 4x & + & 11 & \leq & 2x & - & 5 \\ -2x & - & 11 & & -2x & - & 11 \\ \hline 2x & & & \leq & & - & 16 \end{array}$$

Divide each side by 2, and you get $x\leq -8$ or $(-\infty,-8]$.

3. **7.** First, add 4 to each side. Then divide both sides by 3.

$$\begin{array}{rcrcr} 3x & - & 4 & = & 17 \\ & + & 4 & & 4 \\ \hline 3x & & & = & 21 \end{array} \quad \text{then} \quad \begin{array}{c} \frac{3x}{3}=\frac{21}{3} \\ x=7 \end{array}$$

4. $\boldsymbol{(3,\infty)}$ or $\left(-\infty,-\frac{11}{5}\right)$. Write the two corresponding linear inequalities and solve them for x.

$$5x-2>13 \quad \text{or} \quad 5x-2<-13$$

When $5x-2>13$, $5x>15\rightarrow x>3$, which is written $(3,\infty)$.
When $5x-2<-13$, $5x<-11\rightarrow x<-\frac{11}{5}$, which is written $\left(-\infty,-\frac{11}{5}\right)$.

5. **6.** First, add 13 to each side and subtract $5y$ from each side. Then divide both sides by 4.

$$\begin{array}{rcrcrcr} 5y & + & 11 & = & 9y & - & 13 \\ -5y & + & 13 & & -5y & + & 13 \\ \hline & & 24 & = & 4y & & \end{array} \quad \text{then} \quad \begin{array}{c} \frac{24}{4}=\frac{4y}{4} \\ 6=y \end{array}$$

6. $\boldsymbol{d=\frac{2S_n-2a_1n}{n(n-2)}}$. First, multiply each side of the equation by $\frac{2}{n}$. Then subtract $2a_1$ from each side. Finally, divide each side by $(n-2)$.

$$\frac{2}{n}\cdot S_n=\frac{2}{n}\cdot\frac{n}{2}\left(2a_1+(n-2)d\right)\rightarrow\frac{2S_n}{n}=2a_1+(n-2)d\rightarrow\frac{2S_n}{n}-2a_1=(n-2)d\rightarrow$$

$$\frac{\frac{2S_n}{n}-2a_1}{n-2}=\frac{(n-2)d}{(n-2)}\rightarrow\frac{\frac{2S_n}{n}-2a_1}{n-2}=d$$

The complex fraction can be simplified by writing the numerator as a single fraction with denominator n and then dividing. This just means multiplying by the reciprocal of $(n-2)$.

$$\frac{\frac{2S_n}{n}-2a_1}{n-2}=d \rightarrow \frac{\frac{2S_n}{n}-\frac{2a_1n}{n}}{n-2}=d \rightarrow \frac{\frac{2S_n-2a_1n}{n}}{n-2}=d$$

$$\frac{\frac{2S_n-2a_1n}{n}}{n-2}=\frac{2S_n-2a_1n}{n}\cdot\frac{1}{n-2}=\frac{2S_n-2a_1n}{n(n-2)}=d$$

(7) **−12.** First, multiply each term by the least common denominator, 12.

$$\begin{array}{l}\frac{2x}{\cancel{3}}\cdot\frac{\cancel{12}^4}{1}-\frac{3x}{\cancel{4}}\cdot\frac{\cancel{12}^3}{1}=1\cdot 12\\ 8x \quad - \quad 9x \quad =12\end{array}$$

Simplify on the left, and then multiply each side of the equation by −1.

$-x=12 \rightarrow x=-12$

(8) $\left[-\frac{6}{5},2\right]$. First, subtract 3 from each side of the inequality and divide each side by 2.

$$2|5x-2|+3\le 19 \rightarrow 2|5x-2|\le 16 \rightarrow |5x-2|\le 8$$

Now you can write the absolute value inequality as a compound inequality. Solve for x by first adding 2 to each section and then dividing each section by 5:

$-8\le 5x-2\le 8 \rightarrow -6\le 5x\le 10 \rightarrow -\frac{6}{5}\le x\le 2$. In interval notation, this is $\left[-\frac{6}{5},2\right]$.

(9) $\frac{1}{12}$. First, multiply each term by the least common denominator, $3y$.

$$\begin{array}{l}\frac{8}{\cancel{y}}\cdot\frac{3\cancel{y}}{1}+\frac{4}{\cancel{3y}}\cdot\frac{\cancel{3y}}{1}=112\cdot 3y\\ 24 \quad + \quad 4 \quad =336y\\ \qquad 28=336y\end{array}$$

Divide each side by 336.

$$\frac{28}{336}=\frac{336y}{336} \rightarrow \frac{1}{12}=y$$

(10) $x=4+\frac{4}{3}y$. Add $4y$ to each side, and then divide both sides of the equation by 3.

$$\begin{array}{rcl}3x-4y&=&12\\ +4y&&+4y\\ \hline 3x&=&12+4y\end{array} \quad \text{then} \quad \begin{array}{l}\frac{\cancel{3}x}{\cancel{3}}=\frac{12+4y}{3}\\ x=\frac{12+4y}{3}=4+\frac{4}{3}y\end{array}$$

(11) **1** and **−11.** Solve the two equations: $x+5=6$ and $x+5=-6$.

$$\begin{array}{rcl}x+5&=&6\\ -5&&-5\\ \hline x&=&1\end{array} \quad \text{and} \quad \begin{array}{rcl}x+5&=&-6\\ -5&&-5\\ \hline x&=&-11\end{array}$$

(12) **2 and −1.** First, divide each side of the equation by 3.

$$\frac{\cancel{3}|2y-1|}{\cancel{3}}=\frac{9}{3}$$
$$|2y-1|=3$$

Now solve the two equations $2y-1=3$ and $2y-1=-3$.

$$\begin{array}{rcrcr} 2y & - & 1 & = & 3 \\ & + & 1 & & +1 \\ \hline 2y & & & = & 4 \\ & & y & = & 2 \end{array} \quad \text{and} \quad \begin{array}{rcrcr} 2y & - & 1 & = & -3 \\ & + & 1 & & +1 \\ \hline 2y & & & = & -2 \\ & & y & = & -1 \end{array}$$

(13) $\frac{1}{5}$ **and 3.** Solve the two equations: $8-5y=7$ and $8-5y=-7$.

$$\begin{array}{rcrcr} 8 & - & 5y & = & 7 \\ -8 & & & & -8 \\ \hline & - & 5y & = & -1 \\ & & y & = & \frac{1}{5} \end{array} \quad \text{and} \quad \begin{array}{rcrcr} 8 & - & 5y & = & -7 \\ -8 & & & & -8 \\ \hline & - & 5y & = & -15 \\ & & y & = & 3 \end{array}$$

(14) $(-2,6]$**.** Add 4 to each section, and then divide each section by 3: $-10<3z-4\leq 14 \rightarrow -6<3z\leq 18 \rightarrow -2<z\leq 6$. In interval notation, this is written $(-2,6]$.

(15) **−7.** First, distribute the 4 and −2; then combine the like terms on the left.

$$4z+4-6+2z=7z+5$$
$$6z-2=7z+5$$

Subtract 6z and 5 from each side.

$$\begin{array}{rcrcrcr} 6z & - & 2 & = & 7z & + & 5 \\ -6z & - & 5 & & -6z & - & 5 \\ \hline & - & 7 & = & z & & \end{array}$$

IN THIS CHAPTER

» Rooting and factoring to solve quadratic equations

» Breaking down equations with the quadratic formula

» Squaring up the quadratic

» Conquering advanced quadratics

» Acting rational with the rational root theorem

Chapter 3

Handling Quadratic and Other Polynomial Equations

Quadratic equations are some of the more common equations you see in the mathematics classroom. A quadratic equation contains a term with an exponent of two, and no term with a higher power. The standard form is $ax^2 + bx + c = 0$.

In other words, this equation is a quadratic expression with an equal sign (see Chapter 1 for a short-and-sweet discussion on quadratic expressions). Quadratic equations potentially have two solutions. You may not find two, but you start out assuming that you'll find two and then proceed to prove or disprove your assumption.

Quadratic equations are not only very manageable — because you can always find ways to tackle them — but they also serve as good role models, playing parts in many practical applications. If you want to track the height of an arrow you shoot into the air, for example, you can find your answer with a quadratic equation. The area of a circle is technically a quadratic equation. The profit (or loss) from the production and sales of items often follows a quadratic pattern.

And then, in this chapter, I bring in some polynomial equations! Yes, a quadratic equation is also a polynomial equation — it meets the criteria — and when you leave the quadratic format you need some different procedures. You'll find them here!

Implementing the Square Root Rule

Some quadratic equations are easier to solve than others; half the battle is recognizing which equations are easy and which are more challenging.

The simplest quadratic equations that you can solve quickly are those that allow you to take the square root of both sides of the equation. These lovely equations are made up of a squared term and a number, and start out as $ax^2 + m = n$ or $ax^2 = k$. You solve equations written this way by changing the form (if necessary) to set the squared term equal to a number; then you use the square root rule. When you have $x^2 = k$, you get $x = \pm\sqrt{k}$.

Notice that by using the square root rule, you come up with two solutions: both the positive and the negative. When you square a positive number, you get a positive result, and when you square a negative number, you also get a positive result.

EXAMPLE

Q. Solve for x: $3x^2 + 2 = 50$.

A. ± 4. First, subtract 2 from each side to get $3x^2 = 48$, and then divide each side by 3 to get $x^2 = 16$. Taking the square root of each side, you have $x = \pm\sqrt{16} = \pm 4$.

Q. Solve for y: $y^2 = 40$.

A. $\pm 2\sqrt{10}$. Take the square root of each side and then simplify the radical term: $y^2 = 40 \rightarrow y = \pm\sqrt{40} = \pm\sqrt{4}\sqrt{10} = \pm 2\sqrt{10}$.

Refer to Chapter 1 for more on simplifying radical expressions.

The choice to use the square root rule is pretty obvious when you have an equation with a squared variable and a perfect square number. The decision may seem a bit murkier when the number involved isn't a perfect square. But, as you see in this second example, you can still use the square root rule in these situations.

The number represented by k also has to be positive if you want to find real answers with this rule. If k is negative, you get an imaginary answer, such as $3i$ or $5 - 4i$. (For more on imaginary numbers, check out Chapter 17.)

YOUR TURN

1 Solve for x: $2x^2 - 3 = 47$.

2 Solve for z: $z^2 + 7 = 25$.

Successfully Factoring for Solutions

You can factor many quadratic expressions — one side of a quadratic equation — by rewriting them as products of two or more numbers, variables, terms in parentheses, and so on. The advantage of the factored form is that you can solve quadratic equations by setting the factored expression equal to zero (making it an equation) and then using the multiplication property of zero. How you factor the expression depends on the number of terms in the quadratic and how those terms are related.

Factoring binomials

In the previous section, you saw how to solve a quadratic binomial using the square root rule. Now I get into the factoring possibilities to solve these equations. You can factor a quadratic binomial (which contains two terms, one of them with a variable raised to the power 2) in one of two ways, if you can factor it at all (you may find no common factor, or the two terms may not both be squares).

Here are the two ways you can use to factor a quadratic binomial:

- Divide out a common factor from each of the terms.
- Write the quadratic as the product of two binomials, if the quadratic is the difference of perfect squares.

Taking out a greatest common factor

The greatest common factor (GCF) of two or more terms is the largest number (and variable combination) that divides each of the terms evenly. This is one of the quickest, easiest ways of solving a quadratic equation using factoring.

Q. Solve the equation $4x^2+8x=0$.

EXAMPLE **A.** **0, −2.** Factor out the greatest common factor, which is $4x$. After dividing, you get $4x(x+2)=0$. Using the multiplication property of zero, you can now state three facts about this equation:

- $4=0$, which is false — this isn't a solution.
- $x=0$, which is a solution.
- $x+2=0$, written as $x=-2$, which is a solution.

You find two solutions for the original equation $4x^2+8x=0$: $x=0$ or $x=-2$. If you replace the x's with either of these solutions, you create a true statement.

Q. Solve the quadratic equation $6y^2+18=0$.

A. **No solution.** You can factor this equation by dividing out the 6 from each term: $6y^2+18=6\left(y^2+3\right)=0$.

Unfortunately, this factored form doesn't yield any real solutions for the equation, because it doesn't have any. Applying the multiplication property of zero, you first get $6=0$. No help there. Setting $y^2+3=0$, you can subtract 3 from each side to get $y^2=-3$. The number –3 isn't positive, so you can't apply the square root rule, because you get $y=\pm\sqrt{-3}$. No real number's square is –3. So, you can't find an answer to this problem among real numbers. (For information on complex or imaginary answers, head to Chapter 17.)

Be careful when the GCF of an expression is just *x*, and always remember to set that front factor, *x*, equal to zero so you don't lose one of your solutions. A very common error in algebra is to take a perfectly nice equation such as $x^2+5x=0$, factor it into $x(x+5)=0$, and give the only answer as $x=-5$. For some reason, people often ignore that lonely *x* in front of the parenthesis. Don't forget the solution $x=0$!

YOUR TURN

3 Solve for *x*: $6x^2-3x=0$.

4 Solve for *y*: $4y^2=20y$.

Factoring the difference of squares

If you run across a binomial that you don't think calls for the application of the square root rule, you can factor the difference of the two squares and solve for the solution by using the multiplication property of zero. If any solutions exist, and you can find them with the square root rule, then you can also find them by using the difference of the two squares method.

This method states that if $x^2-a^2=0$, $(x-a)(x+a)=0$, giving you $x=a$ or $x=-a$.

Generally, if $k^2x^2-a^2=0$, $(kx-a)(kx+a)=0$, and $x=\frac{a}{k}$ or $x=-\frac{a}{k}$.

Q. Solve $x^2 - 25 = 0$.

EXAMPLE **A.** **5, −5.** **Factor the equation into $(x-5)(x+5)=0$, and the multiplication property of zero tells you that $x=5$ or $x=-5$.**

Q. Solve $49y^2 - 64 = 0$.

A. $\frac{8}{7}, -\frac{8}{7}$. **Factor into the difference and sum of the square roots. To solve, you can factor the terms on the left into $(7y-8)(7y+8)=0$, and the two solutions are $y=\frac{8}{7}$, $y=-\frac{8}{7}$.**

YOUR TURN

5 Solve for x: $x^2 - 100 = 0$.

6 Solve for y: $9y^2 = 25z^2$.

Factoring trinomials

Like quadratic binomials, a quadratic trinomial can have as many as two solutions — or it may have one solution or no solution at all. If you can factor the trinomial and use the multiplication property of zero to solve for the roots, you're home free. If the trinomial doesn't factor, or if you can't figure out how to factor it, you can utilize the quadratic formula (see the section, "Resorting to the Quadratic Formula," later in this chapter). The rest of this section deals with the trinomials that you can factor.

Finding two solutions in a trinomial

It isn't always apparent whether your trinomial will factor and give you two solutions or one solution or none. You learn more about what you have after you've done the factoring.

Q. Solve $x^2 - 2x - 15 = 0$.

EXAMPLE **A.** **5, −3.** **You can factor the left side of the equation into $(x-5)(x+3)=0$ and then set each factor equal to zero. When $x-5=0$, $x=5$, and when $x+3=0$, $x=-3$. (If you can't remember how to factor these trinomials, see Chapter 1, or, for even more details, see *Algebra I For Dummies* (John Wiley & Sons, Inc.).)**

Q. Solve $24x^2 + 52x - 112 = 0$.

A. $\frac{4}{3}, \frac{7}{2}$. It may not be immediately apparent how you should factor a seemingly complicated trinomial like this one, but before you bail out and go to the quadratic formula, consider factoring 4 out of each term to simplify the picture; you get $4(6x^2 + 13x - 28) = 0$. The quadratic in the parentheses factors into the product of two binomials. If you're using the Box Method (see Chapter 1), then you multiply $6x^2$ times –28 and then find which factors of $168x^2$ have a sum of 13x. You find these factors and write the factorization $4(3x - 4)(2x + 7) = 0$. Setting the binomials equal to 0, you get the two solutions $x = \frac{4}{3}$ and $x = -\frac{7}{2}$. How about the factor of 4? If you set 4 equal to 0, you get a false statement, which is fine; you already have the two numbers that make the equation a true statement.

YOUR TURN

7 Solve for x: $x^2 - 7x - 44 = 0$.

8 Solve for x: $50x^2 + 65x - 150 = 0$.

Doubling up on a trinomial solution

Some trinomials may look like they will give you two solutions, but they can be misleading. There are special instances where you get only one solution, and even that is misleading; there are actually two solutions, but they're the same number. This is called a *double root*.

You need to pay attention to double roots when you're graphing, because they act differently on the axes. This distinction is important when you're graphing any polynomial. Graphs at double roots don't cross the axis — they just touch. You also see these entities when solving inequalities; see Chapter 4 for more on how double roots affect those problems.

EXAMPLE

Q. Solve $x^2 - 12x + 36 = 0$.

A. 6. This equation is a perfect square trinomial, which simply means that it's the square of a single binomial. Assigning this equation that special name points out why the two solutions you find are actually just one. Look at the factoring: $x^2 - 12x + 36 = (x - 6)(x - 6) = (x - 6)^2 = 0$. The two different factors give the same solution: $x = 6$. A quadratic trinomial can have as many as two solutions or roots. This trinomial technically does have two roots, 6 and 6, so you can say that the equation has a double root.

Q. Solve $18x^2 + 84x + 98 = 0$.

A. $-\frac{7}{3}$. First, factor out the GCF of 2. You now have $9x^2 + 14x + 49 = 0$. Both the first and last coefficients are perfect squares, and this factors: $9x^2 + 14x + 49 = (3x+7)(3x+7) = (3x+7)^2 = 0$. Setting the binomial equal to zero, you have the double root $x = -\frac{7}{3}$.

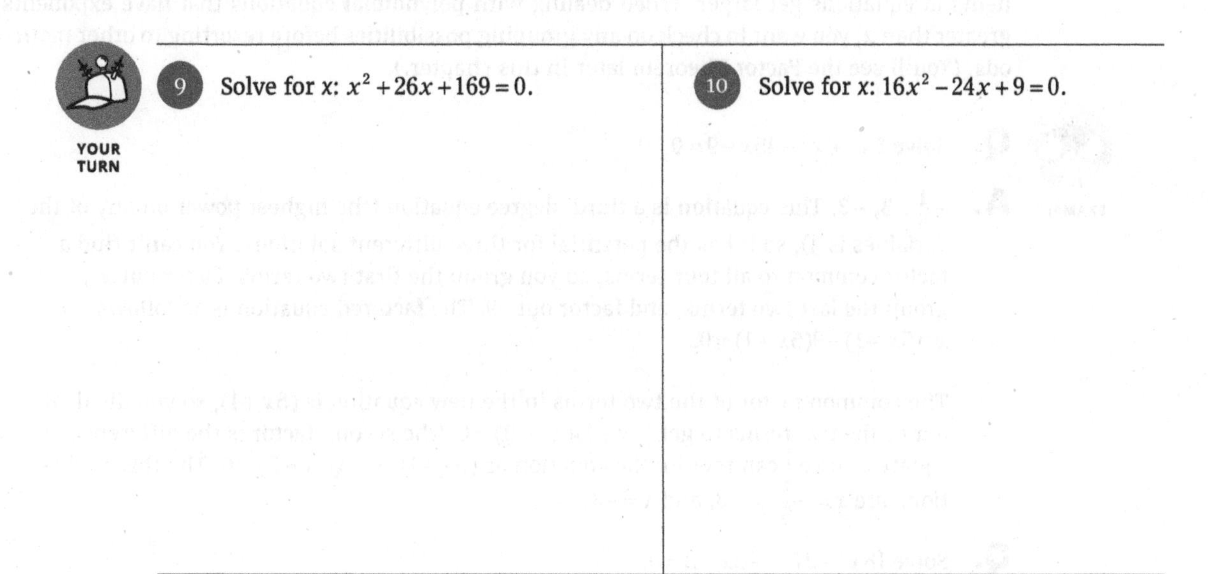

9 Solve for x: $x^2 + 26x + 169 = 0$.

10 Solve for x: $16x^2 - 24x + 9 = 0$.

Factoring by grouping

Factoring by grouping is a great method for rewriting a quadratic equation so that you can use the multiplication property of zero (see Chapter 1) and find all the solutions. The main idea behind factoring by grouping is to arrange the terms into smaller groupings that have a common factor. You go to little groupings because you can't find a greatest common factor for all the terms; however, by taking two terms at a time, you can find something to divide them by.

Grouping terms in a quadratic

A quadratic equation such as $2x^2 + 8x - 5x - 20 = 0$ has four terms. Yes, you can combine the two middle terms on the left but leave them as is for the sake of the grouping process. The four terms in the equation don't share a greatest common factor. You can divide the first, second, and fourth terms evenly by 2, but the third term doesn't comply. The first three terms all have a factor of x, but the last term doesn't. So, you group the first two terms together and take out their common factor, 2x. The last two terms have a common factor of –5. The factored form, therefore, is $2x(x+4) - 5(x+4) = 0$.

The new, factored form has two terms. Each of the terms has an $(x+4)$ factor, so you can divide that factor out of each term. When you divide the first term, you have 2x left. When you divide the second term, you have –5 left. Your new factored form is $(x+4)(2x-5) = 0$. Now you can set each factor equal to zero to get $x = -4$ and $x = \frac{5}{2}$.

Factoring by grouping works only when you can create a new form of the quadratic equation that has fewer terms and a common factor. If the factor $(x+4)$ hadn't shown up in both of the factored terms in the previous example, you would've gone in a different direction.

Finding quadratic factors in a grouping

Solving quadratic equations by grouping and factoring is even more important when the exponents in equations get larger. When dealing with polynomial equations that have exponents greater than 2, you want to check on any grouping possibilities before resorting to other methods. (You'll see the Factor Theorem later in this chapter.)

EXAMPLE

Q. Solve $5x^3 + x^2 - 45x - 9 = 0$.

A. $-\frac{1}{5}$, **3, −3. This equation is a third-degree equation (the highest power on any of the variables is 3), so it has the potential for three different solutions. You can't find a factor common to all four terms, so you group the first two terms, factor out x^2, group the last two terms, and factor out −9. The factored equation is as follows: $x^2(5x+1) - 9(5x+1) = 0$.**

The common factor of the two terms in the new equation is $(5x+1)$, so you divide it out of the two terms to get $(5x+1)(x^2 - 9) = 0$. The second factor is the difference of squares, so you can rewrite the equation as $(5x+1)(x-3)(x+3) = 0$. The three solutions are $x = -\frac{1}{5}$, $x = 3$, and $x = -3$.

Q. Solve $18x^3 + 27x^2 - 2x - 3 = 0$.

A. $-\frac{3}{2}, \frac{1}{3}, -\frac{1}{3}$. **The first two terms have a common factor of $9x^2$, but the best you can do with the last two terms is a common factor of −1. Using grouping, $18x^3 + 27x^2 - 2x - 3 = 9x^2(2x+3) - 1(2x+3) = 0$. Pulling out the common factor, you now have $(2x+3)(9x^2 - 1) = 0$. The difference of squares factors, giving you $(2x+3)(3x-1)(3x+1) = 0$ and the three solutions $x = -\frac{3}{2}, x = \frac{1}{3}, x = -\frac{1}{3}$.**

YOUR TURN

11 Solve for x: $x^3 + 7x^2 - x - 7 = 0$.

12 Solve for x: $12x^3 + 16x^2 - 75x - 100 = 0$.

Resorting to the Quadratic Formula

The quadratic formula is a wonderful tool to use when other factoring methods fail — an algebraic vending machine, of sorts. You take the numbers from a quadratic equation, plug them into the formula, and out come the solutions of the equation. You can even use the formula when the equation does factor, but you haven't been able to figure out how.

The quadratic formula states that when you have a quadratic equation in the form $ax^2 + bx + c = 0$ (with the *a* as the coefficient of the squared term, the *b* as the coefficient of the first-degree term, and the *c* as the constant), the equation has the solutions:

$$x = \frac{-b \pm \sqrt{b^2 - 4ac}}{2a}$$

The process of solving a quadratic equation is almost always faster and more accurate if you can factor the equation. The quadratic formula is wonderful, but like a vending machine that eats your quarters, it has some built-in challenges:

- You have to remember to insert the *opposite* of *b*.
- You have to simplify the numbers under the radical correctly.
- You have to divide the *whole numerator* by the denominator.

REMEMBER

Don't get me wrong, you shouldn't hesitate to use the quadratic formula whenever necessary! It's great. But factoring is usually better, faster, and more accurate.

Finding rational solutions

Quadratics that can be factored have rational solutions. This means that each solution can be written as an integer or fraction, positive or negative. If you use the quadratic formula to find the solutions of a quadratic equation and the solutions are rational, then the quadratic could have been factored. The big plus of this formula is that it also takes care of the irrational solutions. More on that later.

EXAMPLE

Q. Solve $3x^2 + 11x + 10 = 0$ using the quadratic formula.

A. $-\frac{5}{3}$ **or** -2**.** Let $a = 3$, $b = 11$, and $c = 10$. Filling in the values and solving for *x*, you get:

$$x = \frac{-11 \pm \sqrt{11^2 - 4(3)(10)}}{2(3)} = \frac{-11 \pm \sqrt{121 - 120}}{6} = \frac{-11 \pm \sqrt{1}}{6} = \frac{-11 \pm 1}{6}$$

You finish up the answers by dealing with the ± symbol one sign at a time. Considering both signs:

$$x=\frac{-11+1}{6}=\frac{-10}{6}=-\frac{5}{3} \text{ or } x=\frac{-11-1}{6}=\frac{-12}{6}=-2$$

You find two different solutions. The fact that the solutions are rational numbers (numbers that you can write as fractions) tells you that you could've factored the equation. If you end up with a radical in your answer, you know that factorization isn't possible for that equation. Here's the factorization of the quadratic:
$3x^2+11x+10=(3x+5)(x+2)=0$

Q. Solve $12x^2-16x-35=0$ using the quadratic formula.

A. $\frac{5}{2}$ or $-\frac{7}{6}$. Let $a = 12$, $b = -16$, and $c = -35$. Filling in the values and solving for x, you get:

$$x=\frac{-(-16)\pm\sqrt{(-16)^2-4(12)(-35)}}{2(12)}=\frac{16\pm\sqrt{256+1680}}{24}=\frac{16\pm\sqrt{1936}}{24}=\frac{16\pm 44}{24}=\frac{4\pm 11}{6}$$

Considering both signs:

$$x=\frac{4+11}{6}=\frac{15}{6}=\frac{5}{2} \text{ or } x=\frac{4-11}{6}=-\frac{7}{6}$$

Just in case you're curious, here's the factorization of the quadratic:
$12x^2-16x-35=(2x-5)(6x+7)$

Straightening out irrational solutions

The quadratic formula is especially valuable for solving quadratic equations that don't factor. Unfactorable equations, when they do have solutions, have irrational numbers in their answers. Irrational numbers have no fractional equivalent; they feature decimal values that go on forever and never have patterns that repeat.

TIP

Factoring a quadratic equation is almost always preferable to using the quadratic formula. But at certain times, you're better off opting for the quadratic formula, even when you can factor the equation. In cases where the numbers are huge and have many multiplication possibilities, I suggest that you bite the bullet, haul out your calculator, and go for it.

For instance, a great problem in calculus (known as "Finding the largest box that can be formed from a rectangular piece of cardboard" for the curious among you) has an answer that you find when you solve a quadratic equation. This is fine and dandy, but sometimes the quadratic equations can get pretty nasty. If you have a calculator handy, this helps. I've provided an example.

EXAMPLE

Q. Solve $2x^2+5x-6=0$.

A. $\frac{-5\pm\sqrt{73}}{4}$. This quadratic is not factorable, so the quadratic formula is your only option. Let $a = 2$, $b = 5$, and $c = -6$.

$$x=\frac{-5\pm\sqrt{5^2-4(2)(-6)}}{2(2)}=\frac{-5\pm\sqrt{25+48}}{4}=\frac{-5\pm\sqrt{73}}{4}$$

The value under the radical doesn't factor, so there is no more to simplify.

Q. Here's the open box example. In this case, the piece of cardboard was 36 inches by 48 inches, and you cut equal squares out of the corners and folded up the sides. After performing a calculus process (called the derivative), you get to solve a quadratic equation to find the answer. Solve $12x^2-336x+1728=0$.

A. $14\pm2\sqrt{13}$. This quadratic is not factorable. Using the quadratic formula, let $a = 12$, $b = -36$, and $c = 1728$.

$$x=\frac{336\pm\sqrt{336^2-4(12)(1728)}}{2(12)}=\frac{336\pm\sqrt{112{,}896-82{,}944}}{24}=\frac{336\pm\sqrt{29{,}952}}{24}=\frac{336\pm\sqrt{29{,}952}}{24}$$

$$=\frac{336\pm\sqrt{2^8\cdot3^2\cdot13}}{24}=\frac{336\pm2^4\cdot3\sqrt{13}}{24}=14\pm2\sqrt{13}$$

The two solutions come out to be about 21.2 inches and about 6.8 inches. Either cut-out measure will create the best box.

If you get a negative number under the radical when using the quadratic formula, you know that the problem has no real answer. Chapter 17 explains how to deal with these imaginary/complex answers. And if you need some help with simplifying the radical, go to Chapter 1.

YOUR TURN

13 Solve for x: $3x^2-6x-4=0$.

14 Solve for x: $48x^2-155x+125=0$.

Solving Quadratics by Completing the Square

Of all the choices you have for solving a quadratic equation (factoring and the quadratic formula, to name two; see the previous sections in this chapter), completing the square is usually your last resort. But the binomial-squared format is very nice to have when you're working with conic sections (circles, ellipses, hyperbolas, and parabolas) and writing their standard forms (as you can see in Chapter 13).

For instance, using completing the square on the equation of a parabola gives you a visual answer to questions about where the vertex is and how it opens (to the side, up, or down; see Chapter 8). The big payoff is that you have a result for all your work. After all, you do get answers to the quadratic equations by using this process.

Squaring up a quadratic equation

To solve the quadratic equation $ax^2 + bx + c = 0$ by completing the square, follow these steps:

1. **Divide every term in the equation by the coefficient *a*.**

   $$\frac{ax^2}{a} + \frac{bx}{a} + \frac{c}{a} = x^2 + \frac{b}{a}x + \frac{c}{a} = 0$$

2. **Move the constant term (the term without a variable) to the opposite side of the equation by adding or subtracting.**

   $$x^2 + \frac{b}{a}x = -\frac{c}{a}$$

3. **Find half the value of the coefficient on the first-degree term of the variable; then square the result of the halving, and add that amount to each side of the equation.**

   Find half of $\frac{b}{a}$, which is $\frac{b}{2a}$. Square the fraction, giving you $\frac{b^2}{4a^2}$, and add the square to each side of the equation:

   $$x^2 + \frac{b}{a}x + \frac{b^2}{4a^2} = \frac{b^2}{4a^2} - \frac{c}{a}$$

   Now add the two terms on the right:

   $$x^2 + \frac{b}{a}x + \frac{b^2}{4a^2} = \frac{b^2 - 4ac}{4a^2}$$

4. **Factor the side of the equation that's a perfect square trinomial (you just created it) into the square of a binomial.**

   Factor the left side of the equation:

   $x^2 + \frac{b}{a}x + \frac{b^2}{4a^2}$ factors into $\left(x + \frac{b}{2a}\right)^2$

   So now you have $\left(x + \frac{b}{2a}\right)^2 = \frac{b^2 - 4ac}{4a^2}$

5. **Find the square root of each side of the equation.**

$$\sqrt{\left(x+\frac{b}{2a}\right)^2}=\pm\sqrt{\frac{b^2-4ac}{4a^2}}$$
$$x+\frac{b}{2a}=\pm\sqrt{\frac{b^2-4ac}{4a^2}}$$

6. **Isolate the variable term by adding or subtracting to move the constant to the other side. Then simplify the radical and combine terms.**

$$x=-\frac{b}{2a}\pm\sqrt{\frac{b^2-4ac}{4a^2}}=-\frac{b}{2a}\pm\frac{\sqrt{b^2-4ac}}{\sqrt{4a^2}}=-\frac{b}{2a}\pm\frac{\sqrt{b^2-4ac}}{2a}=\frac{-b\pm\sqrt{b^2-4ac}}{2a}$$

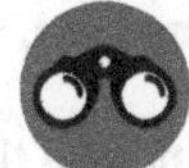

EXAMPLE

Q. Use *completing the square* to solve the quadratic equation $x^2-8x-9=0$.

A. x = 9 and x = −1. Use the steps previously described.

1. **Divide every term in the equation by the coefficient of the squared term.**

   Because the coefficient is 1, there is no change to the equation:
   $x^2-8x-9=0$

2. **Move the constant term to the opposite side of the equation by adding 9 to each side.**

   $x^2-8x=9$

3. **Find half the value of the coefficient on the first-degree term of the variable; then square the result of the halving, and add that amount to each side of the equation.**

   Find half of 8, which is 4. Square the 4, giving you 16, and add the square to each side of the equation:

   $x^2-8x+16=9+16=25$

4. **Factor the side of the equation that's a perfect square trinomial (left side) into the square of a binomial.**

   $x^2-8x+16=25$ factors into $(x-4)^2=25$

5. **Find the square root of each side of the equation.**

   $$\sqrt{(x-4)^2}=\pm\sqrt{25}=\pm5$$
   $$x-4=\pm5$$

6. **Isolate the variable term by adding 4 to each side and solve for the value of x.**

   $x=4\pm5$

   The two answers are $x=9$ and $x=-1$.

Q. Use *completing the square* to solve the quadratic equation $3x^2+10x-8=0$.

A. $\frac{2}{3}$ **and** -4. Use the steps previously described.

1. **Divide every term in the equation by the coefficient of the squared term, 3.**

   $$x^2+\frac{10}{3}x-\frac{8}{3}=0$$

2. **Move the constant term to the opposite side of the equation by adding $\frac{8}{3}$.**

   $x^2+\frac{10}{3}x-\frac{8}{3}+\frac{8}{3}=0+\frac{8}{3}$, which simplifies to $x^2+\frac{10}{3}x=\frac{8}{3}$

3. **Find half the value of the coefficient on the first-degree term of the variable; then square the result of the halving, and add that amount to each side of the equation.**

   Find half of $\frac{10}{3}$, which is $\frac{5}{3}$. Square the fraction, giving you $\left(\frac{5}{3}\right)^2=\frac{25}{9}$, and add the square to each side of the equation:

   $$x^2+\frac{10}{3}x+\frac{25}{9}=\frac{8}{3}+\frac{25}{9}=\frac{49}{9}$$

4. **Factor the side of the equation that's a perfect square trinomial (left side) into the square of a binomial.**

   $x^2+\frac{10}{3}x+\frac{25}{9}=\frac{49}{9}$ factors into $\left(x+\frac{5}{3}\right)^2=\frac{49}{9}$

5. **Find the square root of each side of the equation.**

   $$\sqrt{\left(x+\frac{5}{3}\right)^2}=\pm\sqrt{\frac{49}{9}}$$

   $$x+\frac{5}{3}=\pm\frac{7}{3}$$

6. **Isolate the variable term by subtracting $\frac{5}{3}$ from each side, and solve for the value of x.**

   $$x+\frac{5}{3}-\frac{5}{3}=-\frac{5}{3}\pm\frac{7}{3}$$

   $$x=-\frac{5}{3}\pm\frac{7}{3}$$

   The two answers are $x=-\frac{5}{3}+\frac{7}{3}=\frac{2}{3}$ and $x=-\frac{5}{3}-\frac{7}{3}=-4$.

Completing the square twice over

Completing the square on an equation with both x's and y's leaves you just one step away from what you need to work with conics. Conic sections (circles, ellipses, hyperbolas, and parabolas) have standard equations that give you plenty of information about individual curves, where their centers are, which direction they go in, and so on. Chapter 13 covers this information in detail. In the meantime, you practice completing the square twice over in this section.

When working with conics, you'll see an equation written something like $ax^2 + by^2 + cx + dy + e = 0$. To put it in the standard form for a circle or ellipse or hyperbola, you want it to be the sum of two binomials squared and a constant. Think of the equation as having two separate completing-the-square problems to, well, complete. Follow these steps to give the equation a twice-over:

1. **Rewrite the equation with the x's and y's together, with a space between the x terms and the y terms, and with the constant on the other side of the equation:**

   $$ax^2 + cx + \quad by^2 + dy \quad = f$$

2. **Find any numerical factors for each grouping; you want the coefficient of the squared term to be one. Write the factor outside the parentheses, with the variables inside.**

   $$a\left(x^2 + \frac{c}{a}x \quad\right) + b\left(y^2 + \frac{d}{b}y \quad\right) = f$$

3. **Complete the square on the x's and the y's, and add whatever you used to complete the squares to the other side of the equation, too, to keep the equation balanced.**

   $$a\left(x^2 + \frac{c}{a}x + \frac{c^2}{4a^2}\right) + b\left(y^2 + \frac{d}{b}y + \frac{d^2}{4b^2}\right) = f + \frac{c^2}{4a} + \frac{d^2}{4b}$$

4. **Simplify each side of the equation by writing the trinomials on the left as binomials squared and by combining the terms on the right.**

   $$(x+h)^2 + (y+k)^2 = n^2$$

EXAMPLE

Q. Write the equation $x^2 + 6x + 2y^2 - 8y + 13 = 0$ as the sum of two binomials squared and a constant.

A. $\mathbf{(x+3)^2 + 2(y-2)^2 = 4}$. Use the steps as described.

1. **Rewrite the equation with the x's and y's together, with a space between the x terms and the y terms, and with the constant on the other side of the equation.**

   $$x^2 + 6x \quad + 2y^2 - 8y \quad = -13$$

2. **Find any numerical factors for each grouping; you want the coefficient of the squared term to be one. Write the factor outside the parentheses, with the variables inside.**

   $$x^2 + 6x \quad + 2\left(y^2 - 4y \quad\right) = -13$$

3. **Complete the square on the x's and the y's, and add whatever you used to complete the squares to the other side of the equation, too, to keep the equation balanced.**

   $$x^2 + 6x + 9 + 2\left(y^2 - 4y + 4\right) = -13 + 9 + 8$$

4. **Simplify each side of the equation by writing the trinomials on the left as binomials squared and by combining the terms on the right.**

   $$(x+3)^2 + 2(y-2)^2 = 4$$

Q. Write the equation $9x^2-4y^2-18x-16y-43=0$ as the sum of two binomials squared and a constant.

A. $9(x-1)^2-4(y+2)^2=36$. Use the steps as described.

1. **Rewrite the equation with the x's and y's together, with a space between the x terms and the y terms, and with the constant on the other side of the equation:**

   $9x^2-18x \quad -4y^2-16y \quad =43$

2. **Find any numerical factors for each grouping; you want the coefficient of the squared term to be one. Write the factor outside the parentheses, with the variables inside.**

   $9(x^2-2x \quad)-4(y^2+4y \quad)=43$

3. **Complete the square on the x's and the y's, and add whatever you used to complete the squares to the other side of the equation, too, to keep the equation balanced.**

   $9(x^2-2x+1)-4(y^2+4y+4)=43+9-16$

4. **Simplify each side of the equation by writing the trinomials on the left as binomials squared and by combining the terms on the right.**

   $9(x-1)^2-4(y+2)^2=36$

YOUR TURN

15 Use *completing the square* to solve the quadratic equation $x^2+8x-48=0$.

16 Use *completing the square* to solve the quadratic equation $5x^2+4x-1=0$.

Write the equation $x^2+18x+y^2-8y+52=0$ as the sum of two binomials squared and a constant.

Write the equation $25x^2-50x+4y^2+24y-39=0$ as the sum of two binomials squared and a constant.

Tackling Higher-Powered Polynomials

Solving polynomial equations requires that you know how to count and plan. Okay, so it isn't really that simple. But if you can count up to the number that represents the degree (highest power) of the equation, then you can account for the solutions you find and determine if you have them all. And if you can make a plan to use patterns in binomials or techniques from quadratics, you're well on your way to a solution.

Handling the sum or difference of cubes

As discussed in Chapter 1, you factor an expression that's the difference between two perfect squares into the difference and sum of the roots, $a^2-b^2=(a-b)(a+b)$. If you're solving an equation that represents the difference of two squares, you can apply the multiplication property of zero and solve. However, you can't factor the sum of two squares this way, so you're usually out of luck when it comes to finding any real solution.

In the case of the difference or sum of two cubes, you can factor the binomial, and you do find a solution. Here are the factorizations of the difference and sum of cubes:

$$a^3-b^3=(a-b)(a^2+ab+b^2)$$
$$a^3+b^3=(a+b)(a^2-ab+b^2)$$

Solving cubes by factoring

If you want to solve a cubic equation $x^3 - a^3 = 0$ by using the factorization of the difference of cubes, you get $(x-a)(x^2 + ax + a^2) = 0$. Using the multiplication property of zero when $x - a$ equals 0, x equals a. But when $x^2 + ax + a^2 = 0$, you need to use the quadratic formula, and you won't be too happy with the results.

Applying the quadratic formula, you get

$$x = \frac{-a \pm \sqrt{a^2 - 4(1)(a^2)}}{2(1)} = \frac{-a \pm \sqrt{a^2 - 4a^2}}{2} = \frac{-a \pm \sqrt{-3a^2}}{2}$$

You have a negative number under the radical, which means you find no real root. (Chapter 17 discusses what to do about negatives under radicals.) You deduce from this that the equation $x^3 - a^3 = 0$ has just one real solution, $x = a$.

Be careful when predicting the number of solutions of an equation from its power. In the previous example, the power 3 suggests that you may find as many as three solutions. But in actuality, the exponent just tells you that you can find *no more than* three solutions.

Solving cubes by taking the cube root

You may wonder if the factorization of the difference or sum of two perfect cubes always results in a quadratic factor that has no real roots (like the example I show in the preceding section, "Solving cubes by factoring"). Well, wonder no longer. The answer is a resounding "Yes!" When you factor a difference, $a^3 - b^3 = (a-b)(a^2 + ab + b^2)$, the quadratic $a^2 + ab + b^2 = 0$ doesn't have a real root; when you factor a sum, $a^3 + b^3 = (a+b)(a^2 - ab + b^2)$, the equation $a^2 - ab + b^2 = 0$ doesn't have a real root, either.

TIP

You can make the best of the fact that you find only one real root for equations in the cubic form by deciding how to deal with solving equations of that form. I suggest changing the equations $x^3 - b^3 = 0$ and $x^3 + b^3 = 0$ to $x^3 = b^3$ and $x^3 = -b^3$, respectively, and taking the cube root of each side.

EXAMPLE

Q. Solve $x^3 - 27 = 0$ using factoring.

A. $x = 3$. When factored, $x^3 - 27 = (x-3)\left(x^2 + 3x + 9\right) = 0$. The only real root is found when $x - 3 = 0$, which is $x = 3$.

Q. Solve $8x^3 + 125 = 0$ using a cube root.

A. $2x = -5$ or $-\frac{5}{2}$. Rewrite the equation as $8x^3 = -125$. Take the cube root of each side.

$\sqrt[3]{8x^3} = \sqrt[3]{-125}$, which becomes $2x = -5$ or $x = -\frac{5}{2}$.

YOUR TURN

19 Solve $125x^3 - 1 = 0$ using factoring.

20 Solve $27x^3 - 343 = 0$ using a cube root.

Solving quadratic-like trinomials

A quadratic-like trinomial is a trinomial of the form $ax^{2n} + bx^n + c = 0$. The power on one variable term is twice that of the other variable term, and a constant term completes the picture. The good thing about quadratic-like trinomials is that they're candidates for factoring and then for the application of the multiplication property of zero.

EXAMPLE

Q. Solve $y^4 - 17y^2 + 16 = 0$.

A. **4, −4, 1, −1.** Let $y^2 = x$ and $y^4 = x^2$. Then you have the quadratic $x^2 - 17x + 16 = 0$. You can factor this trinomial and get $(x-1)(x-16) = 0$. Substitute the y values back into this factored form and you have $(y^2 - 16)(y^2 - 1) = 0$. Both of the factors can now be factored: $= (y-4)(y+4)(y-1)(y+1) = 0$. Setting the individual factors equal to zero, you get $y = 4$, $y = -4$, $y = 1$, and $y = -1$.

Q. Solve $z^6 - 26z^3 - 27 = 0$.

A. **27, 3, −1, −1.** Let $z^3 = x$ and $z^6 = x^2$. Then you have the quadratic $x^2 - 26x - 27 = 0$. This factors into $(x-27)(x+1)$. Now you replace the x's in the factorization with z^3, and you have the factorization for the equation with the z's; set it equal to zero: $z^6 - 26z^3 - 27 = (z^3 - 27)(z^3 + 1) = 0$. When $z^3 - 27 = 0$, $z^3 = 27$, and $z = 3$. And when $z^3 + 1 = 0$, $z^3 = -1$, and $z = -1$.

You can just take the cube roots of each side of the equations you form (see the previous section), because when you take that odd root, you know you can find only one real solution.

YOUR TURN

 Solve $9x^4 - 37x^2 + 4 = 0$.

 Solve $x^6 - 9x^3 + 8 = 0$.

Using the Rational Root Theorem and Synthetic Division

Polynomials can take on all shapes and sizes. You have many tools at your disposal to help you factor quadratic and cubic polynomials. But what if the cubic polynomial isn't the sum or difference of cubes? And what if a higher-degree polynomial isn't quadratic-like? This is where the Rational Root Theorem and synthetic division come in.

First, consider the Factor Theorem. If $(x-a)$ is a factor of the polynomial $y = a_n x^n + a_{n-1}x^{n-1} + a_{n-2}x^{n-2} + \ldots + a_1 x^1 + a_0$, then a is a solution or root of $y = a_n x^n + a_{n-1}x^{n-1} + a_{n-2}x^{n-2} + \ldots + a_1 x^1 + a_0 = 0$. So how do you find those factors?

Without further ado, here's the Rational Root Theorem: If the polynomial $y = a_n x^n + a_{n-1}x^{n-1} + a_{n-2}x^{n-2} + \ldots + a_1 x^1 + a_0$ has any rational roots, they all meet the requirement that you can write them as a fraction equal to $\frac{\text{factor of } a_0}{\text{factor of } a_n}$. In other words, according to the theorem, any rational root of a polynomial is formed by dividing a factor of the constant term by a factor of the lead coefficient.

To help determine the possible roots, you also have Descartes' Rule of Signs (with details found in Chapter 9). Count the number of times the sign changes in f, and call that value p. The value of p is the maximum number of positive real roots of f. If the number of positive roots isn't p, then it is $p-2$, $p-4$, or some number less by a multiple of two. To determine the maximum number of negative real roots, you replace each x in the polynomial with $-x$ and then count the changes in sign again. As with the positive number of roots, you decrease by 2s if necessary.

Synthetic division is used to divide polynomials by binomials, and it essentially just uses the roots. So you make a list of possible roots and check them out with synthetic division. The winners are those you use in the factorization of the polynomial. (You can find more information about, and practice problems for, long division and synthetic division in one of my other spine-tingling thrillers, *Algebra I Workbook For Dummies* [John Wiley & Sons, Inc.].)

Q. Find the roots of the polynomial $y = x^4 - 4x^3 - x^2 + 16x - 12$.

EXAMPLE **A.** **1, 2, 3, −2.** First, with the factor theorem, you divide all the factors of 12 by 1 and get: $\pm 1, \pm 2, \pm 3, \pm 4, \pm 6, \pm 12$. There are only 4 possible factors, and you have 12 possibilities. Using Descartes' Rule of Signs, there are 3 changes (positive to negative to positive to negative). Replace the x's with $y = x^4 + 4x^3 - x^2 - 16x - 12$ and the equation becomes $y = x^4 + 4x^3 - x^2 - 16x - 12$, which has just 1 change in sign. So you have either 1 or 3 positive roots and just 1 negative root. Looks like I'll try the positive roots first!

Using synthetic division involves these steps:

1. **Write the polynomial in order of decreasing powers of the exponents. Replace any missing powers with zero to represent the coefficient.**

   In this case, you've lucked out. The polynomial is already in the correct order, and there are no missing terms: $y = x^4 - 4x^3 - x^2 + 16x - 12$.

2. **Write the coefficients in a row, including any zeros, and the number you're trying out as a root in the half-box in front. (I chose the number 1.) Draw a horizontal line below the row of coefficients, leaving room for numbers under the coefficients.**

$$\begin{array}{r|rrrrr} 1 & 1 & -4 & -1 & 16 & -12 \\ & & & & & \\ \hline & & & & & \end{array}$$

3. **Bring the first coefficient straight down below the line. Multiply the number you bring below the line by the number that you're dividing into everything. Put the result under the second coefficient.**

$$\begin{array}{r|rrrrr} 1 & 1 & -4 & -1 & 16 & -12 \\ & & 1 & & & \\ \hline & 1 & & & & \end{array}$$

4. **Add the second coefficient and the product, putting the result below the line. Repeat the multiplication/addition with the rest of the coefficients.**

$$\begin{array}{r|rrrrr} 1 & 1 & -4 & -1 & 16 & -12 \\ & & 1 & -3 & -4 & 12 \\ \hline & 1 & -3 & -4 & 12 & 0 \end{array}$$

   The last entry on the bottom is a zero, so you know $x = 1$ is a root. Now, you can do a modified synthetic division when testing for the next root; you just use the numbers across the bottom except for the bottom-right zero entry. (These values are actually coefficients of the quotient, if you do long division.)

   Now, doing synthetic division on the bottom row:

1. **Write the coefficients in a row, including any zeros, and the number you're trying out as a root in the half-box in front. (I chose the number 2.) Draw a horizontal line below the row of coefficients, leaving room for numbers under the coefficients.**

2	1	–3	–4	12

2. **Bring the first coefficient straight down below the line. Multiply the number you bring below the line by the number that you're dividing into everything. Put the result under the second coefficient.**

2	1	–3	–4	12
		2		
	1			

3. **Add the second coefficient and the product, putting the result below the line. Repeat the multiplication/addition with the rest of the coefficients.**

2	1	–3	–4	12
		2	–2	–12
	1	–1	–6	0

The last entry on the bottom is a zero, so you know $x = 2$ is a root. You can stop with the synthetic division now. Use the bottom row to write the quadratic equation $x^2 - x - 6 = 0$. Factor the equation to find the two remaining roots: $x^2 - x - 6 = (x-3)(x+2) = 0$. You find the roots $x = 3$ and $x = -2$ here. You already had $x = 1$ and $x = 2$. So that's 3 positive roots and 1 negative root.

YOUR TURN

23 Solve $y = x^5 + 5x^4 - 2x^3 - 28x^2 - 8x + 32 = 0$.

24 Solve $y = 2x^3 - 13x^2 + 3x + 18 = 0$.

Practice Questions Answers and Explanations

1. **5,−5.** Add 3 to each side of the equation. Then divide each side by 2. Take the square root of both sides.

$$2x^2-3=47 \rightarrow 2x^2=50 \rightarrow x^2=25$$
$$\sqrt{x^2}=\pm\sqrt{25}$$
$$x=\pm 5$$

2. **$\pm 3\sqrt{2}$.** Subtract 7 from both sides. Take the square root of both sides. Simplify the radical.

$$z^2+7=25 \rightarrow z^2=18$$
$$\sqrt{z^2}=\pm\sqrt{18}$$
$$=\pm\sqrt{9\cdot 2}=\pm 3\sqrt{2}$$

3. **$0, \frac{1}{2}$.** Factor out the $3x$. Set each factor equal to 0.

$$6x^2-3x=3x(2x-1)=0$$
$$3x=0 \rightarrow x=0$$
$$2x-1=0 \rightarrow x=\frac{1}{2}$$

4. **0,5.** Subtract $20y$ from both sides, then factor out the $4y$. Set each factor equal to 0.

$$4y^2=20y \rightarrow 4y^2-20y=0 \rightarrow 4y(y-5)=0$$
$$4y=0 \rightarrow y=0$$
$$y-5=0 \rightarrow y=5$$

5. **10,−10.** Factor and set the two factors equal to 0.

$$x^2-100=(x-10)(x+10)=0$$
$$x-10=0 \rightarrow x=10$$
$$x+10=0 \rightarrow x=-10$$

6. **$\frac{5z}{3}, -\frac{5z}{3}$.** Subtract $25z^2$ from each side. Then factor and set the two factors equal to 0.

$$9y^2=25z^2 \rightarrow 9y^2-25z^2=(3y-5z)(3y+5z)=0$$
$$3y-5z=0 \rightarrow y=\frac{5z}{3}$$
$$3y+5z=0 \rightarrow y=-\frac{5z}{3}$$

7. **11,−4.** Factor and set the factors equal to 0.

$$x^2-7x-44=(x-11)(x+4)=0$$
$$x-11=0 \rightarrow x=11$$
$$x+4=0 \rightarrow x=-4$$

8) $-\frac{5}{2}, \frac{6}{5}$. **First, factor out the 5. Then use FOIL to factor the trinomial. Set the factors equal to 0.**

$$50x^2+65x-150=0 \rightarrow 5\left(10x^2+13x-30\right)=0$$
$$5(2x+5)(5x-6)=0$$
$$2x+5=0 \rightarrow x=-\frac{5}{2}$$
$$5x-6=0 \rightarrow x=\frac{6}{5}$$

9) **−13. Factor as a binomial squared. Then set the factors equal to 0.**

$$x^2+26x+169=(x+13)(x+13)=(x+13)^2=0$$
$$x+13=0 \rightarrow x=-13$$

10) $\frac{3}{4}$. **Factor as a binomial squared. Then set the factors equal to 0.**

$$16x^2-24x+9=0 \rightarrow (4x-3)(4x-3)=(4x-3)^2=0$$
$$4x-3=0 \rightarrow x=\frac{3}{4}$$

11) **−7,1,−1. Use grouping to factor. Then set the factors equal to 0.**

$$x^3+7x^2-x-7=0 \rightarrow x^2(x+7)-1(x+7)=0$$
$$(x+7)\left(x^2-1\right)=0 \rightarrow (x+7)(x-1)(x+1)=0$$
$$x+7=0 \rightarrow x=-7$$
$$x-1=0 \rightarrow x=1$$
$$x+1=0 \rightarrow x=-1$$

12) $-\frac{4}{3}, \frac{5}{2}, -\frac{5}{2}$. **Use grouping to factor. Then set the factors equal to 0.**

$$12x^3+16x^2-75x-100=0 \rightarrow 4x^2(3x+4)-25(3x+4)=0$$
$$(3x+4)\left(4x^2-25\right)=0 \rightarrow (3x+4)(2x-5)(2x+5)=0$$
$$3x+4=0 \rightarrow x=-\frac{4}{3}$$
$$2x-5=0 \rightarrow x=\frac{5}{2}$$
$$2x+5=0 \rightarrow x=-\frac{5}{2}$$

13) $\frac{3 \pm \sqrt{21}}{3}$. **Use the quadratic formula where $a=3$, $b=-6$, $c=-4$.**

$$x=\frac{6 \pm \sqrt{(-6)^2-4(3)(-4)}}{2(3)}=\frac{6 \pm \sqrt{36+48}}{6}=\frac{6 \pm \sqrt{84}}{6}=\frac{6 \pm \sqrt{4 \cdot 21}}{6}=\frac{6 \pm 2\sqrt{21}}{6}=\frac{3 \pm \sqrt{21}}{3}.$$

14) $\frac{5}{3}, \frac{25}{16}$. **Even though this factors, use the quadratic formula where $a=48$, $b=-155$, $c=125$.**

$$x=\frac{155 \pm \sqrt{(-155)^2-4(48)(125)}}{2(48)}=\frac{155 \pm \sqrt{24{,}025-24{,}000}}{96}=\frac{155 \pm \sqrt{25}}{96}=\frac{155 \pm 5}{96}$$

The two answers are: $\frac{155+5}{96}=\frac{160}{96}=\frac{5}{3}$ and $\frac{155-5}{96}=\frac{150}{96}=\frac{25}{16}$

If you had solved this by factoring, you would have gotten:
$48x^2-155x+125=(3x-5)(16x-25)=0$

15. **4,−12.** Add 48 to each side of the equation, then complete the square by adding 16 to each side. Factor on the left and find the square root of both sides. Solve for *x*.

$$x^2+8x-48=0 \rightarrow x^2+8x=48 \rightarrow x^2+8x+16=48+16$$
$$x^2+8x+16=64 \rightarrow (x+4)^2=64 \rightarrow \sqrt{(x+4)^2}=\pm\sqrt{64} \rightarrow x+4=\pm 8$$
$$x=-4\pm 8 \rightarrow x=4 \text{ or } x=-12$$

16. $\mathbf{\frac{1}{5}}$**,−1.** First, divide each term by 5. Then add $\frac{1}{5}$ to each side of the equation. Complete the square by adding $\frac{4}{25}$ to each side. Factor on the left and find the square root of both sides. Solve for *x*.

$$5x^2+4x-1=0 \rightarrow x^2+\frac{4}{5}x-\frac{1}{5}=0 \rightarrow x^2+\frac{4}{5}x=\frac{1}{5} \rightarrow x^2+\frac{4}{5}x+\frac{4}{25}=\frac{1}{5}+\frac{4}{25}$$
$$x^2+\frac{4}{5}x+\frac{4}{25}=\frac{9}{25} \rightarrow \left(x+\frac{2}{5}\right)^2=\frac{9}{25} \rightarrow \sqrt{\left(x+\frac{2}{5}\right)^2}=\pm\sqrt{\frac{9}{25}} \rightarrow x+\frac{2}{5}=\pm\frac{3}{5}$$
$$x=-\frac{2}{5}\pm\frac{3}{5}=\frac{-2\pm 3}{5} \rightarrow x=\frac{1}{5}, \text{ or } x=-1$$

17. $\mathbf{(x+9)^2+(y-4)^2=45}$**.** Subtract 52 from both sides. Then complete the squares of both the *x* terms and *y* terms, adding the amounts needed to both sides of the equation.

$$x^2+18x+y^2-8y+52=0 \rightarrow x^2+18x+y^2-8y=-52$$
$$x^2+18x+81+y^2-8y+16=-52+81+16$$
$$(x+9)^2+(y-4)^2=45$$

18. $\mathbf{5(x-1)^2+4(y+3)^2=80}$**.** Add 39 to both sides. Factor the two sets of terms in the same variable. Then complete the squares of the factored *x* terms and *y* terms, adding the amounts needed to both sides of the equation.

$$25x^2-50x+4y^2+24y-39=0 \rightarrow 25x^2-50x+4y^2+24y=39 \rightarrow 25\left(x^2-2x\right)+4\left(y^2+6y\right)=39$$
$$5\left(x^2-2x+1\right)+4\left(y^2+6y+9\right)=39+5+36$$
$$5(x-1)^2+4(y+3)^2=80$$

19. $\mathbf{\frac{1}{5}}$**.** Using the pattern for the difference of cubes: $125x^3-1=(5x-1)\left(25x^2+5x+1\right)=0$. The only solution is from $5x-1=0$, so $x=\frac{1}{5}$.

20. $\mathbf{\frac{7}{3}}$**.** Add 343 to both sides and then take the cube root of both sides:

$$27x^3-343=0 \rightarrow 27x^3=343 \rightarrow \sqrt[3]{27x^3}=\sqrt[3]{343} \rightarrow 3x=7 \rightarrow x=\frac{7}{3}$$

21) $\frac{1}{3}, -\frac{1}{3}, 1, -1$. **Create a quadratic equation in y's from this quadratic-like version. Factor using FOIL and replace the substituted values.**

$$9x^4 - 37x^2 + 4 = 0 \rightarrow 9y^2 - 37y + 4 = 0 \rightarrow (9y-1)(y-4) = 0 \rightarrow (9x^2-1)(x^2-1) = 0.$$

Now factor the binomials and solve for x: $(9x^2-1)(x^2-1) = (3x-1)(3x+1)(x-1)(x+1) = 0$. Setting the factors equal to 0, you have $x = \frac{1}{3},\ x = -\frac{1}{3},\ x = 1,\ x = -1$.

22) **1,2.** Create a quadratic equation in y's from this quadratic-like version. Factor using FOIL and replace the substituted values.

$$x^6 - 9x^3 + 8 = 0 \rightarrow y^2 - 9y + 8 = 0 \rightarrow (y-8)(y-1) = 0 \rightarrow (x^3-8)(x^3-1) = 0$$

Now factor the binomials and solve for x:
$(x^3-8)(x^3-1) = (x-2)(x^2+2x+4)(x-1)(x^2+x+1) = 0$

There are only two solutions: $x = 2$ and $x = 1$.

23) **1,2,−2,−2,−4.** Create a list of possible roots: $\pm1, \pm2, \pm4, \pm8, \pm16, \pm32$. There are 2 changes in sign, so there are 2 or 0 positive roots. Replacing all x's with $-x$'s, there are 3 changes in sign so there are 3 or 1 negative roots. Making a first guess of −2 and performing synthetic division:

$$\begin{array}{r|rrrrrr} -2 & 1 & 5 & -2 & -28 & -8 & 32 \\ & & -2 & -6 & 16 & 24 & -32 \\ \hline & 1 & 3 & -8 & -12 & 16 & 0 \end{array}$$

This is a root. Now try it again.

$$\begin{array}{r|rrrrr} -2 & 1 & 3 & -8 & -12 & 16 \\ & & -2 & -2 & 20 & -16 \\ \hline & 1 & 1 & -10 & 8 & 0 \end{array}$$

This is a root — again. Now try 1.

$$\begin{array}{r|rrrr} 1 & 1 & 1 & -10 & 8 \\ & & 1 & 2 & -8 \\ \hline & 1 & 2 & -8 & 0 \end{array}$$

This is a root, also. Write the corresponding quadratic from the numbers in the bottom row and factor it.

This last bit of factoring should yield $(x + 4)(x - 2)$: $x^2 + 2x - 8 = (x+4)(x-2) = 0$. This gives you the roots −4 and 2. So the five roots are: 1, 2, −2, −2, −4.

24) **$-1, \frac{3}{2}, 6$.** Create a list of possible roots: $\pm1, \pm2, \pm3, \pm6, \pm9, \pm\frac{1}{2}, \pm\frac{3}{2}, \pm\frac{9}{2}$. That's a lot to choose from! There are 2 changes in sign, so there are 2 or 0 positive roots. Replacing all x's with $-x$'s, there is 1 change in sign, indicating 1 negative root. Making a first guess of −1 and performing synthetic division: $y = 2x^3 - 13x^2 + 3x + 18 = 0$.

$$\begin{array}{r|rrrr} -1 & 2 & -13 & 3 & 18 \\ & & -2 & 15 & -18 \\ \hline & 2 & -15 & 18 & 0 \end{array}$$

This is a root, also. Write the corresponding quadratic from the numbers in the bottom row and factor it.

$2x^2 - 15x + 18 = (2x-3)(x-6) = 0$. This gives you the roots $\frac{3}{2}$ and 6. So the three roots are: $-1, \frac{3}{2}, 6$.

Whaddya Know? Chapter 3 Quiz

Quiz time! Complete each problem to test your knowledge on the various topics covered in this chapter. You can then find the solutions and explanations in the next section.

1. **Solve for** x: $x^2 + 6x - 55 = 0$.

2. **Write the polynomial as the sum of two binomials squared:** $x^2 + y^2 - 6x + 12y - 49 = 0$.

3. **When you solve** $2x^2 + 12x - 11 = 0$ **for** x **by completing the square, your first step is to do what?**

4. **Solve for** x: $6x^2 - 7x - 20 = 0$.

5. **Solve for** y: $12y^2 - 90y + 42 = 0$.

6. **Solve the quadratic-like equation for** x: $x^6 - 19x^3 - 216 = 0$.

7. **Solve for** x: $4x^2 - 7 = 29$.

8. **Solve for** x: $x^3 - 4x^2 - 9x + 36 = 0$.

9. **Solve for** z: $4z^2 - 81 = 0$.

10. **Solve for** x **using the quadratic formula:** $2x^2 - 3x - 4 = 0$.

11. **Solve for** x: $12x^2 - x - \frac{1}{6} = 0$.

12. **Solve for** x **using the quadratic formula:** $2x^2 - 6x + 3 = 0$.

13. **When solving for** x **in the equation** $x^2 - 8x - 9 = 0$ **using completing the square, what is the next step after you get to** $x^2 - 8x \quad = 9$?

14. **Solve the quadratic-like equation for** y: $36y^4 - 13y^2 + 1 = 0$.

Answers to Chapter 3 Quiz

1. **5,−11.** Factor the quadratic: $x^2+6x-55=(x-5)(x+11)=0$. Setting the two factors equal to zero, you have $x=5$ and $x=-11$.

2. **$(x-3)^2+(y+6)^2=94$.** First rearrange the terms, grouping the x's and y's. Add 49 to each side.

$$x^2-6x\quad+y^2+12y\quad=49$$

Now complete the square for both the x's and y's, adding like amounts to the right side of the equation. Then factor the perfect square trinomials.

$$x^2-6x+9+y^2+12y+36=49+9+36$$
$$(x-3)^2+(y+6)^2=94$$

3. **Divide through by 2.** When you solve $2x^2+12x-11=0$ for x by completing the square, your first step is to divide each term by 2, giving you $x^2+6x-\frac{11}{2}=0$. Then you add $\frac{11}{2}$ to each side and complete the square by adding 9 to each side: $x^2+6x+9=\frac{11}{2}+9=\frac{29}{2}$. Factor on the left, then take the square root of both sides: $(x+3)^2=\frac{29}{2}\rightarrow x+3=\sqrt{\frac{29}{2}}$. Subtract 3 from each side, and you've solved the equation using completing the square.

4. **$-\frac{4}{3},\frac{5}{2}$.** Factoring the quadratic: $6x^2-7x-20=(3x+4)(2x-5)$. Setting the two factors equal to zero, you have $x=-\frac{4}{3}$ and $x=\frac{5}{2}$.

5. **$\frac{1}{2}$,7.** First, factor 6 from each term: $12y^2-90y+42=6(2y^2-15y+7)$. Then, factoring the trinomial, you have $6(2y-1)(y-7)=0$. Setting the two factors equal to zero, $y=\frac{1}{2}$ and $y=7$.

6. **3,−2.** Replace the x^6 with y^2 and x^3 with y. Factor the quadratic equation.

$$x^6-19x^3-216=0\rightarrow y^2-19y-216=0\rightarrow(y-27)(y+8)=0$$

Replace the y's with x^3 and factor the two differences of cubes.

$$(y-27)(y+8)=0\rightarrow(x^3-27)(x^3+8)=0\rightarrow(x-3)(x^2+3x+9)(x+2)(x^2-2x+4)=0$$

The two binomials provide the only two solutions: $x=3$ and $x=-2$.

7. **±3.** Add 7 to both sides. Then find the square root of both sides: $4x^2=36\rightarrow\sqrt{4x^2}=\pm\sqrt{36}\rightarrow 2x=\pm6$. Finally, divide each side by 2: $x=\pm3$.

8. **3, −3, 4.** Group the first two and last two terms together using their common factors: $=x^2(x-4)-9(x-4)=0$. Factor again, using the common factor of the two terms: $=(x-4)(x^2-9)$. Finally, factor the difference of squares and solve for x: $=(x-4)(x-3)(x+3)=0$, giving you $x=4$, $x=3$, $x=-3$.

9. **$\pm\frac{9}{2}$.** Factor the quadratic: $4z^2-81=(2z-9)(2z+9)=0$. Solving for z, $z=\frac{9}{2}$ and $z=-\frac{9}{2}$.

(10) $\frac{3}{4}+\frac{\sqrt{41}}{4}$ or $\frac{3}{4}-\frac{\sqrt{41}}{4}$. Applying the quadratic formula:

$$x=\frac{3\pm\sqrt{9-4(2)(-4)}}{2(2)}=\frac{3\pm\sqrt{9+32}}{4}=\frac{3\pm\sqrt{41}}{4}=\frac{3}{4}\pm\frac{\sqrt{41}}{4}$$

(11) $\frac{1}{6},-\frac{1}{12}$. First, multiply the equation through by 6: $6\left(12x^2-x-\frac{1}{6}\right)=72x^2-6x-1$. Factor the quadratic: $=(12x+1)(6x-1)$. Set the factors equal to zero and solve for x: $x=-\frac{1}{12}, x=\frac{1}{6}$.

(12) $\frac{3}{2}+\frac{\sqrt{3}}{2}$ or $\frac{3}{2}-\frac{\sqrt{3}}{2}$. Applying the quadratic formula:

$$x=\frac{6\pm\sqrt{36-4(2)(3)}}{2(2)}=\frac{6\pm\sqrt{36-24}}{4}=\frac{6\pm\sqrt{12}}{4}$$

The 12 under the radical is the product of a perfect square and 3.

$$\frac{6\pm\sqrt{12}}{4}=\frac{6\pm\sqrt{4\cdot 3}}{4}=\frac{6\pm\sqrt{4}\sqrt{3}}{4}=\frac{6\pm 2\sqrt{3}}{4}=\frac{\cancel{6}^{3}\pm\cancel{2}\sqrt{3}}{\cancel{4}^{2}}=\frac{3\pm\sqrt{3}}{2}=\frac{3}{2}\pm\frac{\sqrt{3}}{2}$$

(13) **Complete the square by adding half of –8 squared to each side.** First, add 9 to each side. Then find half of the coefficient –8, square it, and add that to both sides of the equation.

$$\begin{aligned} x^2-8x-9&=0\\ x^2-8x&=9\\ x^2-8x+\left(\frac{-8}{2}\right)^2&=9+\left(\frac{-8}{2}\right)^2\\ x^2-8x+16&=9+16=25 \end{aligned}$$

Factor the trinomial on the left, and then find the square root of both sides of the equation.

$$\begin{aligned} (x-4)^2&=25\\ \sqrt{(x-4)^2}&=\pm\sqrt{25}\\ x-4&=\pm 5 \end{aligned}$$

Add the 4 to each side.

$$\begin{aligned} x&=4\pm 5\\ &=9 \text{ and } -1 \end{aligned}$$

(14) $\pm\frac{1}{3}$ and $\pm\frac{1}{2}$. Substitute x for y^2 and x^2 for y^4 in the equation. Then factor the quadratic.

$$\begin{aligned} &36x^2-13x+1=0\\ &(9x-1)(4x-1)=0 \end{aligned}$$

Solve each factor for x. Then replace each x with y^2 and solve for y.

$9x-1=0$ gives you $x=\frac{1}{9}$ or $y^2=\frac{1}{9}$. Finding the square root of both sides, $y=\pm\frac{1}{3}$.

$4x-1=0$ gives you $x=\frac{1}{4}$ or $y^2=\frac{1}{4}$. Finding the square root of both sides, $y=\pm\frac{1}{2}$.

IN THIS CHAPTER

» Taking on quadratic inequalities

» Incorporating the number line for solutions

» Getting rational with rational inequalities

» Being absolutely sure with absolute value inequalities

» Applying the techniques

Chapter 4 Controlling Quadratic and Rational Inequalities

An algebraic inequality is just what it says: an inequality (<, >, ≤, or ≥) that involves a linear, polynomial, rational, or absolute value expression. I begin here with quadratic inequalities, and once you have this type of inequality mastered, you see how to employ the same method to solve high-degree inequalities, rational inequalities (which contain variables in fractions), and absolute value inequalities. To finish off the topic, I offer several types of real-world applications for using inequalities.

Checking Out Quadratic Inequalities

In Chapter 2, you discover the techniques used to solve linear inequalities. The next step is to tackle quadratic inequalities. You need to be able to solve quadratic equations in order to solve quadratic inequalities. (For a refresher on quadratic equations, check out Chapter 3.) With quadratic equations, you set the expressions equal to zero; inequalities deal with what's on either side of the zero (positives and negatives).

To solve a quadratic inequality, follow these steps:

1. **Move all the terms to one side of the inequality sign; the sign should now point at a zero.**
2. **Factor, if possible.**
3. **Determine all zeros (roots or solutions). Zeros are the values of the variable that make factors equal to zero and, by multiplying, that make the expression equal to zero.**
4. **Put the zeros in order on a number line.**
5. **Create a sign-line to show where the expression in the inequality is positive or negative.**

   A sign-line shows the signs of the different factors in each interval. If the expression is factored, show the signs of the individual factors.

6. **Determine the solution, writing it in inequality notation or interval notation (I cover interval notation in Chapter 1).**

What are you actually doing when you *solve* an inequality? The following example asks you to solve $x^2 - x > 12$. You need to determine what values of x you can square so that when you subtract the original number, your answer will be bigger than 12. For instance, when $x = 5$, you get $25 - 5 = 20$. That's certainly bigger than 12, so the number 5 works; $x = 5$ is a solution. How about the number 2? When $x = 2$, you get $4 - 2 = 2$, which isn't bigger than 12. You can't use $x = 2$ in the solution. Do you then conclude that smaller numbers don't work? No. When you try $x = -10$, you get $100 + 10 = 110$, which is most definitely bigger than 12. You can actually find an infinite amount of numbers that make this inequality a true statement. Rather than keep guessing the answers, the following examples give you a procedure.

EXAMPLE

Q. Solve the inequality $x^2 - x > 12$.

A. $x < -3$ or $x > 4$. Solve the quadratic inequality by using the preceding steps:

1. **Subtract 12 from each side of the inequality to move all the terms to one side.**

   You end up with $x^2 - x - 12 > 0$.

2. **Factoring on the left side of the inequality, you get $(x-4)(x+3) > 0$.**
3. **Determine that the zeros for the inequality are $x = 4$ and $x = -3$.**
4. **Put the zeros in order on a number line, as shown here.**

5. **Create a sign-line to show the signs of the different factors in each interval.**

   Between -3 and 4, try letting $x = 0$ (you can use any number between -3 and 4). When $x = 0$, the factor $(x-4)$ is negative, and the factor $(x+3)$ is positive. Put those signs on the sign-line to correspond to the factors. Do the same for the interval of numbers to the left of -3 and to the right of 4.

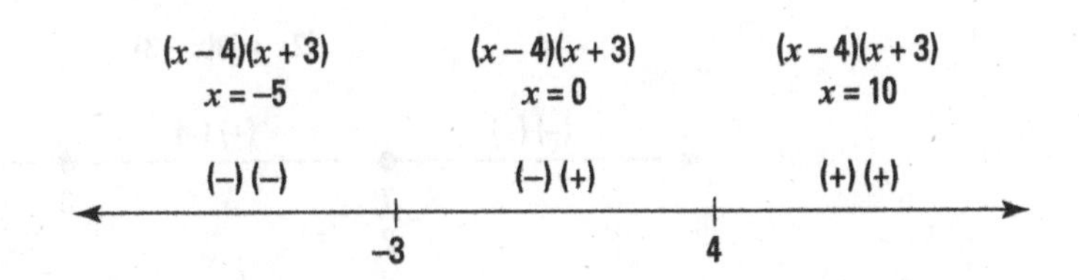

The x-values in each interval are really random choices (as you can see from my choice of $x = -5$ and $x = 10$). Any number in each of the intervals gives the same positive or negative value to the factor.

6. **To determine the solution, look at the signs of the factors; you want the expression to be positive, corresponding to the inequality greater than zero.**

   The interval to the left of –3 has a negative times a negative, which is positive. So, any number to the left of –3 works. You can write that part of the solution as $x < -3$ or, in interval notation (see Chapter 1), $(-\infty, -3)$. The interval to the right of 4 has a positive times a positive, which is positive. So, $x > 4$ is a solution; you can write it as $(4, \infty)$. The interval between –3 and 4 is always negative; you have a negative times a positive. The complete solution lists both intervals that have working values in the inequality.

   The solution of the inequality $x^2 - x > 12$, therefore, is $x < -3$ or $x > 4$. Writing this result in interval notation, you replace the word "or" with the symbol $\cup$ and write it as $(-\infty, -3) \cup (4, \infty)$.

Q. Solve the inequality $2x^2 + x - 21 \le 0$.

A. $-\frac{7}{2} \le x \le 3$. Solve this quadratic inequality by using the preceding steps:

1. **The entire expression is already on the left side.**
2. **Factoring on the left side of the inequality, you get $(2x + 7)(x - 3) \le 0$.**
3. **Determine that the zeros for the inequality are $x = -\frac{7}{2}$ and $x = 3$.**
4. **Put the zeros in order on a number line, as shown here.**

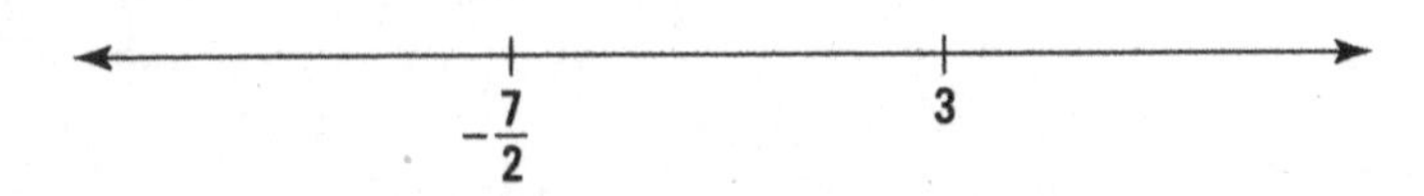

5. **Create a sign-line to show the signs of the different factors in each interval. Place solid dots on the two zeros to indicate that these values will be included in the solution.**

   Between $-\frac{7}{2}$ and 3, try letting $x = 0$. When $x = 0$, the factor $(2x + 7)$ is positive, and the factor $(x - 3)$ is negative. Put those signs on the sign-line to correspond to the factors. Do the same for the interval of numbers to the left of $-\frac{7}{2}$ and to the right of 3.

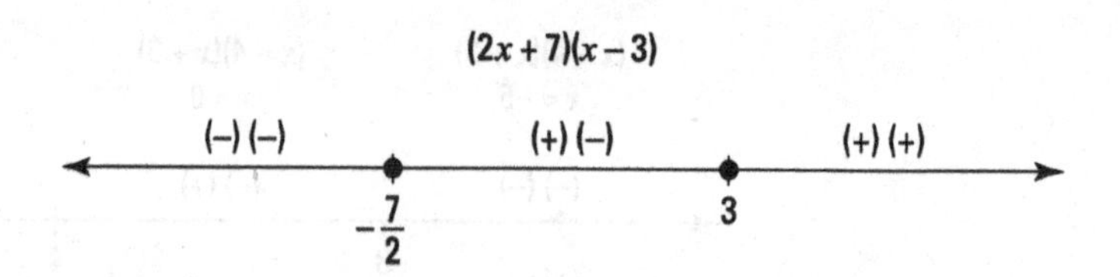

The x-values in each interval are really random choices. Any number in each of the intervals gives the same positive or negative value to the factor.

6. **To determine the solution, look at the signs of the factors; you want the expression to be negative or zero, corresponding to the inequality less-than-or-equal-to zero.**

   The interval to the left of $-\frac{7}{2}$ has a negative times a negative, which is positive. So, any number to the left of $-\frac{7}{2}$ is not a solution. The interval to the right of 3 has a positive times a positive, which is positive, so this is still not part of the solution. The interval between $-\frac{7}{2}$ and 3 gives you a negative times a positive, so there's the solution.

   The solution of the inequality $2x^2 + x - 21 \leq 0$, therefore, is $-\frac{7}{2} \leq x \leq 3$. Writing this result in interval notation, you get $\left[-\frac{7}{2}, 3\right]$.

YOUR TURN

1 Solve for x: $x^2 - x - 30 \geq 0$.

2 Solve for x: $x^2 < 25$.

3 Solve for x: $3 > x^2 - 2x$.

4 Solve for x: $x^2 + 20 \le 9x$.

Taking on Rational Inequalities and Non-Solutions

The sign-line process is great for solving rational inequalities, such as $\frac{x-2}{x+6} \le 0$. The signs of the results of multiplication and division use the same rules, so to determine your answer, you can treat the numerator and denominator the same way you treat two different factors in multiplication. The only thing to watch for, though, is not to use any solution value that creates a zero in the denominator.

EXAMPLE

Q. Solve $\frac{x-2}{x+6} \le 0$.

A. $(-6, 2]$. Follow the steps from the previous section.

1. **Every term in $\frac{x-2}{x+6} \le 0$ is already to the left of the inequality sign.**
2. **Neither the numerator nor the denominator factors any further.**
3. **The two zeros are $x = 2$ and $x = -6$.**
4. **Place the two numbers on a number line.**

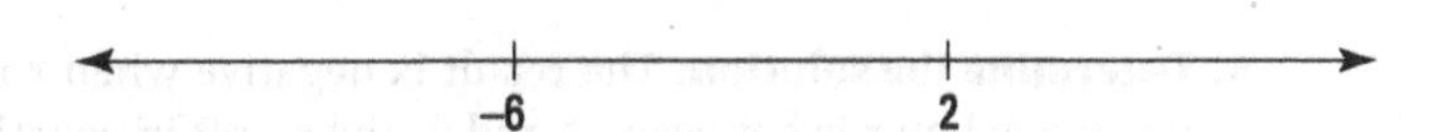

5. **Create a sign-line for the two zeros; you can see that the numerator is positive when x is greater than 2, and the denominator is positive when x is greater than –6.**

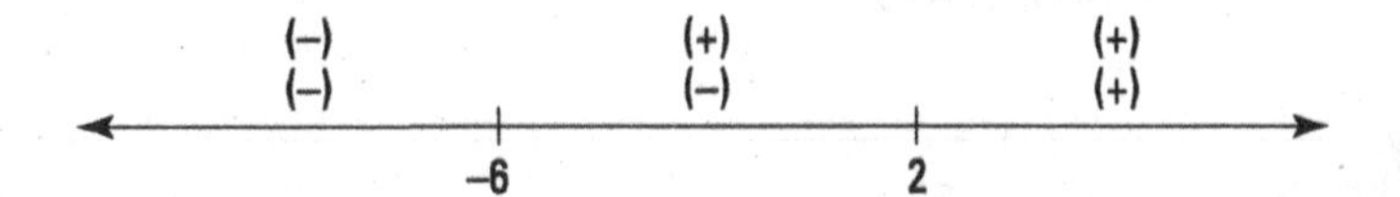

6. **When determining the solution, keep in mind that the inequality calls for something less than or equal to zero.**

   The fraction is a negative number when you choose an x between –6 and 2. You get a negative numerator and a positive denominator, which gives a negative result. Another solution to the original inequality is 2. Letting $x = 2$, you get a numerator equal to 0, which you want because the inequality is less than or equal to zero. You can't let the denominator be zero, though. Having a zero in the denominator isn't allowed because no such number exists. So, the solution of $\frac{x-2}{x+6} \le 0$ is $-6 < x \le 2$. In interval notation, you write the solution as $(-6, 2]$.

Q. Solve $x > \frac{9}{x}$.

A. $(-3,0) \cup (3,\infty)$. Follow the steps from the preceding section.

1. **Subtract $\frac{9}{x}$ from each side; find a common denominator and combine the two terms.**

$$x - \frac{9}{x} > 0 \rightarrow \frac{x^2}{x} - \frac{9}{x} > 0 \rightarrow \frac{x^2 - 9}{x} > 0$$

2. **Factor the numerator: $\frac{(x-3)(x+3)}{x} > 0$.**
3. **There are three zeros: $x = 3$, $x = -3$, and $x = 0$.**
4. **Place the three numbers on a number line.**

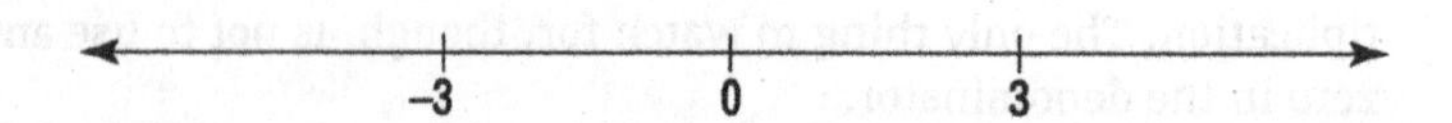

5. **Create a sign-line for the three zeros.** The numerator is positive and the denominator is negative when x is less than –3; the numerator is negative and the denominator is negative when x is between –3 and 0; the numerator is negative and the denominator is positive when x is between 0 and 3; and the numerator is positive and the denominator is positive when x is greater than 3.

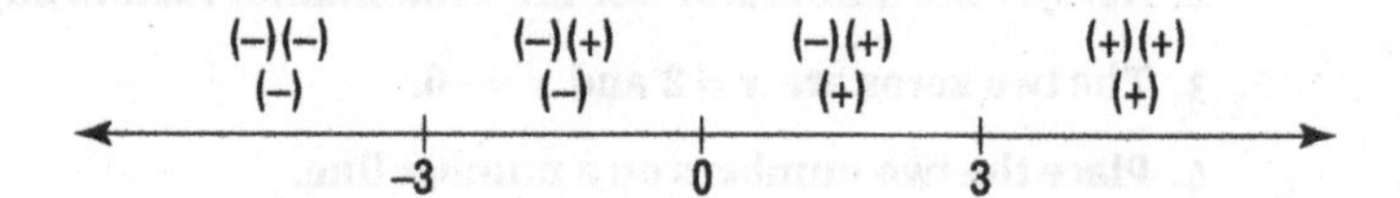

6. **Determine the solution.** The result is negative when x is less than –3; the result is positive when x is between –3 and 0; the result is negative when x is between 0 and 3; and the result is positive when x is greater than 3. So, the solution of $x > \frac{9}{x}$ or $x - \frac{9}{x} > 0$ is $-3 < x < 0$ or $x > 3$. In interval notation, you write the solution as $(-3,0) \cup (3,\infty)$.

YOUR TURN

5 Solve for x: $\frac{5-x}{x+3} < 0$.

6 Solve for x: $\frac{x^2}{x+2} \geq 1$.

Increasing the Factors in Polynomial Inequalities

The method you use to solve a quadratic inequality works nicely with fractions and higher-degree expressions. You aren't limited to just two factors when solving inequalities involving polynomials with higher degrees than 2. Just factor the polynomial, identify the roots, and create a sign-line to help you determine the sign of each interval. The main thing to watch for is whether or not you use a particular endpoint.

EXAMPLE

Q. Solve $(x+2)(x-4)(x+7)(x-5)^2 \geq 0$.

A. $[-7,-2] \cup [4,\infty)$. **The inequality is already factored, so you can move directly to determining the zeros. The zeros are $-2, 4, -7$, and 5 (the 5 is a double root and the factor is always positive or 0). Here is the sign-line with the roots identified and the signs of the factors:**

$(x+2)(x-4)(x+7)(x-5)^2$

(–)(–)(–)(+) | (–)(–)(+)(+) | (+)(–)(+)(+) | (+)(+)(+)(+) | (+)(+)(+)(+)

–7 | –2 | 4 | 5

You want the expression on the left to be positive or zero, given the original language of the inequality. You find an even number of negative factors between –7 and –2 and for numbers greater than 4 (zero is an even number). You include the zeros, so the solution you find is $-7 \leq x \leq -2$ or $x \geq 4$. In interval notation, you write the solution as $[-7,-2] \cup [4,\infty)$.

Q. Solve $x^5 + 12x^4 + 35x^3 - 12x^2 - 36x < 0$.

A. $x < -6$ **or** $-6 < x < -1$. **First, factor the polynomial, starting with the common factor of x. Then, using the Rational Root Theorem and synthetic division, first check x = 1 and x = –1. After determining that they are factors, you're left with the trinomial $x^2 + 12x + 36$ which is a perfect square trinomial.**

$$x^5 + 12x^4 + 35x^3 - 12x^2 - 36x < 0$$
$$x\left(x^4 + 12x^3 + 35x^2 - 12x - 36\right) < 0$$
$$x\left(x^2 - 1\right)\left(x^2 + 12x + 36\right) < 0$$
$$x(x-1)(x+1)(x+6)^2 < 0$$

The roots are 0, 1, –1, and the double root of –6. The expression is negative. So, a sign-line will help determine where this occurs.

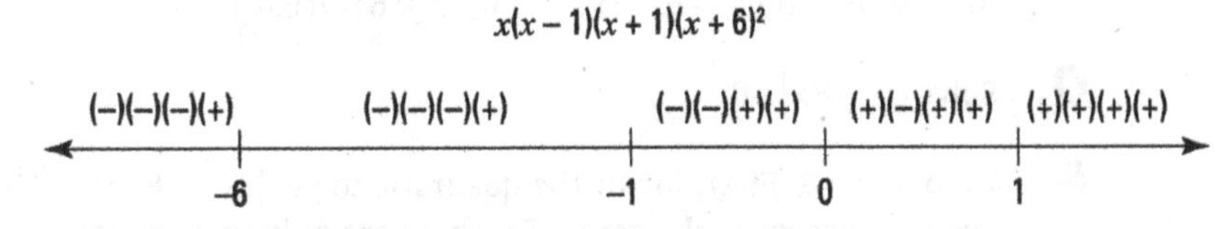

There are three negative factors less than –6, and the same three negative factors between –6 and –1. There is one negative factor between 0 and 1. So, the solution is $x < -6$ or $-6 < x < -1$ or $0 < x < 1$. In interval notation, this is written $(-\infty,-6) \cup (-6,-1) \cup (0,1)$.

YOUR TURN

7 Solve for x: $x(x-1)(x+4)(x-6)^2 \leq 0$.

8 Solve for x: $x^4 - 20x^2 + 64 > 0$.

Considering Absolute Value Inequalities

Another type of inequality that you can solve either using this sign-line method or just considering the roots is an absolute value inequality involving a quadratic or higher-degree variable. In Chapter 2, you discover how to deal with linear inequalities. Here, I show you how to deal with absolute value inequalities of a higher degree. The process is pretty much the same as with quadratic or polynomial inequalities. The only difference comes in the evaluation of the intervals — you refer to the original statement.

EXAMPLE

Q. Solve $|x^3 - 5x^2 - 6x| > 0$.

A. **$x = -1$, $x = 0$, or $x = 6$.** First, factor the cubic to get $|x(x-6)(x+1)| > 0$. The three zeros are $x = 0$, $x = 6$, and $x = -1$. Put them on a sign-line and establish the signs of the intervals you determined.

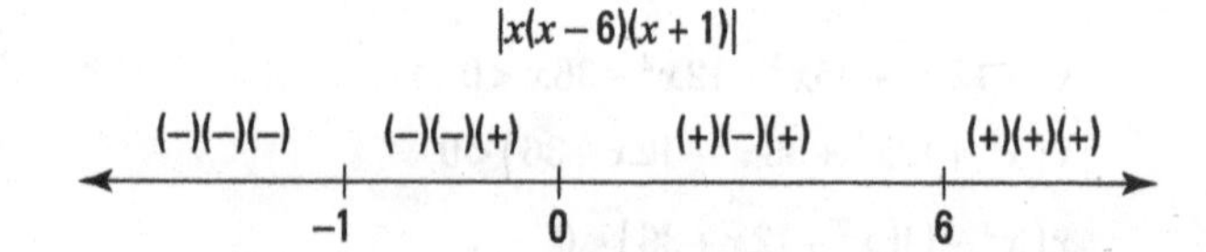

This part is really quite simple to evaluate; the expression is *always* positive (except at the zeros). The absolute value operation creates all positive or 0 results. So, the solution is $x < -1$ or $-1 < x < 0$ or $0 < x < 6$ or $x > 6$. An easier way to read the solution is to just say that the solution is all real numbers except for $x = -1$, $x = 0$, or $x = 6$. Or there's the interval notation: $(-\infty, -1) \cup (-1, 0) \cup (0, 6) \cup (6, \infty)$.

Q. Solve $|x^2 - 8x| \leq 0$.

A. **$x = 0$ or $x = 8$.** First, factor the quadratic to get $|x(x-8)| \leq 0$. This expression is always positive, except at the zeros. So, there are only two solutions, at the two roots: $x = 0$ or $x = 8$.

YOUR TURN

9 Solve for x: $|x(x+11)| \le 0$.

10 Solve for x: $|x^3 - 3x^2 + 3x - 1| > 0$.

11 Solve for x: $|x^2 - 9x - 10| \ge 0$.

Compounding the Situation with Compound Inequalities

A compound inequality is one that considers two or more algebraic expressions, comparing them with one another and placing them in terms of values. You often see these problems using the connecting words "and" or "or." There are also situations where the "and" is not used, and the statement is written with just inequality symbols.

EXAMPLE

Q. Solve for x if: $2x+1>7$ or $5x-6\le 14$.

A. $(-\infty, 4] \cup (3, \infty)$. Solve the two inequalities separately, and then combine their solutions: $2x+1>7 \rightarrow 2x>6 \rightarrow x>3$ or $5x-6\le 14 \rightarrow 5x \le 20 \rightarrow x \le 4$. In interval notation, this is $(-\infty, 4] \cup (3, \infty)$.

Q. Solve for x if: $3x-7\le 11$ and $x+3>5$.

A. $(2, 6]$. Solve the two inequalities separately, and then combine their solutions: $3x-7\le 11 \rightarrow 3x \le 18 \rightarrow x \le 6$ and $x+3>5 \rightarrow x>2$. The solution includes all values greater than 2 but less than or equal to 6. In a single, compound statement, this is written $2<x\le 6$, and in interval notation, $(2, 6]$.

Q. Solve for x if: $-8 \le 3x+1 \le 19$.

A. $[-3,6]$. Subtract 1 from each section; then divide each section by 3: $-8 \le 3x+1 \le 19 \rightarrow -9 \le 3x \le 18 \rightarrow -3 \le x \le 6$. In interval notation, this is written $[-3,6]$.

Q. Solve for x if: $5 < 2-x < 12$.

A. $(-10,-3)$. Subtract 2 from each section. Then multiply each section by –1. When you do this, you have to reverse all the inequality symbols: $5 < 2-x < 12 \rightarrow 3 < -x < 10 \rightarrow -3 > x > -10$. The numbers are in the opposite order of those on the number line, so reverse the numbers and reverse the inequalities again: $-10 < x < -3$. In interval notation, this is written $(-10,-3)$.

YOUR TURN

12 Solve for x if: $5x-3<12$ or $3x+1 \ge 16$.

13 Solve for x if: $15 \ge 2x+11$ and $4x+21 \ge 9$.

14 Solve for x: $-5 \le 4x-3 < 17$.

15 Solve for x: $-13 > 5-3x \ge -40$.

Solving Inequality Word Problems

The whole point of learning how to handle inequality problems is to have something practical to do with them. And that's what most mathematics is about. You often encounter situations where you have to make a decision based on the information given. It could be something like, "You need at least 14 hamburgers to make George happy, but you can't afford more than 12 hamburgers." Oh, so sad. But, to deal with problems like this, you can write an inequality statement and solve the problem (or borrow money for two more hamburgers).

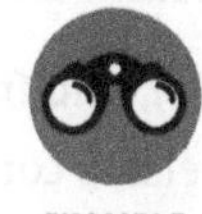

EXAMPLE

Q. You are trying out a new video game and have downloaded it with a 30-minute free use of the game to see if you really want to buy it. It'll take 4 minutes to set up the game, and then you're expecting to be able to complete each level you find in 5 minutes. How many levels will you be able to complete for free, before committing?

A. Write the inequality $4+5m\le 30$ and solve for m: $4+5m\le 30 \to 5m\le 26 \to m\le 5.2$. It looks like you'll have time enough to complete five levels before the free period runs out.

Q. The formula for the height of a projectile shot up into the air at 96 feet per second is $H=-16t^2+96t$, where t is the number of seconds. Casey can shoot down the projectile as long as it's at least 80 feet in the air. What is the range of seconds in which he has to be successful with his shot?

A. Write the inequality $-16t^2+96t\ge 80$ and solve for the value of t. Rewrite the inequality with all the terms on the right. Then factor the quadratic: $-16t^2+96t\ge 80 \to 0\ge 16t^2-96t+80 \to 0\ge 16\left(t^2-6t+5\right) \to 0\ge 16(t-1)(t-5)$. The two roots are 1 and 5. Put these root values on a number line and determine when the product is less than or equal to 0.

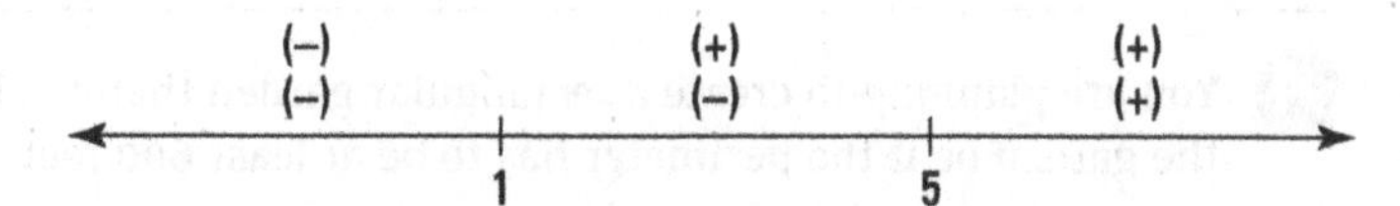

The expression is negative between 0 and 1, and zero at those two values. So, the time range that Casey has is $1\le t\le 5$ seconds.

YOUR TURN

16 The Marine recruits are practicing parachute jumping; they are lined up at a point on a bridge that is 600 feet above the water. Their height after jumping is determined with the formula $H=600-16t^2$, where t is the number of seconds. They need to pull the cord on their parachute between 408 and 280 feet above the water. Between how many seconds is that?

17 **A balance scale has a 40-pound weight and dozens of 9-pound weights that can be placed on one side of the scale. You need to weigh a large dog on the other side and have it be positioned even with or just above the weights on the opposite side. Using the 40-pound weight and some 9-pound weights, how many 9-pound weights will you need if the dog weighs 165 pounds?**

18 **To change from Fahrenheit to Celsius degree measures, you use $C = \frac{5}{9}(F - 32)$, and from Celsius to Fahrenheit, you use $F = \frac{9}{5}C + 32$. What temperatures in degrees Celsius correspond to 250° to 550° Fahrenheit?**

19 **One leg of a right triangle is 5 inches shorter than the other leg. How long should the shorter leg be to create a triangle whose hypotenuse is at least 25 inches in length?**

20 **You are planning to create a rectangular garden that can be 30 feet wide. How long can the garden be if the perimeter has to be at least 600 feet?**

21 **A cylindrical tank is being built with a radius of 20 feet. If the height of the tank can be between 25 and 60 feet, then what is the range of the capacity (volume) of the tank? Use the formula $V = \pi r^2 h$.**

Practice Questions Answers and Explanations

1. **$(-\infty,-5]\cup[6,\infty)$.** Factor the quadratic and identify the two zeros: $(x-6)(x+5)\geq 0$ has zeros $x=6$ and $x=-5$. Place the zeros on a number line, placing solid dots on the points to indicate that the zeros are part of the solution. Then, testing points to the left of –5, between –5 and 6, and to the right of 6, indicate the signs of the factors.

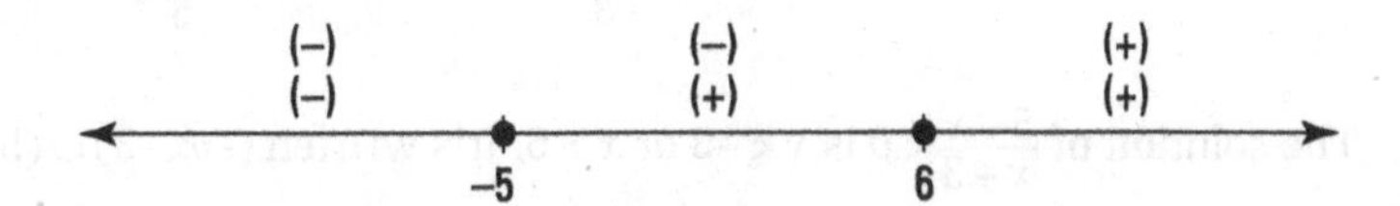

The product is positive to the left of –5 and to the right of 6. It's equal to 0 at those two points. The solution is $x\leq -5$ or $x\geq 6$, written $(-\infty,-5]\cup[6,\infty)$ in interval notation.

2. **$(-5,5)$.** Subtract 25 from each side, factor, and determine the zeros: $x^2<25\rightarrow x^2-25<0\rightarrow(x-5)(x+5)<0$. The zeros are ±5. Place the zeros on a number line, using open circles on the points to indicate that they are not part of the solution.

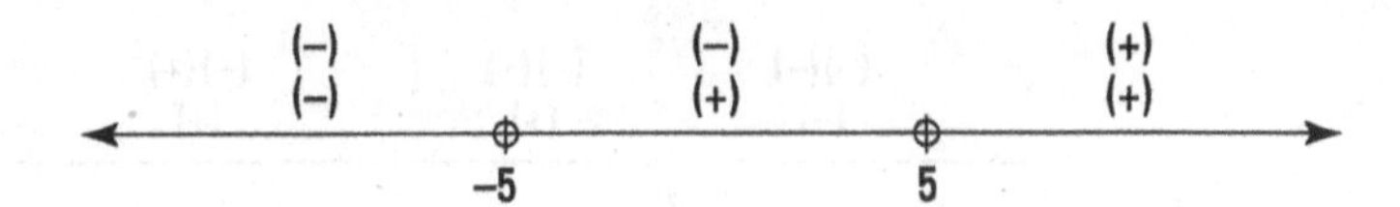

The product is positive to the left of –5 and to the right of 5. The product is negative between the two zeros. The solution is $-5<x<5$, written $(-5,5)$ in interval notation.

3. **$(-1,3)$.** Subtract x^2-2x from each side. Then multiply each side by –1 and reverse the direction of the inequality. Factor, and determine the zeros: $3>x^2-2x\rightarrow -x^2+2x+3>0\rightarrow x^2-2x-3<0\rightarrow(x-3)(x+1)<0$. The zeros are 3 and –1.

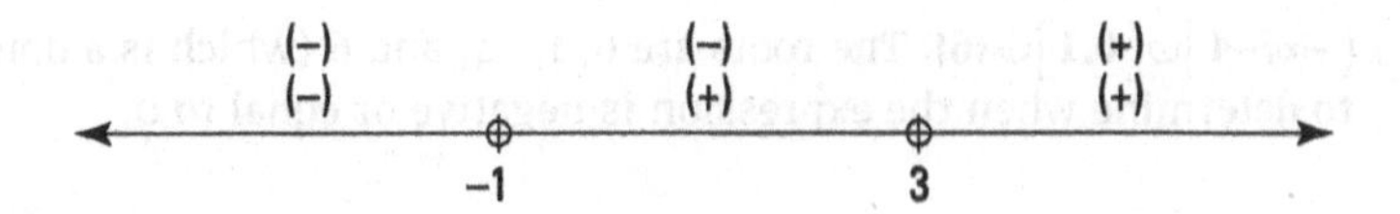

The product is positive to the left of –1 and to the right of 3. The product is negative between the two zeros. The solution is $-1<x<3$, written $(-1,3)$ in interval notation.

4. **$[4,5]$.** Subtract 9x from each side, factor, and determine the zeros: $x^2+20\leq 9x\rightarrow x^2-9x+20\leq 0\rightarrow(x-4)(x-5)\leq 0$. The zeros are 4 and 5. Place the zeros on a number line, using solid dots on the points to indicate that they are part of the solution.

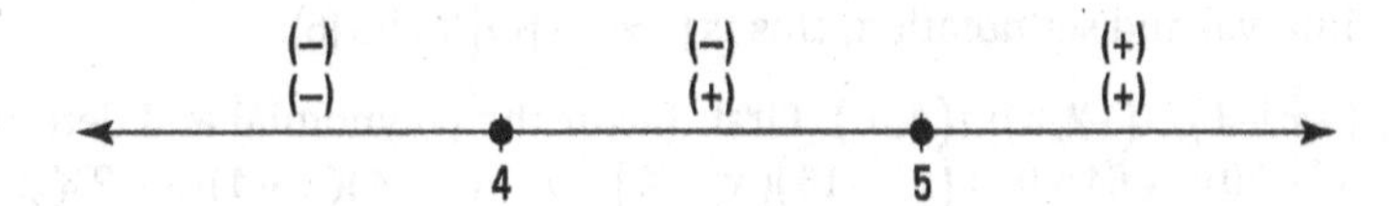

The product is positive to the left of 4 and to the right of 5. The product is negative between the two zeros. The solution is $4\leq x\leq 5$ to include the two values that create a 0. It's written $[4,5]$ in interval notation.

5. $(-\infty,-3)\cup(5,\infty)$. The two zeros are 5 and −3. Placing the zeros on a number line and determining the signs of the numerator and denominator, you determine that the fraction is negative when x is less than −3 or greater than 5. The fraction is positive between those two values.

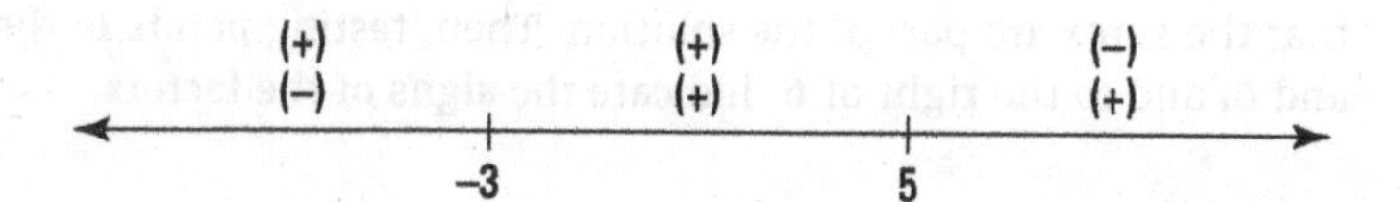

The solution of $\frac{5-x}{x+3}<0$ is $x<-3$ or $x>5$. It's written $(-\infty,-3)\cup(5,\infty)$ in interval notation.

6. $(-2,1]\cup[2,\infty)$. First, subtract 1 from each side, find a common denominator, and subtract the fractions: $\frac{x^2}{x+2}\geq 1\rightarrow\frac{x^2}{x+2}-1\geq 0\rightarrow\frac{x^2}{x+2}-\frac{x+2}{x+2}\geq 0\rightarrow\frac{x^2-x-2}{x+2}\geq 0$. Factor the numerator, and you have $\frac{(x-2)(x+1)}{x+2}\geq 0$.

The three zeros are −2, −1, and 2. Place the zeros on a number line and determine the signs of the factors.

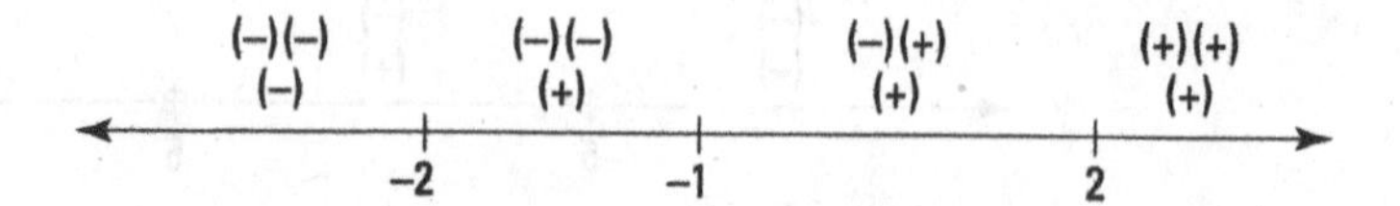

The fraction is negative when x is less than −2 or between 1 and 2. The fraction is positive when x is between −2 and −1 or greater than 2. The fraction is equal to 0 when x is equal to 2 or −1. The value of x cannot be −2, because that would create a 0 in the denominator.

The solution of $\frac{x^2}{x+2}\geq 1\rightarrow\frac{x^2}{x+2}-1\geq 0$ is $-2<x\leq 1$ or $x\geq 2$. This is written $(-2,1]\cup[2,\infty)$ in interval notation.

7. $(-\infty,-4]\cup[0,1]\cup\{6\}$. The roots are 0, 1, −4, and 6 (which is a double root). Create a sign-line to determine when the expression is negative or equal to 0.

$x(x-1)(x+4)(x-6)^2$

(−)(−)(−)(+) (−)(−)(+)(+) (+)(−)(+)(+) (+)(+)(+)(+) (+)(+)(+)(+)

−4 0 1 6

The expression is negative when x is less than −4 or between 0 and 1. The expression is 0 at all the roots, 0, 1, −4, and 6. So the solution is $x\leq 4$ or $0\leq x\leq 1$ or $x=6$. Written in interval and set notation, this is $(-\infty,-4]\cup[0,1]\cup\{6\}$.

8. $(-\infty,-4)\cup(-2,2)\cup(4,\infty)$. First, factor the polynomial and determine the roots: $x^4-20x^2+64>0\rightarrow(x^2-16)(x^2-4)>0\rightarrow(x-4)(x+4)(x-2)(x+2)>0$. The roots are 4, −4, 2, and −2. Create a sign-line to determine when the expression is positive.

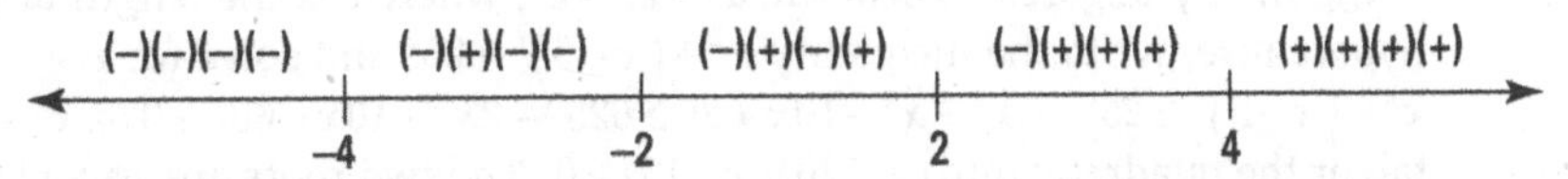

The expression is positive when there is an even number of negative signs. This occurs when $x < -4$ or $-2 < x < 2$ or $x > 4$. This is written $(-\infty,-4)\cup(-2,2)\cup(4,\infty)$ in interval notation.

9. **$x = 0$ or $x = -11$.** The expression is always positive or 0, so the only two possible solutions are the roots, when it's equal to 0. The roots are $x = 0$ or $x = -11$.

10. **$(-\infty,1)\cup(1,\infty)$.** First, factor the expression: $\left|x^3 - 3x^2 + 3x - 1\right| > 0 \rightarrow \left|(x-1)^3\right| > 0$. This expression is always positive, except at the root. So the solution is $x < 1$ or $x > 1$, which, written in interval notation, is $(-\infty,1)\cup(1,\infty)$.

11. **$(-\infty,\infty)$.** You could factor this quadratic, but it really isn't necessary. The expression is always positive or 0, so all values of x are solutions: $x \in \{\text{all real numbers}\}$, written in interval notation as $(-\infty,\infty)$.

12. **$(-\infty,3)\cup[5,\infty)$.** Solve the inequalities separately and then combine their solutions: $5x - 3 < 12 \rightarrow 5x < 15 \rightarrow x < 3$ or $3x + 1 \geq 16 \rightarrow 3x \geq 15 \rightarrow x \geq 5$. So, $x < 3$ or $x \geq 5$. In interval notation, this is written $(-\infty,3)\cup[5,\infty)$.

13. **$[-3,2]$.** Solve the inequalities separately and then combine their solutions: $15 \geq 2x + 11 \rightarrow 4 \geq 2x \rightarrow 2 \geq x$ and $4x + 21 \geq 9 \rightarrow 4x \geq -12 \rightarrow x \geq -3$. The single compound statement combining these results is $-3 \leq x \leq 2$, and in interval notation, this is written $[-3,2]$.

14. **$\left[-\frac{1}{2},5\right)$.** Add 3 to each section, and then divide by 4: $-5 \leq 4x - 3 < 17 \rightarrow -2 \leq 4x < 20 \rightarrow -\frac{1}{2} \leq x < 5$. In interval notation, this is written $\left[-\frac{1}{2},5\right)$.

15. **$(6,15]$.** First, subtract 5 from each section, and then divide each section by −3. When you do this, you reverse all the inequalities: $-13 > 5 - 3x \geq -40 \rightarrow -18 > -3x \geq -45 \rightarrow 6 < x \leq 15$. In interval notation, this is written $(6,15]$.

16. **$3.5 \leq t \leq 4.5$.** Write the inequality $280 \leq 600 - 16t^2 \leq 408$. Solving for t, first subtract 600 from each section, and then divide by −16. Be sure to reverse the inequalities when you divide by the negative number: $280 \leq 600 - 16t^2 \leq 408 \rightarrow -320 \leq -16t^2 \leq -192 \rightarrow 20 \geq t^2 \geq 12$. Finding the square roots, you have $t \approx 4.5$ and $t \approx 3.5$ seconds.

17. **13.** Write the inequality $40 + 9n \leq 165$, where n is the number of 9-pound weights. Subtract 40 from each side and divide by 9: $40 + 9n \leq 165 \rightarrow 9n \leq 125 \rightarrow n \leq 14\frac{8}{9}$. To keep the dog above the weights on the other side, just use 13 9-pound weights.

18. **$121\frac{1}{9} \leq {}^\circ C \leq 287\frac{7}{9}$.** Write the inequality $250 \leq F \leq 550$ and replace the degrees Fahrenheit with $F = \frac{9}{5}C + 32$. Then solve for degrees Celsius:

$$250 \leq \frac{9}{5}C + 32 \leq 550 \rightarrow 218 \leq \frac{9}{5}C \leq 518 \rightarrow 218 \cdot \frac{5}{9} \leq C \leq 518 \cdot \frac{5}{9} \rightarrow 121\frac{1}{9} \leq C \leq 287\frac{7}{9}$$

19. **At least 15 in.** Let the lengths of the two legs be represented by x and $x-5$. Using the Pythagorean Theorem, $a^2+b^2=c^2$, where c is the length of the hypotenuse, write the inequality $x^2+(x-5)^2 \geq 25^2$ and solve for x: $x^2+(x-5)^2 \geq 25^2 \to x^2+x^2-10x+25 \geq 625 \to 2x^2-10x-600 \geq 0 \to x^2-5x-300 \geq 0$. Now factor the quadratic into $(x-20)(x+15) \geq 0$. The two roots are 20 and −15. The measure of the side of a triangle can't be negative, so the number you want is 20. The shorter side is then 15 inches, and this gives you a 15-20-25 right triangle.

20. **270 ft.** The perimeter of a rectangle is found with the formula $P=2(l+w)$. Write the inequality $2(l+30) \geq 600$ and solve for l: $2(l+30) \geq 600 \to l+30 \geq 300 \to l \geq 270$. The length must be at least 270 feet.

21. **$31{,}400 \leq V \leq 75{,}360$ cu. ft.** First, solve for h in the volume formula and write the inequality using the two height constraints: $25 \leq \frac{V}{\pi r^2} \leq 60$. Replace the radius symbol with 20, and then multiply each section by the denominator of the fraction: $25 \leq \frac{V}{\pi 20^2} \leq 60 \to 25 \leq \frac{V}{400\pi} \leq 60 \to 25 \cdot 400\pi \leq V \leq 60 \cdot 400\pi \to 10{,}000\pi \leq V \leq 24{,}000\pi$. Replacing π with 3.14, you have $31{,}400 \leq V \leq 75{,}360$ cubic feet of water.

Whaddya Know? Chapter 4 Quiz

Quiz time! Complete each problem to test your knowledge on the various topics covered in this chapter. You can then find the solutions and explanations in the next section.

1. **You need \$4,000 to buy a used car. You currently have \$500 and can save \$330 per week. What is the minimum number of weeks it will take you to accumulate at least \$4,000?**

2. **Solve: $4x - 3 > 13$.**

3. **Solve: $\left|x^2 - 5x - 14\right| > 0$.**

4. **Solve: $5 - \frac{x}{2} \le 9$.**

5. **Solve: $-6 \le 4x - 2 < 10$.**

6. **Solve: $x^2 - x \ge 12$.**

7. **Solve $\left|x(x+3)\right| \le 0$.**

8. **Solve: $\frac{6}{x+1} < 2$.**

9. **Solve: $x^3 + x^2 - 16x > 16$.**

10. **A fireworks rocket is being launched from a platform that's 40 feet off the ground. The initial velocity of the rocket is 128 feet per second. The height of the rocket t seconds after launching is found with the formula $H = -16t^2 + 128t + 40$. The rocket is set to explode when it reaches 152 feet or more aboveground (for the safety of the spectators). After at least how many seconds will this occur?**

11. **Solve for x: $\frac{1}{x} + \frac{x}{x-6} \ge 0$.**

12. **Solve for x if $3x + 4 \ge 22$ or $5x - 2 < 13$.**

13. **In boxing, to be in the welterweight division you can't weigh more than 152 pounds. To be in the middleweight division, you can't weigh more than 165 pounds. Henry currently weighs 148 pounds. How much weight can he gain to classify as a middleweight boxer? (A boxer may not qualify for welterweight and middleweight at the same time.)**

Answers to Chapter 4 Quiz

1. **11 weeks.** Write the inequality: $4{,}000 \le 500 + 330w$, where w represents the number of weeks. Solve for w.

$$4{,}000 \le 500 + 330w \rightarrow 3500 \le 330w \rightarrow \frac{3500}{330} \le \frac{330w}{330} \rightarrow 10\frac{20}{33} \le w$$

It will take 11 weeks to save enough money.

2. **$x > 4$.** Add 3 to each side and then divide both sides by 4.

$$\begin{array}{rcrcr} 4x & - & 3 & > & 13 \\ & + & 3 & & +3 \\ \hline 4x & & & > & 16 \\ & & x & > & 4 \end{array}$$

Using interval notation, this is written $(4, \infty)$.

3. **$(-\infty, -2) \cup (-2, 7) \cup (7, \infty)$.** Factoring, the inequality becomes $|(x-7)(x+2)| > 0$. The expression is always positive — when not equal to 0. So, the only numbers not in the solution are the roots: 7 and −2. The solution is $x < -2$ or $-2 < x < 7$ or $x > 7$. Written in interval notation, this is $(-\infty, -2) \cup (-2, 7) \cup (7, \infty)$.

4. **$x \ge -8$.** First, multiply each term by 2. Then subtract 10 from each side of the inequality.

$$\begin{array}{rcrcr} 5 & - & \frac{x}{2} & \le & 9 \\ 10 & - & x & \le & 18 \\ -10 & & & & -10 \\ \hline & - & x & \le & 8 \end{array}$$

Multiply each side of the inequality by −1 and reverse the inequality sign.

$$\begin{array}{rcr} -x & \le & 8 \\ x & \ge & -8 \end{array}$$

Using interval notation, this is written $[-8, \infty)$.

5. **$-1 \le x < 3$.** First, add 2 to each section of the inequality. Then divide each term by 4.

$$\begin{array}{rcrcrcr} -6 & \le & 4x & - & 2 & < & 10 \\ +2 & & & + & 2 & & +2 \\ \hline -4 & \le & 4x & & & < & 12 \\ -1 & \le & x & & & < & 3 \end{array}$$

Using interval notation, this is written $[-1, 3)$.

6) $x \le -3$ or $x \ge 4$. **First, subtract 12 from each side of the inequality. Then factor the quadratic.**

$$x^2 - x \ge 12 \rightarrow x^2 - x - 12 \ge 0 \rightarrow (x-4)(x+3) \ge 0$$

Draw a number line identifying the numbers 4 and −3. Use solid dots to indicate that those two numbers are included in the solution. Determine the sign of each factor for values of x less than −3, between −3 and 4, and then greater than 4.

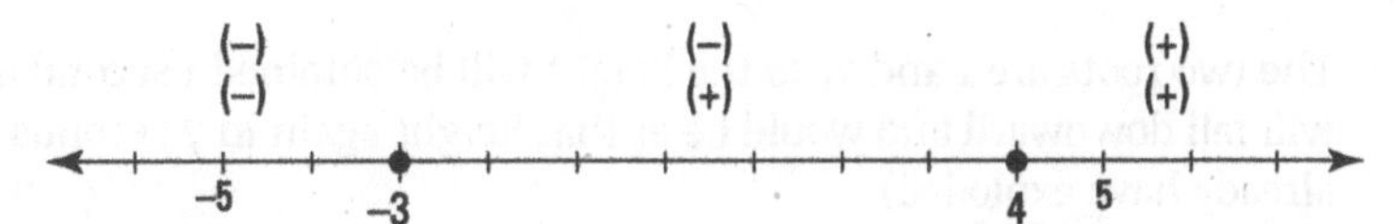

The product is positive or equal to zero from −3 and to the left, and it is positive or equal to zero from 4 and to the right: $x \le -3$ or $x \ge 4$. In interval notation, this is written $(-\infty,-3] \cup [4,\infty)$.

7) $x = 0$ or $x = -3$. **Having the inequality inside the absolute value guarantees that the expression will always be either positive or 0. Thus, the only solutions will be the two roots, 0 and −3.**

8) $x < -1$ or $x > 2$. **Subtract 2 from each side of the inequality. Then find a common denominator and write the left side as one fraction.**

$$\frac{6}{x+1} < 2 \rightarrow \frac{6}{x+1} - 2 < 0 \rightarrow \frac{6}{x+1} - \frac{2(x+1)}{x+1} < 0 \rightarrow \frac{6-2x-2}{x+1} < 0 \rightarrow \frac{4-2x}{x+1} < 0$$

The numerator is equal to 0 when $x = 2$, and the denominator is 0 when $x = -1$. You can't use the −1 as part of the solution, but it is used when determining signs of the expression. Put those values on the number line and determine when the quotient is negative. Use open circles for the two numbers to indicate that they are not included in the solution.

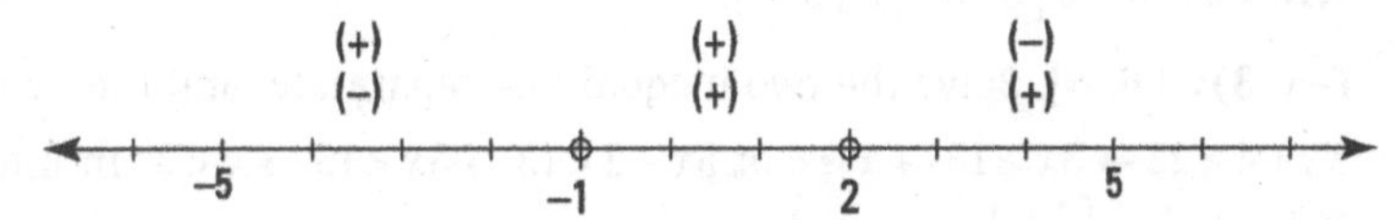

The fraction is negative when $x < -1$ or $x > 2$. In interval notation, this is written $(-\infty,-1) \cup (2,\infty)$.

9) $(-4,-1) \cup (4,\infty)$. **First, subtract 16 from each side, and then factor the polynomial using grouping.**

$$x^3 + x^2 - 16x > 16 \rightarrow x^3 + x^2 - 16x - 16 > 0 \rightarrow x^2(x+1) - 16(x+1) > 0 \rightarrow (x^2-16)(x+1) > 0 \rightarrow$$
$$(x-4)(x+4)(x+1) > 0$$

The roots are 4, −4, and −1. Put the roots on a number line to determine where the expression is positive.

(x − 4)(x + 4)(x + 1)

(−)(−)(−) (−)(+)(−) (−)(+)(+) (+)(+)(+)

−4 −1 4

The product is positive when x is between –4 and –1 or when it is greater than 4: $-4 < x < -1$ or $x > 4$. In interval notation, this is written $(-4,-1) \cup (4,\infty)$.

(10) **1 second.** First, write the inequality describing the height to be obtained: $-16t^2 + 128t + 40 \geq 152$. Subtract 152 from both sides and factor the quadratic.

$$-16t^2 + 128t + 40 \geq 152 \rightarrow -16t^2 + 128t - 112 \geq 0 \rightarrow -16\left(t^2 - 8t + 7\right) \geq 0 \rightarrow -16(t-1)(t-7) \geq 0$$

The two roots are 1 and 7, so the height will be obtained 1 second after launch. The rocket will fall downward and would be at that height again in 7 seconds (except that it would already have exploded).

(11) $x \leq -3$, $0 < x \leq 2$, or $x > 6$. First, find a common denominator and write the two fractions as one.

$$\frac{1}{x} + \frac{x}{x-6} \geq 0 \rightarrow \frac{x-6}{x(x-6)} + \frac{x^2}{x(x-6)} \geq 0 \rightarrow \frac{x^2 + x - 6}{x(x-6)} \geq 0 \rightarrow \frac{(x+3)(x-2)}{x(x-6)} \geq 0$$

Create a number line indicating the values: –3, 2, 0, and 6. Then determine whether the value of the fraction is positive or negative in each section. Use solid dots for –3 and 2; use empty dots for 0 and 6.

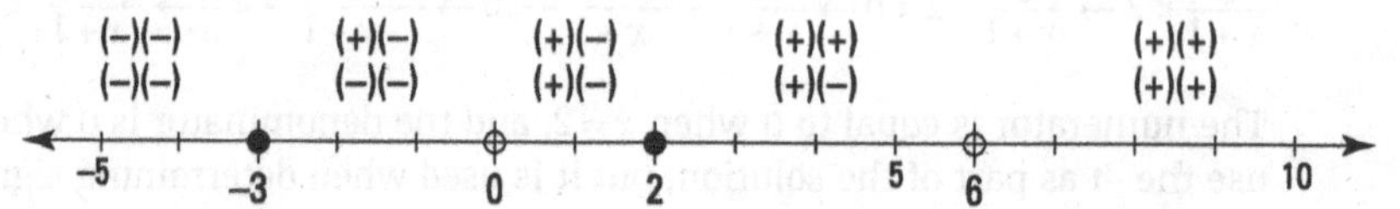

The expression is positive when $x < -3$, $0 < x < 2$, and $x > 6$. The expression equals 0 when $x = -3$ or 2. Thus, the solution is $x \leq -3$, $0 < x \leq 2$, or $x > 6$. In interval notation, this is written $(-\infty,-3] \cup (0,2] \cup (6,\infty)$.

(12) $(-\infty,3) \cup [6,\infty)$. Solve the two inequalities separately, and then write the combined solution: $3x + 4 \geq 22 \rightarrow 3x \geq 18 \rightarrow x \geq 6$ or $5x - 2 < 13 \rightarrow 5x < 15 \rightarrow x < 3$. In interval notation, this is written $(-\infty,3) \cup [6,\infty)$.

(13) $4 < x \leq 17$. Write the inequality: $152 < 148 + x \leq 165$. Then solve for x.

$$\begin{aligned} 152 < 148 + &x \leq 165 \\ 4 < \quad &x \leq 17 \end{aligned}$$

He should gain more than 4 pounds and no more than 17 pounds.

IN THIS CHAPTER

» Working through rational equations

» Dealing with radicals in equations

» Flipping and factoring negative exponents

» Combining and factoring fractional exponents

Chapter 5
Soothing the Rational, the Radical, and the Negative

Solving an algebraic equation requires some know-how. You need the basic mathematical tools, and you need to know what is and isn't allowed. You don't want to take a perfectly good equation and change it into drivel. You need a game plan to solve equations with fractions, radicals, and negative or fractional exponents — one that involves careful planning and a final check of your answers. In this chapter, you find out how to tackle equations by changing them into new equations that are more familiar and easier to solve. You also see a recurring theme of checking your answers, because changing equations into different forms can introduce mysterious strangers into the mix in the form of false answers.

Systematically Solving Rational Equations

A rational term in an equation is a fraction, and an equation with one or more terms, some of which are rational, is called a *rational equation*. You probably hope that all your problems (and the people you associate with) are rational, but an equation that contains fractions isn't always easy to handle.

A general plan for solving a rational equation is to get rid of the fraction or fractions by changing the equation into an equivalent form with the same answer as the original — a form that makes it easier to solve. Two of the most common ways to get rid of the fractions are multiplying through by the least common denominator (LCD) and cross-multiplying proportions. I just happen to discuss both of these techniques in the sections that follow.

But you must be cautious! This mathematical sleight of hand, using alternate equations to solve more complicated problems, isn't without its potential problems. At times, the new equation produces an extraneous solution (also referred to as an *extraneous root*). This is a false solution that pops up because you messed around with the original format of the equation. To guard against including extraneous solutions in your answers, you need to check the solutions you come up with in the original equations. Don't worry; I have you covered in the following sections.

You can solve rational equations such as $\frac{x+1}{4} - \frac{2x+1}{x-2} = \frac{2-x}{5}$ without as much hassle if you simply get rid of all the denominators. To do so, you work with an old friend, the least common denominator. The least common denominator is also known as the least common multiple — the smallest number or factor that two or more numbers divide into evenly (such as 2, 3, and 4 all dividing the LCD 12 evenly).

To solve an equation with the LCD, you find a common denominator, write each fraction with that common denominator, and then multiply each side of the equation by that same denominator to get a nice quadratic or higher-degree polynomial equation (see Chapter 3 for a full discussion of quadratic and polynomial equations). Quadratic and polynomial equations can have two or more solutions, so they present more opportunities for extraneous solutions. Be on the lookout!

Here's a general method to use when dealing with rational equations:

1. **Find a common denominator.**

   Each of the denominators has to be able to divide into the common denominator evenly. In other words, the LCD is a multiple of each of the original denominators.

2. **Write each fraction with the common denominator.**

   Multiply each of the terms in the original equation by the specific value such that, after the multiplication, each resulting term has the same denominator. The "specific value" I speak of is equal to 1, because each of the fractions multiplying the terms is the same in the numerator and denominator. But you carefully select the fractions that serve as multipliers — the numerators and denominators must consist of all the factors necessary to complete the LCD. You could just divide the LCD by the current denominator to determine what more you need in order to create the common denominator in that term.

3. **Multiply each side of the equation by that same denominator.** You multiply each term in the equation by the least common denominator to reduce each term and get rid of the denominators.

4. **Solve the new equation.**

5. **Check your answers to avoid extraneous solutions.**

The most common indication that you have an extraneous solution is that you end up with a zero in the denominator after replacing all the variables with that answer. Occasionally, you get a nonsense equation such as $4 = 7$ when checking — and that tells you that the solution is extraneous — but those are very special cases. You should always check your answers after solving equations. Make sure that the values you find create true statements.

Q. Solve for x in $\frac{x+1}{4} - \frac{2x+1}{x-2} = \frac{2-x}{5}$.

EXAMPLE **A.** $x = -\frac{2}{9}$, $x = 7$. Follow the preceding steps:

1. **Find a common denominator.**

   The first step in solving the rational equation is to find the least common denominator for all the terms in the equation.

   The common denominator of all three fractions in the equation $\frac{x+1}{4} - \frac{2x+1}{x-2} = \frac{2-x}{5}$ consists of the product of all the factors in the three denominators, $4 \cdot (x-2) \cdot 5 = 20(x-2)$.

   You determine the common denominator to be $20(x-2)$ because it's a multiple of 4 (you multiply by $5(x-2)$ to get it), it's a multiple of $(x-2)$ (you multiply by 20 to get it), and it's a multiple of 5 (you multiply by $4(x-2)$ to get it). All three denominators divide this product evenly.

2. **Write each fraction with the common denominator.**

   $$\frac{x+1}{4} \cdot \frac{5(x-2)}{5(x-2)} - \frac{2x+1}{x-2} \cdot \frac{20}{20} = \frac{2-x}{5} \cdot \frac{4(x-2)}{4(x-2)}$$

   Now, multiply each fraction and simplify.

   $$\frac{5(x^2-x-2)}{20(x-2)} - \frac{20(2x+1)}{20(x-2)} = \frac{4(-x^2+4x-4)}{20(x-2)}$$
   $$\frac{5x^2-5x-10}{20(x-2)} - \frac{40x+20}{20(x-2)} = \frac{-4x^2+16x-16}{20(x-2)}$$

3. **Multiply each side of the equation by that same denominator.**

   $$\frac{\cancel{20(x-2)}}{1} \cdot \frac{5x^2-5x-10}{\cancel{20(x-2)}} - \frac{\cancel{20(x-2)}}{1} \cdot \frac{40x+20}{\cancel{20(x-2)}} = \frac{\cancel{20(x-2)}}{1} \cdot \frac{-4x^2+16x-16}{\cancel{20(x-2)}}$$

   Now simplify what's left.

   $$\begin{aligned} 5x^2-5x-10-(40x+20) &= -4x^2+16x-16 \\ 5x^2-45x-30 &= -4x^2+16x-16 \\ 9x^2-61x-14 &= 0 \end{aligned}$$

   One pitfall of multiplying both sides of an equation by a variable is that you may have to multiply both sides by zero, which may introduce an extraneous solution. Be sure to check your answer in the original equation when you're finished to make sure that your answer doesn't make one or more of the denominators equal to zero.

4. **Solve the new equation.**

You have produced a quadratic equation (if you don't know what to do with those, turn to Chapter 3). To solve the new quadratic equation, you can either factor or use the quadratic formula.

This equation factors into $(9x+2)(x-7)=0$. After factoring, you set each factor equal to zero and solve for x. When $9x+2=0$, $x=-\frac{2}{9}$, and when $x-7=0$, $x=7$.

5. **Check your answers to avoid extraneous solutions.**

You now have to check to be sure that both your solutions work in the original equation. As I discuss in the introduction to this section, one or both solutions may be extraneous. First, checking the solution $x=-\frac{2}{9}$, you have

$$\frac{-\frac{2}{9}+1}{4}-\frac{2\left(-\frac{2}{9}\right)+1}{-\frac{2}{9}-2}=\frac{2-\left(-\frac{2}{9}\right)}{5}\rightarrow\frac{\frac{7}{9}}{4}-\frac{\frac{5}{9}}{-\frac{20}{9}}=\frac{\frac{20}{9}}{5}\rightarrow\frac{7}{36}+\frac{1}{4}=\frac{4}{9}\rightarrow\frac{7}{36}+\frac{9}{36}=\frac{16}{36}$$

And then, checking $x = 7$:

$$\frac{7+1}{4}-\frac{2(7)+1}{7-2}=\frac{2-7}{5}\rightarrow\frac{8}{4}-\frac{15}{5}=\frac{-5}{5}\rightarrow 2-3=-1$$

When replacing the x in the original equation with the two solutions, they both work.

Q. Solve for x in $\frac{x}{x+2}+\frac{3x+2}{x(x+2)}=\frac{3}{x}$.

A. $x = 2$. Follow the preceding steps:

1. **Find a common denominator.**

The two factors found in the three fractions are $(x+2)$ and x. The common denominator is $x(x+2)$, which is already found in the middle fraction.

2. **Write each fraction with the common denominator.**

$$\frac{x}{x+2}\cdot\frac{x}{x}+\frac{3x+2}{x(x+2)}=\frac{3}{x}\cdot\frac{x+2}{x+2}\rightarrow\frac{x^2}{x(x+2)}+\frac{3x+2}{x(x+2)}=\frac{3(x+2)}{x(x+2)}$$

3. **Multiply each side of the equation by that same denominator.**

$$\cancel{x(x+2)}\cdot\frac{x^2}{\cancel{x(x+2)}}+\cancel{x(x+2)}\cdot\frac{3x+2}{\cancel{x(x+2)}}=\cancel{x(x+2)}\cdot\frac{3(x+2)}{\cancel{x(x+2)}}$$

$$x^2+3x+2=3(x+2)$$

4. **Solve the new equation.**

$$x^2+3x+2=3(x+2)\rightarrow x^2=4\rightarrow x^2-4=0\rightarrow(x+2)(x-2)=0$$

The solutions for this equation are $x=-2$ and $x=2$.

5. **Check your answers to avoid extraneous solutions.**

When you try $x=2$ in the original equation, it works out:

$$\frac{2}{2+2}+\frac{3(2)+2}{2(2+2)}=\frac{3}{2}$$

which becomes $\frac{2}{4}+\frac{8}{8}=\frac{3}{2}$ or $\frac{1}{2}+1=\frac{3}{2}$.

However, when you substitute $x=-2$ into the original equation, you get

$$\frac{-2}{-2+2}+\frac{3(-2)+2}{-2(-2+2)}=\frac{3}{-2}$$

which becomes $\frac{-2}{0}+\frac{-4}{0}=\frac{3}{-2}$.

Stop right there! You can't have a zero in the denominator. The solution $x=2$ works just fine in the quadratic equation, but it isn't a solution of the rational equation; $x=-2$ is extraneous.

YOUR TURN

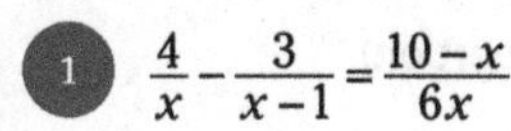

1 $\frac{4}{x}-\frac{3}{x-1}=\frac{10-x}{6x}$

2 $\frac{x+3}{4}+\frac{10}{x}=\frac{2x+2}{3}$

3 $\frac{10}{x^2-4}-\frac{x-13}{x+2}=\frac{x+1}{x-2}$

4 $\frac{3}{x^2-7x+10}-\frac{1}{x-5}=\frac{x}{x-2}$

Simplifying and Solving Proportions

A proportion is an equation in which one fraction is set equal to another. For example, the equation $\frac{a}{b}=\frac{c}{d}$ is a proportion. Proportions have several very nice features that make them desirable to work with when you're solving rational equations because you can eliminate the fractions or change them so that they feature better denominators. Also, they factor in four different ways.

When you have the proportion $\frac{a}{b}=\frac{c}{d}$, the following are also true:

- *ad* and *bc*, the cross-products, are equal, giving you $ad=bc$.
- $\frac{b}{a}=\frac{d}{c}$, so the reciprocals are equal (you can flip the proportion).

Reducing every which way but loose

Another wonderful feature of proportions is that you can reduce the fractions in a proportion by finding common factors in four different directions: top, bottom, left, and right. The ability to reduce a proportion comes in handy when you have large numbers in the equation.

Here are the rules for reducing proportions across the top (numerators), bottom (denominators), left, and right, and an example for each:

Numerators	*Denominators*	*Left*	*Right*
$\frac{\cancel{m}\cdot a}{b}=\frac{\cancel{m}\cdot c}{d}$	$\frac{a}{\cancel{m}\cdot b}=\frac{c}{\cancel{m}\cdot d}$	$\frac{\cancel{m}\cdot a}{\cancel{m}\cdot b}=\frac{c}{d}$	$\frac{a}{b}=\frac{\cancel{m}\cdot c}{\cancel{m}\cdot d}$

The reduced forms of the proportions make cross-multiplication much easier and more manageable.

EXAMPLE

Q. Solve for *x* in $\frac{80x}{16}=\frac{30}{x-5}$.

A. **x = 6, x = −1.** Reduce the two numerators by dividing by 10: $\frac{{}^{8}\cancel{80}x}{16}=\frac{\cancel{30}^{3}}{x-5}\rightarrow\frac{8x}{16}=\frac{3}{x-5}$.

Now reduce the left fractions by dividing by 8: $\frac{{}^{1}\cancel{8}x}{{}_{2}\cancel{16}}=\frac{3}{x-5}\rightarrow\frac{x}{2}=\frac{3}{x-5}$.

Cross-multiply and solve the quadratic equation:

$$\frac{x}{2}=\frac{3}{x-5}\rightarrow x(x-5)=6\rightarrow x^2-5x-6=(x-6)(x+1)=0$$

When $x-6=0$, $x=6$, and when $x+1=0$, $x=-1$. Both solutions check out.

Q. Solve for x in $\frac{x-4}{27}=\frac{35}{21x}$.

A. **x = 9, x = −5.** Reduce the right fractions by dividing by 7: $\frac{x-4}{27}=\frac{\cancel{35}^5}{\cancel{21}^3x}\rightarrow\frac{x-4}{27}=\frac{5}{3x}$.

Next, reduce the denominators by dividing by 3: $\frac{x-4}{\cancel{27}^9}=\frac{5}{\cancel{3}^1x}\rightarrow\frac{x-4}{9}=\frac{5}{x}$.

Cross-multiply and solve the equation:

$$\frac{x-4}{9}=\frac{5}{x}\rightarrow x(x-4)=45\rightarrow x^2-4x=45\rightarrow x^2-4x-45=0\rightarrow(x-9)(x+5)=0$$

When $x-9=0$, $x=9$, and when $x+5=0$, $x=-5$. Both solutions work.

YOUR TURN

5 Solve for x in $\frac{5x}{120}=\frac{9}{4x-12}$.

6 Solve for x in $\frac{15}{4x-2}=\frac{4x}{24}$.

Using cross-products to solve a rational equation

So far, I've shown you how to solve equations written as proportions, taking full advantage of the properties of proportions to make the solution quicker and easier. You sometimes find rational equations that can be quickly turned into proportions. Go for it!

EXAMPLE

Q. Solve $\frac{x+5}{2}-\frac{3}{x}=\frac{9}{x}$ by first changing the format to a proportion.

A. **x = −8, x = 3.** Move the second fraction to the right side of the equation by adding it to each side. Then the two fractions on the right can be added together, forming a proportion.

$$\frac{x+5}{2}-\frac{3}{x}+\frac{3}{x}=\frac{9}{x}+\frac{3}{x} \text{ simplifies to } \frac{x+5}{2}=\frac{12}{x}$$

Now cross-multiply and solve for x:

$$\frac{x+5}{2}=\frac{12}{x} \to x(x+5)=24 \to x^2+5x=24 \to x^2+5x-24=0 \to (x+8)(x-3)=0$$

The solutions are $x=-8$ or $x=3$. Both work.

Q. Solve $\frac{x}{3}-\frac{6}{x+6}=\frac{7}{x+6}+\frac{4}{3}$ for x.

A. **x = −9, x = 7.** First, subtract $\frac{4}{3}$ from each side and add $\frac{6}{x+6}$ to each side. Combine the like-terms.

$$\frac{x}{3}-\frac{4}{3}-\frac{6}{x+6}+\frac{6}{x+6}=\frac{7}{x+6}+\frac{6}{x+6}+\frac{4}{3}-\frac{4}{3}$$
$$\frac{x-4}{3}=\frac{13}{x+6}$$

Cross-multiply and solve for x:

$$\frac{x-4}{3}=\frac{13}{x+6} \to (x-4)(x+6)=39 \to x^2+2x-24=39 \to x^2+2x-63=0 \to (x+9)(x-7)=0$$

The solutions are $x=-9$ or $x=7$. Both work.

YOUR TURN

7 Solve for x in $\frac{x-11}{2}+\frac{2}{x+2}=\frac{9}{x+2}$.

8 Solve for x in $\frac{x+1}{5}-\frac{7}{x-6}=\frac{4}{x-6}+\frac{1}{5}$.

Ridding Yourself of a Radical

The radical symbol indicates that you want to perform the operation of finding a root — a square root of a number, a cube root, and so on. A radical in an equation gives the same message, but it adds a whole new dimension to what could've been a perfectly nice equation to solve. In general, you deal with radicals in equations the same way you deal with fractions in equations — you get rid of them. But watch out: The extraneous answers that first rear their ugly heads in the earlier section, "Systematically Solving Rational Equations," pop up here as well. So — you guessed it — you have to check your answers.

Squaring both sides of a radical equation

If you have an equation in the form $\sqrt{ax+b}=c$, you square both sides of the equation to get rid of the radical. The only problem arises when you end up with an extraneous root.

Consider the non-equation $-3=3$. You know that the equation isn't correct, but what happens when you square both sides of this statement? You get $(-3)^2=(3)^2$, or $9=9$. Now you have an equation. Squaring both sides can mask or hide an incorrect statement.

Much like the process of getting rid of fractions in equations, the method of squaring both sides is the easiest way to deal with radicals in equations. You just accept that you always have to watch for extraneous roots when solving equations by squaring.

EXAMPLE

Q. Solve the equation $\sqrt{4x+21}-6=x$.

A. **x = −3** or **x = −5.** First, change the equation so that the radical term is by itself on the left. Just add 6 to both sides.

$$\sqrt{4x+21}-6=x \rightarrow \sqrt{4x+21}=x+6$$

Now square both sides of the equation.

$$\left(\sqrt{4x+21}\right)^2=(x+6)^2 \rightarrow 4x+21=x^2+12x+36$$

Simplify the equation and solve for x.

$$4x+21=x^2+12x+36 \rightarrow 0=x^2+8x+15 \rightarrow 0=(x+3)(x+5)$$

The two solutions are $x=-3$ or $x=-5$.

You need to check to be sure that there are no extraneous solutions. Always go back to the original equation.

$$x=-3 \rightarrow \sqrt{4(-3)+21}-6=-3 \rightarrow \sqrt{9}-6=-3 \rightarrow 3-6=-3 \text{ Check!}$$

$$x=-5 \rightarrow \sqrt{4(-5)+21}-6=-5 \rightarrow \sqrt{1}-6=-5 \rightarrow 1-6=-5 \text{ Check!}$$

Q. Solve the equation $\sqrt{5x-1}+2=x-1$.

A. **$x=10$. First, isolate the radical on the left.**

$\sqrt{5x-1}+2=x-1 \rightarrow \sqrt{5x-1}=x-3$

Now square both sides of the equation.

$\left(\sqrt{5x-1}\right)^2=(x-3)^2 \rightarrow 5x-1=x^2-6x+9$

Simplify the equation and solve for x.

$5x-1=x^2-6x+9 \rightarrow 0=x^2-11x+10 \rightarrow 0=(x-1)(x-10)$

The two solutions are $x=1$ or $x=10$.

You need to check to be sure that there are no extraneous solutions. $x=1 \rightarrow \sqrt{5\cdot 1-1}+2=1-1 \rightarrow \sqrt{4}+2=0 \rightarrow 4 \neq 0$ **Extraneous!**

$x=10 \rightarrow \sqrt{5\cdot 10-1}+2=10-1 \rightarrow \sqrt{49}+2=9 \rightarrow 7+2=9$ **Check!**

Only the 10 is a solution of the original equation.

YOUR TURN

9 Solve for x: $\sqrt{3x+7}+1=x$.

10 Solve for x: $\sqrt{2x+7}-2x=1$.

11 Solve for x: $\sqrt{14-x}+3=x-5$.

Calming two radicals

Some equations that contain radicals call for more than one application of squaring both sides. For example, you have to square both sides more than once when you can't isolate a radical term by itself on one side of the equation. And you usually need to square both sides more than once when you have three terms in the equation — two of them with radicals.

Q. Solve for x: $\sqrt{3x+19}-\sqrt{5x-1}=2$.

A. **$x = 2$. First, isolate one radical by itself. Then square both sides of the equation.**

$$\sqrt{3x+19}=2+\sqrt{5x-1}\rightarrow\left(\sqrt{3x+19}\right)^2=\left(2+\sqrt{5x-1}\right)^2\rightarrow 3x+19=4+4\sqrt{5x-1}+5x-1$$

Isolate the radical term again, simplify the other terms, and divide through by the common factor. Then square both sides again.

$$3x+19-4-5x+1=4\sqrt{5x-1}\rightarrow -2x+16=4\sqrt{5x-1}\rightarrow -x+8=2\sqrt{5x-1}\rightarrow$$
$$(-x+8)^2=\left(2\sqrt{5x-1}\right)^2\rightarrow x^2-16x+64=4(5x-1)$$

Simplify and solve for x.

$$x^2-16x+64=4(5x-1)\rightarrow x^2-16x+64=20x-4\rightarrow x^2-36x+68=0\rightarrow(x-2)(x-34)=0$$

The two solutions are $x=2$ or $x=34$.

Don't forget to check each solution in the original equation.

$$x=2\rightarrow\sqrt{3(2)+19}-\sqrt{5(2)-1}=2\rightarrow\sqrt{25}-\sqrt{9}=5-3=2 \text{ Checks!}$$
$$x=34\rightarrow\sqrt{3(34)+19}-\sqrt{5(34)-1}=2\rightarrow\sqrt{121}-\sqrt{169}=11-13\neq-2 \text{ Extraneous!}$$

The solution $x=2$ works. The other solution doesn't work in the equation. The number 34 is an extraneous solution.

YOUR TURN

12 Solve for x: $\sqrt{2x-3}-\sqrt{x-2}=1$.

13 Solve for x: $\sqrt{x+4}+\sqrt{21-x}=7$.

Changing Negative Attitudes toward Negative Exponents

Equations with negative exponents offer some unique challenges. The first challenge deals with the fact that you're working with negative numbers and so you have to keep track of the rules needed to add, subtract, multiply, and divide those negative numbers. Another challenge deals with the solution — if you find one — and checking to see if it works in the original form of the equation. The original form will take you back to those negative exponents, so it's round and round you go with number challenges.

Flipping negative exponents out of the picture

In general, negative exponents are easier to work with if they disappear. Yes, as wonderful as negative exponents are in the world of mathematics, solving equations that contain them is just easier if you change the format to positive exponents and fractions and then deal with solving the fractional equations.

EXAMPLE

Q. Solve for x in: $x^{-1} = 4$.

A. $4x = 1$ **or** $x = \frac{1}{4}$**.** Write the variable x in the denominator of a fraction and then solve for x. A nice way to solve for x is to write the 4 as a fraction, creating a proportion, and then cross-multiply:

$x^{-1} = 4$ is written $\frac{1}{x} = \frac{4}{1}$, which, when cross-multiplied, becomes $4x = 1$ or $x = \frac{1}{4}$.

Q. Solve for x in: $(x-3)^{-1} - x^{-1} = \frac{3}{10}$.

A. $x = 5$ **or** $x = -2$**.** Rewrite the equation, changing the terms with negative exponents into rational or fractional terms.

$$\frac{1}{x-3} - \frac{1}{x} = \frac{3}{10}$$

Then find the common denominator for the three fractions and rewrite each fraction as an equivalent fraction with that common denominator. Now multiply through to get rid of all the denominators (whew!), and solve the resulting equation.

$$\frac{1}{x-3} - \frac{1}{x} = \frac{3}{10} \rightarrow \frac{10x}{10x(x-3)} - \frac{10(x-3)}{10x(x-3)} = \frac{3x(x-3)}{10x(x-3)} \rightarrow$$

$$10x - 10(x-3) = 3x(x-3) \rightarrow 10x - 10x + 30 = 3x^2 - 9x \rightarrow$$

$$0 = 3x^2 - 9x - 30 = 3(x^2 - 3x - 10) = 3(x-5)(x+2)$$

The solutions are $x = 5$ or $x = -2$.

Factoring out negatives to solve equations

Negative exponents don't have to have the same power within a particular equation. In fact, it may be more common to have a mixture of powers in an equation. Here are two useful methods for solving equations with negative exponents:

- Factoring out a greatest common factor (GCF)
- Solving the equation as if it's a quadratic (quadratic-like; see Chapter 3)

Factoring out a negative GCF

Factoring out a negative exponent is an alternative to changing the terms to fractions and finding common denominators. You get to choose either your favorite procedure or just what will work best in a particular situation.

EXAMPLE

Q. Solve for x: $3x^{-3} - 5x^{-2} = 0$.

A. $x = \frac{3}{5}$. First, factor out the greatest common factor.

$$3x^{-3} - 5x^{-2} = x^{-3}(3 - 5x) = 0$$

Now set each factor equal to 0 and solve for x.

$$x^{-3}(3 - 5x) = 0$$

When $x^{-3} = 0$, $\frac{1}{x^3} = 0$; but that can never be true. The numerator has to be 0 to have a fraction be equal to 0.

When $3 - 5x = 0$, $x = \frac{3}{5}$.

Q. Solve for x: $x^{-4} - x^{-2} = 0$.

A. $x = 1$ **or** $x = -1$. First, factor out the greatest common factor.

$$x^{-4} - x^{-2} = x^{-4}(1 - x^2) = 0$$

Now set each factor equal to 0 and solve for x.

$$x^{-4}(1 - x^2) = 0$$

When $x^{-4} = 0$, $\frac{1}{x^4} = 0$; but that can never be true. The numerator has to be 0 to have a fraction be equal to 0.

When $1 - x^2 = 0$ you have $(1 - x)(1 + x) = 0$ and $x = 1$ or $x = -1$.

YOUR TURN

14 Solve: $x^{-2} - 9 = 0$.

15 Solve: $2x^{-1} - (x-1)^{-1} = \frac{1}{6}$.

16 Solve: $x^{-5} - x^{-4} = 0$.

17 Solve: $x^{-3} - 8x^{-4} + 7x^{-5} = 0$.

Solving quadratic-like trinomials

Trinomials are expressions with three terms, and if the terms are raised to the second degree, the expression is quadratic. You can simplify quadratic trinomials by factoring them into two binomial factors. (See Chapter 3 for details on factoring trinomials.)

Often, you can factor trinomials with negative powers into two binomials if they have the following pattern: $ax^{-2n} + bx^{-n} + c$. The exponents on the variables have to have a special arrangement, where one of the exponents is twice the other. When this factors, you produce two solutions and both work when substituted into the original equation. You haven't changed the format of the equation, but you still have to be sure that you aren't putting a zero in the denominator for an answer.

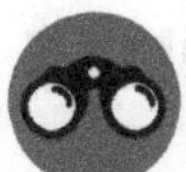

EXAMPLE

Q. Solve $3x^{-2}+5x^{-1}-2=0$.

A. $x=3,\ x=-\frac{1}{2}$. Factoring:

$$3x^{-2}+5x^{-1}-2=\left(3x^{-1}-1\right)\left(x^{-1}+2\right)=0$$

When $3x^{-1}-1=0,\ \frac{3}{x}=1,\ x=3$, and when $x^{-1}+2=0,\ \frac{1}{x}=-2,\ x=-\frac{1}{2}$.

Q. Solve: $x^{-4}-15x^{-2}-16=0$.

A. $x=\frac{1}{4},\ x=-\frac{1}{4}$. Factor: $x^{-4}-15x^{-2}-16=\left(x^{-2}-16\right)\left(x^{-2}+1\right)=0$

The first factor can be factored again: $\left(x^{-2}-16\right)\left(x^{-2}+1\right)=\left(x^{-1}-4\right)\left(x^{-1}+4\right)\left(x^{-2}+1\right)=0$. Setting the factors equal to 0, only the first two yield solutions.

$$x^{-1}-4=0\rightarrow\frac{1}{x}=4\rightarrow x=\frac{1}{4}$$

$$x^{-1}+4=0\rightarrow\frac{1}{x}=-4\rightarrow x=-\frac{1}{4}$$

$x^{-2}+1=0\rightarrow\frac{1}{x^2}=-1$, and this has no solution.

YOUR TURN

18 Solve $x^{-2}-9=0$.

19 Solve: $x^{-4}-5x^{-2}+4=0$.

Solving Equations with Fractional Exponents

You use fractional exponents ($x^{1/2}$, for example) to replace radicals and powers under radicals. Writing terms with fractional exponents allows you to perform operations on terms more easily when they have the same base or variable.

You write the radical expression $\sqrt[3]{x^4}$, for example, as $x^{4/3}$. The power of the variable under the radical goes in the numerator of the fraction, and the root of the radical goes in the denominator of the fraction.

Combining terms with fractional exponents

Fortunately, the rules of exponents stay the same when the exponents are fractional:

- **You can add or subtract terms with the same base and exponent:**

  $4x^{2/3}+5x^{2/3}-2x^{2/3}=7x^{2/3}$

- **You can multiply terms with the same base by adding their exponents:**

  $\left(5x^{3/4}\right)\left(9x^{2/3}\right)=45x^{3/4+2/3}=45x^{9/12+8/12}=45x^{17/12}$

- **You can divide terms with the same base by subtracting their exponents:**

  $\frac{45x^{3/5}}{9x^{1/5}}=5x^{3/5-1/5}=5x^{2/5}$

- **You can raise a fractional power to a power by multiplying the two powers:**

  $\left(x^{3/4}\right)^{-5/2}=x^{3/4(-5/2)}=x^{-15/8}$

Fractional exponents may not look that much better than the radicals they represent, but can you imagine trying to simplify $5\sqrt[4]{x^3}\cdot 9\sqrt[3]{x^2}$ without changing the format? You can always refer to the second entry in the previous list to see how fractional powers make doing the multiplication possible.

Factoring fractional exponents

You can easily factor expressions that contain variables with fractional exponents if you know the rule for dividing numbers with the same base (see the previous section): Subtract their exponents. Of course, you have the challenge of finding common denominators when you subtract fractions. Other than that, it's smooth sailing.

EXAMPLE

Q. Factor the expression $2x^{1/2} - 3x^{1/3}$.

A. You note that the smaller of the two exponents is the fraction $\frac{1}{3}$. Factor out x raised to that lower power, changing to a common denominator where necessary:

$$2x^{1/2} - 3x^{1/3} = x^{1/3}\left(2x^{1/2-1/3} - 3x^{1/3-1/3}\right)$$
$$= x^{1/3}\left(2x^{3/6-2/6} - 3x^{0}\right) = x^{1/3}\left(2x^{1/6} - 3\right)$$

Q. Factor the expression $4x^{-1/2} - x^{-1/4}$.

A. Factor out $x^{-1/2}$: $4x^{-1/2} - x^{-1/4} = x^{-1/2}\left(4 - x^{1/4}\right)$

Solving equations by working with fractional exponents

Fractional exponents represent radicals and powers, so when you can isolate a term with a fractional exponent, you can raise each side to an appropriate power to get rid of the exponent and eventually solve the equation. When you can't isolate the fractional exponent, you must resort to other methods for solving equations, such as factoring.

Raising each side to a power

When you can isolate a term that has a fractional exponent attached to it, go for it. The goal is to make the exponent equal to 1 so that you can solve the equation for x. You accomplish your goal by raising each side of the equation to a power that's equal to the reciprocal of the fractional exponent.

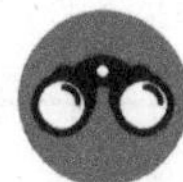

EXAMPLE

Q. Solve the equation $x^{4/3} = 16$.

A. 8. Raise each side of the equation to the $\frac{3}{4}$ power, because multiplying a number and its reciprocal always gives you a product of 1:

$\left(x^{4/3}\right)^{3/4} = (16)^{3/4}$, giving you $x^{1} = \sqrt[4]{16^{3}}$.

You finish the problem by evaluating the radical:

$$x = \sqrt[4]{16^{3}} = \left(\sqrt[4]{16}\right)^{3} = (2)^{3} = 8$$

The evaluation is easier if you take the fourth root first and then raise the answer to the third power. You can do this because powers and roots are on the same level in the order of operations (see Chapter 1 for more on this topic), so you can calculate them in either order — whatever is most convenient.

Q. Solve the equation $x^{-1/2} = 2$.

A. $\frac{1}{4}$. Raise each side of the equation to the –2 power.

$\left(x^{-1/2}\right)^{-2} = (2)^{-2}$, giving you $x = \frac{1}{2^{2}} = \frac{1}{4}$.

YOUR TURN

20 Simplify: $2x^{\frac{3}{4}} - 3x^{\frac{3}{4}} + 5\left(x^{-3}\right)^{-\frac{1}{4}}$.

21 Factor: $3x^{-\frac{5}{6}} + 6x^{\frac{1}{6}}$.

22 Solve for *x*: $x^{\frac{1}{3}} = 2$.

23 Solve for *x*: $x^{-\frac{1}{2}} = 10$.

Factoring out variables with fractional exponents

You don't always have the luxury of being able to raise each side of an equation to a power to get rid of the fractional exponents. Your next best plan of attack involves factoring out the variable with the smaller exponent and setting the two factors equal to zero.

Often, you can factor trinomials with fractional exponents into the product of two binomials. After the factoring, you set the two binomials equal to zero to determine if you can find any solutions.

EXAMPLE

Q. Solve $x^{5/6} - 3x^{1/2} = 0$.

A. **0, 27.** First, factor out an *x* with an exponent of $\frac{1}{2}$:

$$x^{5/6} - 3x^{1/2} = x^{1/2}\left(x^{5/6-1/2} - 3\right) = x^{1/2}\left(x^{5/6-3/6} - 3\right) = x^{1/2}\left(x^{2/6} - 3\right) = x^{1/2}\left(x^{1/3} - 3\right) = 0$$

Now you can set the two factors equal to zero and solve for *x*. When $x^{1/2} = 0$, $x = 0$; and when $x^{1/3} - 3 = 0$, $x^{1/3} = 3$, $\left(x^{1/3}\right)^3 = (3)^3$, $x = 27$.

Q. Solve $x^{1/2} - 6x^{1/4} + 5 = 0$.

A. **1, 625. First, factor the trinomial into two binomials. The exponent of the first variable is twice that of the second, which should indicate to you that the trinomial has factoring potential. After you factor, you set the expression equal to zero and solve for *x*:**

$$x^{1/2} - 6x^{1/4} + 5 = \left(x^{1/4} - 1\right)\left(x^{1/4} - 5\right) = 0$$

When $x^{1/4} - 1 = 0$, $x^{1/4} = 1$, $\left(x^{1/4}\right)^4 = (1)^4$, $x = 1$; and when $\left(x^{1/4}\right)^4 = (5)^4$, $x = 625$, $x^{1/4} - 5 = 0$, $x^{1/4} = 5$. Check your answers in the original equation; you find that both $x = 1$ and $x = 625$ work.

YOUR TURN

24 Solve $x^{8/15} + 64x^{1/5} = 0$.

25 Solve $x^{1/3} - 3x^{1/6} - 4 = 0$.

Putting fractional and negative exponents together

This chapter wouldn't be complete without an explanation of how you can combine fractional and negative exponents into one big equation. Creating this mega problem isn't something you do just to see how exciting an equation can be. The following is an example of a situation that occurs in calculus problems. The derivative (a calculus process) has already been performed, and now you have to solve the equation. The hardest part of calculus is often the algebra, so I feel I should address this topic before you get to calculus.

Q. Factor and solve: $x^{-1/4}+2x^{-1/8}-3=0$.

EXAMPLE

A. **1. The first exponent is twice the second, so this can be factored as a quadratic: $x^{-1/4}+2x^{-1/8}-3=\left(x^{-1/8}+3\right)\left(x^{-1/8}-1\right)=0$. Setting each factor equal to zero, when $x^{-1/8}+3=0, x^{-1/8}=-3$ and you have no solution, because the even-powered root can't have a negative result. Next, setting $x^{-1/8}-1=0$ you have $x^{-1/8}=1$, which is true when $x=1$.**

Q. Factor and solve: $x\left(x^3+8\right)^{1/3}\left(x^2-4\right)^{-1/2}+x^2\left(x^3+8\right)^{-2/3}\left(x^2-4\right)^{1/2}=0$.

A. **0, 2, −2. You have a common factor of $x\left(x^3+8\right)^{-2/3}\left(x^2-4\right)^{-1/2}$. Notice that both terms have a power of x, a power of $\left(x^3+8\right)$, and a power of $\left(x^2-4\right)$ in them. (You have to choose the lower of the powers on a factor and use that on the greatest common factor.)**

Dividing each term by the greatest common factor, you have $x(x^3+8)^{-2/3}(x^2-4)^{-1/2}\left[(x^3+8)^1+x(x^2-4)^1\right]=0$

Now you simplify the terms inside the brackets:

$$x(x^3+8)^{-2/3}(x^2-4)^{-1/2}\left[(x^3+8)^1+x(x^2-4)^1\right]=0\rightarrow x(x^3+8)^{-2/3}(x^2-4)^{-1/2}\left[x^3+8+x^3-4x\right]=0\rightarrow$$
$$x(x^3+8)^{-2/3}(x^2-4)^{-1/2}\left[2x^3-4x+8\right]=0\rightarrow 2x(x^3+8)^{-2/3}(x^2-4)^{-1/2}\left[x^3-2x+4\right]=0$$

You can set each of the four factors equal to zero to find any solutions for the equation:

When $2x=0$, $x=0$. When $\left(x^3+8\right)^{-2/3}=0$, $x^3+8=0$, $x^3=-8$, $x=-2$. When $\left(x^2-4\right)^{-1/2}=0$, $x^2-4=0$, $x=\pm 2$. And, finally, when $x^3-2x+4=\left(x+2\right)\left(x^2-2x+2\right)=0$, $x=-2$. The solutions you find are $x=0$, $x=2$, and $x=-2$. You see repeated roots, but I name each of them only once here. The trinomial that remains in the factorization, x^3-2x+4, doesn't have any real solution when set equal to 0.

YOUR TURN

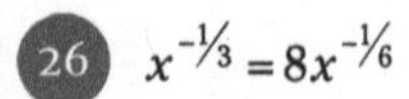

26 $x^{-1/3}=8x^{-1/6}$

27 $x\left(x^3-1\right)^{-1/3}-25x^{-1}\left(x^3-1\right)^{-1/3}=0$

Practice Questions Answers and Explanations

1 **7, -2.** 1) Find the common denominator. 2) Rewrite the fractions. 3) Multiply by the LCD. 4) Solve for x. 5) Check.

1) $x(x-1)(6)=6x(x-1)$

2) $\frac{4}{x}\cdot\frac{6(x-1)}{6(x-1)}-\frac{3}{x-1}\cdot\frac{6x}{6x}=\frac{10-x}{6x}\cdot\frac{x-1}{x-1}\rightarrow\frac{24x-24}{6x(x-1)}-\frac{18x}{6x(x-1)}=\frac{-x^2+11x-10}{6x(x-1)}$

3) $24x-24-18x=-x^2+11x-10$

4) $x^2-5x-14=0\rightarrow(x-7)(x+2)=0\rightarrow x=7, x=-2$

5) $x=7\rightarrow\frac{4}{7}-\frac{3}{7-1}=\frac{10-7}{6\cdot 7}\rightarrow\frac{4}{7}-\frac{3}{6}=\frac{3}{42}\rightarrow\frac{24}{42}-\frac{21}{42}=\frac{3}{42}$ Check!

$x=-2\rightarrow\frac{4}{-2}-\frac{3}{-2-1}=\frac{10-(-2)}{6(-2)}\rightarrow\frac{4}{-2}-\frac{3}{-3}=\frac{12}{-12}\rightarrow-2+1=-1$ Check!

2 $-\frac{24}{5}$**, 5.** 1) Find the common denominator. 2) Rewrite the fractions. 3) Multiply by the LCD. 4) Solve for x. 5) Check.

1) $4\cdot x\cdot 3=12x$

2) $\frac{x+3}{4}\cdot\frac{3x}{3x}+\frac{10}{x}\cdot\frac{12}{12}=\frac{2x+2}{3}\cdot\frac{4x}{4x}\rightarrow\frac{3x^2+9x}{12x}+\frac{120}{12x}=\frac{8x^2+8x}{12x}$

3) $3x^2+9x+120=8x^2+8x$

4) $0=5x^2-x-120\rightarrow(5x+24)(x-5)=0\rightarrow x=-\frac{24}{5}, x=5$

5) $x=-\frac{24}{5}\rightarrow\frac{-\frac{24}{5}+3}{4}+\frac{10}{-\frac{24}{5}}=\frac{2\left(-\frac{24}{5}\right)+2}{3}\rightarrow\frac{-9}{20}+\frac{-25}{12}=\frac{-38}{15}\rightarrow\frac{-27}{60}+\frac{-125}{60}=\frac{-152}{60}$ Check!

$x=5\rightarrow\frac{5+3}{4}+\frac{10}{5}=\frac{2(5)+2}{3}\rightarrow 2+2=4$ Check!

3 **3.** 1) Find the common denominator. 2) Rewrite the fractions. 3) Multiply by the LCD. 4) Solve for x. 5) Check.

1) $(x+2)(x-2)=x^2-4$

2) $\frac{10}{x^2-4}-\frac{x-13}{x+2}\cdot\frac{x-2}{x-2}=\frac{x+1}{x-2}\cdot\frac{x+2}{x+2}\rightarrow\frac{10}{x^2-4}-\frac{x^2-15x+26}{x^2-4}=\frac{x^2+3x+2}{x^2-4}$

3) $10-x^2+15x-26=x^2+3x+2$

4) $0=2x^2-12x+18=2(x^2-6x+9)\rightarrow 2(x-3)^2=0\rightarrow x=3$

5) $x=3\rightarrow\frac{10}{3^2-4}-\frac{3-13}{3+2}=\frac{3+1}{3-2}\rightarrow\frac{10}{5}-\frac{-10}{5}=\frac{4}{1}\rightarrow 2+2=4$ Check!

4 **-1.** 1) Find the common denominator. 2) Rewrite the fractions. 3) Multiply by the LCD. 4) Solve for x. 5) Check.

1) $(x-5)(x-2)=x^2-7x+10$

2) $\frac{3}{x^2-7x+10}-\frac{1}{x-5}\cdot\frac{x-2}{x-2}=\frac{x}{x-2}\cdot\frac{x-5}{x-5}\rightarrow\frac{3}{x^2+x-20}-\frac{x-2}{x^2+x-20}=\frac{x^2-5x}{x^2+x-20}$

3) $3-x+2=x^2-5x$

4) $0=x^2-4x-5\rightarrow(x-5)(x+1)=0\rightarrow x=5, x=-1$

5) $x = 5 \rightarrow \frac{3}{5^2 - 7\cdot 5 + 10} - \frac{1}{5-5} = \frac{5}{5-2} \rightarrow \frac{3}{0} - \frac{1}{0} = \frac{5}{3}$ Extraneous!

$x = -1 \rightarrow \frac{3}{(-1)^2 - 7(-1) + 10} - \frac{1}{-1-5} = \frac{-1}{-1-2} \rightarrow \frac{3}{18} - \frac{1}{-6} = \frac{1}{3} \rightarrow \frac{1}{6} + \frac{1}{6} = \frac{1}{3} \rightarrow \frac{2}{6} = \frac{1}{3}$ Check!

(5) **9, −6.** First, reduce the left fractions by dividing by 5. Then reduce the denominators by dividing by 4.

$$\frac{\cancel{5}^1 x}{\cancel{120}^{24}} = \frac{9}{4x-12} \rightarrow \frac{x}{24} = \frac{9}{4x-12} \rightarrow \frac{x}{24} = \frac{9}{4x-12} \rightarrow \frac{x}{\cancel{24}^6} = \frac{9}{\cancel{4}^1(x-3)} \rightarrow \frac{x}{6} = \frac{9}{x-3}$$

Cross-multiply and solve for x.

$$\frac{x}{6} = \frac{9}{x-3} \rightarrow x(x-3) = 54 \rightarrow x^2 - 3x - 54 = 0 \rightarrow (x-9)(x+6) = 0$$

The two solutions are $x = 9$ or $x = -6$.

(6) $-\frac{9}{2}$**, 5.** First, reduce the right fractions by dividing by 4. Then reduce the denominators by dividing by 2.

$$\frac{15}{4x-2} = \frac{\cancel{4}^1 x}{\cancel{24}^6} \rightarrow \frac{15}{4x-2} = \frac{x}{6} \rightarrow \frac{15}{4x-2} = \frac{x}{6} \rightarrow \frac{15}{\cancel{2}^1(2x-1)} = \frac{x}{\cancel{6}^3} \rightarrow \frac{15}{2x-1} = \frac{x}{3}$$

Cross-multiply and solve for x.

$$\frac{15}{2x-1} = \frac{x}{3} \rightarrow 45 = x(2x-1) \rightarrow 0 = 2x^2 - x - 45 \rightarrow 0 = (2x+9)(x-5)$$

The two solutions are $x = -\frac{9}{2}$ or $x = 5$.

(7) **12, −3.** First, subtract the second term on the left from each side and combine the newly created like-terms on the right.

$$\frac{x-11}{2} + \frac{2}{x+2} - \frac{2}{x+2} = \frac{9}{x+2} - \frac{2}{x+2} \rightarrow \frac{x-11}{2} = \frac{7}{x+2}$$

Now cross-multiply and solve for x.

$$\frac{x-11}{2} = \frac{7}{x+2} \rightarrow (x-11)(x+2) = 14 \rightarrow x^2 - 9x - 22 = 14 \rightarrow x^2 - 9x - 36 = 0 \rightarrow (x-12)(x+3) = 0$$

The two solutions are $x = 12$ or $x = -3$.

(8) **11, −5.** First, add the second term on the left to each side, subtract the $\frac{1}{5}$ from each side, and combine the newly created like-terms.

$$\frac{x+1}{5} - \frac{1}{5} - \frac{7}{x-6} + \frac{7}{x-6} = \frac{4}{x-6} + \frac{7}{x-6} + \frac{1}{5} - \frac{1}{5} \rightarrow \frac{x+1-1}{5} = \frac{4+7}{x-6} \rightarrow \frac{x}{5} = \frac{11}{x-6}$$

Now cross-multiply and solve for x.

$$\frac{x}{5} = \frac{11}{x-6} \rightarrow x(x-6) = 55 \rightarrow x^2 - 6x - 55 = 0 \rightarrow (x-11)(x+5) = 0$$

The two solutions are $x = 11$ or $x = -5$.

(9) **6. First, isolate the radical term, and then square both sides.**

$$\sqrt{3x+7}+1=x \to \sqrt{3x+7}=x-1 \to \left(\sqrt{3x+7}\right)^2=(x-1)^2 \to 3x+7=x^2-2x+1$$

Solve for x and check the solutions obtained.

$$3x+7=x^2-2x+1 \to 0=x^2-5x-6 \to 0=(x-6)(x+1)$$

The two solutions are $x=6$ or $x=-1$.

Checking to be sure that there are no extraneous solutions:

$x=6 \to \sqrt{3\cdot6+7}+1=6 \to \sqrt{25}+1=6 \to 6=6$ Check!

$x=-1 \to \sqrt{3(-1)+7}+1=-1 \to \sqrt{4}+1=-1 \to 3\neq-1$ Extraneous!

(10) **1. First, isolate the radical term, and then square both sides.**

$$\sqrt{2x+7}-2x=1 \to \sqrt{2x+7}=2x+1 \to \left(\sqrt{2x+7}\right)^2=(2x+1)^2 \to 2x+7=4x^2+4x+1$$

Solve for x and check the solutions obtained.

$$2x+7=4x^2+4x+1 \to 0=4x^2+2x-6 \to 0=2\left(2x^2+x-3\right)=2(2x+3)(x-1)$$

The two solutions are $x=-\frac{3}{2}$ or $x=1$.

Checking to be sure that there are no extraneous solutions: $x=-\frac{3}{2} \to \sqrt{2\left(-\frac{3}{2}\right)+7}-2\left(-\frac{3}{2}\right)=1 \to \sqrt{-3+7}+3=1 \to 2+3\neq1$ Extraneous!

$x=1 \to \sqrt{2(1)+7}-2(1)=1 \to \sqrt{9}-2=1 \to 3-2=1$ Check!

(11) **10. First, isolate the radical term, and then square both sides.**

$$\sqrt{14-x}+3=x-5 \to \sqrt{14-x}=x-8 \to \left(\sqrt{14-x}\right)^2=(x-8)^2 \to 14-x=x^2-16x+64$$

Solve for x and check the solutions obtained.

$$14-x=x^2-16x+64 \to 0=x^2-15x+50 \to 0=(x-5)(x-10)$$

The two solutions are $x=5$ or $x=10$.

Checking to be sure that there are no extraneous solutions:

$x=5 \to \sqrt{14-5}+3=5-5 \to \sqrt{9}+3=0 \to 3+3\neq0$ Extraneous!

$x=10 \to \sqrt{14-10}+3=10-5 \to \sqrt{4}+3=5 \to 5=5$ Check!

(12) **2, 6. First, isolate the left radical; then square both sides, simplify, and isolate the remaining radical.**

$$\sqrt{2x-3}-\sqrt{x-2}=1 \to \sqrt{2x-3}=1+\sqrt{x-2} \to \left(\sqrt{2x-3}\right)^2=\left(1+\sqrt{x-2}\right)^2 \to$$

$$2x-3=1+2\sqrt{x-2}+x-2 \to x-2=2\sqrt{x-2}$$

Square both sides, simplify, and solve the resulting equation.

$$(x-2)^2=\left(2\sqrt{x-2}\right)^2 \rightarrow x^2-4x+4=4(x-2)\rightarrow x^2-4x+4=4x-8\rightarrow$$
$$x^2-8x+12=0\rightarrow(x-2)(x-6)=0$$

The two solutions are $x=2$ or $x=6$.

Checking in the original equation:

$x=2\rightarrow\sqrt{2\cdot2-3}-\sqrt{2-2}=1\rightarrow\sqrt{1}-\sqrt{0}=1\rightarrow1-0=1$ Checks!

$x=6\rightarrow\sqrt{2\cdot6-3}-\sqrt{6-2}=1\rightarrow\sqrt{9}-\sqrt{4}=1\rightarrow3-2=1$ Checks!

(13) **12.** First, isolate the left radical; then square both sides, simplify, and isolate the remaining radical.

$$\sqrt{x+4}+\sqrt{21-x}=7\rightarrow\sqrt{x+4}=7-\sqrt{21-x}\rightarrow\left(\sqrt{x+4}\right)^2=\left(7-\sqrt{21-x}\right)^2\rightarrow$$
$$x+4=49-14\sqrt{21-x}+21-x\rightarrow2x-66=-14\sqrt{21-x}$$

Now, divide both sides by 2. Then square both sides, simplify, and solve the resulting equation.

$$x-33=-7\sqrt{21-x}\rightarrow(x-33)^2=\left(-7\sqrt{21-x}\right)^2\rightarrow x^2-66x+1089=49(21-x)\rightarrow$$
$$x^2-66x+1089=1029-49x\rightarrow x^2-17x+60=0\rightarrow(x-5)(x-12)=0$$

The two solutions are $x=5$ or $x=12$.

Checking in the original equation:

$x=5\rightarrow\sqrt{5+4}+\sqrt{21-5}=7\rightarrow\sqrt{9}+\sqrt{16}=7\rightarrow3+4=7$ Check!

$x=12\rightarrow\sqrt{12+4}+\sqrt{21-12}=7\rightarrow\sqrt{16}+\sqrt{9}=7\rightarrow4+3=7$ Check!

(14) $\pm\frac{1}{3}$**.** Rewrite the equation by changing the negative exponent term to a fraction and adding 9 to each side: $\frac{1}{x^2}=9$. Write the 9 as a fraction, cross-multiply, and solve for x using the square-root rule: $\frac{1}{x^2}=\frac{9}{1}\rightarrow9x^2=1\rightarrow x^2=\frac{1}{9}\rightarrow x=\pm\frac{1}{3}$

(15) **3, 4.** Rewrite the equation by changing the negative exponent terms to fractions: $\frac{2}{x}-\frac{1}{x-1}=\frac{1}{6}$. Find a common denominator, change the fractions, and multiply through by the common denominator to solve for x.

$$\frac{2}{x}\cdot\frac{6(x-1)}{6(x-1)}-\frac{1}{x-1}\cdot\frac{6x}{6x}=\frac{1}{6}\cdot\frac{x(x-1)}{x(x-1)}\rightarrow\frac{12(x-1)}{6x(x-1)}-\frac{6x}{6x(x-1)}=\frac{x(x-1)}{6x(x-1)}\rightarrow$$
$$12(x-1)-6x=x(x-1)\rightarrow12x-12-6x=x^2-x\rightarrow0=x^2-7x+12\rightarrow0=(x-3)(x-4)$$

The two solutions are $x=3$ or $x=4$. They both check.

(16) **1.** Factor out x^{-5} and solve for x: $x^{-5}-x^{-4}=0\rightarrow x^{-5}(1-x)=0$. The only solution is $x=1$ because $\frac{1}{x^5}=0$ has no solution.

17 **7, 1.** First, factor out x^{-5}: $x^{-3}-8x^{-4}+7x^{-5}=0 \rightarrow x^{-5}\left(x^2-8x+7\right)=0$. The GCF with the negative exponent gives you no solution but the quadratic factors.

$x^{-5}(x-7)(x-1)=0$ has two solutions: $x=7$ or $x=1$.

18 $\frac{1}{3},-\frac{1}{3}$. Factor and set the factors equal to 0: $x^{-2}-9=0 \rightarrow \left(x^{-1}-3\right)\left(x^{-1}+3\right)=0$

$x^{-1}-3=0 \rightarrow \frac{1}{x}=3 \rightarrow x=\frac{1}{3}$

$x^{-1}+3=0 \rightarrow \frac{1}{x}=-3 \rightarrow x=-\frac{1}{3}$

19 $\frac{1}{2},-\frac{1}{2},1,-1$. Factor and set the factors equal to 0:

$x^{-4}-5x^{-2}+4=0 \rightarrow \left(x^{-2}-4\right)\left(x^{-2}-1\right)=0 \rightarrow \left(x^{-1}-2\right)\left(x^{-1}+2\right)\left(x^{-1}-1\right)\left(x^{-1}+1\right)=0$

These factors give you the solutions $x=\frac{1}{2}, x=-\frac{1}{2}, x=1, x=-1$, respectively.

20 $4x^{3/4}$. First, raise the power to a power, and then combine like terms:

$2x^{3/4}-3x^{3/4}+5\left(x^{-3}\right)^{-1/4}=2x^{3/4}-3x^{3/4}+5x^{3/4}=4x^{3/4}$

21 $3x^{-5/6}(1+2x)$. The GCF is $3x^{-5/6}$: $3x^{-5/6}+6x^{1/6}=3x^{-5/6}\left(1+2x^{1}\right)$

22 **8.** Cube each side of the equation: $\left(x^{1/3}\right)^3=2^3 \rightarrow x=8$

23 $\frac{1}{100}$. Raise each side of the equation to the –2 power: $\left(x^{-1/2}\right)^{-2}=10^{-2} \rightarrow x=\frac{1}{10^2}=\frac{1}{100}$

24 **0, –262,144.** The GCF is $x^{1/5}$. First, divide by that factor and then determine any solutions from the factors: $x^{8/15}+64x^{1/5}=x^{1/5}\left(x^{8/15-3/15}+64\right)=x^{1/5}\left(x^{1/3}+64\right)=0$. When $x^{1/5}=0$, the solution is $x=0$. When $x^{1/3}+64=0$, you subtract 64 from each side and then cube each side: $x^{1/3}=-64$ becomes $\left(x^{1/3}\right)^3=(-64)^3$ or $x=-262{,}144$.

25 **4096.** Factor as a quadratic and set the factors equal to zero to determine the solutions: $x^{1/3}-3x^{1/6}-4=\left(x^{1/6}-4\right)\left(x^{1/6}+1\right)=0$.

When $x^{1/6}=4$, you raise each side to the sixth power, and the solution is $x=4096$. When $x^{1/6}+1=0$, you subtract 1 from each side and have $x^{1/6}=-1$. There is no solution here, because an even power doesn't result in a negative answer.

26 **262,144.** Set the equation equal to zero and factor: $x^{-1/3}-8x^{-1/6}=x^{-1/6}\left(x^{-1/6}-8\right)=0$. There is no solution with the first factor, because $\frac{1}{x^{1/6}}=0$ is not possible. When $x^{1/6}=8$, you raise each side to the sixth power and get $x=262{,}114$.

27 **1, –5, 5.** The GCF is $x^{-1}\left(x^3-1\right)^{-1/3}$. Factoring: $x\left(x^3-1\right)^{-1/3}-25x^{-1}\left(x^3-1\right)^{-1/3}=x^{-1}\left(x^3-1\right)^{-1/3}\left(x^2-25\right)=0$.

Now set each factor equal to zero to determine the solutions. When $x^{-1}=0$, there is no solution, because $\frac{1}{x}=0$ has no solution. When $\left(x^3-1\right)^{-1/3}=0$, there is one solution when $x=1$. And when $x^2-25=0$, you have $(x+5)(x-5)=0$, which simplifies to $x=-5$ or $x=5$.

Whaddya Know? Chapter 5 Quiz

Quiz time! Complete each problem to test your knowledge on the various topics covered in this chapter. You can then find the solutions and explanations in the next section.

1. **Solve for y: $y^{1/2} - 6 = 4$.**
2. **Find the least common denominator of the fractions: $\frac{3}{16}, \frac{5}{36}, \frac{9}{40}$.**
3. **Solve for x: $\frac{8}{x^2-1} - \frac{x-5}{x-1} = \frac{x+5}{x+1}$.**
4. **Find the least common denominator of the fractions and rewrite the fractions with that common denominator: $\frac{7}{y+3}, \frac{y-4}{y^2}$.**
5. **Solve for x: $\frac{120}{375} = \frac{24}{x}$.**
6. **Solve for y: $\frac{27}{126} = \frac{2y+7}{14}$.**
7. **Solve for x: $\sqrt{9x-8} + 7 = 15$.**
8. **Solve for x: $\frac{10}{x^2+x-6} - \frac{x}{x-2} = \frac{2}{x+3}$.**
9. **Rewrite the expression, eliminating all negative exponents: $4x^{-2}y^{-3}z$.**
10. **Solve for x: $x^{-3} - 4 = 23$.**
11. **Solve for y: $\sqrt{y-3} + y = 15$.**
12. **Solve for z: $2z^{-2} - 7z^{-1} - 4 = 0$.**
13. **Solve for x: $x^{2/3} + x^{1/3} = 20$.**
14. **Solve for z: $\sqrt{4z+1} - \sqrt{10-z} = 3$.**
15. **Solve for x: $x^{-1} - 3x = 2$.**
16. **Solve for x: $x^{-1/3}\left(x^{1/3}-1\right)\left(x^{-1/3}-1\right) + x^{-1/3}\left(x^{1/3}-1\right)\left(x^{-1/2}+1\right) = 0$.**

Answers to Chapter 5 Quiz

1. **100.** Add 6 to each side of the equation. Then square both sides.

$$y^{1/2}-6=4\to y^{1/2}=10\to\left(y^{1/2}\right)^2=(10)^2\to y=100$$

2. **720.** Writing the prime factorizations of the denominators:

$$16=2^4,\quad 36=2^2\cdot 3^2,\quad 40=2^3\cdot 5$$

The LCD must contain four factors of 2, two factors of 3, and one factor of 5: $2^4\cdot 3^2\cdot 5=720$. Thus, $\frac{3}{16}=\frac{135}{720}$, $\frac{5}{36}=\frac{100}{720}$, $\frac{9}{40}=\frac{162}{720}$.

3. **3, −3.** First, find the common denominator and write each fraction with that denominator. Multiply through by the LCD to create the equation to be solved.

$$\frac{8}{x^2-1}-\frac{x-5}{x-1}=\frac{x+5}{x+1}\to\frac{8}{x^2-1}-\frac{x-5}{x-1}\cdot\frac{x+1}{x+1}=\frac{x+5}{x+1}\cdot\frac{x-1}{x-1}\to\frac{8}{x^2-1}-\frac{(x-5)(x+1)}{x^2-1}=\frac{(x+5)(x-1)}{x^2-1}\to$$
$$8-(x-5)(x+1)=(x+5)(x-1)\to 8-x^2+4x+5=x^2+4x-5\to 0=2x^2-18$$

Factor the quadratic to determine the solutions: $2\left(x^2-9\right)=2(x-3)(x+3)=0$ gives you the solutions $x=3$ or $x=-3$. Checking the solutions:

$$x=3\to\frac{8}{3^2-1}-\frac{3-5}{3-1}=\frac{3+5}{3+1}\to\frac{8}{8}-\frac{-2}{2}=\frac{8}{4}\to 1+1=2\text{ Check!}$$

$$x=-3\to\frac{8}{(-3)^2-1}-\frac{-3-5}{-3-1}=\frac{-3+5}{-3+1}\to\frac{8}{8}-\frac{-8}{-4}=\frac{2}{-2}\to 1-2=-1\text{ Check!}$$

4. $\frac{\mathbf{7y^2}}{\mathbf{y^2(y+3)}}$ **and** $\frac{\mathbf{y^2-y-12}}{\mathbf{y^2(y+3)}}$. The LCD is the product of the two denominators, because they have no factor in common.

$$\frac{7}{y+3}\cdot\frac{y^2}{y^2}=\frac{7y^2}{y^2(y+3)}\quad\text{and}\quad\frac{y-4}{y^2}\cdot\frac{y+3}{y+3}=\frac{y^2-y-12}{y^2(y+3)}$$

5. **75.** Reduce the fraction on the left by dividing numerator and denominator by 15.

$$\frac{\cancel{120}^{8}}{\cancel{375}^{25}}=\frac{24}{x}$$

Now reduce the two numerators by dividing each one by 8.

$$\frac{\cancel{8}^{1}}{25}=\frac{\cancel{24}^{3}}{x}$$

Cross-multiply to find x.

$$\frac{1}{25}=\frac{3}{x}$$
$$1\cdot x=3\cdot 25$$
$$x=75$$

(6) **−2.** Reduce the fraction on the left by dividing the numerator and denominator by 9.

$$\frac{\cancel{27}^{3}}{\cancel{126}^{14}}=\frac{2y+7}{14}$$

Reduce across the bottom. Then solve for *y*.

$$\frac{3}{\cancel{14}_1}=\frac{2y+7}{\cancel{14}_1}\rightarrow 3=2y+7\rightarrow -4=2y\rightarrow -2=y$$

(7) **8.** Subtract 7 from each side of the equation. Then square both sides.

$$\sqrt{9x-8}+7=15\rightarrow\sqrt{9x-8}=8\rightarrow\left(\sqrt{9x-8}\right)^2=8^2\rightarrow 9x-8=64$$

Solve for *x*.

$$9x=72\rightarrow x=8$$

(8) **−7.** First, find a common denominator and write each term with that denominator. Multiply through to remove the fractions. Then solve the resulting equation.

$$\frac{10}{x^2+x-6}-\frac{x}{x-2}=\frac{2}{x+3}\rightarrow\frac{10}{(x-2)(x+3)}-\frac{x}{x-2}\cdot\frac{x+3}{x+3}=\frac{2}{x+3}\cdot\frac{x-2}{x-2}\rightarrow$$

$$\frac{10}{(x-2)(x+3)}-\frac{x(x+3)}{(x-2)(x+3)}=\frac{2(x-2)}{(x-2)(x+3)}\rightarrow 10-x(x+3)=2(x-2)\rightarrow 10-x^2-3x=2x-4\rightarrow 0$$

$$=x^2+5x-14$$

Factor the quadratic to determine the solutions: $x^2+5x-14=(x+7)(x-2)=0$ gives you the solutions $x=-7$ or $x=2$. Checking the solutions:

$$x=-7\rightarrow\frac{10}{(-7)^2+(-7)-6}-\frac{-7}{-7-2}=\frac{2}{-7+3}\rightarrow\frac{10}{36}-\frac{-7}{-9}=\frac{2}{-4}\rightarrow\frac{10}{36}-\frac{28}{36}=-\frac{18}{36}\text{ Check!}$$

$x=2\rightarrow\frac{10}{(2)^2+(2)-6}-\frac{2}{2-2}=\frac{2}{2+3}\rightarrow\frac{10}{36}-\frac{2}{0}=\frac{2}{5}$. The 0 in the denominator of the second fraction shows that this solution is extraneous.

(9) $\boldsymbol{\frac{4z}{x^2y^3}}$**.** Move the factors with the negative exponents to the denominator of a fraction and change their exponents to positive numbers.

(10) $\boldsymbol{\frac{1}{3}}$**.** Add 4 to each side. Then rewrite the equation with fractions and no negative exponents.

$$x^{-3}-4=23$$

$$x^{-3}=27$$

$$\frac{1}{x^3}=\frac{27}{1}$$

Flip the proportion. Then take the cube root of each side of the equation.

$$\frac{x^3}{1}=\frac{1}{27}\rightarrow\sqrt[3]{x^3}=\sqrt[3]{\frac{1}{27}}\rightarrow x=\frac{1}{3}$$

(11) **12.** Subtract y from each side of the equation. Then square both sides.

$$\sqrt{y-3}+y=15 \rightarrow \sqrt{y-3}=15-y \rightarrow \left(\sqrt{y-3}\right)^2=(15-y)^2 \rightarrow y-3=225-30y+y^2$$

Write the quadratic equation in standard form and solve for y.

$$y^2-31y+228=0 \rightarrow (y-12)(y-19)=0 \rightarrow y=12 \text{ or } y=19$$

Check both solutions.

$$\sqrt{12-3}+12\stackrel{?}{=}15 \rightarrow \sqrt{9}+12\stackrel{?}{=}15 \rightarrow 9+12=15 \text{ Checks}$$

$$\sqrt{19-3}+19\stackrel{?}{=}15 \rightarrow \sqrt{16}+19\stackrel{?}{=}15 \rightarrow 4+19 \neq 15 \text{ Extraneous}$$

(12) **−2** and $\mathbf{\frac{1}{4}}$. Replace z^{-2} with x^2 and z^{-1} with x. Then solve the quadratic equation for x.

$$2x^2-7x-4=0 \rightarrow (2x+1)(x-4)=0 \rightarrow x=-\frac{1}{2} \text{ or } x=4$$

Now replace the x's with z^{-1} and solve for z.

$$z^{-1}=-\frac{1}{2} \quad \text{and} \quad z^{-1}=4$$

$$z=-2 \qquad\qquad z=\frac{1}{4}$$

(13) **−125** or **64**. Subtract 20 from both sides of the equation. Then replace $x^{2/3}$ with y^2 and $x^{1/3}$ with y.

$$x^{2/3}+x^{1/3}-20=0 \rightarrow y^2+y-20=0$$

Solve the quadratic equation for y.

$$(y+5)(y-4)=0 \rightarrow y=-5 \text{ or } y=4$$

Replace the y's with their equivalent x values and solve for x.

$$y=-5 \qquad\qquad y=4$$

$$x^{1/3}=-5 \qquad\qquad x^{1/3}=4$$

$$\left(x^{1/3}\right)^3=(-5)^3 \quad \text{or} \quad \left(x^{1/3}\right)^3=(4)^3$$

$$x=-125 \qquad\qquad x=64$$

(14) **6.** Add the second radical to both sides of the equation. Then square both sides.

$$\sqrt{4z+1}=3+\sqrt{10-z} \rightarrow \left(\sqrt{4z+1}\right)^2=\left(3+\sqrt{10-z}\right)^2 \rightarrow 4z+1=9+6\sqrt{10-z}+10-z$$

Isolate the radical term on the right by subtracting 19 and adding z to each side. Then square both sides.

$$5z-18=6\sqrt{10-z} \rightarrow (5z-18)^2=\left(6\sqrt{10-z}\right)^2 \rightarrow$$

$$25z^2-180z+324=36(10-z) \rightarrow 25z^2-180z+324=360-36z \rightarrow 25z^2-144z-36=0$$

Solve the quadratic for z.

$$25z^2-144z-36=0 \rightarrow (25z+6)(z-6)=0 \rightarrow z=-\frac{6}{25} \text{ or } z=6$$

Check both solutions.

$$\sqrt{4\left(-\frac{6}{25}\right)+1}-\sqrt{10-\left(-\frac{6}{25}\right)}\stackrel{?}{=}3 \rightarrow \sqrt{\frac{1}{25}}-\sqrt{\frac{256}{25}} \rightarrow \frac{1}{5}-\frac{16}{5} \neq 3 \text{ Extraneous}$$

$$\sqrt{4(6)+1}-\sqrt{10-(6)}\stackrel{?}{=}3 \rightarrow \sqrt{25}-\sqrt{4}\stackrel{?}{=}3 \rightarrow 5-2=3 \text{ Checks}$$

(15) **−1 and $\frac{1}{3}$.** Multiply each term in the equation by x. (This eliminates 0 as being a possible solution.)

$$x\left(x^{-1}-3x\right)=2x \rightarrow 1-3x^2=2x$$

Solve the quadratic equation for x.

$$1-x^2=2x \rightarrow 0=3x^2+2x-1 \rightarrow 0=(3x-1)(x+1) \rightarrow x=\frac{1}{3} \text{ or } x=-1$$

Check both solutions.

$$\left(\frac{1}{3}\right)^{-1}-3\left(\frac{1}{3}\right)\stackrel{?}{=}2 \rightarrow 3-1=2 \text{ Checks}$$

$$(-1)^{-1}-3(-1)\stackrel{?}{=}2 \rightarrow -1+3=2 \text{ Checks}$$

(16) **1.** First, factor out the greatest common factor:

$$x^{-1/3}\left(x^{1/3}-1\right)\left(x^{-1/3}-1\right)+x^{-1/3}\left(x^{1/3}-1\right)\left(x^{-1/2}+1\right)=x^{-1/3}\left(x^{1/3}-1\right)\left(x^{-1/3}-1+x^{-1/2}+1\right)=0$$

Simplify the terms in the last factor. Then set each factor equal to zero to determine solutions.

$$x^{-1/3}\left(x^{1/3}-1\right)\left(x^{-1/3}+x^{-1/2}\right)=0$$

When $x^{-1/3}=0$, there is no solution, because $\frac{1}{x^{1/3}}=0$ does not exist.

When $x^{1/3}-1=0$, there is the solution $x=1$.

When $x^{-1/3}+x^{-1/2}=0$, there is no solution. The x has to be positive for the even root, and the sum of two positive numbers cannot be 0.

IN THIS CHAPTER

» Establishing the coordinate system

» Using intercepts and symmetry to your advantage

» Plotting lines on a graph

» Reviewing the ten most common graphs in algebra

» Embracing technology with graphing calculators

Chapter 6
Giving Graphing a Gander

An algebraic graph is a drawing that illustrates an algebraic relationship or equation in a two-dimensional plane (like a piece of graph paper). A graph allows you to see the characteristics of an algebraic statement immediately.

The graphs in algebra are unique because they reveal relationships that you can use to model situations. The graph of a mathematical curve has a ton of information crammed into an elegant package. For example, parabolas can model daily temperature, and a flat, S-shaped curve can model the number of people infected with the flu.

In this chapter, you find out how lines, parabolas, other polynomials, and radical curves all fit into the picture. I start off with a quick refresher on graphing basics. The rest of the chapter helps you become a more efficient grapher so that you don't have to plot a million points every time you need to create a graph. Along the way, I discuss using intercepts and symmetry and working with linear equations. I also provide you with a quick overview of ten basic forms and equations you run into again and again in Algebra II. After you recognize a few landmarks, you can quickly sketch a reasonable graph of an equation. And to top it all off, you get some pointers on using a graphing calculator.

Coordinating Axes, Points, and Quadrants

You do most graphing in algebra in the Cartesian coordinate system — a gridlike system where you plot points depending on the magnitude and signs of numbers. You may be familiar with the basics of graphing coordinates — (x, y) points — from Algebra I or middle school math. Within the Cartesian coordinate system (which is named for the philosopher and mathematician René Descartes), you can plug-and-plot points to draw a curve, or you can take advantage of knowing a little something about what the graphs should look like. In either case, the coordinates and their respective plotted points fit together to give you a picture.

Identifying the parts of the coordinate plane

A coordinate plane (shown in Figure 6-1) features two intersecting axes — two perpendicular lines that cross at a point called the *origin*. The axes divide up the coordinate plane into four quadrants, usually numbered with Roman numerals starting in the upper right-hand corner and moving counterclockwise.

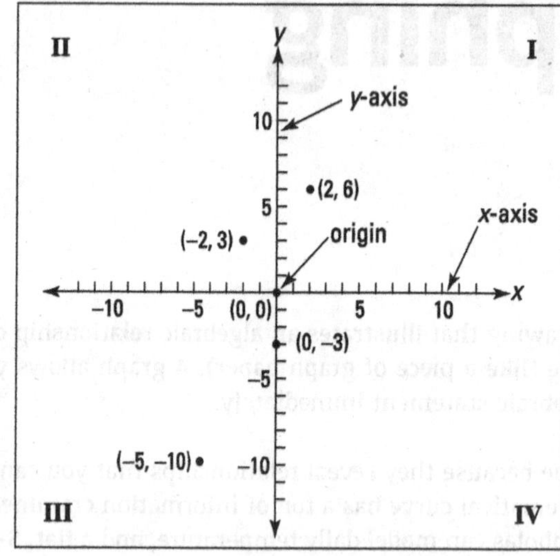

FIGURE 6-1: Identifying all the players in the coordinate plane.

REMEMBER

The horizontal line is called the x-axis, and the vertical line is called the y-axis. The scales on the axes are shown with little tick marks. Usually, you have the same scale on both axes — each tick mark may represent one unit or five units — but sometimes the plane needs to have two different scales. In any case, the scale on any one axis is the same all along that axis.

You identify points on the coordinate plane with numbers called *coordinates*, which come in ordered pairs — where the order matters. In the ordered pair (x, y), the first number is the x-coordinate; it tells you how far and in which direction the point is from the origin along the x-axis. The second number is the y-coordinate; it tells you how far from the origin and in which direction the point is along the y-axis. In Figure 6-1, you can see several points drawn in with their corresponding coordinates.

Plotting from dot to dot

Graphing in algebra isn't quite as simple as a quick game of connect-the-dots but the main concept is the same: You connect the points in the correct order, and you see the desired picture.

In the section, "Graphing the Ten Basic Forms," later in this chapter, you discover the advantage of having at least a hint of what a graph is supposed to look like. But even when you have a pretty good idea of what you're going to get, you still need the basic ability to plot points.

In algebra, you usually have to create your set of points to graph. A problem gives you a formula or equation, and you determine the coordinates of the points that work in that equation. Your main goal is to sketch the graph after finding as few points as necessary. If you know what the general shape of the graph should be, all you need are a few anchor points to be on your way.

EXAMPLE

Q. Plot the points and connect them in order: $(0,3),(-2,4),(-4,3),(-5,0),(-2,-3),(0,-2.5),(2,-3),(5,0),(4,3),(2,4),(0,3),(1.5,5),(1,6),(0,3)$

A. Connecting the points in order, you get a picture. You can see the points and the corresponding picture in Figure 6-2.

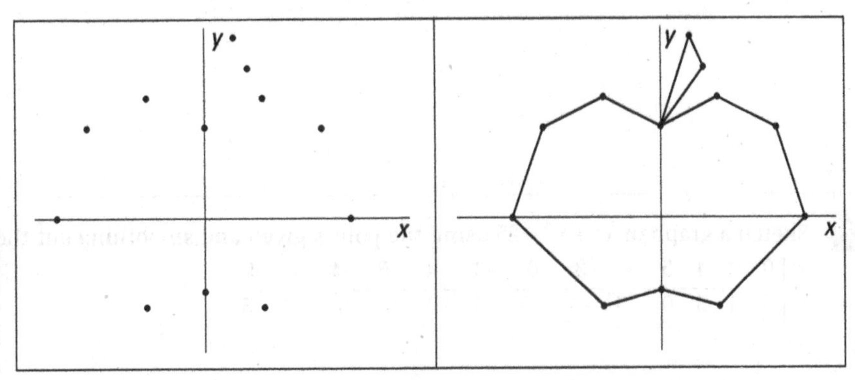

FIGURE 6-2: Connecting the points in order creates a picture.

Q. Sketch the graph of $y = x^2 - x - 6$.

A. Find some points (x, y) that make the equation true. Some points that work (picked at random, making sure they work in the equation) include $(4,6),(3,0),(2,-4),(1,-6),(0,-6),(-1,-4),(-2,0)$, and $(-3,6)$. You can find many, many more points that work, but these points should give you a pretty good idea of what's going on. In Figure 6-3a, you can see the points graphed; I connect them with a smooth curve in Figure 6-3b. The order of connecting the points goes by the order of the x-coordinates.

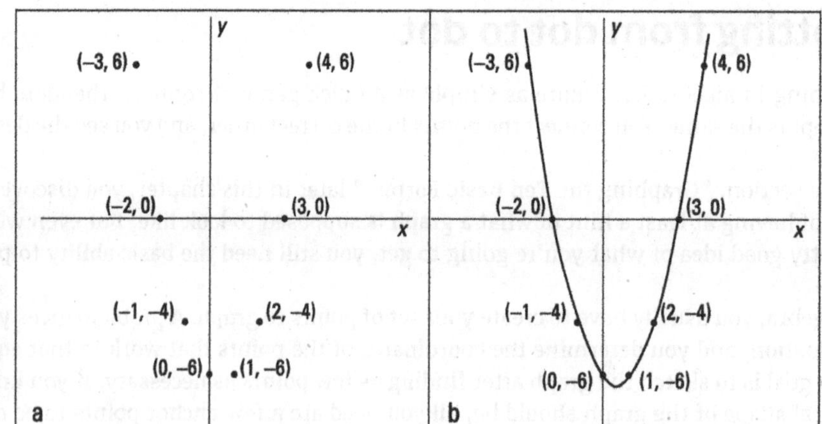

FIGURE 6-3: Creating a set of points to fit the graph of an equation.

YOUR TURN

1 **Plot the following points and connect them in this order: (2,−2),(−2,−2),(2,2),(−2,2),(2,−2),(2,2),(0,5),(−2,2),(−2,−2).**

2 **Sketch a graph of $x^2 + y^2 = 25$ using the points given and smoothing out the curve.**

x	0	3	4	5	4	3	0	−3	−4	−5	−4	−3	0
y	5	4	3	0	−3	−4	−5	−4	−3	0	3	4	5

Crossing the Line Using Intercepts and Symmetry

Graphing curves can take as long as you like or be as quick as you like. If you take advantage of the characteristics of the curves you're graphing, you can cut down on the time it takes to graph and improve your accuracy. Two features that you can quickly recognize and solve for are the intercepts and symmetry of the graphs.

Finding x- and y-intercepts

The intercepts of a graph appear at the points where the lines of the graph cross the axes. The graph of a curve may never cross an axis, but when it does, knowing the points that represent the intercepts is very helpful.

The x-intercepts always have the format $(h,0)$ — the y-coordinate of an x-intercept is 0 because the point is on the x-axis. The y-intercepts have the form $(0,k)$ — the x-coordinate is zero because the point is on the y-axis. You find the x- and y-intercepts by letting y and x, respectively, equal zero and solving for the other coordinate.

EXAMPLE

Q. Find the intercepts of the graph of $y = -x^2 + x + 6$.

A. The graph has two x-intercepts and one y-intercept:

To find the x-intercepts, let $y = 0$; you then have the quadratic equation $0 = -x^2 + x + 6 = -(x^2 - x - 6)$. You solve this equation by factoring it into $0 = -(x-3)(x+2)$. You find two solutions, $x = 3$ and $x = -2$, so the two x-intercepts are $(3,0)$ and $(-2,0)$. (For more on factoring, see Chapters 1 and 3.)

To find the y-intercept, let $x = 0$. This gives you the equation $y = -0 + 0 + 6 = 6$. The y-intercept, therefore, is $(0,6)$.

Figure 6-4a shows the intercepts of the parabola placed on the axes. The graph doesn't tell you a whole lot unless you're aware that the equation is a parabola. If you know that you have a parabola (find out more about them in Chapter 8), then you know that a U-shaped curve goes through the intercepts. For now, you can just plug-and-plot to find a few more points to help you with the graph. Using the equation to find other points that work with the graph, you get $(1,6)$, $(2,4)$, $(4,-6)$, and $(-1,4)$. Figure 6-4b shows the completed graph.

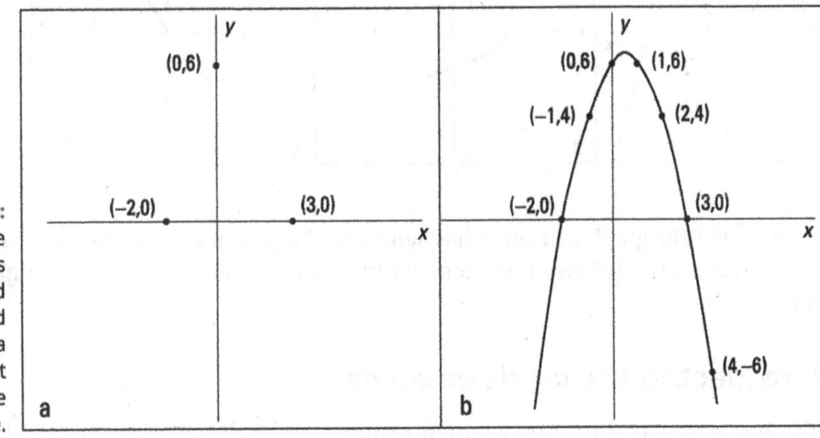

FIGURE 6-4: Plotting the intercepts and calculated points on a graph to get the whole picture.

YOUR TURN

3 **Find the intercepts of the function $y = x^2 - 7x - 8$.**

4 **Find the intercepts of the ellipse $\frac{x^2}{4} + y^2 = 1$.**

Reflecting on a graph's symmetry

When an item is symmetric, you can see a sameness or pattern to it. A graph that's symmetric with respect to one of the axes appears to be a mirror image of itself on either side of the axis. A graph symmetric about the origin appears to be the same after a 180-degree turn. Figure 6-5 shows three curves and three symmetries: symmetry with respect to the y-axis (a), symmetry with respect to the x-axis (b), and symmetry with respect to the origin (c).

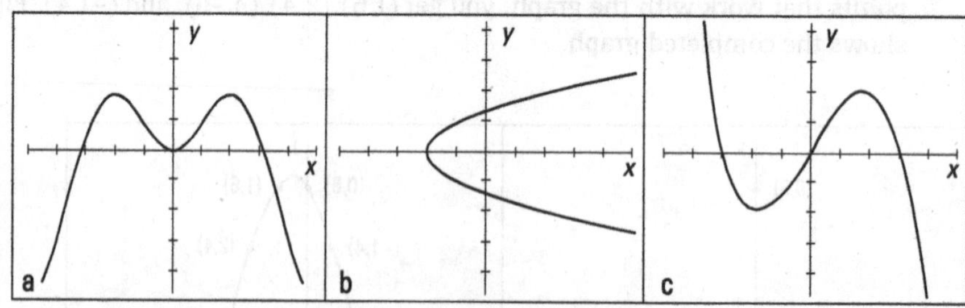

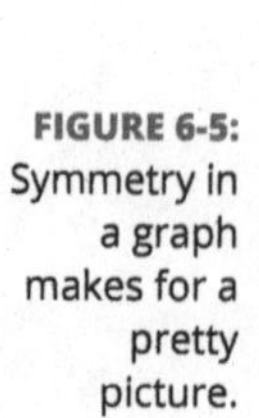

FIGURE 6-5: Symmetry in a graph makes for a pretty picture.

Recognizing that the graph of a curve has symmetry helps you sketch the graph and determine its characteristics. The following sections outline ways to determine from a graph's equation if symmetry exists.

With respect to the *y*-axis or *x*-axis

Consider an equation in this form: y = some expression in x. If replacing every x with $-x$ doesn't change the value of y, then the curve is the mirror image of itself over the y-axis. The graph contains the points (x, y) and $(-x, y)$.

Now consider an equation in this form: $x =$ some expression in y. If replacing every y with $-y$ doesn't change the value of x, then the curve is the mirror image of itself over the x-axis. The graph contains the points (x, y) and $(x, -y)$.

EXAMPLE

Q. Determine whether the graph of the equation $y = x^4 - 3x^2 + 1$ is symmetric with respect to the y-axis.

A. If you replace each x with $-x$, the equation remains unchanged. Replacing each x with $-x$, $y = (-x)^4 - 3(-x)^2 + 1 = x^4 - 3x^2 + 1$. The y-intercept is $(0, 1)$. Some other points include $(1, -1), (-1, -1)$ and $(2, 5), (-2, 5)$. Notice that the y is the same for both positive and negative x's. With symmetry about the y-axis, for every point (x, y) on the graph, you also find $(-x, y)$. It's easier to find points when an equation has symmetry because of the pairs. Figure 6-6 shows the graph of $y = x^4 - 3x^2 + 1$.

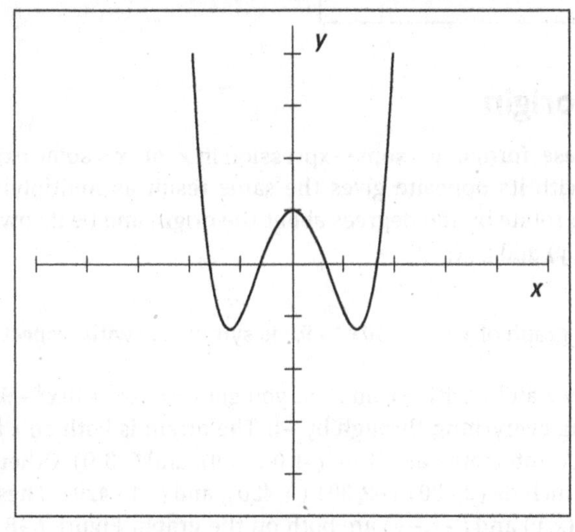

FIGURE 6-6: A graph's reflection over a vertical line.

Q. Determine if the graph of $x = \frac{10}{y^2 + 1}$ is symmetric with respect to the x-axis.

A. When you replace each y with $-y$, the x value remains unchanged. The x-intercept is $(10, 0)$. Some other points on the graph include the following: $(5, 1), (5, -1)$; $(2, 2), (2, -2)$; and $(1, 3), (1, -3)$. Notice the pairs of points that have positive and negative values for y but the same value for x. This is where the symmetry comes in: The points have both positive and negative values — on both sides of the x-axis — for each x coordinate. Symmetry about the x-axis means that for every point (x, y) on the curve, you also find the point $(x, -y)$. This symmetry makes it easy to find points and plot the graph. The graph of $x = \frac{10}{y^2 + 1}$ is shown in Figure 6-7.

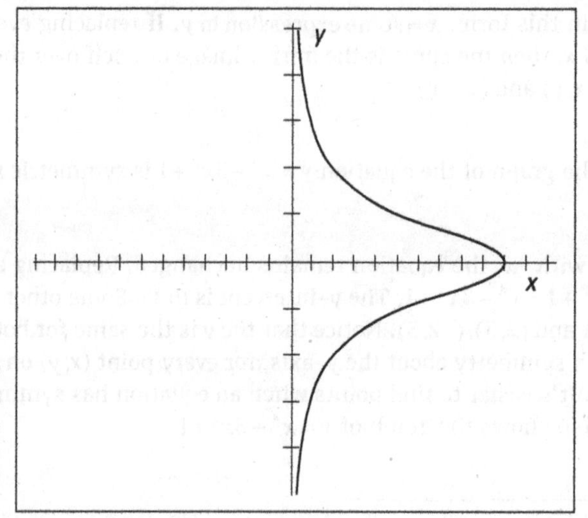

FIGURE 6-7: A graph's reflection over a horizontal line.

With respect to the origin

Consider an equation in these forms: $y =$ some expression in x or $x =$ some expression in y. If replacing every variable with its opposite gives the same result as multiplying the entire equation by –1, the curve can rotate by 180 degrees about the origin and be its own image. The graph contains the points (x, y) and $(-x, -y)$.

EXAMPLE

Q. Determine whether the graph of $y = x^5 - 10x^3 + 9x$ is symmetric with respect to the origin.

A. When you replace every x and y with $-x$ and $-y$, you get $-y = -x^5 + 10x^3 - 9x$, which is the same as multiplying everything through by –1. The origin is both an x and a y-intercept. The other x-intercepts are $(1, 0)$, $(-1, 0)$, $(3, 0)$, and $(-3, 0)$. Other points on the graph of the curve include $(2, -30)$, $(-2, 30)$, $(4, 420)$, and $(-4, -420)$. These points illustrate the fact that (x, y) and $(-x, -y)$ are both on the graph. Figure 6-8 shows the graph of $y = x^5 - 10x^3 + 9x$. (I changed the scale on the y-axis to make each tick-mark equal 10 units).

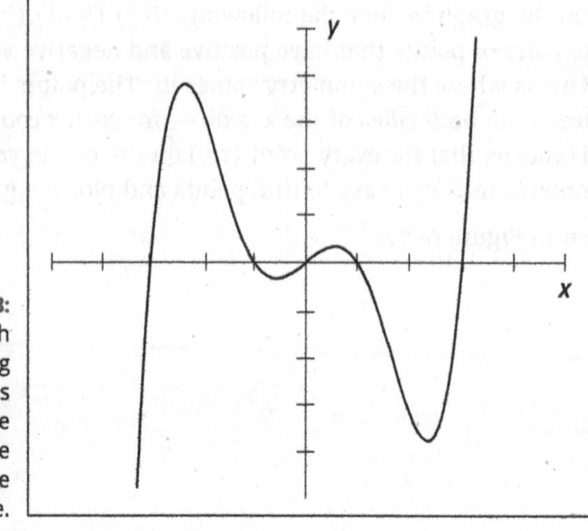

FIGURE 6-8: A graph revolving 180 degrees about the origin of the coordinate plane.

YOUR TURN

Determine any symmetry of the graph of $y = 4x^3$.

6 **Determine any symmetry of the graph of $x^2 + (y-1)^2 = 4$.**

7 **Determine any symmetry of the graph of $x = y^2 - 1$.**

Preparing to Graph Lines

Lines are some of the simplest graphs to sketch. It takes only two points to determine the one and only line that exists in that place on the coordinate plane, so one simple method for graphing lines is to find two points — any two points — on the line. Another useful method is to use a point and the slope of the line. The method you choose is often just a matter of personal preference.

The slope of a line also plays a big role in comparing it with other lines that run parallel or perpendicular to it. The slopes of parallel and perpendicular lines are closely related to one another.

Finding the slope of a line

The slope of a line, m, has a complicated math definition, but it's basically a number — positive or negative, large or small — that highlights some of the characteristics of the line, such as its steepness and direction. The numerical value of the slope tells you if the line slowly rises or drops from left to right or dramatically soars or falls from left to right. In order to find the slope and discover the properties of the line, you can use the equation of the line to solve for the information you need, or you can look at the graph of the line to get the general picture. If you opt to concentrate on the equation of the line, you can either look at a specific format of the line's equation or you can solve for points on the line, which allows you to solve for the slope.

Identifying characteristics of a line's slope

A line can have a positive or a negative slope. If a line's slope is positive, the line rises from left to right. If the slope is negative, the line falls from left to right. The greater the absolute value (the value of the number without regard to the sign; in other words, the distance of the number from zero) of a line's slope, the steeper the line is. For example, if the slope is a number between –1 and 1, the line is rather flat. A slope of zero means that the line is absolutely horizontal.

A vertical line doesn't have a slope. This is tied to the fact that numbers go infinitely high and math doesn't have a highest number — you just say infinity. Only an infinitely high number can represent a vertical line's slope, but usually, if you're talking about a vertical line, you just say that the slope doesn't exist.

Formulating the value of a line's slope

You can determine the slope of a line, m, if you know two points on the line. The formula for finding the slope with this method involves finding the difference between the y-coordinates of the points and dividing that difference by the difference of the x-coordinates of the points.

You find the slope of the line that goes through the points (x_1, y_1) and (x_2, y_2) with the formula $m = \frac{y_2 - y_1}{x_2 - x_1}$.

EXAMPLE

Q. Find the slope of the line that crosses the points $(-3, 2)$ and $(-4, -12)$.

A. Use the formula to get $m = \frac{y_2 - y_1}{x_2 - x_1} = \frac{-12 - 2}{4 - (-3)} = \frac{-14}{7} = -2$. This line is fairly steep — the absolute value of –2 is 2 — and it falls as it moves from left to right, which makes it negative.

Q. Find the slope of the line that crosses the points $(4, -1)$ and $(6, -1)$.

A. Use the formula to get $m = \frac{y_2 - y_1}{x_2 - x_1} = \frac{-1 - (-1)}{6 - 4} = \frac{0}{2} = 0$. When a line has a slope of 0, it means there's no rise or fall of the line. It's horizontal.

If the 0 had been in the denominator of the slope formula, you would know that there is no slope. This indicates that the line is vertical. When you use the slope formula, it doesn't matter which point you choose to be (x_1, y_1) — the order of the points doesn't matter — but you can't mix up the order of the two different coordinates. You can run a quick check by seeing if the coordinates of each point are above and below one another. Also, be sure that the y-coordinates are in the numerator; a common error is to have the difference of the y-coordinates in the denominator.

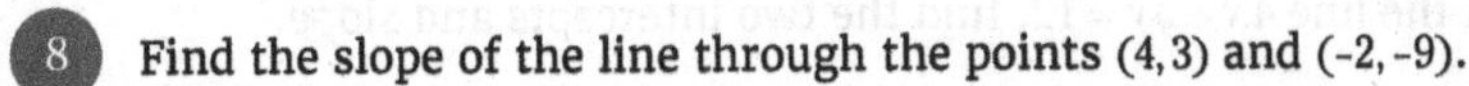

YOUR TURN

8 Find the slope of the line through the points (4, 3) and (-2, -9).

9 Find the slope of the line through the points (7, 0) and (2, -5).

Facing two types of equations for lines

Algebra offers two different forms for the equation of a line. The first is the standard form, written $Ax + By = C$, with the two variable terms on one side and the constant on the other side. The other form is the slope-intercept form, written $y = mx + b$; the y value is set equal to the product of the slope, m, and x, and the product is added to the y-coordinate of the y-intercept, b.

Meeting high standards with the standard form

The standard form for the equation of a line is written $Ax + By = C$. It takes only two points to graph a line, so you can choose any two points and plot them on the coordinate plane to create your line. However, the standard form has more information than may be immediately apparent. You can determine, just by looking at the numbers in the equation, the intercepts and slope of the line. The intercepts, in particular, are great for graphing the line, and you can find them easily because they fall right on the axes.

The line $Ax + By = C$ has:

» An x-intercept of $\left(\frac{C}{A}, 0\right)$
» A y-intercept of $\left(0, \frac{C}{B}\right)$
» A slope of $-\frac{A}{B}$

Q. Given the line $4x + 3y = 12$, find the two intercepts and slope.

EXAMPLE **A.** The x-intercept $\left(\frac{C}{A}, 0\right) = \left(\frac{12}{4}, 0\right) = (3,0)$; the y-intercept $\left(0, \frac{C}{B}\right) = \left(0, \frac{12}{3}\right) = (0,4)$; and the slope $m = -\frac{A}{B} = -\frac{4}{3}$. Figure 6-9 shows the two intercepts and the graph of the line. Note that the line falls as it moves from left to right, confirming the negative value of the slope from the formula $m = -\frac{A}{B} = -\frac{4}{3}$.

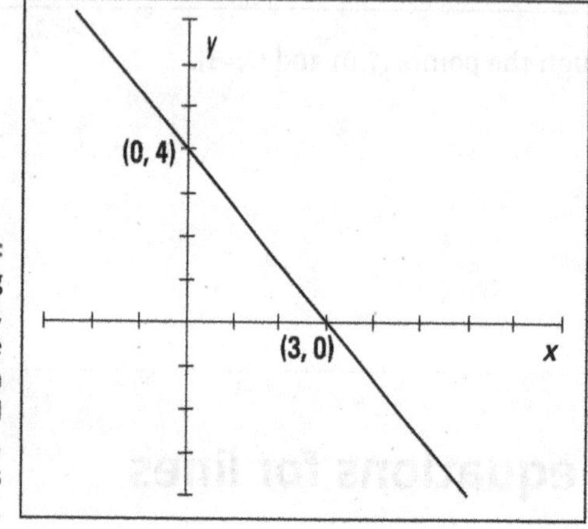

FIGURE 6-9: Graphing $4x + 3y = 12$, a line written in standard form, using its intercepts.

Picking off the slope-intercept form

When the equation of a line is written in the slope-intercept form, $y = mx + b$, you have good information right at your fingertips. The coefficient (a number multiplied with a variable or unknown quantity) of the x term, m, is the slope of the line. And the constant, b, is the y value of the y-intercept (the coefficient of the y term must be 1). With these two bits of information, you can quickly sketch the line.

Q. Graph the line $y = 2x + 5$.

EXAMPLE **A.** First, plot the y-intercept, (0,5), and then count off the slope from that point. The slope of the line $y = 2x + 5$ is 2; think of the 2 as the slope fraction, with the y-coordinate on top and the x-coordinate on the bottom. Thus, the slope is $\frac{2}{1} = 2$.

Counting off the slope means to start at the y-intercept, move one unit to the right (for the 1 in the denominator of the previous fraction), and, from the point one unit to the right, count two units up (for the 2 in the numerator). This process gets you to another point on the line.

You connect the point you count off with the y-intercept to create your line. Figure 6-10 shows the intercept, an example of counting off, and the new point that appears one unit to the right and two units up from the intercept — the point (1,7).

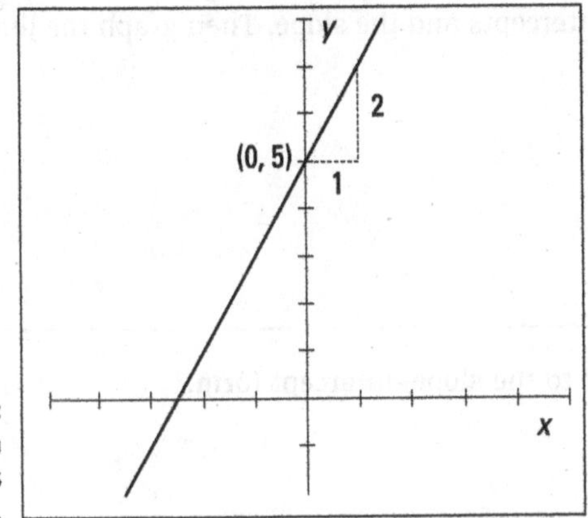

FIGURE 6-10: A line with a slope of 2 is fairly steep.

Changing from one form to the other

You can graph lines by using the standard form or the slope-intercept form of the equations. If you prefer one form to the other — or if you need a particular form for an application you're working on — you can change the equations to your preferred form by performing simple algebra.

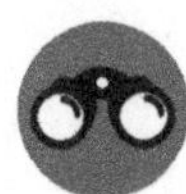
EXAMPLE

Q. Change the equation $2x - 5y = 8$ to the slope-intercept form.

A. First, subtract $2x$ from each side and then divide both sides by the -5 multiplier on the y term:

$$2x - 5y = 8 \rightarrow -5y = -2x + 8 \rightarrow \frac{-5}{-5}y = \frac{-2}{-5}x + \frac{8}{-5} \rightarrow y = \frac{2}{5}x - \frac{8}{5}$$

In the slope-intercept form, you can quickly determine the slope and y-intercept, $m = \frac{2}{5}, b = -\frac{8}{5}$.

Q. Change the equation $y = -\frac{3}{4}x + 5$ to the standard form.

A. You first multiply each side by 4 and then add $3x$ to each side of the equation:

$$4(y) = 4\left(-\frac{3}{4}x + 5\right) \rightarrow 4y = \cancel{4}\left(-\frac{3}{\cancel{4}}x\right) + 4(5) \rightarrow 4y = -3x + 20 \rightarrow 3x + 4y = 20$$

YOUR TURN

10 **Given the line $5x - 3y = 10$, find the intercepts and slope. Then graph the line.**

11 **Given the line $y = 4x$, find the intercepts and the slope. Then graph the line.**

12 **Change the equation $3x - 4y = 12$ to the slope-intercept form.**

13 **Change the equation $y = -\frac{1}{2}x + 11$ to the standard form.**

Identifying Parallel and Perpendicular Lines

Lines on the same coordinate plane are parallel if they never touch — no matter how far out you draw them. Lines are perpendicular when they intersect at a 90-degree angle. Both of these instances are fairly easy to spot when you see the lines graphed, but how can you be sure that the lines are truly parallel or that the angle is really 90 degrees and not 89.9 degrees? The answer lies in the slopes.

Consider two lines, $y = m_1x + b_1$ and $y = m_2x + b_2$.

Two lines are parallel when their slopes are equal ($m_1 = m_2$). Two lines are perpendicular when their slopes are negative reciprocals of one another $\left(m_2 = -\frac{1}{m_1}\right)$.

Q. Determine whether the lines $y = 3x + 7$ and $y = 3x - 2$ are parallel or perpendicular.

EXAMPLE

A. **Both lines have slopes of 3, so they are parallel.** Their y-intercepts are different — one crosses the y-axis at (0,7) and the other at (0,–2). Figure 6-11a shows these two lines.

Q. Determine whether the lines $y = -\frac{3}{8}x + 4$ and $y = \frac{8}{3}x - 2$ are parallel or perpendicular.

A. **The lines $y = -\frac{3}{8}x + 4$ and $y = \frac{8}{3}x - 2$ are perpendicular.** The slopes are negative reciprocals of one another. You see these two lines graphed in Figure 6-11b. A quick way to determine if two numbers are negative reciprocals is to multiply them together; you should get a result of –1. For the previous example problem, you get $-\frac{3}{8} \cdot \frac{8}{3} = -1$.

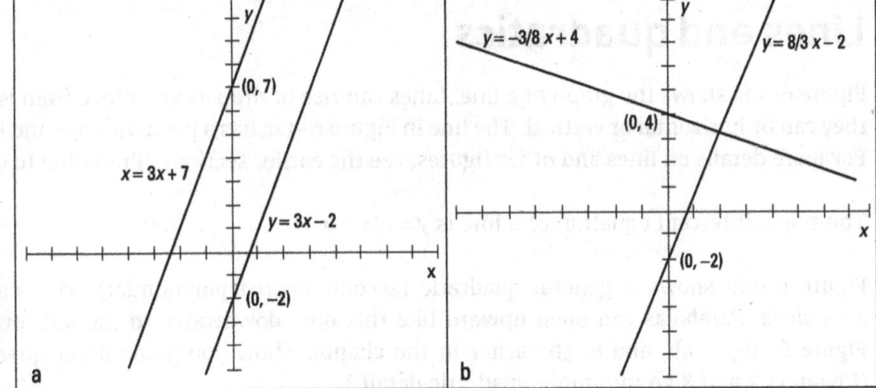

FIGURE 6-11: Parallel lines have equal slopes, and perpendicular lines have slopes that are negative reciprocals.

YOUR TURN

14 Determine whether the two lines are parallel, perpendicular, or neither: $y = 2x - 7$ and $2x + 4y = 11$.

15 Determine whether the two lines are parallel, perpendicular, or neither: $4x + 3y = 16$ and $3x + 4y = 8$.

Graphing the Ten Basic Forms

You study many types of equations and graphs in Algebra II. You find very specific details on graphs of quadratics, other polynomials, radicals, rationals, exponentials, and logarithms in the various chapters of this book that are devoted to the different types. What I present in this section is a general overview of some of the types of graphs, designed to help you distinguish one type from another as you get ready to ferret out all the gory details. Knowing the basic graphs is the starting point for graphing variations on those graphs or more complicated curves.

Ten basic graphs seem to occur most often in Algebra II. I cover the first, a line, earlier in this chapter, but I also include it in the following sections with the other nine basic graphs.

Lines and quadratics

Figure 6-12a shows the graph of a line. Lines can rise or drop as you move from left to right, or they can be horizontal or vertical. The line in Figure 6-12a has a positive slope and is fairly steep. For more details on lines and other figures, see the earlier section, "Preparing to Graph Lines."

The slope-intercept equation of a line is $y = mx + b$.

Figure 6-12b shows a general quadratic (second-degree polynomial). This curve is called a *parabola*. Parabolas can open upward like this one, downward, to the left, or to the right. Figure 6-3b, 6-4b, and 6-5b earlier in the chapter show you some other quadratic curves. (Chapters 3 and 8 go into more quadratic detail.)

Two general equations for quadratics are $y = ax^2 + bx + c$ and $x = ay^2 + by + c$.

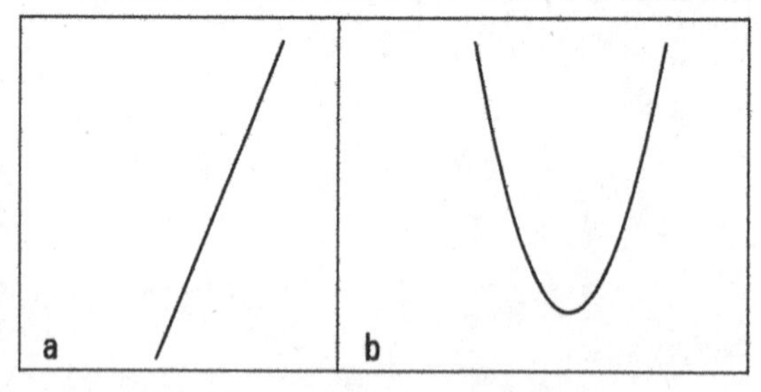

FIGURE 6-12: Graphs of a steep line and an upward-facing quadratic.

Cubics and quartics

Figure 6-13a shows the graph of a cubic (third-degree polynomial). A cubic curve (a polynomial with degree 3) can be an S-shaped curve as shown, or it can flatten out in the middle and not feature those turns. A cubic can appear to come from below and end up high and to the right, or it can drop from the left, take some turns, and continue on down (for more on polynomials, check out Chapter 9).

The general equation for a cubic polynomial is $y = ax^3 + bx^2 + cx + d$.

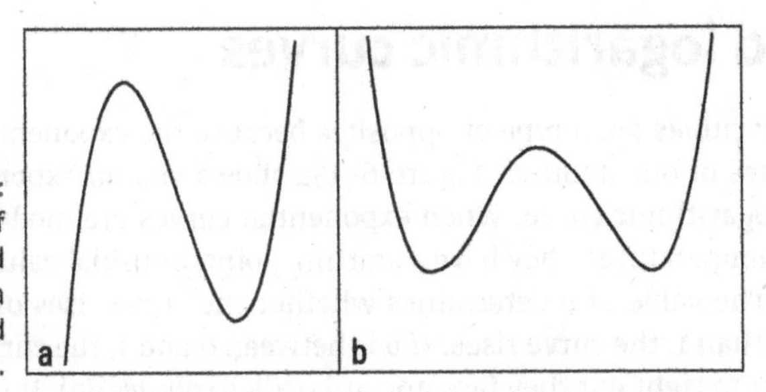

FIGURE 6-13: Graphs of an S-shaped cubic and a W-shaped quartic.

Figure 6-13b shows the graph of a quartic (fourth-degree polynomial). The graph has a distinctive W-shape but that shape can flatten out, depending on how many terms appear in its equation. Similar to the quadratic (parabola), the quartic can open downward, in which case it looks like an M rather than a W.

The quartic polynomial has the following general equation: $y = ax^4 + bx^3 + cx^2 + dx + e$.

Radicals and rationals

In Figure 6-14a, you see a radical curve, and in Figure 6-14b, you see a rational curve. They seem like opposites, don't they? But a major characteristic radical and rational curves have in common is that you can't draw them everywhere. A radical curve can have an equation such as $y = \sqrt{x-4}$, where the square root doesn't allow you to put values under the radical that give you a negative result. In this case, you can't use any number smaller than four, so you have no graph when x is smaller than four. The radical curve in Figure 6-14a shows an abrupt stop at the left end, which is typical of a radical curve. (You can find more on radicals in Chapter 5.)

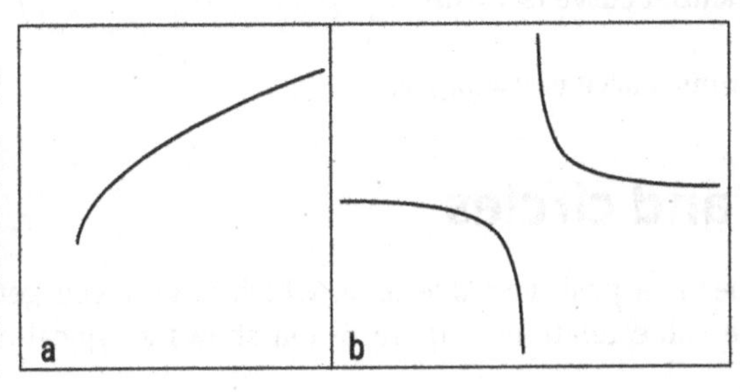

FIGURE 6-14: Graphs of radicals often have abrupt stops, and graphs of rationals have gaps.

A general equation for some radical curves is $y = \sqrt[n]{ax^n + bx^{n-1} + \ldots}$, where n is a natural number.

Rational curves have a different type of restriction. When the equation of a curve involves a fraction, it opens itself up to the possibility of having a zero in the denominator of the fraction — a big no-no. You can't have a zero in the denominator because math offers no such number, which is why the graphs of rational curves have spaces in them, places where the graph has a disconnect.

In Figure 6-14b, you see the graph of $y = \frac{1}{x}$. The graph doesn't have any value when $x = 0$. (See Chapter 10 for information on rational functions.) A general equation for some rational curves is $y = \frac{k}{ax^n + bx^{n-1} + \ldots}$.

Exponential and logarithmic curves

Exponential and logarithmic curves are a type of opposites because the exponential and logarithmic functions are inverses of one another. Figure 6-15a shows you an exponential curve, and Figure 6-15b shows a logarithmic curve. When exponential curves are modeling physical situations (money, matter, temperature), they have a starting point, or initial value, of a, which is actually the y-intercept. The value of b determines whether the curve rises or falls in that gentle C shape. If b is larger than 1, the curve rises. If b is between 0 and 1, the curve falls. They generally rise or fall from left to right but they face upward (called *concave up*). It helps to think of them collecting snow or sand in their gentle curves. Logarithmic curves can also rise or fall as you go from left to right but they face downward.

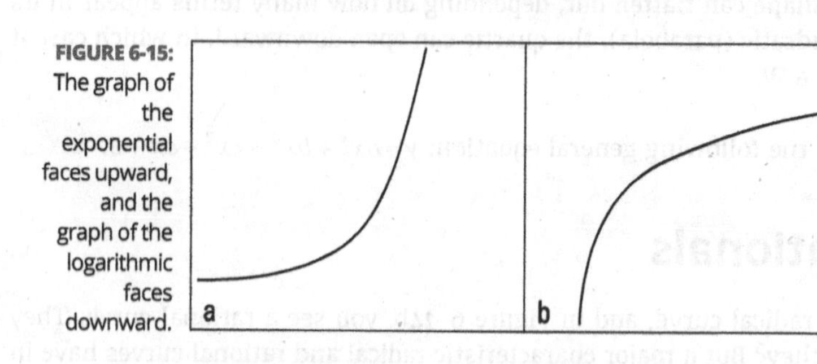

FIGURE 6-15: The graph of the exponential faces upward, and the graph of the logarithmic faces downward.

Both exponential and logarithmic curves can model growth (or decay), where a previous amount is multiplied by some factor. The change continues on at the same rate throughout, so you produce steady curves up or down. (See Chapter 11 for more info on exponential and logarithmic functions.)

The general form for an exponential curve is $y = ab^{kx}$.

The general form for a logarithmic curve is $y = \log_b x$.

Absolute values and circles

The absolute value of a number is a positive value or 0, which is why you get a distinctive V-shaped curve from absolute value relations. Figure 6-16a shows a typical absolute value curve.

A general equation for an absolute value curve is $y = |ax + b|$.

The graph in Figure 6-16b is probably the most recognizable shape, except for the line, of the ten. A circle goes round and round a fixed distance from its center. (I cover circles and other conics in detail in Chapter 13.)

Circles with their centers at the origin have equations $x^2 + y^2 = r^2$.

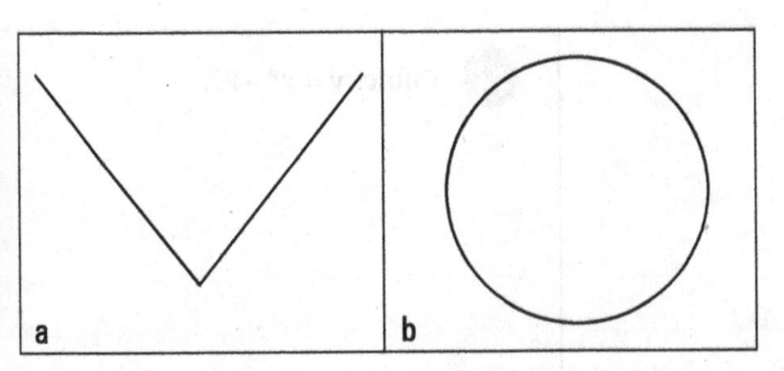

FIGURE 6-16: Graphs of absolute values have distinctive V-shapes, and graphs of circles go round and round.

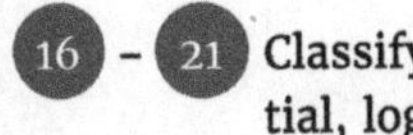

Classify each graph as a line, quadratic, cubic, quartic, radical, rational, exponential, logarithmic, absolute value, or circle.

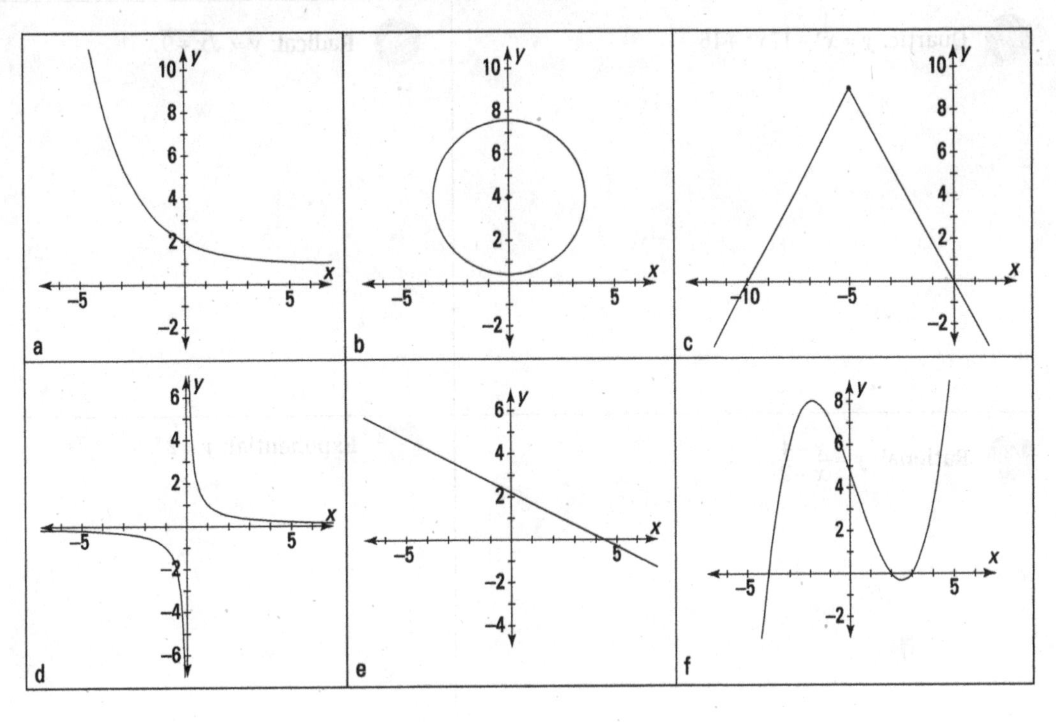

22–30 *Find the x and y intercepts of each graph.*

Line: $y = 6x - 12$.

23 Quadratic: $y = x^2 - 9x - 10$.

24 Cubic: $y = x^3 - 27$.

25 Quartic: $y = x^4 - 17x^2 + 16$.

26 Radical: $y = \sqrt{x+9}$.

27 Rational: $y = \frac{x-4}{x+2}$.

28 Exponential: $y = 2^x$.

29 Absolute value: $y = |x-3| + 2$.

30 Circle: $x^2 + y^2 = 9$.

Enter the Machines: Using a Graphing Calculator

You may think that discussions on graphing aren't necessary if you have a calculator that does the graphing for you. Graphing calculators are wonderful. They take much of the time-consuming drudgery out of graphing lines, curves, and all sorts of intricate functions. But you have to know how to use the calculator correctly and effectively. Two of the biggest stumbling blocks students and teachers face when using a graphing calculator to solve problems have to do with entering a function correctly and then finding the correct graphing window. You can use your calculator's manual or online help to get all the other nitty-gritty details.

Entering equations into graphing calculators correctly

The trick to entering equations correctly into your graphing calculator is to enter what you mean (and mean what you insert). Your calculator will follow the order of operations, so you need to use parentheses to create the equation that you really want. (See Chapter 1 if you need a review of the order of operations.) The four main trouble areas are fractions, exponents, radicals, and negative numbers.

Facing fractions

Say, for example, that you want to graph the equations $y=\frac{1}{x+2}$, $y=\frac{x+3}{x}$, and $y=\frac{1}{4}x+3$. You may mistakenly enter them into your graphing calculator as $y_1=1\div x+2$, $y_2=x+3\div x$, and $y_3=1\div 4x+3$. They don't appear at all as you expected. Why is that?

When you enter $1\div x+2$, your calculator (using order of operations) reads that as "Divide 1 by x and then add 2." Instead of having a gap in the graph when $x=-2$, like you expect, the gap or discontinuity appears when $x=0$ — your hint that you have a problem. If you want the numerator divided by the entire denominator, you use parentheses around the terms in the denominator. Enter the first equation as $y_1=1\div(x+2)$.

When you enter a division symbol, ÷, in your calculator, the screen will show a slanted line, /. When you enter multiplication, your calculator will show an asterisk, *.

Figure 6-17 shows you how the three equations in this discussion will look in a typical graphing calculator.

```
Plot1 Plot2 Plot3
\Y1■1/(X+2)
\Y2■X/(X+3)
\Y3■(1/4)X+3
\Y4=
\Y5=
\Y6=
\Y7=
```

FIGURE 6-17: Equations entered in the *y*-menu of a graphing calculator.

Expressing exponents

Graphing calculators have some built-in exponents that make it easy to enter the expressions. You have a button for squared,^{2}, and, if you look hard, you can find one for cubed,^{3}. Calculators also provide an alternative to these options and to all the other exponents: You can enter an exponent with a caret, ^ (pronounced "carrot," like what Bugs Bunny eats).

You have to be careful when entering exponents with carets. When graphing $y = x^{1/2}$ or $y = 2^{x+1}$, you don't get the correct graphs if you type $y_1 = x \wedge 1 \div 2$ and $y_2 = 2 \wedge x + 1$. Instead of getting a radical curve (raising to the 1/2 power is the same as finding the square root) for y_1, you get a line. You need to put the fractional exponent in parentheses: $y_1 = x \wedge (1 \div 2)$. The same goes for the second curve. When you have more than one term in an exponent, you put the whole exponent in parentheses: $y_2 = 2 \wedge (x+1)$.

Raising Cain over radicals

Graphing calculators often have keys for square roots — and even other roots. The main problem comes if you don't put every term that falls under the radical where it belongs.

If you want to graph $y = \sqrt{4-x}$ and $y = \sqrt[6]{4+x}$, for example, you use parentheses around what falls under the radical and parentheses around any fractional exponents. Figure 6-18 shows two ways to enter the first and one way to enter the second equation.

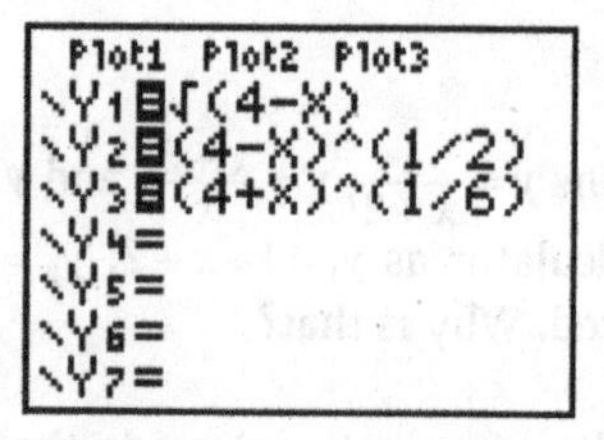

FIGURE 6-18: Radicals can be represented by fractional exponents.

Negating or subtracting

In all phases of algebra, you treat *subtract, minus, opposite*, and *negative* the same way, and they all have the same symbol, "–". Graphing calculators aren't quite as forgiving. They have a special button for *negative*, and they have another button that means *subtract*.

If you're performing the operation of subtraction, such as in $4-3$, you use the subtract button found between the addition and multiplication buttons. If you want to type in the number -3, you use the button with parentheses around the negative sign, (–). You get an error message if you enter the wrong one, but it helps to know why the action is considered an error.

Another problem with negatives has to do with the order of operations (see Chapter 1). If you want to square -4, you can't enter -4^2 into your calculator — if you do, you get the wrong answer. Your calculator squares the 4 and then takes its opposite, giving you -16. To square the -4, you put parentheses around both the negative sign and the 4: $(-4)^2 = 16$ or $(-4) \wedge 2 = 16$.

Looking through the graphing window

Graphing calculators usually have a standard window to graph in — from –10 to +10, both side-to-side and up-and-down. The standard window is a wonderful starting point and does the job for many graphs, but you need to know when to change the window or view of your calculator to solve problems.

Using *x*-intercepts

The graph of $y = x(x-11)(x+12)$ appears only as a set of axes if you use the standard setting. The funky picture shows up because the two intercepts — $(11, 0)$ and $(-12, 0)$ — don't show up if you just go from –10 to 10. (In Chapter 8, you find out how to determine these intercepts, if you're not familiar with that process.) Knowing where the x-intercepts are helps you to set the window properly. You need to change the view or window on your calculator so that it includes all the intercepts. One possibility for the window of the example curve is to have the x-values go from –13 to 12 (one lower than the lowest and one higher than the highest).

If you change the window in this fashion, you have to adjust the window upward and downward for the heights of the graph, or you can use *Fit*, a graphing capability of most graphing calculators. After you set how far to the left and right the window must go, you set the calculator to adjust the upward and downward directions with *Fit*. You have to tell the calculator how wide you want the graph to be. After you set the parameters, the *Fit* button automatically adjusts the height (up and down) so that the window includes the entire graph in that region.

Disconnecting curves

Your calculator, bless its heart, tries to be very helpful. But sometimes helpful isn't accurate. For example, when you graph rational functions that have gaps, or graph piecewise functions that have holes, your calculator tries to connect the two pieces. You can just ignore the connecting pieces of the graph, or you can change the mode of your calculator to dot mode. Most calculators have a Mode menu where you can change decimal values, angle measures, and so on. Go to that general area and toggle from connected to dot mode.

The downside to using dot mode is that sometimes you lose some of the curve — it won't be nearly as complete as with the connected mode. As long as you recognize what's happening with your calculator, you can adjust for its quirks.

YOUR TURN

Choose the entry you use to create the algebraic expression.

31 To enter $y=\dfrac{x+3}{x-2}$, use:

$y_1 = x+3 \div x-2$
$y_2 = (x+3) \div x-2x$
$y_3 = x+3 \div (x-2)$
$y_4 = (x+3) \div (x-2)$

32 To enter $y = x^{-1/2}$, use:

$y_1 = x \wedge (-1 \div 2)$
$y_2 = (x \wedge -1) \div 2$
$y_3 = x \wedge (-1) \div 2$
$y_4 = x \wedge -(1) \div 2$

33 To enter $y=\sqrt{x+4}$, use:

$y_1 = (x+4) \wedge 1 \div 2$
$y_2 = x+4 \wedge (1 \div 2)$
$y_3 = (x+4) \wedge (1 \div 2)$
$y_4 = x+4 \wedge (1 \div 2)$

Practice Questions Answers and Explanations

1. **(See the figure.)** The points (2,−2),(−2,−2),(2,2),(−2,2),(2,−2),(2,2),(0,5),(−2,2),(−2,−2) connected in this order show you how to draw this figure without lifting your pencil off the paper.

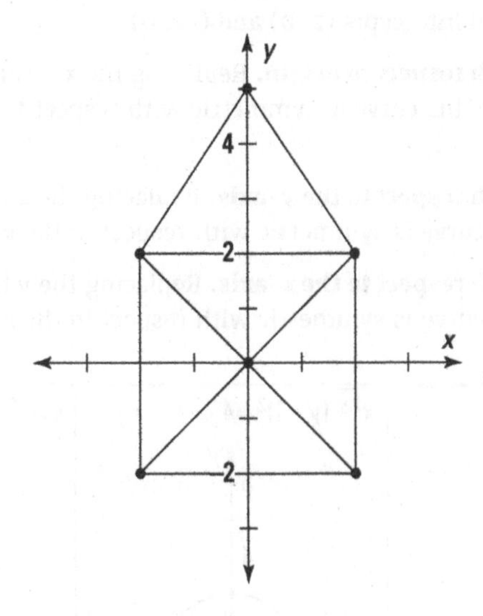

2. **(See the figure.)** The graph has been smoothed out.

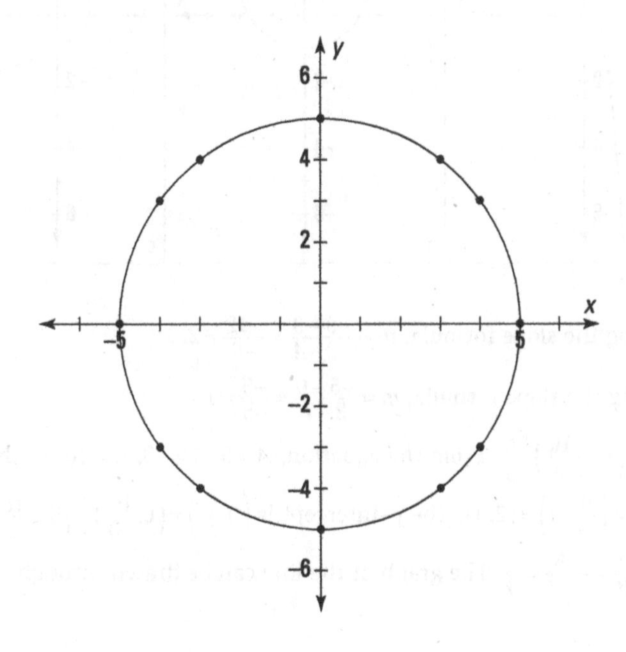

(3) **(0,-8), (8, 0), (-1, 0).** Setting $x = 0$, you have $y = 0^2 - 7 \cdot 0 - 8 = -8$, so the y-intercept is (0,–8). When $y = 0$, you have $0 = x^2 - 7x - 8 = (x-8)(x+1)$, giving you the two zeros 8 and –1. Those intercepts are written (8, 0) and (–1, 0).

(4) **(0, 1), (0, -1), (2, 0), (-2, 0).** Setting $x = 0$, you have $\frac{0^2}{4} + y^2 = 1$ or $y^2 = 1$ with solutions ±1 and intercepts (0, 1) and (0, –1). Setting $y = 0$, $\frac{x^2}{4} + 0^2 = 1$, and you have $\frac{x^2}{4} + 0^2 = 1$ or $x^2 = 4$ with solutions ±2 and intercepts (2, 0) and (–2, 0).

(5) **Symmetric with respect to origin.** Replacing the x term with $-x$ is the same as multiplying the term by –1. This curve is symmetric with respect to the origin. Refer to the following figure.

(6) **Symmetric with respect to the y-axis.** Replacing the x term with $-x$ does not change the equation. This curve is symmetric with respect to the y-axis. Refer to the following figure.

(7) **Symmetric with respect to the x-axis.** Replacing the y term with $-y$ does not change the equation. This curve is symmetric with respect to the x-axis. Refer to the following figure.

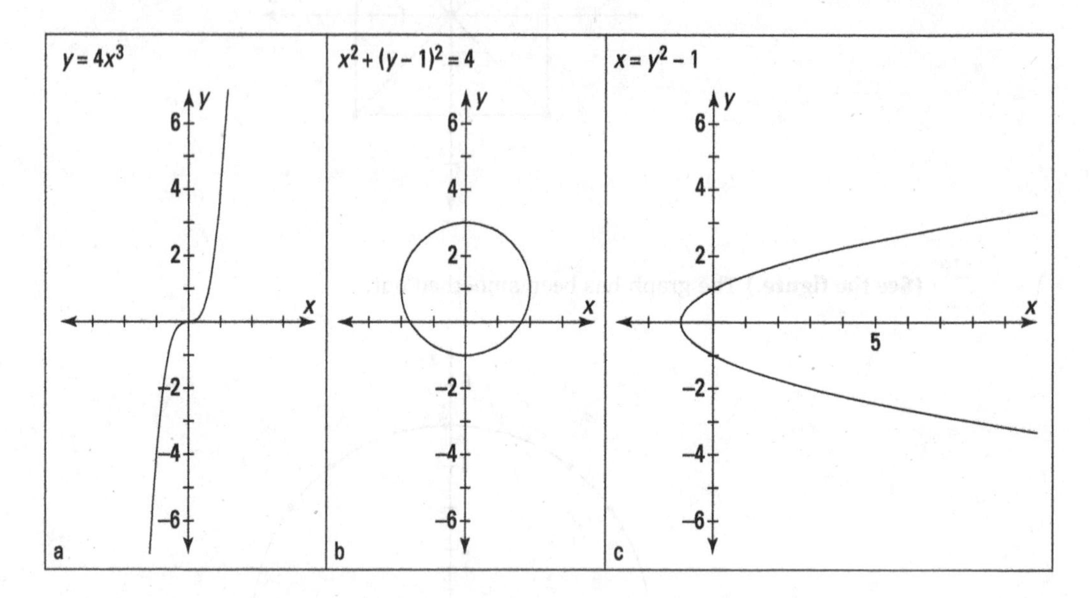

(8) **2.** Using the slope formula, $m = \frac{-9-3}{-2-4} = \frac{-12}{-6} = 2$.

(9) **1.** Using the slope formula, $m = \frac{-5-0}{2-7} = \frac{-5}{-5} = 1$.

(10) **(2,0), $\left(0, -\frac{10}{3}\right)$, $\frac{5}{3}$.** From the equation, $A = 5$, $B = -3$, $C = 10$, so the x-intercept is $\left(\frac{C}{A}, 0\right) = \left(\frac{10}{5}, 0\right) = (2,0)$. The y-intercept is $\left(0, \frac{C}{B}\right) = \left(0, \frac{10}{-3}\right) = \left(0, -\frac{10}{3}\right)$, and the slope is $m = -\frac{A}{B} = -\frac{5}{-3} = \frac{5}{3}$. The graph of the line can be drawn through the two intercepts.

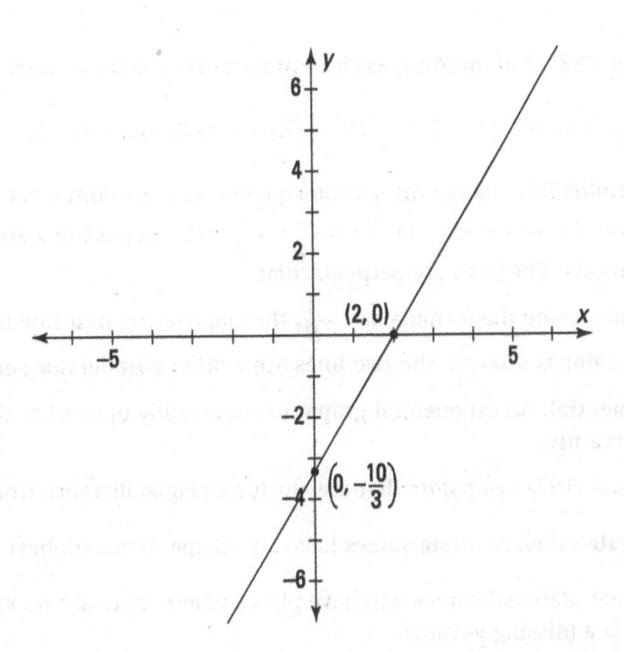

11. **(0,0), $m = 4$. From the slope-intercept form, $m = 4$ and $b = 0$, so the y-intercept is (0,0). This is also the x-intercept. Given the line $y = 4x$, find the intercepts and the slope. The graph of the line can be drawn starting at the intercept (0,0) and then using the slope — counting one unit to the right and four units up.**

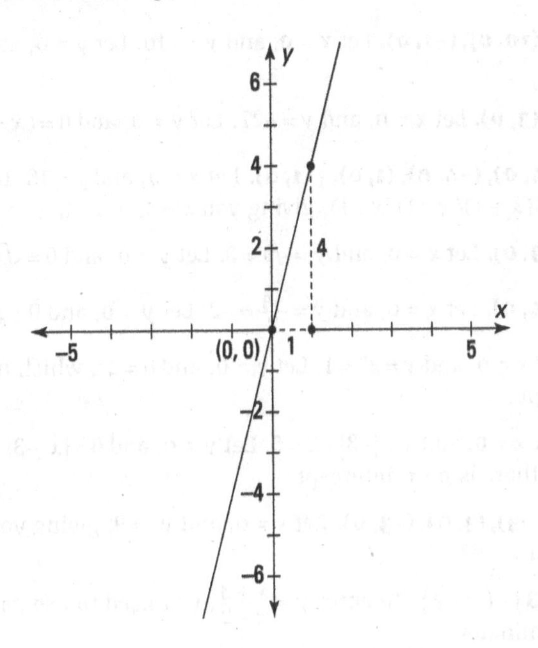

12. **$y = \frac{3}{4}x - 3$. Subtract $3x$ from each side, and then divide each term by −4.**

$$3x - 4y = 12 \rightarrow -4y = -3x + 12 \rightarrow \frac{-4y}{-4} = \frac{-3x}{-4} + \frac{12}{-4} \rightarrow y = \frac{3}{4}x - 3$$

13. $x+2y=22$**.** First, multiply each term by 2. Then add x to each side.

$$y=-\frac{1}{2}x+11 \rightarrow 2\cdot y=-\frac{1}{2}\cdot 2x+11\cdot 2 \rightarrow 2y=-x+22 \rightarrow x+2y=22$$

14. **Perpendicular.** Change the second equation to slope-intercept form: $2x+4y=11 \rightarrow 4y=-2x+11 \rightarrow y=-\frac{1}{2}x+\frac{11}{4}$. The slopes are 2 and $-\frac{1}{2}$, which are negative reciprocals. The lines are perpendicular.

15. **Neither.** Using the formula $m=-\frac{A}{B}$, the slope of the first line is $m=-\frac{4}{3}$ and the slope of the second line is $m=-\frac{3}{4}$. The two lines are neither parallel nor perpendicular.

16. **Exponential.** An exponential graph moves steadily upward or downward and doesn't change its curvature.

17. **Circle.** A circle has points that are always an equal distance from the center.

18. **Absolute value.** Absolute values have a V-shape with a highest point or a lowest point.

19. **Rational.** Rational curves will have places where there are no x-values and a place where there is a missing y-value.

20. **Line.** A line is steadily straight — no curves or bends.

21. **Cubic.** The cubic polynomial has an S-shaped characteristic.

22. **(0, −12), (2, 0).** Let $x = 0$, then $y = -12$. Let $y = 0$, and $0=6(x-2)$, giving you $x=2$.

23. **(0, −10), (10, 0), (−1, 0).** Let $x = 0$, and $y=-10$. Let $y = 0$, and $0=(x-10)(x+1)$, giving you $x=10,-1$.

24. **(0, −27), (3, 0).** Let $x = 0$, and $y=-27$. Let $y = 0$, and $0=(x-3)\left(x^2+3x+9\right)$, giving you $x=3$.

25. **(0, 16), (4, 0), (−4, 0), (1, 0), (−1, 0).** Let $x = 0$, and $y=16$. Let $y = 0$, and $0=(x-4)(x+4)(x-1)(x+1)$, giving you $x=4,-4,1,-1$.

26. **(0, 3), (−9, 0).** Let $x = 0$, and $y=\sqrt{9}=3$. Let $y = 0$, and $0=\sqrt{x+9}$, giving you $x=-9$.

27. **(0, −2), (4, 0).** Let $x = 0$, and $y=\frac{-4}{2}=-2$. Let $y = 0$, and $0=x-4$, giving you $x=4$.

28. **(0, 1).** Let $x = 0$, and $y=2^0=1$. Let $y = 0$, and $0=2^x$, which has no solution; there is no x-intercept.

29. **(0, 5).** Let $x = 0$, and $y=|-3|+2=5$. Let $y = 0$, and $0=|x-3|+2$ or $-2=|x-3|$, which has no solution; there is no x-intercept.

30. **(0, 3), (0, −3), (3, 0), (−3, 0).** Let $x = 0$, and $y^2=9$, giving you $y=\pm 3$. Let $y = 0$, and $x^2=9$, giving you $x=\pm 3$.

31. $y_4=(x+3)\div(x-2)$**.** To enter $y=\frac{x+3}{x-2}$, you need to use parentheses on both the numerator and denominator.

32. $y_1=x\wedge(-1\div 2)$**.** To enter $y=x^{-1/2}$, you need to use parentheses around the numbers and operation in the exponent.

33. $y_3=(x+4)\wedge(1\div 2)$**.** To enter $y=\sqrt{x+4}$, you need to put the terms under the radical in parentheses and the numbers in the fractional exponent in parentheses.

Whaddya Know? Chapter 6 Quiz

Quiz time! Complete each problem to test your knowledge on the various topics covered in this chapter. You can then find the solutions and explanations in the next section.

1. Find the x- and y-intercepts of the parabola $y = x^2 - 6x + 8$.
2. The point $(-4,5)$ lies in which quadrant?
3. Change the equation of the line $x - 3y = 8$ to slope-intercept form.
4. Determine the intercepts of the line $3x - 2y = 6$, and graph the line.
5. Determine the symmetry of the graph of $x = -2y^2$.
6. Find the slope and y-intercept of $y = -\frac{2}{5}x + 4$.
7. Find the slopes of the lines that are parallel and perpendicular to the line $y = \frac{1}{2}x - 3$.
8. Determine the symmetry of the graph of $y = x^2 + 6$.
9. Which function is best represented by the curve shown in the following figure?

   (A) $y = (x+4)(x-2)(x+5)$

   (B) $y = (x+4)(x-2)(x-5)$

   (C) $y = (x+4)(x+2)(x-5)$

   (D) $y = (x-4)(x+2)(x-5)$

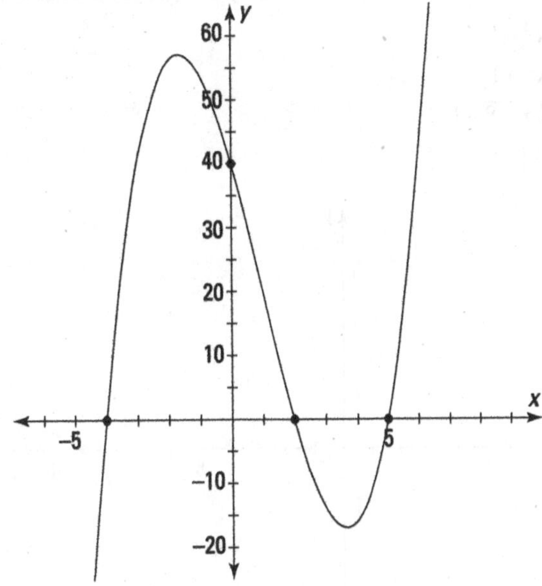

10. Change the equation of the line $y = -\frac{1}{3}x + 3$ to standard form.
11. Which equation best matches the graph shown here?

(A) $y=-x^4+5x^2-4$

(B) $y=-x^4-5x^2-4$

(C) $y=x^4+5x^2-4$

(D) $y=-x^4+5x^2+4$

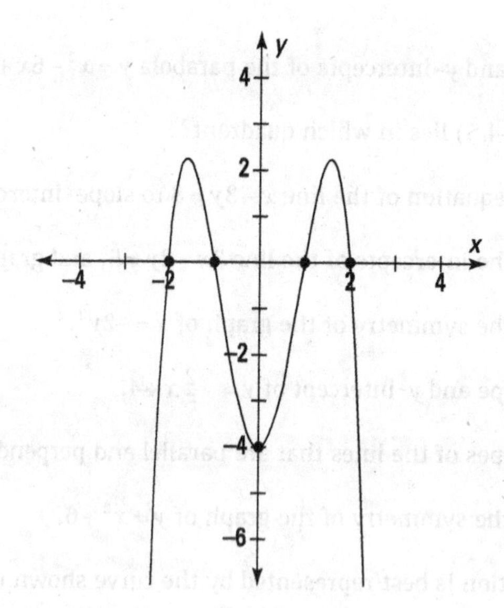

12 **Determine the symmetry of the graph of $y=x^3-x$.**

13 **Find the slope of a line that goes through both (5,–2) and (–5,8).**

14 **Which function is best represented by the graph shown here?**

(A) $y=x^3-1$

(B) $y=x^2-1$

(C) $y=\sqrt[3]{x-8}-1$

(D) $y=|x|-1$

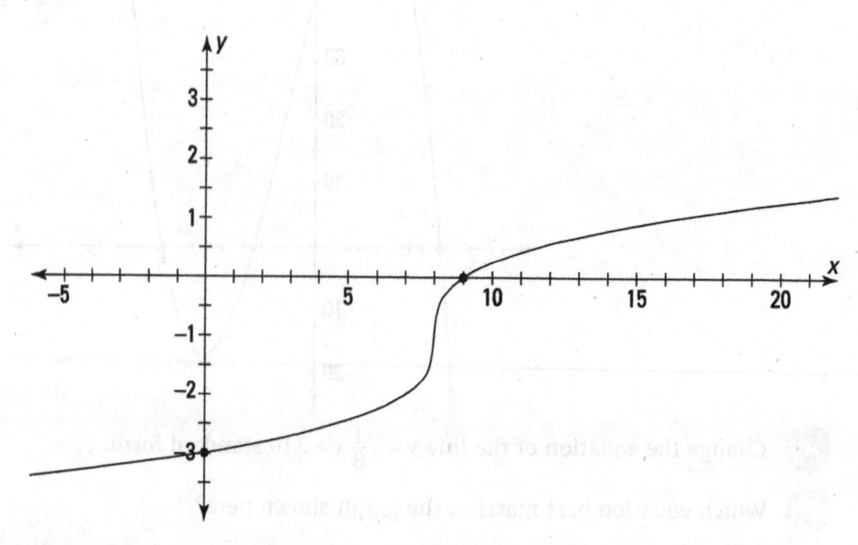

Answers to Chapter 6 Quiz

1. **(0,8), (4,0), (2,0).** To find the y-intercept, set x equal to 0: $y = 0^2 - 6 \cdot 0 + 8 = 8$, so the y-intercept is (0,8). To find the x-intercepts, set y equal to 0, factor the quadratic, and solve for x: $0 = x^2 - 6x + 8 = (x-4)(x-2)$. The two solutions are 4 and 2, so the two intercepts are (4,0) and (2,0).

2. **Quadrant II.** Moving from the origin, (0,0), first to the left and then up 5, you end up in the second quadrant.

3. $y = \frac{1}{3}x - \frac{8}{3}$**.** Solving for y, subtract x from each side and then divide both sides by –3.

   $x - 3y = 8 \rightarrow -3y = -x + 8 \rightarrow \frac{-3y}{-3} = \frac{-x}{-3} + \frac{8}{-3} \rightarrow y = \frac{1}{3}x - \frac{8}{3}$

4. **(2,0),(0,–3).** To find the x-intercept, let y = 0 and solve for x: $3x - 2 \cdot 0 = 6 \rightarrow 3x = 6 \rightarrow x = 2$. The intercept is (2,0). To find the y-intercept, let x = 0 and solve for y: $3 \cdot 0 - 2y = 6 \rightarrow -2y = 6 \rightarrow y = -3$. The intercept is (0,–3).

   Draw the line through the two intercepts.

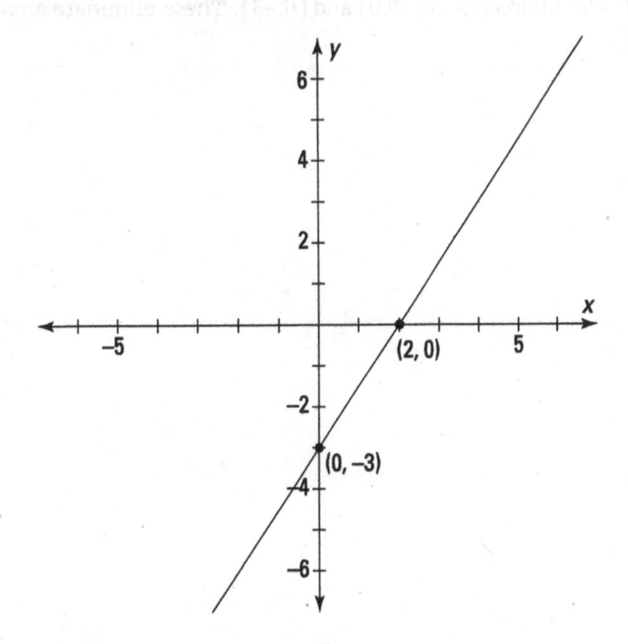

5. **Symmetric with respect to the x-axis.** Replacing y with $-y$ in the equation $x = -2y^2$, the equation does not change.

6. $m = -\frac{2}{5}$**,(0,4).** Referring to the slope-intercept form, $y = mx + b$, the slope is the coefficient m and the y-intercept is the constant b.

7. **Parallel:** $m = \frac{1}{2}$**, perpendicular:** $m = -2$**.** Lines that are parallel have equal slopes. The slope of a line perpendicular to another line has a slope that is the negative reciprocal.

8. **Symmetric with respect to the y-axis.** Replacing x with $-x$ in the equation $y = x^2 + 6$, the equation does not change.

(9) **B.** Referring to the graph, the x-intercepts are: $(2,0),(5,0)$, and $(-4,0)$. The y-intercept is 40. Answer choice B, $y=(x+4)(x-2)(x-5)$, has solutions $x=-4,2,5$, which match the x-intercepts, and the product produces a constant of 40, which matches the y-intercept.

(10) $\boldsymbol{x+3y=9}$. First, add $-\frac{1}{3}x$ to each side, and then multiply through by 3.
$y=-\frac{1}{3}x+3 \rightarrow \frac{1}{3}x+y=3 \rightarrow 3\cdot\frac{1}{3}x+3\cdot y=3\cdot 3 \rightarrow x+3y=9$

(11) **A.** The y-intercept is -4, so you only need to consider A, B, or C. The graph rises from negative infinity from the left and falls to negative infinity to the right, so you can eliminate C. Factoring A, you have $y=-x^4+5x^2-4=-(x-1)(x+1)(x-2)(x+2)$, which corresponds to the x-intercepts: $(1,0),(-1,0),(2,0)$, and $(-2,0)$. So the equation in answer choice A matches the graph.

(12) **Symmetric with respect to the origin.** Replacing x with $-x$ and y with $-y$ in the equation $y=x^3-x$ yields opposite results.

(13) **-1.** Using the slope formula, $m=\frac{8-(-2)}{-5-5}=\frac{10}{-10}=-1$.

(14) **C.** The curve moves from negative infinity to positive infinity with a "flattening" at the point $(8,-1)$. The intercepts are $(9,0)$ and $(0,-3)$. These eliminate answer choices A, B, and D.

2
Figuring on Functions

In This Unit . . .

IN THIS CHAPTER

» Freeze-framing function characteristics

» Concentrating on domain and range

» Tackling one-to-one functions

» Applying rules to piecewise functions

» Handling composition duties

» Working with inverses

Chapter 7 Formulating Functions

Is your computer functioning well? Are you going to that company function? Are your kidneys functioning properly? What is all this business about functions? In algebra, the word *function* is very specific. You reserve it for certain math expressions that meet the tough standards of input and output values, as well as other mathematical rules of relationships. Therefore, when you hear that a certain relationship is a function, you know that the relationship meets some stringent requirements.

In this chapter, you find out more about these requirements. I also cover topics ranging from the domain and range of functions to the inverses of functions, and I show you how to deal with piecewise functions and do composition of functions. After grazing through these topics, you can confront a function equation with great confidence and a plan of attack.

Featuring Functions

In mathematics, a *function* is a relationship between two variables that designates exactly one output value for every input value — in other words, exactly one answer for every number inserted.

Q. Determine whether or not the equation $y = x^2 + 5x - 4$ is a function.

EXAMPLE **A.** The x is the input variable, and the y is the output variable. If you input the number 3 for each of the x's, you get $y = 3^2 + 5(3) - 4 = 9 + 15 - 4 = 20$. The output is 20, the only possible answer. You won't get another number if you input the 3 again. So, yes, this is a function.

Q. Determine whether or not the equation $x = y^2$ is a function.

A. The x is the input variable, and the y is the output variable. If you input 9 for x, the value of y could be either 3 or -3. Either one satisfies the equation. With two possible answers, this can't be a function.

You'll be encountering plenty of strange math equations out there. You have to watch out that if you want a function, you have a function.

Introducing function notation

Functions feature some special notation that makes them much easier to work with, more recognizable, and more understandable. The notation doesn't change any of the properties; it just allows you to identify different functions quickly and indicate various operations and processes more efficiently.

The variables x and y are pretty standard in functions and come in handy when you're creating their graphs. But mathematicians also use another format called *function notation*. For instance, say you have these three functions: $y = x^2 + 5x - 4$, $y = \sqrt{3x-8}$, and $y = 6xe^x - 2e^{2x}$. Assume you want to call them by name. Instead, you can write them as $f(x) = x^2 + 5x - 4$, $g(x) = \sqrt{3x-8}$, and $h(x) = 6xe^x - 2e^{2x}$.

REMEMBER

The names of these functions are f, g, and h. You read them as follows: "f of x is x squared plus 5x minus 4," and so on. When you see a bunch of functions written together, you can simply refer to individual functions as f or g or h rather than "the middle one" or "the first one," and so on.

Evaluating functions

When you see a written function that uses function notation, you can easily identify the input variable, the output variable, and what operations you need in order to evaluate the function for some input (or replace the variables with numbers and simplify). You can do so because the input value is placed in the parentheses right after the function name or output value. When you replace the input variable with a number, you know to replace every one of the variables in the function expression or rule with the number and simplify.

Q. Evaluate $g(x) = \sqrt{3x-8}$ when x is 3.

EXAMPLE **A.** 1. Change the function name to $g(3)$. This means you substitute a 3 for every x in the function expression and perform the operations to get the output answer: $g(3) = \sqrt{3(3)-8} = \sqrt{9-8} = \sqrt{1} = 1$. Now you can say that $g(3) = 1$, or "g of 3 equals 1." The output of the function g is 1 if the input is 3.

Q. Evaluate $k(x) = 2x^2 - 3x + 1$ when x is -1.

A. **6.** Change the function name to $k(-1)$. This means you substitute a -1 for every x in the function expression and perform the operations to get the output answer: $k(-1) = 2(-1)^2 - 3(-1) + 1 = 2 \cdot 1 + 3 + 1 = 6$.

1 Given the function $f(x) = |2x - 1| - 4$, evaluate it when $x = 2$.

2 Given the function $g(x) = \frac{x^2 - 4x + 7}{x - 3}$, evaluate it when $x = 2$.

Homing In on Domain and Range

The input and output values of a function are of major interest to people working in algebra. If these terms don't yet strum your guitar, then allow me to pique your interest. The words *input* and *output* describe what's happening in the function (namely, what number you put in and what result comes out), but the official designations for the input and output are domain and range, respectively.

Determining a function's domain

The domain of a function consists of all the input values of the function (think of a monarch's domain of all their servants entering their kingdom). In other words, the domain is the set of all numbers that you can input without creating an unwanted or impossible situation. Such situations can occur when operations appear in the definition of the function, such as fractions, radicals, logarithms, and so on.

Many functions have no exclusions of values, but fractions are notorious for causing trouble if zeros appear in the denominators. Radicals have restrictions when the roots are even numbers, and logarithms can deal only with positive numbers.

REMEMBER

You need to be prepared to determine the domain of a function so that you can tell where you can use the function — in other words, for what input values it does any good. You can determine the domain of a function from its equation or function definition. You look at the domain in terms of which real numbers you can use for input and which ones you have to eliminate. You can express the domain by using the following table.

Function	Domain: Words	Domain: Inequality	Domain: Interval Notation
$f(x)=x^2+2$	All real numbers	$-\infty<x<\infty$	$(-\infty,\infty)$
$g(x)=\sqrt{x}$	All numbers greater than or equal to 0	$x\geq 0$	$[0,\infty)$
$h(x)=ln(x-1)$	All numbers greater than 1	$x>1$	$(1,\infty)$

The way you express the domain depends on what's required in the task you're working on, such as evaluating functions, graphing, or determining a good fit as a model, to name a few.

EXAMPLE

Q. Describe the domain of $f(x)=\sqrt{x-11}$ using inequality notation.

A. $x\geq 11$**.** You can't use numbers smaller than 11 because you'd be taking the square root of a negative number, which isn't a real number. You write this domain as $x\geq 11$.

Q. Describe the domain of $g(x)=\frac{x}{x^2-4x-12}=\frac{x}{(x-6)(x+2)}$ using interval notation.

A. **The domain consists of all real numbers except 6 and –2.** In interval notation, this is written $(-\infty,-2)\cup(-2,6)\cup(6,\infty)$. The reason you can't use –2 or 6 is because these numbers result in a 0 in the denominator of the fraction, and a fraction with 0 in the denominator creates a number that doesn't exist.

Describing a function's range

The range of a function is all its output values — every value you get by inputting the domain values into the rule (the function equation) for the function. You may be able to determine the range of a function from its equation, but sometimes you have to graph it to get a good idea of what's going on.

A range may consist of all real numbers, or it may be restricted because of the way a function equation is constructed. You have no easy way to describe ranges — at least, not as easy as describing domains — but you can discover clues with some functions by looking at their graphs and with others by knowing the characteristics of those kinds of curves.

The following are some examples of functions and their ranges. Like domains, you can express ranges in words, inequalities, or interval notation.

Function	Range: Words	Range: Inequality	Range: Interval Notation
$k(x)=x^2+3$	3 and any number greater than 3	$k(x)\geq 3$	$[3,\infty)$
$m(x)=\sqrt{x+7}$	All positive numbers and zero	$m(x)\geq 0$	$[0,\infty)$
$p(x)=\frac{2}{x-5}$	All numbers except 0	$p(x)<0$ or $p(x)>0$	$(-\infty,0)\cup(0,\infty)$

- The range of $k(x)=x^2+3$ consists of the number 3 and any number greater than 3 because the squares of any input numbers are always positive or 0. The smallest number you can add to the 3 is 0. You never get anything smaller than 3.
- In $m(x)=\sqrt{x+7}$, the sum under the radical can never be negative, and all the square roots come out positive or zero. The smallest sum comes from adding –7 to the 7.
- Some functions' equations, such as $p(x)=\frac{2}{x-5}$, don't give an immediate clue to the range values. It often helps to sketch the graphs of these functions. Figure 7-1 shows the graph of the function p.

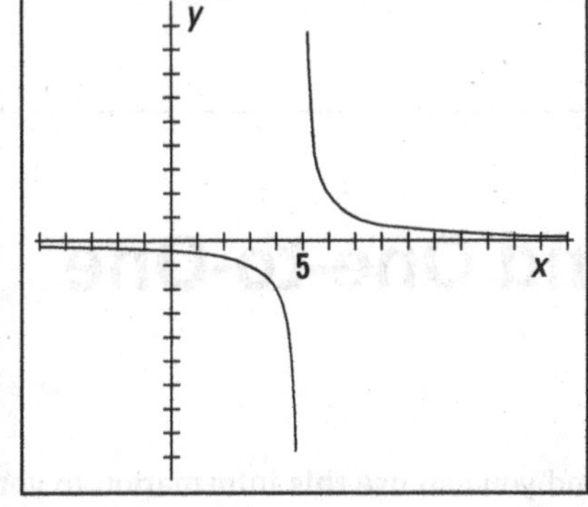

FIGURE 7-1: Try graphing equations that don't have an obvious range.

The graph of this function never touches the x-axis, but it gets very close. For the numbers in the domain bigger than 5, the graph has some really high y-values and some y-values that get really close to zero. But the graph never touches the x-axis, so the function value never really reaches zero. For numbers in the domain smaller than 5, the curve is below the x-axis. These function values are negative — some really small. But, again, the y-values never reach zero. So, if you guessed that the range of the function is every real number except zero, you're right! You write the range as $y\neq 0$, or $(-\infty,0)\cup(0,\infty)$. Did you also notice that the function doesn't have a value when $x=5$? This happens because 5 isn't in the domain.

For some tips on how to graph functions, head to Chapters 8 through 11, where I discuss the graphs of the different kinds of functions.

EXAMPLE

Q. Describe the range of $f(x)=\frac{1}{x^2}$ using inequality notation.

A. $f(x)>0$. Squaring the x input value always gives you a positive number — as long as x is not 0. You can't divide by x. So the range is all positive numbers: $f(x)>0$.

Q. **Describe the range of $f(x) = -|x+4|$ using inequality notation.**

A. **$f(x) \le 0$. The absolute value function presents you with all positive numbers or 0. The negative sign makes all the positive results negative. So the range is $f(x) \le 0$.**

YOUR TURN

Given the function $g(x) = x^2 - 2$, determine the domain and range of the function.

Given the function $h(x) = \sqrt{x^2 - 9}$, determine the domain and range of the function.

Recognizing Even, Odd, and One-to-One Functions

You can classify numbers as even or odd (and you can use this information to your advantage; for example, you know you can divide even numbers by 2 and come out with a whole number). You can also classify some functions as even or odd. The even and odd integers (like 2, 4, 6, and 1, 3, 5) play a role in this classification, but they aren't the be-all and end-all. You have to put a bit more calculation work into classifying the functions. But the effort will be worth it as you apply their classifications in your work.

Classifying even and odd functions

An even function is one in which both a domain value (an input) and its opposite result in the same range value (output) — so, $f(-x) = f(x)$. An odd function is one in which a domain value and its opposite produce opposite results in the range — meaning, $f(-x) = -f(x)$.

To determine if a function is even or odd (or neither), you replace every x in the function equation with $-x$ and simplify. If the function is even, the result in the simplified version looks exactly like the original. If the function is odd, the result in the simplified version looks like what you get after multiplying the original function equation by –1.

The descriptions for even and odd functions may remind you of how even and odd numbers for exponents act. If you raise –2 to an even power, you get a positive number: $(-2)^4 = 16$. If you raise –2 to an odd power, you get a negative (opposite) result: $(-2)^5 = -32$.

EXAMPLE

Q. Determine if the function $f(x) = x^4 - 3x^2 + 6$ is even or odd.

A. $f(x) = x^4 - 3x^2 + 6$ **is even.** This is because whether you input 2 or –2, you get the same output:

$$f(2) = (2)^4 - 3(2)^2 + 6 = 16 - 12 + 6 = 10$$

$$f(-2) = (-2)^4 - 3(-2)^2 + 6 = 16 - 3(4) + 6 = 10$$

This is just a demonstration of how one pair of numbers works. To show that the function is even for any input value, you replace x with $-x$ in the function rule: $f(-x) = (-x)^4 - 3(-x)^2 + 6 = x^4 - 3x^2 + 6$. This result is the original function — it must be even.

Q. Determine if the function $g(x) = \frac{12x}{x^2 + 2}$ is even or odd.

A. $g(x) = \frac{12x}{x^2 + 2}$ **is odd.** This is because the inputs 2 and –2 give you opposite answers:

$$g(2) = \frac{12(2)}{(2)^2 + 2} = \frac{24}{4 + 2} = \frac{24}{6} = 4$$

$$g(-2) = \frac{12(-2)}{(-2)^2 + 2} = \frac{-24}{4 + 2} = \frac{-24}{6} = -4$$

To show that this result happens for all pairs of input values, replace x with $-x$ in the function rule:

$g(-x) = \frac{12(-x)}{(-x)^2 + 2} = \frac{-12x}{x^2 + 2} = -\left(\frac{12x}{x^2 + 2}\right)$. This is the same as multiplying the original function rule by –1. The function must be odd.

You can't say that a function is even just because it has even exponents and coefficients, and you can't say that a function is odd just because the exponents and coefficients are odd numbers. If you do make these assumptions, you classify the functions incorrectly, which messes up your graphing. You have to apply the rules to determine which label a function has.

Applying even and odd functions to graphs

The biggest distinction between even and odd functions is how their graphs look.

- **Even functions:** The graphs of even functions are symmetric with respect to the *y*-axis (the vertical axis). You see what appears to be a mirror image to the left and right of the vertical axis. For an example of this type of symmetry, see Figure 7-2a, which is the graph of the even function,

  $$f(x) = \frac{5}{x^2+1}$$

- **Odd functions:** The graphs of odd functions are symmetric with respect to the origin. The symmetry is radial, or circular, so it looks the same if you rotate the graph by 180 degrees. The graph in Figure 7-2b, which is the odd function $g(x) = x^3 - 8x$, displays origin symmetry.

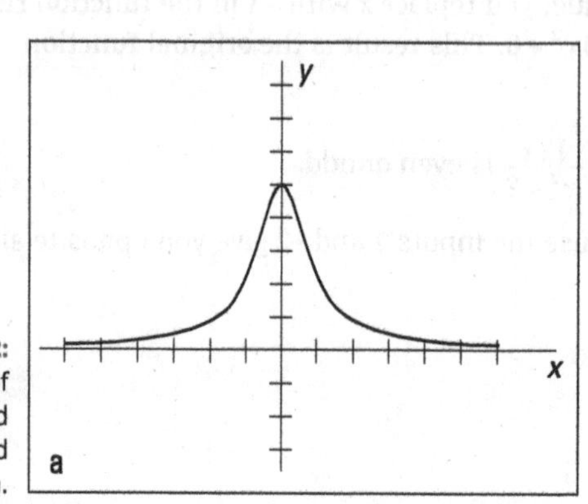

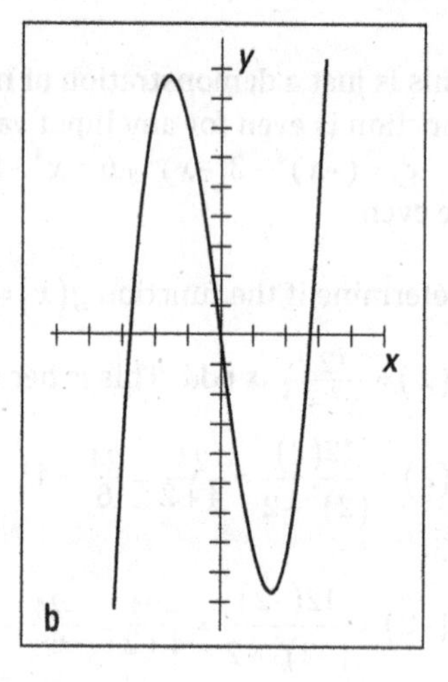

FIGURE 7-2: Graphs of an even and an odd function.

You may be wondering whether you can have symmetry with respect to the *x*-axis. After all, is the *y*-axis all that better than the *x*-axis? I'll leave that up to you, but yes, *x*-axis symmetry does exist — just not in the world of functions. By its definition, a function can have only one *y* value for every *x* value. If you have points on either side of the *x*-axis, above and below an *x* value, you don't have a function. Head to Chapter 13 if you want to see some pictures of curves that are symmetric all over the place.

YOUR TURN

5 Given the function $f(x) = \sqrt{x^2+3}$, determine whether the function is even, odd, or neither.

6 Given the function $g(x) = \sqrt[3]{x+5}$, determine whether the function is even, odd, or neither.

7 Given the function $h(x) = x - x^5$, determine whether the function is even, odd, or neither.

Defining one-to-one functions

Functions can have many classifications or names, depending on the situation (maybe you want to model a business transaction or use functions to figure out payments and interest) and what you want to do with them (put the formulas or equations in spreadsheets or maybe just graph them, for example). One very important classification is deciding whether a function is one-to-one.

A function is one-to-one if it has exactly one output value for every input value and exactly one input value for every output value. Formally, you write this definition as follows:

If $f(x_1) = f(x_2)$, then $x_1 = x_2$

In simple terms, if the two output values of a function are the same, then the two input values must also be the same.

One-to-one functions are important because they're the only functions that can have inverses, and functions with inverses aren't all that easy to come by. If a function has an inverse, you can work backward and forward — find an answer if you have a question and find the original question if you know the answer (sort of like *Jeopardy!*). For more on inverse functions, see the section, "Dealing with Inverse Functions," later in this chapter.

EXAMPLE

Q. Determine if the function $f(x) = x^3$ is a one-to-one function.

A. The rule for the function involves cubing the variable. The cube of a positive number is positive, and the cube of a negative number is negative. Therefore, every input has a unique output — no other input value gives you that output. This function is one-to-one.

Q. Determine if the function $g(x) = x^3 - x$ is a one-to-one function.

A. Some functions without the one-to-one designation may look like the previous example, which is one-to-one. The function $g(x) = x^3 - x$ counts as a function because only one output comes with every input. However, the function isn't one-to-one, because you can create many outputs or function values from more than one input. For instance, $g(1) = (1)^3 - (1) = 1 - 1 = 0$, and $g(-1) = (-1)^3 - (-1) = -1 + 1 = 0$. You have two inputs, 1 and −1, that result in the same output of 0.

Functions that don't qualify for the one-to-one label can be hard to spot, but you can rule out any functions with all even-numbered exponents right away. Functions with absolute value usually don't cooperate either.

Eliminating one-to-one violators

You can determine which functions are one-to-one and which are violators by sleuthing (guessing and trying), using algebraic techniques, and graphing. Most mathematicians prefer the graphing technique because it gives you a nice, visual answer. The basic graphing technique is the horizontal line test. But, to better understand this test, you need to meet its partner, the vertical line test. (I show you how to graph various functions in Chapters 8 through 11.)

Vertical line test

The graph of a function always passes the vertical line test. The test stipulates that any vertical line drawn through the graph of the function passes through that function no more than once. This is a visual illustration that only one y value (output) exists for every x value (input), which is a rule of functions. Figure 7-3a shows a function that passes the vertical line test, and Figure 7-3b contains a curve that isn't a function and therefore flunks the vertical line test.

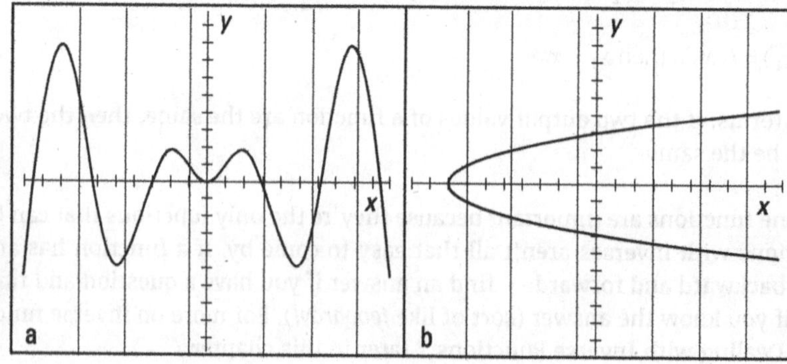

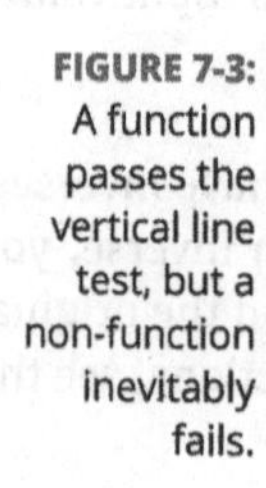
FIGURE 7-3: A function passes the vertical line test, but a non-function inevitably fails.

Horizontal line test

All functions pass the vertical line test but only one-to-one functions pass the horizontal line test. With this test, you can see if any horizontal line drawn through the graph cuts through the function more than one time. If the line passes through the function more than once, the function fails the test and therefore isn't a one-to-one function. Figure 7-4a shows a function that passes the horizontal line test and Figure 7-4b shows a function that flunks it.

Both graphs in Figure 7-4 are functions, however, so they both pass the vertical line test.

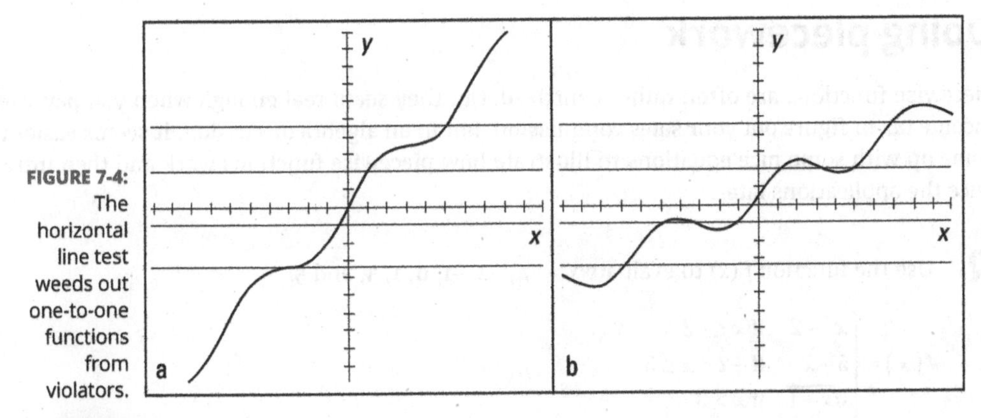

FIGURE 7-4: The horizontal line test weeds out one-to-one functions from violators.

YOUR TURN

8 Determine whether or not $f(x) = 3x^2 + 5$ is one-to-one.

9 Determine whether or not $g(x) = \sqrt{x+2}$ is one-to-one.

Going to Pieces with Piecewise Functions

A piecewise function consists of two or more function rules (function equations) pieced together (listed separately for different x-values) to form one bigger function. A change in the function equation occurs for different values in the domain. For example, you may have one rule for all the negative numbers, another rule for numbers bigger than three, and a third rule for all the numbers between those two rules.

Piecewise functions have their place in situations where you don't want to use the same rule for everyone or everything. Should a restaurant charge a 3-year-old the same amount for a meal as it does an adult? Do you put on the same amount of clothing when the temperature is 20 degrees as you do in hotter weather? No, you place different rules on different situations. In mathematics, the piecewise function allows for different rules to apply to different numbers in the domain of a function.

Doing piecework

Piecewise functions are often rather contrived. Oh, they seem real enough when you pay your income tax or figure out your sales commission. But in an algebra discussion, it seems easier to come up with some nice equations to illustrate how piecewise functions work and then introduce the applications later.

EXAMPLE

Q. Use the function $f(x)$ to evaluate $x = -4, -2, -1, 0, 1, 3,$ and 5.

$$f(x)=\begin{cases} x^2-2 & \text{if } x\le -2 \\ 5-x & \text{if } -2<x\le 3 \\ \sqrt{x+1} & \text{if } x>3 \end{cases}$$

A. With this function, you use one rule for all numbers smaller than or equal to −2, another rule for numbers between −2 and 3 (including 3), and a final rule for numbers larger than 3. You still have only one output value for every input value. You use the different rules depending on the input value:

$$f(-4)=(-4)^2-2=16-2=14$$
$$f(-2)=(-2)^2-2=4-2=2$$
$$f(-1)=5-(-1)=6$$
$$f(0)=5-0=5$$
$$f(1)=5-1=4$$
$$f(3)=5-3=2$$
$$f(5)=\sqrt{5+1}=\sqrt{6}$$

Figure 7-5 shows you the graph of the piecewise function with these function values.

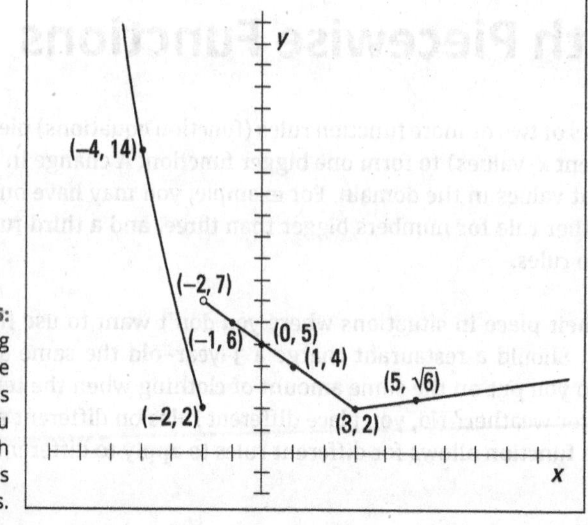

FIGURE 7-5: Graphing piecewise functions shows you both connections and gaps.

Notice the three different sections to the graph. The left curve and the middle line don't connect because a discontinuity exists when $x = -2$. A discontinuity occurs when a gap or hole appears in the graph. Also, notice that the left line falling toward the x-axis ends with a solid dot, and the middle section has an open circle just above it. These features preserve the definition of a function — only one output for each input. The dot tells you to use the rule on the left when $x = -2$.

The middle section connects at point (3, 2) because the rule on the right gets really, really close to the same output value as the rule in the middle when $x = 3$. Technically, you should draw both a hollow circle and a dot, but you really can't spot this feature just by looking at it.

Applying piecewise functions

Why in the world would you need to use a piecewise function? Do you have any good reason to change the rules right in the middle of things? I have two examples that aim to ease your mind — examples that you may relate to very well. One involves utility companies and the other is a taxing situation.

Utility companies can use piecewise functions to charge different rates for users based on consumption levels. A big factory uses a ton of electricity and rightfully gets a different rate than a homeowner.

As April 15 rolls around, many people are faced with their annual struggle with income tax forms. The rate at which you pay tax is based on how much your adjusted income is — a graduated scale where (supposedly) people who make more money pay more income tax. The income values are the inputs (values in the domain), and the government determines the tax paid by putting numbers into the correct formula.

Q. Here's what the Lightning Strike Utility company uses to determine the charges for its customers:

$$C(h) = \begin{cases} 0.0747900h & \text{if } h < 1{,}000 \\ 74.79 + 0.052500(h - 1{,}000) & \text{if } 1{,}000 \le h < 5{,}000 \\ 284.79 + 0.033300(h - 5{,}000) & \text{if } h \ge 5{,}000 \end{cases}$$

where h is the number of kilowatt-hours and C is the cost in dollars.

So, how much does a homeowner who uses 750 kilowatt-hours pay? How much does a business that uses 10,000 kilowatt-hours pay?

A. You use the top rule for the homeowner and the bottom rule for the business, determined by what interval the input value lies in. The homeowner has the following equation:

$$C(h) = 0.0747900h$$
$$C(750) = 0.0747900(750) = 56.0925$$

The person using 750 kilowatt-hours pays a little more than $56 per month. And the business?

$$C(h) = 284.79 + 0.033300(h - 5{,}000)$$

$$C(10{,}000) = 284.79 + 0.033300(10{,}000 - 5{,}000)$$

$$= 284.79 + 0.033300(5{,}000) = 284.79 + 166.50 = 451.29$$

The company pays a little more than $451.

Q. In 2014, a single taxpayer paid their income tax based on their taxable income, according to the following rules (laid out in a piecewise function):

$$T(n) = \begin{cases} 0.10 & n \le 9{,}075 \\ 0.15(n - 9{,}075) + 907.50 & 9{,}075 < n \le 36{,}900 \\ 0.25(n - 36{,}900) + 5{,}081.25 & 36{,}900 < n \le 89{,}350 \\ 0.28(n - 89{,}350) + 18{,}193.75 & 89{,}350 < n \le 186{,}350 \\ 0.33(n - 186{,}350) + 45{,}353.75 & 186{,}350 < n \le 405{,}100 \\ 0.35(n - 405{,}100) + 117{,}541.25 & 405{,}100 < n \le 406{,}750 \\ 0.396(n - 406{,}750) + 118{,}118.75 & n > 406{,}750 \end{cases}$$

where n is the taxable income and T is the tax paid.

If the taxable income was $45,000, how much did they pay in taxes?

A. Insert the value and follow the third rule, because 45,000 is between 36,900 and 89,350:

$$T(45{,}000) = 0.25(45{,}000 - 36{,}900) + 5{,}081.25$$

$$= 2{,}025 + 5{,}081.25 = 7{,}106.25$$

This person paid a little more than $7,100 in income tax.

YOUR TURN

10 Given the piecewise function $h(x)$, evaluate it for $x = -3$, 0, and 5.

$$h(x) = \begin{cases} -2x, & x < -1 \\ x^2 - 2, & -1 \le x < 4 \\ 1 - \sqrt{x - 1}, & x \ge 4 \end{cases}$$

11 A business rewards its employees with year-end raises by giving them a bonus equaling a percentage of their year's sales. Determine the bonuses awarded to employees A, B, and C, who had their sales totals listed — A: \$43,000; B: \$102,000; C: \$140,000.

$$\text{Bonus}(x) = \begin{cases} 0.3\%x, & x \le \$20{,}000 \\ 0.6\%x, & \$20{,}000 < x \le \$140{,}000 \\ 1\%x, & x > \$140{,}000 \end{cases}$$

Composing Functions and Applying the Difference Quotient

You can perform the basic mathematical operations of addition, subtraction, multiplication, and division on the equations used to describe functions. (You can also perform whatever simplification is possible on the different parts of the expression and write the result as a new function.) For example, you can take the two functions $f(x) = x^2 - 3x - 4$ and $g(x) = x + 1$ and perform the four basic operations on them:

$$f + g = \left(x^2 - 3x - 4\right) + \left(x + 1\right) = x^2 - 2x - 3$$

$$f - g = \left(x^2 - 3x - 4\right) - \left(x + 1\right) = x^2 - 4x - 5$$

$$f \cdot g = \left(x^2 - 3x - 4\right)\left(x + 1\right) = x^3 - 2x^2 - 7x - 4$$

$$\frac{f}{g} = \frac{x^2 - 3x - 4}{x + 1} = \frac{(x-4)\cancel{(x+1)}}{\cancel{x+1}} = x - 4, \text{ where } x \ne -1$$

Well done, but you have another operation at your disposal, an operation special to functions, called composition.

Performing compositions

The composition of functions is an operation in which you use one function as the input into another function and perform the operations on that input function. You indicate the composition of functions *f* and *g* with a small circle between the function names, $\circ$, and you define the composition as $f \circ g = f(g)$. This means that you perform the operations and rules assigned to function *f* onto the terms, given the function definition of function *g*.

EXAMPLE

Q. Given functions $f(x) = x^2 - 3x - 4$ and $g(x) = x + 1$, perform the composition $f \circ g$.

A. Using the rules from the functions f and g:

$$\begin{aligned} f \circ g &= f(g) = g^2 - 3g - 4 \\ &= (x+1)^2 - 3(x+1) - 4 = x^2 + 2x + 1 - 3x - 3 - 4 = x^2 - x - 6 \end{aligned}$$

Q. Given functions $f(x) = x^2 - 3x - 4$ and $g(x) = x + 1$, perform the composition $g \circ f$.

A. Using the rules from the functions g and f:

$$\begin{aligned} g \circ f &= g(f) = f + 1 \\ &= x^2 - 3x - 4 + 1 = x^2 - 3x - 3 \end{aligned}$$

The composition of functions isn't commutative (addition and multiplication are commutative, because you can switch the order and not change the result). The order in which you perform the composition — which function comes first — matters. The composition $f \circ g$ isn't the same as $g \circ f$, save for one exception: when the two functions are inverses of one another (see the section, "Dealing with Inverse Functions," later in this chapter).

YOUR TURN

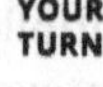

12 Given $f(x) = 2x^2 - 3x + 2$ and $g(x) = 5 - x$, find $f \circ g$.

13 Given $h(x) = \sqrt{x^2 - 3x}$ and $k(x) = x^2 + 11$, find $k \circ h$.

Simplifying the difference quotient

The difference quotient shows up in most high school Algebra II classes as an exercise you do after your instructor shows you the composition of functions. You perform this exercise because the difference quotient is the basis of the definition of the derivative. The difference quotient allows you to find the derivative, which allows you to be successful in calculus (because everyone wants to be successful in calculus, of course). So, where does the composition of functions come in? With the difference quotient, you do the composition of some designated function $f(x)$ and the function $g(x) = x + h$ or $g(x) = x + \Delta x$, depending on what calculus book you use.

The difference quotient for the function f is $\frac{f(x+h)-f(x)}{h}$. You do the composition performing the function rules of f on the binomial $x+h$. Then you subtract the original function f from the composition and divide that result by h.

Now, for an example, perform the difference quotient on the same function f from the previous section.

EXAMPLE

Q. Perform the difference quotient process on $f(x)=x^2-3x-4$.

A. Using the formula:

$$\frac{f(x+h)-f(x)}{h}=\frac{(x+h)^2-3(x+h)-4-(x^2-3x-4)}{h}$$

Notice that you find the expression for $f(x+h)$ by putting $x+h$ in for every x in the function — $x+h$ is the input variable. Now, continuing on with the simplification:

$$=\frac{x^2+2xh+h^2-3x-3h-4-x^2+3x+4}{h}=\frac{2xh+h^2-3h}{h}$$

Did you notice that x^2, $3x$, and 4 all appear in the numerator with their opposites? That's why they disappear with the simplification. Now, to finish:

$$=\frac{h(2x+h-3)}{h}=2x+h-3$$

Although this may not look like much to you, you've created a wonderful result. You're one step away from finding the derivative. Tune in next week at the same time no, I lied. You need to look at *Calculus For Dummies*, by Mark Ryan (John Wiley & Sons, Inc.), if you can't stand the wait and really want to find the derivative. For now, you've just done some really decent algebra.

Q. Perform the difference quotient process on $f(x)=\sqrt{x+1}$.

A. Using the formula:

$$\frac{f(x+h)-f(x)}{h}=\frac{\sqrt{x+h+1}-\sqrt{x+1}}{h}$$

The numerator can't be simplified, so multiply both numerator and denominator by the conjugate of the numerator. (Recall: the *conjugate* of a binomial is another binomial with the opposite sign between the terms. So the conjugate of $x+2$ is $x-2$.)

$$\frac{\sqrt{x+h+1}-\sqrt{x+1}}{h}\cdot\frac{\sqrt{x+h+1}+\sqrt{x+1}}{\sqrt{x+h+1}+\sqrt{x+1}}=\frac{x+h+1-(x+1)}{h\left(\sqrt{x+h+1}+\sqrt{x+1}\right)}$$

$$=\frac{x+h+1-x-1}{h\left(\sqrt{x+h+1}+\sqrt{x+1}\right)}=\frac{h}{h\left(\sqrt{x+h+1}+\sqrt{x+1}\right)}$$

Now the fraction can be reduced.

$$\frac{\cancel{h}}{\cancel{h}\left(\sqrt{x+h+1}+\sqrt{x+1}\right)}=\frac{1}{\left(\sqrt{x+h+1}+\sqrt{x+1}\right)}$$

I know that this doesn't look all that exciting to you, but this is a wonderful result. Calculus loves it, too.

YOUR TURN

14 Perform the difference quotient on $f(x)=3x^2-x-1$.

15 Perform the difference quotient on $f(x)=2x$.

Dealing with Inverse Functions

So why do we need to worry about inverse functions? They can be so useful. You will often come across a process or formula where you want to undo something or find out what it takes to create a value. Functions with inverses can really help out here.

Some functions are inverses of one another, but a function can have an inverse only if it's one-to-one. If two functions are inverses of one another, each function undoes what the other creates. In other words, you use an inverse to get back where you started.

The notation for inverse functions is the exponent –1 written after the function name. The inverse of function $f(x)$, for example, is $f^{-1}(x)$. These aren't going to work the same as inverse operations, but the idea is the same.

EXAMPLE

Q. Demonstrate, with numerical values, that the functions $f(x)=\frac{x+3}{x-4}$ and $f^{-1}(x)=\frac{4x+3}{x-1}$ are inverses of one another. (This isn't a guarantee — just using numbers. The "being sure" part comes next.)

A. If you put 5 into function f, you get 8 as a result. If you put 8 into f^{-1}, you get 5 as a result — you're back where you started:

$$f(5)=\frac{5+3}{5-4}=\frac{8}{1}=8$$

$$f^{-1}(8)=\frac{4(8)+3}{8-1}=\frac{32+3}{7}=\frac{35}{7}=5$$

Q. Demonstrate, with numerical values, that the functions $f(x)=2x^3-1$ and $f^{-1}(x)=\sqrt[3]{\frac{x+1}{2}}$ are inverses of one another.

A. Put the number 3 into f and you get $f(3)=2\cdot 3^3-1=54-1=53$. Now put the number 53 into f^{-1} and you have $f^{-1}(53)=\sqrt[3]{\frac{53+1}{2}}=\sqrt[3]{\frac{54}{2}}=\sqrt[3]{27}=3$.

Determining if functions are inverses

In an example from the introduction to this section, I tell you that two functions are inverses and then demonstrate how they work. You can't really prove that two functions are inverses by plugging in numbers, however. You may face a situation where a couple of numbers work, but, in general, the two functions aren't really inverses.

The only way to be sure that two functions are inverses of one another is to use the following general definition: Functions f and f^{-1} are inverses of one another only if $f\left(f^{-1}(x)\right)=x$ and $f^{-1}\left(f(x)\right)=x$.

In other words, you have to do the composition in both directions (do $f\circ g$ and then do $g\circ f$, which is the opposite order) and show that both functions result in the single value x.

EXAMPLE

Q. Show that $f(x)=\sqrt[3]{2x-3}+4$ and $g(x)=\frac{(x-4)^3+3}{2}$ are inverses of one another.

A. First, you perform the composition $f\circ g$:

$$f\circ g=f(g)=\sqrt[3]{2g-3}+4=\sqrt[3]{2\left(\frac{(x-4)^3+3}{2}\right)-3}+4$$

$$=\sqrt[3]{\cancel{2}\left(\frac{(x-4)^3+3}{\cancel{2}}\right)-3}+4=\sqrt[3]{(x-4)^3+3-3}+4=\sqrt[3]{(x-4)^3}+4=x-4+4=x$$

Now you perform the composition in the opposite order:

$$g\circ f=g(f)=\frac{(f-4)^3+3}{2}=\frac{\left(\sqrt[3]{2x-3}+4-4\right)^3+3}{2}=\frac{\left(\sqrt[3]{2x-3}\right)^3+3}{2}=\frac{2x-3+3}{2}=\frac{2x}{2}=x$$

Both come out with a result of x, so the functions are inverses of one another.

Q. Show that $f(x)=\dfrac{1}{x+1}$ and $g(x)=\dfrac{1-x}{x}$ are inverses of one another.

A. First, you perform the composition $f\circ g$:

$$f\circ g=f(g)=\frac{1}{g+1}=\frac{1}{\frac{1-x}{x}+1}=\frac{1}{\frac{1-x}{x}+\frac{x}{x}}=\frac{1}{\frac{1-x+x}{x}}=\frac{1}{\frac{1}{x}}=1\cdot\frac{x}{1}=x$$

Now you perform the composition in the opposite order:

$$g\circ f=g(f)=\frac{1-f}{f}=\frac{1-\frac{1}{x+1}}{\frac{1}{x+1}}=\frac{\frac{x+1}{x+1}-\frac{1}{x+1}}{\frac{1}{x+1}}=\frac{\frac{x+1-1}{x+1}}{\frac{1}{x+1}}=\frac{\frac{x}{x+1}}{\frac{1}{x+1}}=\frac{x}{\cancel{x+1}}\cdot\frac{\cancel{x+1}}{1}=x$$

Both come out with a result of x, so the functions are inverses of one another.

16 Determine whether $f(x)=2x+7$ and $g(x)=\dfrac{x-7}{2}$ are inverses of one another.

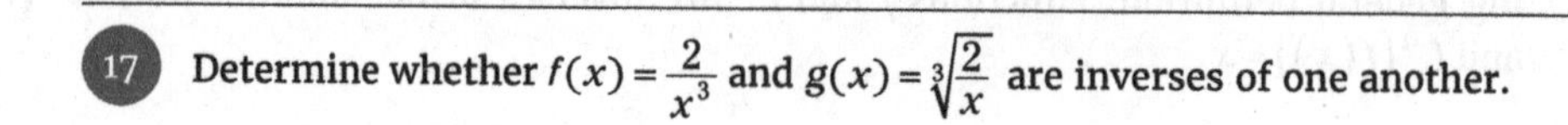

17 Determine whether $f(x)=\dfrac{2}{x^3}$ and $g(x)=\sqrt[3]{\dfrac{2}{x}}$ are inverses of one another.

Solving for the inverse of a function

Up until now in this section, I've been giving you two functions and telling you that they're inverses of one another or asking you to prove that they're inverses. How did I know they were inverses? Was it magic? Did I pull the functions out of a hat? No, you have a nice process to use. I can now show you my secret so you can create all sorts of inverses for all sorts of functions. Here is a step-by-step process that you can use.

To find the inverse of a one-to-one function $f(x)$, follow these steps:

1. **Rewrite the function, replacing $f(x)$ with y to simplify the notation.**
2. **Change each y to an x and each x to a y.**

3. Solve for *y*.

4. Rewrite the function, replacing the *y* with $f^{-1}(x)$.

EXAMPLE

Q. Find the inverse of the function $f(x)=\frac{x}{x-5}$.

A. Use the previous steps:

1. Rewrite the function, replacing *f* (*x*) with *y* to simplify the notation: $y=\frac{x}{x-5}$
2. Change each *y* to an *x* and each *x* to a *y*: $x=\frac{y}{y-5}$
3. Solve for *y*. Start by multiplying each side of the equation by the denominator and distributing:

   $$x=\frac{y}{y-5}\rightarrow x(y-5)=y\rightarrow xy-5x=y$$

   Now subtract *y* from each side and add 5*x* to each side to get the terms with *y* in them on one side. Factor out that *y*:

   $$xy-y=5x\rightarrow y(x-1)=5x$$

   Divide each side by $(x-1)$:

   $$y=\frac{5x}{x-1}$$

4. Rewrite the function, replacing the *y* with $f^{-1}(x)$:

   $$f^{-1}(x)=\frac{5x}{x-1}$$

Q. Find the inverse of the function $f(x)=x^5+4$.

A. Use the previous steps:

1. Rewrite the function, replacing *f* (*x*) with *y* to simplify the notation: $y=x^5+4$
2. Change each *y* to an *x* and each *x* to a *y*: $x=y^5+4$
3. Solve for *y*. Start by subtracting 4 from each side:

   $$x=y^5+4$$

   $$x-4=y^5$$

Now take the fifth root of each side:

$$\sqrt[5]{x-4}=\sqrt[5]{y^5}\rightarrow\sqrt[5]{x-4}=y$$

4. Rewrite the function, replacing the *y* with $f^{-1}(x)$:

   $$f^{-1}(x)=\sqrt[5]{x-4}$$

This process helps you find the inverse of a function, if it has one. If you can't solve for the inverse, the function may not be one-to-one to begin with. For instance, if you try to solve for the inverse of the function $f(x) = x^2 + 3$, you get stuck when you have to take a square root and don't know if you want a positive or negative root. These are the kinds of roadblocks that alert you to the fact that the function has no inverse — and that the function isn't one-to-one.

YOUR TURN

18 Find the inverse of the function $f(x) = 7x - 3$.

19 Find the inverse of the function $f(x) = \frac{2x+1}{x-4}$.

Practice Questions Answers and Explanations

1. **-1.** Replace the x with 2: $f(2)=|2\cdot 2-1|-4=|4-1|-4=3-4=-1$.

2. **-3.** Replace the x with 2: $g(2)=\frac{2^2-4\cdot 2+7}{2-3}=\frac{4-8+7}{-1}=\frac{3}{-1}=-3$.

3. **Domain:** $(-\infty,\infty)$, **range**: $[-2,\infty)$. The domain has no limitation, because all real numbers can be squared and result in another real number. The results from performing the function operations is that –2 is added to all positive numbers and 0. The smallest result will come from –2 + 0, and all other results will be larger.

4. **Domain:** $(-\infty,-3]\cup[3,\infty)$, **range:** $[0,\infty)$. To keep a positive number or 0 under the radical, the value of x has to be –3 and smaller, or 3 and larger. The result of finding the value of the radical is always 0 or a positive number.

5. **Even.** Replacing x with –x in the function rule, $f(-x)=\sqrt{(-x)^2+3}=\sqrt{x^2+3}$. The function is even.

6. **Neither.** Replacing x with –x in the function rule, $g(-x)=\sqrt[3]{-x+5}=-\sqrt[3]{x-5}\neq-\sqrt[3]{x+5}$. The function is neither even nor odd.

7. **Odd.** Replacing x with –x in the function rule, $h(-x)=-x-(-x)^5=-x+x^5=-(x-x^5)$. The function is odd.

8. **Not one-to-one.** The graph of $f(x)=3x^2+5$ is a quadratic or U-shaped curve. Each x value and its corresponding –x value give you the same y-value. And it doesn't pass the horizontal line test. This is not one-to-one.

9. **One-to-one.** The graph of $g(x)=\sqrt{x+2}$ constantly moves upward in a slow curve. There are no repeated y-values. It passes the horizontal line test. This is one-to-one.

10. **6, –2, –1.** When x = –3, use the first rule: $h(-3)=-2(-3)=6$. When x = 0, use the second rule: $h(0)=0^2-2=-2$. And when $x=5$, use the third rule: $h(5)=1-\sqrt{5-1}=1-2=-1$.

$$h(x)=\begin{cases}-2x, & x<-1\\ x^2-2, & -1\le x<4\\ 1-\sqrt{x-1}, & x\ge 4\end{cases}$$

11. **\$258, \$612, \$840.** For each of the employees, use the second rule. A: Bonus(\$43,000) = 0.006(\$43,000) = \$258; B: Bonus(\$102,000) = 0.006(\$102,000) = \$612; C: Bonus(\$140,000) = 0.006(\$140,000) = \$840. A: \$43,000; B: \$102,000; C: \$140,000.

12. $2x^2-17x+37$. Let function g be the input:

$$f\circ g=f(g)=2g^2-3g+2=2(5-x)^2-3(5-x)+2=2\left(25-10x+x^2\right)-15+3x+2$$
$$=50-20x+2x^2-13+3x=2x^2-17x+37$$

13. $x^2-3x+11$. Let function h be the input:

$$k\circ h=k(h)=h^2+11=\left(\sqrt{x^2-3x}\right)^2+11=x^2-3x+11$$

(14) **$6x + 3h - 1$.** Using the formula:

$$\frac{f(x+h)-f(x)}{h}=\frac{3(x+h)^2-(x+h)-1-(3x^2-x-1)}{h}=\frac{3(x^2+2xh+h^2)-x-h-1-3x^2+x+1}{h}$$

$$=\frac{3x^2+6xh+3h^2-x-h-1-3x^2+x+1}{h}=\frac{\cancel{3x^2}+6xh+3h^2-\cancel{x}-h-\cancel{1}-\cancel{3x^2}+\cancel{x}+\cancel{1}}{h}$$

$$=\frac{h(6x+3h-1)}{h}=6x+3h-1$$

(15) **2.** Using the formula:

$$\frac{f(x+h)-f(x)}{h}=\frac{2(x+h)-2x}{h}=\frac{2x+2h-2x}{h}=\frac{2h}{h}=2$$

(16) **They are inverses.** First, perform the composition $f(g)$:

$f(g)=2g+7=\cancel{2}\left(\frac{x-7}{\cancel{2}}\right)+7=x-7+7=x$. Now perform $g(f)$: $g(f)=\frac{f-7}{2}=\frac{2x+7-7}{2}=\frac{2x}{2}=x$.

With both functions resulting in x, they are inverses.

(17) **They are inverses.** First, perform the composition $f(g)$: $f(g)=\frac{2}{g^3}=\frac{2}{\left(\sqrt[3]{\frac{2}{x}}\right)^3}=\frac{2}{\frac{2}{x}}=\frac{\cancel{2}}{1}\cdot\frac{x}{\cancel{2}}=x$.

Now perform $g(f)$; $g(f)=\sqrt[3]{\frac{2}{f}}=\sqrt[3]{\frac{2}{\frac{2}{x^3}}}=\sqrt[3]{\frac{\cancel{2}}{1}\cdot\frac{x^3}{\cancel{2}}}=\sqrt[3]{x^3}=x$. With both functions resulting in x, they are inverses.

(18) **$f^{-1}(x)=\frac{x}{7}+\frac{3}{7}$.** Replacing $f(x)$ with y, you have $y=7x-3$. Switching the y and the x, $x=7y-3$. Solving for y, first add 3 to each side, and then divide each side by 7:

$x=7y-3\rightarrow x+3=7y\rightarrow\frac{x}{7}+\frac{3}{7}=\frac{7y}{7}\rightarrow\frac{x}{7}+\frac{3}{7}=y$. Now rewrite using the inverse function notation: $f^{-1}(x)=\frac{x}{7}+\frac{3}{7}$

(19) **$f^{-1}(x)=\frac{4x+1}{x-2}$.** Replacing $f(x)$ with y, you have $y=\frac{2x+1}{x-4}$. Changing the y to x and each x to y:

$x=\frac{2y+1}{y-4}$. To solve for y, first multiply each side of the equation by the denominator and distribute the terms: $x(y-4)=2y+1\rightarrow xy-4x=2y+1$. Subtract $2y$ from each side and add $4x$ to each side: $xy-4x=2y+1\rightarrow xy-2y=4x+1$. Factor out the y on the left, and then divide each side by the binomial factor: $y(x-2)=4x+1\rightarrow y=\frac{4x+1}{x-2}$. Now rewrite using the inverse function notation: $f^{-1}(x)=\frac{4x+1}{x-2}$.

Whaddya Know? Chapter 7 Quiz

Quiz time! Complete each problem to test your knowledge on the various topics covered in this chapter. You can then find the solutions and explanations in the next section.

1. **Given $f(x) = -3x^3 - 2x + 7$, evaluate $f(-1)$.**

2. **Determine whether the function $f(x) = x^3 + 4x$ is even, odd, one-to-one, a combination of these, or none of these.**

3. **Find the domain and range of the function $f(x) = x^2 - 7$.**

4. **Given the piecewise function $f(x) = \begin{cases} x^2 & \text{if } x < -2 \\ 2 - x & \text{if } -2 \le x < 0 \\ 3x & \text{if } x \ge 0 \end{cases}$, find $f(1)$, $f(-2)$, and $f(3)$.**

5. **Determine whether the function $f(x) = -2x^2 - 7$ is even, odd, one-to-one, a combination of these, or none of these.**

6. **Find the domain and range of the function $g(x) = \dfrac{x}{1-x}$.**

7. **Perform the difference quotient on the function $f(x) = \dfrac{2}{x+3}$.**

8. **Determine whether the function $f(x) = \sqrt{x+1}$ is even, odd, one-to-one, a combination of these, or none of these.**

9. **Find the domain and range of the function $g(x) = \sqrt{x+3}$.**

10. **Find the inverse of the function $f(x) = x^3 - 3$.**

11. **Given $g(y) = \dfrac{4y-2}{y+1}$, evaluate $g(-2)$.**

12. **Given the piecewise function $f(x) = \begin{cases} x+2 & \text{if } x \le 0 \\ x^2 & \text{if } x > 0 \end{cases}$, find $f(0)$ and $f(3)$.**

13. **Perform the function composition: $f \circ g$ when $f(x) = 3x^2 - x + 2$ and $g(x) = x - 3$.**

14. **Determine the difference quotient when $f(x) = 3x^2 - x + 2$.**

15. **Find the inverse of the function $g(x) = \dfrac{x+1}{4-x}$ and determine the domain and range of both the function and its inverse.**

Answers to Chapter 7 Quiz

1. **12.** $f(-1)=-3(-1)^3-2(-1)+7=3+2+7=12$

2. **Odd and one-to-one.** When replacing x with $-x$, $f(-x)=(-x)^3+4(-x)=-x^3-4x=-(x^3+4)$, which makes it odd. And when graphed, the curve moves constantly upward with no repeated y values.

3. **Domain: x all reals, range: $y \geq -7$, or written domain: $(-\infty,\infty)$, range: $[-7,\infty)$.** The domain of a quadratic function consists of all real numbers. This is a parabola that opens upward, so the range goes from the vertex upward. The vertex (lowest point) of the parabola is at $(0,-7)$.

4. **$f(1)=3$, $f(-2)=4$, $f(3)=9$.** For $f(1)$, use the third rule: $f(1)=3(1)=3$. For $f(-2)$, use the second rule: $f(-2)=2-(-2)=4$. For $f(3)$, use the third rule: $f(3)=3(3)=9$.

5. **Even.** When replacing x with $-x$, $f(-x)=-2(-x)^2-7=-2x^2-7$. The function rule does not change. The function is not one-to-one.

6. **Domain: $x \neq 1$, range: $y \neq -1$, or written domain: $(-\infty,1)\cup(1,\infty)$, range: $(-\infty,-1)\cup(-1,\infty)$.** You can't have 0 in the denominator, so x cannot be 1. As the value of x gets closer and closer to 1 from both the right and the left, the value of y gets closer and closer to -1, but it can never be that value.

7. $\dfrac{-2}{(x+h+3)(x+3)}$. Using $\dfrac{f(x+h)-f(x)}{h}$,

$$\frac{\frac{2}{x+h+3}-\frac{2}{x+3}}{h}=\frac{\frac{2}{x+h+3}\cdot\frac{x+3}{x+3}-\frac{2}{x+3}\cdot\frac{x+h+3}{x+h+3}}{h}=\frac{\frac{2(x+3)}{(x+h+3)(x+3)}-\frac{2(x+h+3)}{(x+h+3)(x+3)}}{h}$$

$$=\frac{\frac{2(x+3)-2(x+h+3)}{(x+h+3)(x+3)}}{h}=\frac{\frac{2x+6-2x-2h-6}{(x+h+3)(x+3)}}{h}=\frac{\frac{-2h}{(x+h+3)(x+3)}}{h}$$

$$=\frac{-2\cancel{h}}{(x+h+3)(x+3)}\cdot\frac{1}{\cancel{h}}=\frac{-2}{(x+h+3)(x+3)}$$

8. **None of these.** When replacing x with $-x$, $f(-x)=\sqrt{-x+1}$. This is a different function and cannot be written as $f(x)$ or $-f(x)$. It's not even, odd, or one-to-one.

9. **Domain: $x \geq -3$, range: $y \geq 0$, or written domain: $[-3,\infty)$, range: $[0,\infty)$.** The value under the radical has to be positive or 0, so the smallest possible value for x is -3. The results of finding the square root of numbers 0 and greater are also 0 and greater.

10. **$f^{-1}(x)=\sqrt[3]{x+3}$.** Replace the x term with y and $f(x)$ with x. Then solve for y:

$$f(x)=x^3-3 \rightarrow x=y^3-3 \rightarrow x+3=y^3 \rightarrow \sqrt[3]{x+3}=\sqrt[3]{y^3} \rightarrow \sqrt[3]{x+3}=y$$

11. **10.** $g(-2)=\dfrac{4(-2)-2}{(-2)+1}=\dfrac{-10}{-1}=10$

12. **$f(0)=2$, $f(3)=9$.** For $f(0)$, use the first rule: $f(0)=0+2=0$. For $f(3)$, use the second rule: $f(3)=3^2=9$.

(13) $\mathbf{3x^2 - 19x + 32}$. Replace each x in the function $f(x)$ with $g(x) = x - 3$ and simplify:

$$(f \circ g)(x) = 3(g)^2 - (g) + 2 = 3(x-3)^2 - (x-3) + 2 = 3(x^2 - 6x + 9) - (x-3) + 2$$
$$= 3x^2 - 18x + 27 - x + 3 + 2 = 3x^2 - 19x + 32$$

(14) $\mathbf{6x + 3h - 1}$. Use the function to evaluate the expression: $\frac{f(x+h) - f(x)}{h}$

$$\frac{f(x+h) - f(x)}{h} = \frac{3(x+h)^2 - (x+h) + 2 - (3x^2 - x + 2)}{h}$$
$$= \frac{3(x^2 + 2xh + h^2) - (x+h) + 2 - (3x^2 - x + 2)}{h}$$
$$= \frac{3x^2 + 6xh + 3h^2 - x - h + 2 - 3x^2 + x - 2}{h}$$
$$= \frac{\cancel{3x^2} + 6xh + 3h^2 - \cancel{x} - h + \cancel{2} - \cancel{3x^2} + \cancel{x} - \cancel{2}}{h}$$
$$= \frac{6xh + 3h^2 - h}{h} = 6x + 3h - 1$$

(15) $g^{-1}(x) = \frac{4x-1}{x+1}$; **domain and range of g(x): domain** $x \neq 4$, **range** $y \neq -1$; **domain and range of** $g^{-1}(x)$: **domain** $x \neq -1$, **range** $y \neq 4$.

Write the function rule as $y = \frac{x+1}{4-x}$. Then replace the x terms with y's and y with x. Then solve for y:

$$x = \frac{y+1}{4-y} \rightarrow x(4-y) = y + 1 \rightarrow 4x - xy = y + 1 \rightarrow 4x - 1 = y + xy \rightarrow$$
$$4x - 1 = y(1+x) \rightarrow \frac{4x-1}{1+x} = y$$

IN THIS CHAPTER

» Mastering the standard form of quadratics

» Locating the *x*- and *y*-intercepts

» Reaching the extremes of quadratics

» Putting the axis of symmetry into play

» Piecing together all kinds of quadratic puzzles

» Watching quadratics at work in the real world

Chapter 8 Specializing in Quadratic Functions

A quadratic function is one of the more recognizable and useful polynomial (multi-termed) functions found in all of algebra. The function describes a graceful U-shaped curve called a parabola that you can quickly sketch and easily interpret. People use quadratic functions to model economic situations, physical training progress, and the paths of comets. How much more useful can math get?

The most important features to recognize in order to sketch a parabola are the opening (up or down, steep or wide), the intercepts, the vertex, and the axis of symmetry. In this chapter, I show you how to identify all these features within the standard form of the quadratic function. I also show you some equations of parabolas that model events.

Setting the Standard to Create a Parabola

A parabola is the graph of a quadratic function. The graph is a nice, gentle, U-shaped curve that has points located an equal distance on either side of a line running up through its middle — called its *axis of symmetry*. Parabolas can be turned upward, downward, left, or right, but parabolas that represent functions only turn up or down. (In Chapter 13, you find out more about the other types of parabolas in the general discussion of conics.) The standard form for the quadratic function is $f(x) = ax^2 + bx + c$. This form follows the pattern of polynomial equations. You'll see a different version when discussing parabolas as part of the family of conic sections.

The coefficients (multipliers of the variables) a, b, and c are real numbers; the coefficient a can't be equal to zero because you'd no longer have a quadratic function. You have plenty to discover from the simple standard form equation. The coefficients a and b are important, and some equations may be missing the second or third term — or both. As you can see, there's meaning in everything (or nothing)!

Starting with "a" in the standard form

As the lead coefficient of the standard form of the quadratic function $f(x) = ax^2 + bx + c$, a gives you two bits of information: the direction in which the graphed parabola opens and whether the parabola is steep or flat. Here's the breakdown of how the sign and size of the lead coefficient, a, affect the parabola's appearance:

- If a is positive, the graph of the parabola opens upward (see Figure 8-1a and 8-1b).
- If a is negative, the graph of the parabola opens downward (see Figure 8-1c and 8-1d).
- If a has an absolute value greater than 1, the graph of the parabola is steep (see Figure 8-1a and 8-1c). (See Chapter 2 for a refresher on absolute values.)
- If a has an absolute value less than 1, the graph of the parabola flattens out (see Figure 8-1b and 8-1d).

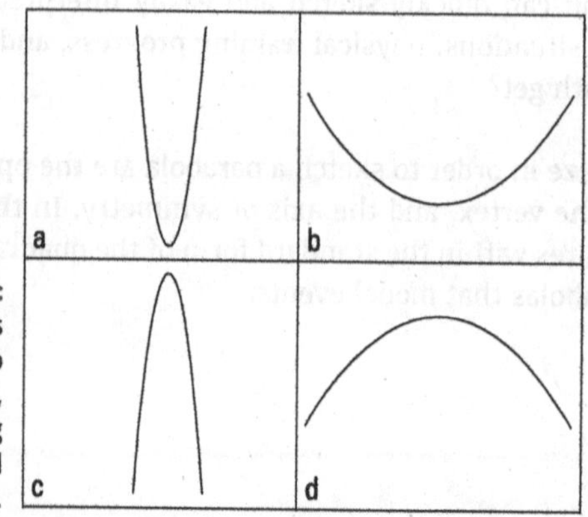

FIGURE 8-1: Parabolas opening up and down, appearing steep and flat.

If you remember the four rules that identify the lead coefficient of a parabola, you don't even have to graph the equation to give a general description as to how the parabola looks.

EXAMPLE

Q. Describe the steepness and direction of the parabola $y = 4x^2 - 3x + 2$.

A. You say that this parabola is steep and opens upward because the lead coefficient is positive and greater than 1.

Q. Describe the steepness and direction of the parabola $y = -\frac{1}{3}x^2 + x - 11$.

A. You say that this parabola is flattened out and opens downward because the lead coefficient is negative and the absolute value of the fraction is less than 1.

Q. Describe the steepness and direction of the parabola $y = -9x^2$.

A. You say that this parabola is very steep and opens downward because the lead coefficient is negative and the absolute value is greater than 1.

Q. Describe the steepness and direction of the parabola $y = 0.002x^2 + 3$.

A. You say that this parabola is flattened out and opens upward because the lead coefficient is positive and the decimal value is less than 1. In fact, the coefficient is so small that the flattened parabola almost looks like a horizontal line.

Following up with "b" and "c"

Much like the lead coefficient in the quadratic function, the values of *b* and *c* give you plenty of information. Mainly, these values tell you a lot if they're *not* there. In the next section, you find out how to use these values to find intercepts (or zeros). For now, you concentrate on their presence or absence.

The lead coefficient, *a*, can never be equal to zero. If that happens, you no longer have a quadratic function, and this discussion is finished. As for the other two terms:

- If the second coefficient, *b*, is zero, the parabola straddles the *y*-axis. The parabola's vertex — the highest or lowest point on the curve, depending on which way it faces — is on that axis, and the parabola is symmetric about the axis (see Figure 8-2a, a graph of a quadratic function where $b = 0$). The second term in the standard form is the *x* term, so if the coefficient *b* is zero, the second term disappears. The standard equation becomes $y = ax^2 + c$, which makes finding intercepts very easy (see the following section).
- If the last term, *c*, is zero, the graph of the parabola goes through the origin — in other words, one of its intercepts is the origin (see Figure 8-2b, a graph of a quadratic function where $c = 0$). The standard equation becomes $y = ax^2 + bx$, which you can easily factor into $y = x(ax + b)$. (See Chapters 1 and 3 for more on factoring.)

FIGURE 8-2: Graphs of $y = x^2 - 3$ ($b = 0$) and $y = x^2 - 3x$ ($c = 0$).

EXAMPLE

Q. **Describe how the parabola $y = x^2 - 4$ is situated in terms of the axes.**

A. **Because the middle term is missing, the parabola straddles the y-axis.**

Q. **Describe how the parabola $y = x^2 - 4x$ is situated in terms of the axes or origin.**

A. **Because the constant term is missing, the parabola passes through the origin. This is just one of its intercepts.**

YOUR TURN

Given the quadratic, describe it in terms of opening, steepness, axis, and origin.

1. $y = -2x^2 + 1$

2. $y = x^2$

3. $y = \frac{1}{4}x^2 + x$

4 $y = -0.2x^2 + 3x - 6$

Recognizing the Intercepts and Vertex

The intercepts of a quadratic function (or any function) are the points where the graph of the function crosses the x-axis or y-axis. The graph of a function can cross the x-axis any number of times, but it can cross the y-axis only once.

Why be concerned about the intercepts of a parabola? In real-life situations, the intercepts occur at points of interest — for instance, at the initial value of an investment or at the break-even point for a business.

REMEMBER

Intercepts are also very helpful when you're graphing a parabola. The points are easy to find because one of the coordinates is always zero. If you have the intercepts, the vertex, and what you know about the symmetry of the parabola, you have a good idea of what the graph looks like.

Finding the one and only y-intercept

The y-intercept of the quadratic function $f(x) = ax^2 + bx + c$ is $(0,c)$. A parabola with this standard equation is a function, so by definition, only one y value can exist for every x value. When $x = 0$, as it does at the y-intercept, the equation becomes $y = a(0)^2 + b(0) + c = 0 + 0 + c = c$, or $y = c$. The statements $x = 0$ and $y = c$ combine to become the y-intercept, $(0,c)$.

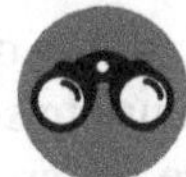
EXAMPLE

Q. Find the y-intercept of $y = 4x^2 - 3x + 2$.

A. Let $x = 0$. When $x = 0$, $y = 2$ (or $c = 2$). The y-intercept is $(0,2)$.

Q. Find the y-intercept of $y = \frac{1}{2}x^2 + \frac{3}{2}x$.

A. When $x = 0$, $y = 0$. The equation provides no constant term; you could also say the missing constant term is zero. The y-intercept is $(0,0)$.

People can model many situations with quadratic functions, and the places where the input variables or output variables equal zero are important. For example, a candle-making company has figured out that its profit is based on the number of candles it produces and sells. The company uses the function $P(x) = -0.05x^2 + 8x - 140$ — where x represents the number of candles — to determine P, the profit. As you can see from the equation, the graph of this parabola opens downward (because a is negative; see the section, "Starting with 'a' in the standard form"). Figure 8-3 is a sketch of the graph of the profit, with the y-axis representing profit and the x-axis representing the number of candles.

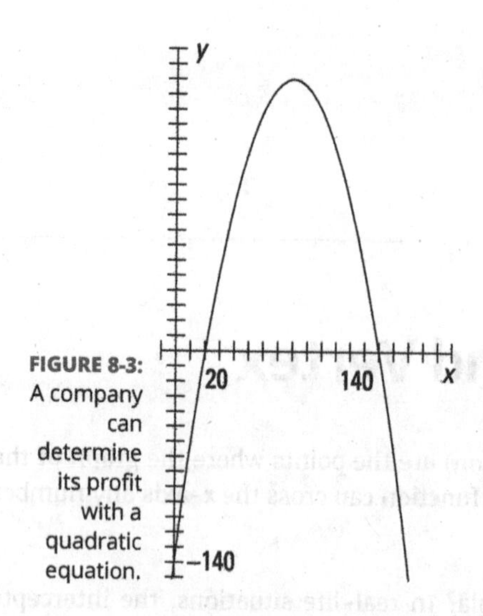

FIGURE 8-3: A company can determine its profit with a quadratic equation.

Does it make sense to use a quadratic function to model profit? Why would the profit decrease after a certain point? Does that make business sense? It does if you consider that, perhaps, when you make too many candles, the cost of overtime and the need for additional machinery play a part.

What about the y-intercept? What part does it play, and what does it mean in this candle-making case? You can say that $x = 0$ represents not producing or selling any candles. According to the equation and graph, the y-intercept has a y-coordinate of -140. It makes sense to find a negative profit if the company has costs that it has to pay no matter what (even if it sells no candles): insurance, salaries, mortgage payments, and so on. With some interpretation, you can find a logical explanation for the y-intercept being negative in this case.

Finding the x-intercepts

You find the x-intercepts of quadratics when you solve for the zeros, or solutions, of the quadratic equation. The method you use to solve for the zeros is the same method you use to solve for the intercepts, because they're really just the same thing. The names change (intercept, zero, solution) depending on the application, but you find the intercepts the same way.

Parabolas with an equation of the standard form $y = ax^2 + bx + c$ open upward or downward and may or may not have x-intercepts. Look at Figure 8-4, for example. You see a parabola with two x-intercepts (Figure 8-4a), one with a single x-intercept (Figure 8-4b), and one with no x-intercept (Figure 8-4c). Notice, however, that they all have a y-intercept.

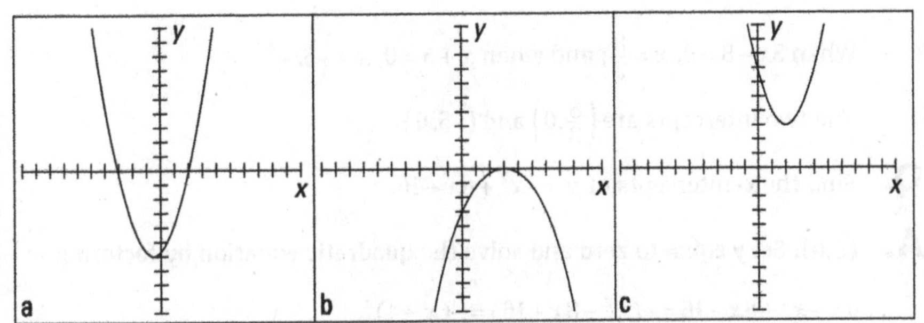

FIGURE 8-4: Parabolas can intercept the x-axis two times, a single time, or not at all.

The coordinates of all x-intercepts have zeros in them. An x-intercept's y value is zero, and you write it in the form (h,0). How do you find the value of h? You let $y = 0$ in the general equation and then solve for x. You have two options when solving the equation $0 = ax^2 + bx + c$:

- Use the quadratic formula (see Chapter 3 for a refresher on the formula).
- Try to factor the expression and use the multiplication property of zero (you can find more on this property in Chapter 1).

Regardless of the path you take, you have some guidelines at your disposal to help you determine the number of x-intercepts you should find.

REMEMBER

When finding x-intercepts by solving $0 = ax^2 + bx + c$

- You find two x-intercepts if
  - The expression factors into two different binomials.
  - The quadratic formula gives you a value greater than zero under the radical.
- You find one x-intercept (a double root) if
  - The expression factors into the square of a binomial.
  - The quadratic formula gives you a value of zero under the radical.
- You find no x-intercept if both
  - The expression doesn't factor; and
  - The quadratic formula gives you a value less than zero under the radical (indicating an imaginary root; Chapter 17 deals with imaginary and complex numbers).

EXAMPLE

Q. Find the x-intercepts of $y = 3x^2 + 7x - 40$.

A. $\left(\frac{8}{3}, 0\right)$ and $(-5, 0)$. Set y equal to zero and solve the quadratic equation by factoring:

$$0 = 3x^2 + 7x - 40 = (3x - 8)(x + 5)$$

When $3x-8=0$, $x=\frac{8}{3}$; and when $x+5=0$, $x=-5$.

The two intercepts are $\left(\frac{8}{3},0\right)$ and $(-5,0)$.

Q. Find the x-intercepts of $y=-x^2+8x-16$.

A. **(4,0).** Set y equal to zero and solve the quadratic equation by factoring:

$$0=-x^2+8x-16=-(x^2-8x+16)=-(x-4)^2$$

So when $x-4=0$, $x=4$.

The only intercept is (4,0).

Q. Find the x-intercepts of $y=x^2+2x-1$.

A. $\left(-1+\sqrt{2},0\right)$ **and** $\left(-1-\sqrt{2},0\right)$. Set y equal to zero and solve the quadratic equation. This doesn't factor, so you need the quadratic formula:

$$x=\frac{-2\pm\sqrt{2^2-4(1)(-1)}}{2(1)}=\frac{-2\pm\sqrt{4+4}}{2}=\frac{-2\pm\sqrt{8}}{2}=\frac{-2\pm2\sqrt{2}}{2}=-1\pm\sqrt{2}$$

The two x intercepts are $\left(-1+\sqrt{2},0\right)$ and $\left(-1-\sqrt{2},0\right)$

Q. Find the x-intercepts of $y=-2x^2+4x-7$.

A. **No x-intercepts.** Set y equal to zero and try to factor the quadratic equation. You can't do it; the equation has no factors that give you this quadratic. When you try the quadratic formula, you see that the value under the radical is less than zero; a negative number under the radical is an imaginary number:

$$x=\frac{-4\pm\sqrt{4^2-4(-2)(-7)}}{2(-2)}=\frac{-4\pm\sqrt{16-(56)}}{-4}=\frac{-4\pm\sqrt{-40}}{-4}$$

Alas, you find no x-intercept for this parabola.

YOUR TURN

Find the x- and y-intercepts of the parabolas.

5 $y=x^2-25$

6. $y = 2x^2 - 5x - 12$

7. $y = 3x^2 + 2x - 2$

8. $y = x^2 + 2x + 5$

Going to the extreme: Finding the vertex

Quadratic functions, or parabolas, that have the standard form $y = ax^2 + bx + c$ are gentle, U-shaped curves that open either upward or downward. When the lead coefficient, a, is a positive number, the parabola opens upward, creating a minimum value for the function — the function values never go lower than that minimum. When a is negative, the parabola opens downward, creating a maximum value for the function — the function values never go higher than that maximum.

The two extreme values, the minimum and maximum, occur at the parabola's vertex. The y-coordinate of the vertex gives you the numerical value of the extreme — its highest or lowest point. The vertex of a parabola is very useful for finding the extreme value, and algebra certainly provides an efficient way of finding it. Right? Well, sure it does! The vertex serves as a sort of anchor for the two parts of the curve to flare out from. The axis of symmetry (see the following section) runs through the vertex. The y-coordinate of the vertex is the function's maximum or minimum value — again, depending on which way the parabola opens.

The parabola $y = ax^2 + bx + c$ has its vertex where $x = \frac{-b}{2a}$. You plug in the a and b values from the equation to come up with x, and then you find the y-coordinate of the vertex by plugging this x value into the equation and solving for y.

EXAMPLE

Q. Find the vertex of $y = -3x^2 + 12x - 7$.

A. **(2,5).** Substitute the coefficients a and b into the equation $x = \frac{-b}{2a}$:

$$x = \frac{-12}{2(-3)} = \frac{-12}{-6} = 2$$

You solve for y by putting the x value back into the equation:

$$y = -3(2)^2 + 12(2) - 7 = -12 + 24 - 7 = 5$$

The coordinates of the vertex are (2,5). You find a maximum value, because a is a negative number, which means the parabola opens downward from this point. The graph of the parabola never goes higher than five units above the x-axis.

Q. Find the vertex of $y = 4x^2 - 19$.

A. **(0,–19).** Substitute the coefficients a and b into the equation $x = \frac{-b}{2a}$:

$$x = \frac{-0}{2(4)} = 0$$

You solve for y by putting the x value into the equation:

$$y = 4(0)^2 - 19 = -19$$

The coordinates of the vertex are $(0,-19)$. You have a minimum value, because a is a positive number, meaning the parabola opens upward from the minimum point.

YOUR TURN

9 Find the vertex of $y = x^2 - 1$.

10 Find the vertex of $y = 3x^2 - 9x - 5$.

Making Symmetry Work with an Axis

The axis of symmetry of a quadratic function is a vertical line that runs through the vertex of the parabola (see the previous section) and acts as a mirror — half the parabola rests on one side of the axis, and half rests on the other. The *x*-value in the coordinates of the vertex appears in the equation for the axis of symmetry. For instance, if a vertex has the coordinates (2,3), the axis of symmetry is $x = 2$. All vertical lines have an equation of the form $x = h$. In the case of the axis of symmetry, the *h* is always the *x*-coordinate of the vertex.

TIP

The axis of symmetry is useful because when you're sketching a parabola and finding the coordinates of a point that lies on it, you know you can find another point that exists as a partner to your point, in that

- It lies on the same horizontal line.
- It lies on the other side of the axis of symmetry.
- It covers the same distance from the axis of symmetry as your point.

Maybe I should just show you what I mean with a sketch! Figure 8-5 shows points on a parabola that lie on the same horizontal line and on either side of the axis of symmetry.

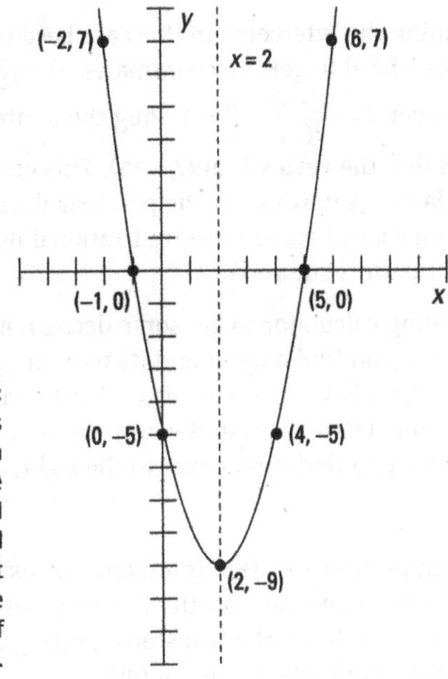

FIGURE 8-5: Points resting on the same horizontal line and equidistant from the axis of symmetry.

The points (4,−5) and (0,−5) are each two units from the axis — the line $x = 2$. The points (−2,7) and (6,7) are each four units from the axis. And the points (−1,0) and (5,0), the *x*-intercepts, are each three units from the axis.

Graphing Parabolas

You have all sorts of information available when it comes to a parabola and its graph. You can use the intercepts, the opening, the steepness, the vertex, the axis of symmetry, or just some random points to plot the parabola; you don't really need all the pieces. As you practice sketching these curves, it becomes easier to figure out which pieces you need for different situations. Sometimes the x-intercepts are hard to find, so you concentrate on the vertex, direction, and axis of symmetry. Other times you find it more convenient to use the y-intercept, a point or two on the parabola, and the axis of symmetry. This section provides you with a couple of examples. Of course, you can go ahead and check off all the information that's possible. Some people are very thorough that way.

EXAMPLE

Q. Sketch the graph of $y = x^2 - 8x + 1$.

A. Use the most convenient information to create this sketch.

- First, notice that the equation represents a parabola that opens upward, because the lead coefficient, a, is positive (+1). The y-intercept is (0,1), which you get by plugging in zero for x.
- If you set y equal to zero to solve for the x-intercepts, you get $0 = x^2 - 8x + 1$, which doesn't factor. You could whip out the quadratic formula — but wait. You have other possibilities to consider.
- The vertex is more helpful than finding the intercepts in this case because of its convenience — you don't have to work so hard to get the coordinates. Use the formula for the x-coordinate of the vertex to get $x = \frac{-(-8)}{2(1)} = \frac{8}{2} = 4$. Plug the 4 into the formula for the parabola, and you find that the vertex is at $(4, -15)$. This coordinate is below the x-axis, and the parabola opens upward, so the parabola does have x-intercepts; you just can't find them easily because they're irrational numbers (square roots of numbers that aren't perfect squares).
- You can try whipping out your graphing calculator to get some decimal approximations of the intercepts. Or, instead, you can find a point and its partner point on the other side of the axis of symmetry, which is $x = 4$. If you let $x = 1$, for example, you find that $y = -6$. This point is three units from $x = 4$, to the left; you find the distance by subtracting $4 - 1 = 3$. Use this distance to find three units to the right, $4 + 3 = 7$. The corresponding point is $(7, -6)$.

If you sketch all that information in a graph first — the y-intercept, vertex, axis of symmetry, and the points $(1, -6)$ and $(7, -6)$ — you can identify the shape of the parabola and sketch in the whole thing. Figure 8-6 shows the two steps: putting in the information (Figure 8-6a), and sketching in the parabola (Figure 8-6b).

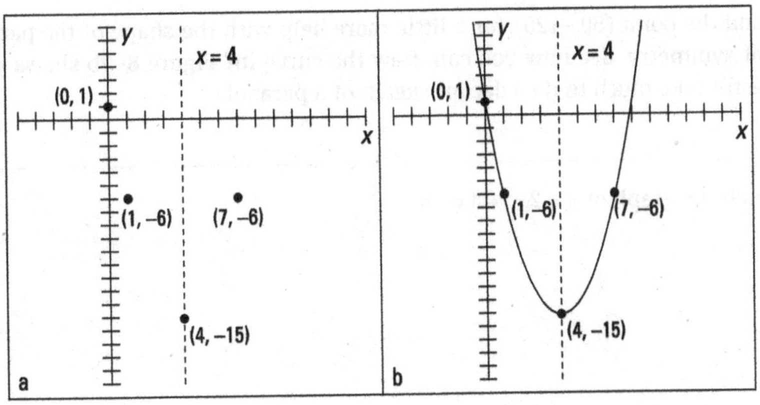

FIGURE 8-6: Using the various pieces of a quadratic as steps for sketching a parabola.

Q. Sketch the graph of $y = -0.01x^2 - 2x$.

A. Look for the hints.

- The parabola opens downward (because a is negative) and is pretty flattened out (because the absolute value of a is less than zero).
- The graph goes through the origin because the constant term (c) is missing. Therefore, the y-intercept and one of the x-intercepts is (0,0).
- The vertex lies at $(-100, 100)$. You get this by solving for the x-coordinate — $x = \frac{-(-2)}{2(-0.01)} = \frac{2}{-0.02} = -100$ — and then substituting back into the equation for the y-coordinate.
- To solve for the other x-intercept, let $y = 0$ and factor:

  $$0 = -0.01x(x + 200)$$

  The second factor tells you that the other x-intercept occurs when $x = -200$.

The intercepts and vertex are sketched in Figure 8-7a.

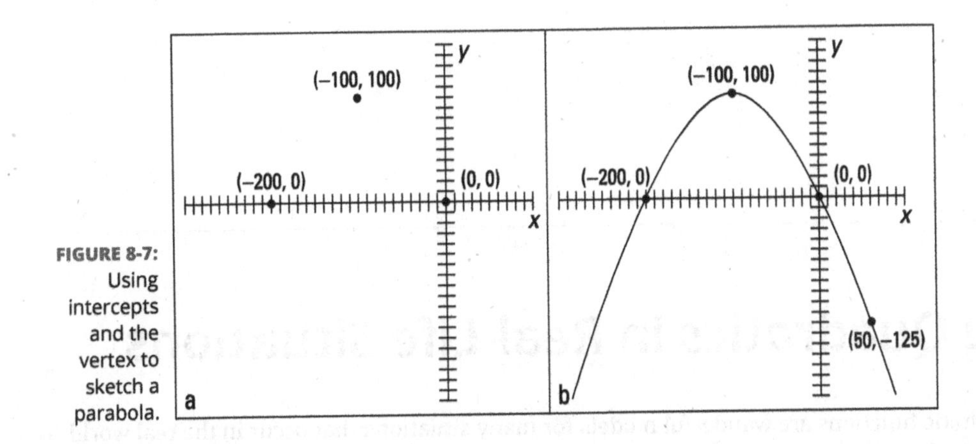

FIGURE 8-7: Using intercepts and the vertex to sketch a parabola.

You can add the point (50,–125) for a little more help with the shape of the parabola by using the axis of symmetry. See how you can draw the curve in? Figure 8-7b shows you the way. It really doesn't take much to do a decent sketch of a parabola.

YOUR TURN

11 Sketch the graph of $y = 2x^2 - 4x - 6$.

12 Sketch the graph of $y = -\frac{1}{2}x^2 - x + 12$.

13 Sketch the graph of $y = 4x^2 - 25$.

14 Sketch the graph of $y = -3x^2 + 12x$.

Applying Quadratics in Real-Life Situations

Quadratic functions are wonderful models for many situations that occur in the real world. You can see them at work in financial and physical applications, just to name a couple. This section provides a few applications for you to consider.

Selling candles

A candle-making company has figured out that its profit is based on the number of candles it produces and sells. The function $P(x) = -0.05x^2 + 8x - 140$ applies to the company's situation, where x represents the number of candles, and P represents the profit. You may recognize this function from the section, "Finding the one and only y-intercept," earlier in the chapter. You can use the function to find out how many candles the company has to produce to garner the greatest possible profit.

You find the two x-intercepts by letting $y = 0$ and solving for x by factoring:

$$0 = -0.05x^2 + 8x - 140 = -0.05(x^2 - 160x + 2800) = -0.05(x - 20)(x - 140)$$

When $x - 20 = 0$, you have $x = 20$; and when $x - 140 = 0$, $x = 140$. These two numbers give you the two x-intercepts: (20,0) and (140,0).

The intercept (20,0) represents where the function (the profit) changes from negative values to positive values. You know this because the graph of the profit function is a parabola that opens downward (because a is negative), so the beginning and ending of the curve appear below the x-axis. The intercept (140,0) represents where the profit changes from positive values to negative values. So, the maximum value, the vertex, lies somewhere between and above the two intercepts. The x-coordinate of the vertex lies between 20 and 140. Refer to Figure 8-3 if you want to see the graph again.

You now use the formula for the x-coordinate of the vertex to find $x = \frac{-8}{2(-.05)} = \frac{-8}{-.1} = 80$. The number 80 lies between 20 and 140; in fact, it rests halfway between them. The nice, even number is due to the symmetry of the graph of the parabola and the symmetric nature of these functions. Now you can find the P value (the y-coordinate of the vertex): $P(80) = -0.05(80)^2 + 8(80) - 140 = 180$.

Your findings say that if the company produces and sells 80 candles, the maximum profit will be \$180. That seems like an awful lot of work for \$180, but maybe the company runs a small business. Work such as this shows you how important it is to have models for profit, revenue, and cost in business so you can make projections and adjust your plans.

Shooting basketballs

A local youth group recently raised money for charity by having a Throw-A-Thon. Participants prompted sponsors to donate money based on a promise to shoot baskets over a 12-hour period. This was a very successful project, both for charity and for algebra, because you can find some interesting bits of information about shooting the basketballs and the number of misses that occurred.

Participants shot baskets for 12 hours, attempting about 200 baskets each hour. The quadratic equation $M(t) = \frac{17}{6}t^2 - \frac{77}{3}t + 100$ models the number of baskets they missed each hour, where t is the time in hours (numbered from 0 through 12) and M is the number of misses.

This quadratic function opens upward (because a is positive), so the function has a minimum value. Figure 8-8 shows a graph of the function.

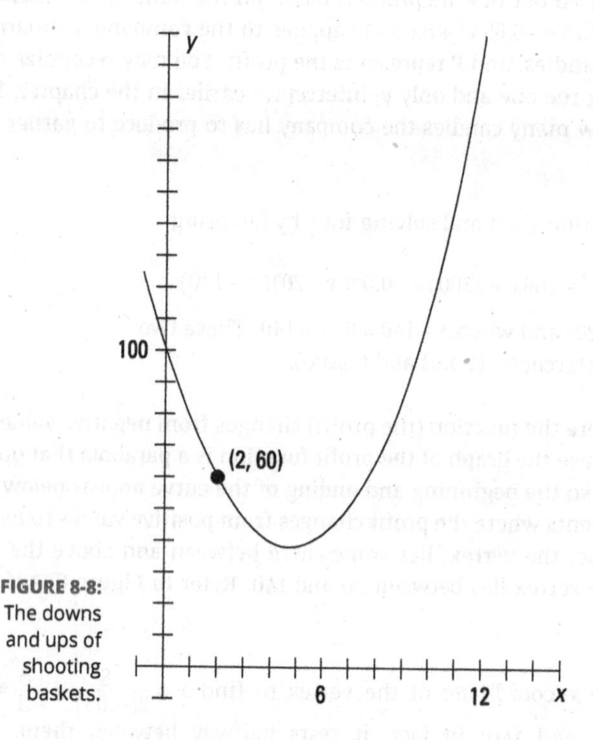

FIGURE 8-8: The downs and ups of shooting baskets.

From the graph, you see that the initial value, the y-intercept, is 100. At the beginning, participants were missing about 100 baskets per hour. The good news is that they got better with practice. $M(2)=60$ shows 60 misses when they're 2 hours into the project; the participants were missing only 60 baskets per hour. The number of misses goes down and then goes back up again. How do you interpret this? Even though the participants got better with practice, they let the fatigue factor take over.

What's the fewest number of misses per hour? When did the participants shoot their best? To answer these questions, find the vertex of the parabola by using the formula for the x-coordinate:

$$h=\frac{\left(\frac{77}{3}\right)}{2\left(\frac{17}{6}\right)}=\frac{77}{3}\cdot\frac{3}{17}=\frac{77}{17}=4\frac{9}{17}$$

The best shooting happened about 4.5 hours into the project. How many misses occurred then? The number you get represents what's happening the entire hour — although that's fudging a bit:

$$M\left(\frac{77}{17}\right)=\frac{17}{6}\left(\frac{77}{17}\right)^2-\frac{77}{3}\left(\frac{77}{17}\right)+100=\frac{4{,}271}{102}\approx 41.87$$

The fraction is rounded to two decimal places. The best shooting is about 42 misses that hour.

Launching a water balloon

One of the favorite springtime activities of the engineering students at a certain university is to launch water balloons from the top of the engineering building so that they hit the statue of the school's founder, which stands 25 feet from the building. The launcher sends the balloons up in an arc to clear a tree that sits next to the building. To hit the statue, the initial velocity and angle of the balloon have to be just right. Figure 8-9 shows a successful launch.

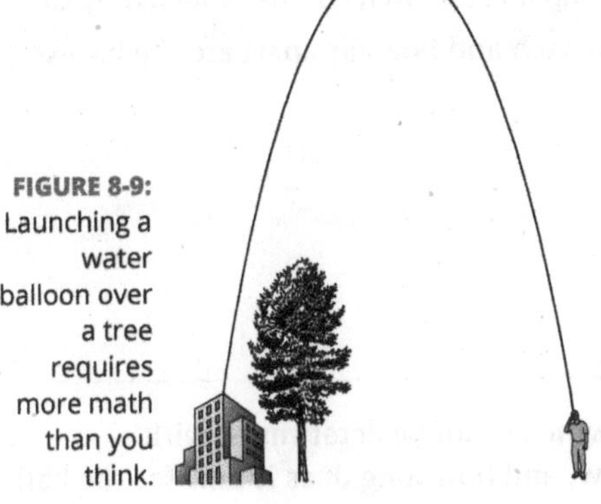

FIGURE 8-9: Launching a water balloon over a tree requires more math than you think.

This year's launch was successful. The students found that by launching the water balloons at 48 feet per second with a precise angle, they could hit the statue. Here's the equation they worked out to represent the path of the balloons: $H(t) = -2t^2 + 48t + 60$. The t represents the number of seconds, and H is the height of the balloon in feet. From this quadratic function, you can answer the following questions:

1. **How high is the building?**

   Solving the first question is probably easy for you. The launch occurs at time $t = 0$, the initial value of the function. When $t = 0$, $H(0) = -2(0)^2 + 48(0) + 60$. The building is 60 feet high.

2. **How high did the balloon travel?**

   You answer the second question by finding the vertex of the parabola: $t = \frac{-48}{2(-2)} = \frac{-48}{-4} = 12$. This gives you t, the number of seconds it takes the balloon to get to its highest point — 12 seconds after launch. Substitute the answer into the equation to get the height: $H(12) = -2(12)^2 + 48(12) + 60$. The balloon went 348 feet into the air.

3. **If the statue is 10 feet tall, how many seconds did it take for the balloon to reach the statue after the launch?**

   To solve the third question, use the fact that the statue is 10 feet high; you want to know when $H = 10$. Replace H with 10 in the equation and solve for t by factoring (see Chapters 1 and 3):

   $$10 = -2t^2 + 48t + 60$$
   $$0 = -2t^2 + 48t + 50 = -2(t^2 - 24t - 25) = -2(t - 25)(t + 1)$$

When $t-25=0$, $t=25$. And when $t+1=0$, $t=-1$.

According to the equation, the amount of time is either 25 seconds or −1 second. The −1 doesn't really make any sense because you can't go back in time; if the balloon had started at ground level, it would have taken that 1 second to reach that initial 60 feet in the air. The 25 seconds, however, tells you how long it took the balloon to reach the statue. Imagine the anticipation!

YOUR TURN

15 The St. Louis Gateway Arch has a parabolic shape. If the arch can be modeled by the equation $y=630-\frac{2}{315}x^2$, then how tall is the Arch and how far apart are the bases?

16 The height of a ball tossed in the air from a window can be determined with $h(t)=-16t^2+32t+48$. How high is the window, and how long does it take for the ball to hit the ground?

17 The profit function telling you how much you can expect to have when you sell x units is $P(x)=-21.875x^2+3500x-40{,}000$, where P is in dollars. How many units have to be sold to reach a maximum profit, and what is that profit? And at what point does the profit become negative again because of the cost of producing so many units?

Practice Questions Answers and Explanations

1. **Down, steep, y-axis straddle.** The negative coefficient on the first term causes the parabola to open downward. The absolute value of that coefficient is greater than 1, so it will be relatively steep. The missing middle term creates a parabola that straddles the y-axis.

2. **Up, y-axis straddle, origin.** The positive coefficient on the first term creates a parabola opening upward; there is no extra steepness or flatness, because the value is 1. The missing middle term causes a straddling of the y-axis, and the missing last term creates an intercept at the origin.

3. **Up, flat, origin.** The positive coefficient on the first term creates a parabola opening upward; the shape will be flattened because the value of the coefficient is less than 1. The missing last term creates an intercept at the origin.

4. **Down, flat.** The negative coefficient on the first term causes the parabola to open downward. The absolute value of that coefficient is less than 1, so it will be relatively flat. It won't straddle the y-axis or go through the origin.

5. $\mathbf{(0,-25), (5,0), (-5,0)}$. When $x = 0$, $y = -25$. So the y-intercept is $(0,-25)$. Let $y = 0$, and $0=(x-5)(x+5)$, giving you the two solutions 5 and -5 and the two x-intercepts $(5,0)$ and $(-5,0)$.

6. $\mathbf{(0,-12), \left(-\frac{3}{2},0\right), (4,0)}$. When $x = 0$, $y = -12$. So the y-intercept is $(0,-12)$. Let $y = 0$, and $0=(2x+3)(x-4)$, giving you the two solutions $-\frac{3}{2}$ and 4, and the two x-intercepts $\left(-\frac{3}{2},0\right)$ and $(4,0)$.

7. $\mathbf{(0,-2), \left(\frac{-1+1\sqrt{7}}{3},0\right), \left(\frac{-1-1\sqrt{7}}{3},0\right)}$. When $x = 0$, $y = -2$. So the y-intercept is $(0,-2)$.

   Let $y=0$, and $0=3x^2+2x-2$ doesn't factor. Using the quadratic formula,
   $x=\frac{-2\pm\sqrt{2^2-4(3)(-2)}}{2(3)}=\frac{-2\pm\sqrt{28}}{6}=\frac{-2\pm2\sqrt{7}}{6}=\frac{-1\pm1\sqrt{7}}{3}$ and the two x-intercepts are
   $\left(\frac{-1+1\sqrt{7}}{3},0\right)$ and $\left(\frac{-1-1\sqrt{7}}{3},0\right)$.

8. $\mathbf{(0,5)}$. When $x = 0$, $y = 5$. So the y-intercept is $(0,5)$. Let $y = 0$, and $0=x^2+2x+5$, which does not factor. Using the quadratic formula, $x=\frac{-2\pm\sqrt{2^2-4(1)(5)}}{2(1)}=\frac{-2\pm\sqrt{-16}}{2}$. There is no real solution, so there are no x-intercepts.

9. $\mathbf{(0,-1)}$. Using $x=\frac{-b}{2a}$, $x=\frac{-0}{2(1)}=0$ and $y=0^2-1=-1$. So the vertex is $(0,-1)$.

10. $\mathbf{\left(\frac{3}{2},-\frac{47}{4}\right)}$. Using $x=\frac{-b}{2a}$, $x=\frac{-(-9)}{2(3)}=\frac{9}{6}=\frac{3}{2}$ and $y=3\left(\frac{3}{2}\right)^2-9\left(\frac{3}{2}\right)-5=\frac{27}{4}-\frac{27}{2}-5$
    $=\frac{27-54-20}{4}=-\frac{47}{4}$. So the vertex is $\left(\frac{3}{2},-\frac{47}{4}\right)$.

11 **See the following sketch.** The parabola opens upward and is steep. The y-intercept is $(0,-6)$. Solving for the x-intercepts, $0=2\left(x^2-2x-3\right)=(x-3)(x+1)$, giving you $(3,0)$ and $(-1,0)$. The x-coordinate of the vertex is $x=\frac{-(-4)}{2(2)}=\frac{4}{4}=1$, and the y-coordinate of the vertex is $y=2(1)^2-4(1)-6=-8$. The axis of symmetry is $x=1$. Choose what you need to complete the graph.

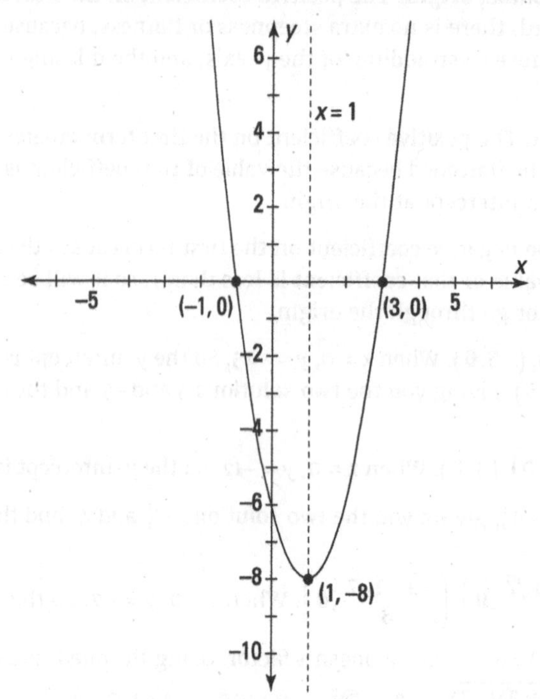

12 **See the following sketch.** The parabola opens downward and is flattened. The y-intercept is (0,12). Solving for the x-intercepts, $0=-\frac{1}{2}\left(x^2+2x-24\right)=-\frac{1}{2}(x+6)(x-4)$, giving you $(-6,0)$ and $(4,0)$. The x-coordinate of the vertex is $x=\frac{-(-1)}{2\left(-\frac{1}{2}\right)}=\frac{1}{-1}=-1$, and the y-coordinate of the vertex is $y=-\frac{1}{2}(-1)^2-(-1)+12=\frac{25}{2}$. The axis of symmetry is $x=-1$. Choose what you need to complete the graph.

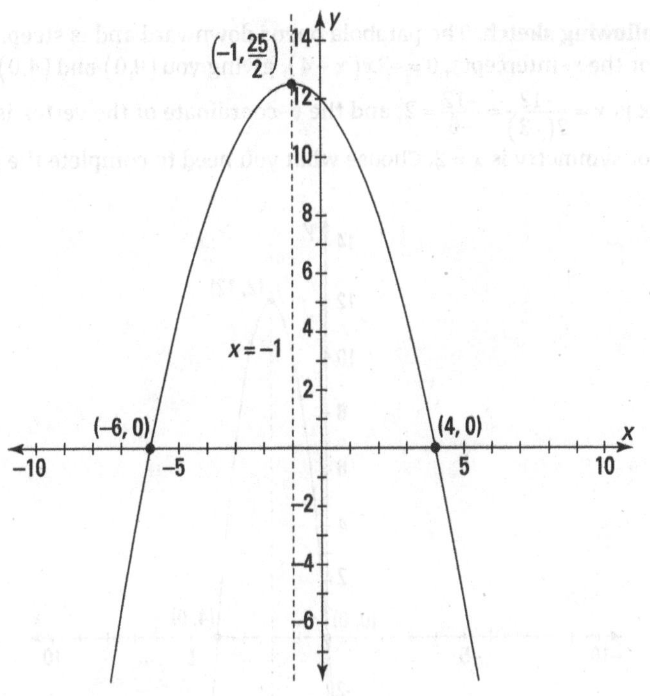

(13) **See the following sketch.** The parabola opens upward and is steep. The y-intercept is $(0,-25)$. Solving for the x-intercepts, $0=(2x-5)(2x+5)$, giving you $\left(\frac{5}{2},0\right)$ and $\left(-\frac{5}{2},0\right)$. The x-coordinate of the vertex is $x=\frac{-0}{2(4)}=0$, and the y-coordinate of the vertex is $(0,-25)$. The axis of symmetry is $x=0$. Choose what you need to complete the graph.

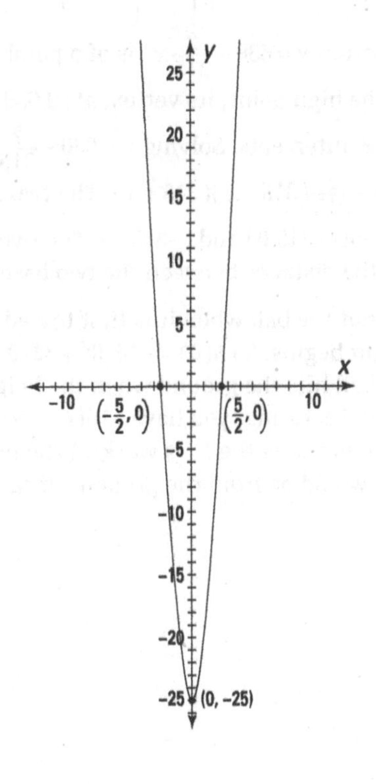

(14) **See the following sketch.** The parabola opens downward and is steep. The y-intercept is $(0,0)$. Solving for the x-intercepts, $0 = -3x(x-4)$, giving you $(0,0)$ and $(4,0)$. The x-coordinate of the vertex is $x = \frac{-12}{2(-3)} = \frac{-12}{-6} = 2$, and the y-coordinate of the vertex is 12, giving you $(2,12)$. The axis of symmetry is $x = 2$. Choose what you need to complete the graph.

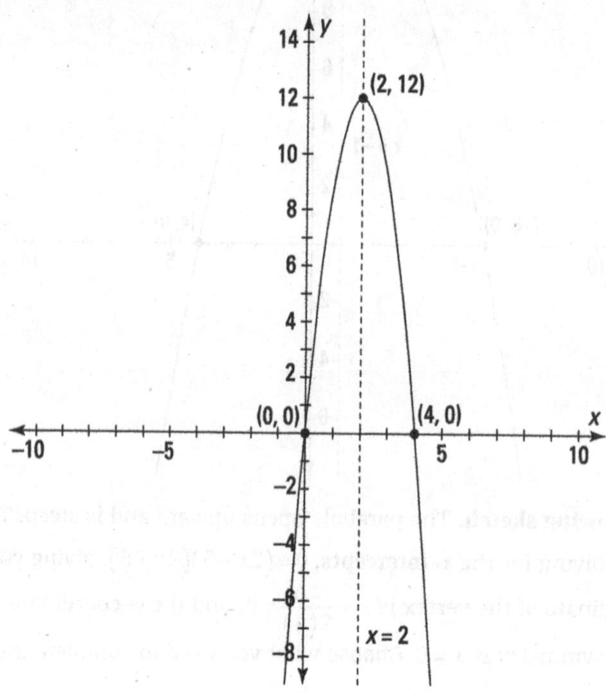

(15) **630 ft., 630 ft.** The equation $y = 630 - \frac{2}{315}x^2$ is of a parabola that is symmetric about the y-axis with the high point, its vertex, at $(0,630)$. To find out how far apart the bases are, you find the x-intercepts. Solving $y = 630 - \frac{2}{315}x^2$, you first factor and get $0 = -\frac{2}{315}(99{,}225 - x^2) = -\frac{2}{315}(315 - x)(315 + x)$. The two solutions are 315 and –315, which give you the two intercepts $(315,0)$ and $(-315,0)$. The two intercepts are 630 feet apart. The height of the arch and the distance between the two bases is the same.

(16) **48 ft., 3 sec.** The height of the ball when it is first tossed is found when x = 0, which is the time that the flight begins. So $h(0) = -16 \cdot 0^2 + 32 \cdot 0 + 48 = 48$. The window is 48 feet above the ground. The ball hits the ground when the height is equal to 0, so you solve $h(t) = -16t^2 + 32t + 48 = 0$. Factoring, you have $-16(t^2 - 2t - 3) = (t-3)(t+1) = 0$. The two solutions are t = 3 and t = –1, but only the t = 3 works. (The negative solution backs the beginning up to where the launch would be from the ground.) It takes 3 seconds for the ball to hit the ground.

(17) **80 units, $100,000 profit; 147 units.** For the maximum profit, you want the vertex of the parabola. Using $x = \frac{-b}{2a}$, you have $x = \frac{-3500}{2(-21.875)} = 80$, which is the x-coordinate. You need to sell 80 units. To get the y-coordinate and the maximum, substitute 80 into the function equation and you have $P(80) = -21.875 \cdot (80)^2 + 3500 \cdot 80 - 40{,}000 = 100{,}000$ or a profit of $100,000. The profit is 0 at the x-intercepts. Setting $P(x) = -21.875x^2 + 3500x - 40{,}000$ equal to 0 and solving for x, you have $-21.875x^2 + 3500x - 40{,}000 = 0$. Using the quadratic formula, $x = \frac{-3500 \pm \sqrt{3500^2 - 4(21.875)(-40{,}000)}}{2(-21.875)}$, you get two values for x: about 12.388 and about 147.612. This indicates that the profit happens after you've produced 13 units (assuming you can create part of a unit). And the profit ceases just after you produce 147 units (rounding up to 148 puts you in the negative profit).

Whaddya Know? Chapter 8 Quiz

Quiz time! Complete each problem to test your knowledge on the various topics covered in this chapter. You can then find the solutions and explanations in the next section.

1. **Find the intercepts of the parabola $y = x^2 - 6x + 5$.**

2. **Describe the steepness and direction of opening of the parabola $y = -3x^2 - 4x + 11$.**

3. **Find the vertex of the parabola $y = -3x^2 + 2x + 1$.**

4. **What is the equation of the axis of symmetry of the parabola $y = 3x^2 - 12x - 17$?**

5. **Graph the parabola: $y = -x^2 - 6x - 5$.**

6. **Find the vertex of the parabola $y = x^2 - 6x + 5$.**

7. **Find the extreme *y* value for the function $y = -2x^2 + 4x - 7$.**

8. **A ball is thrown in the air at a rate of 20 feet per second from a window 50 feet off the ground. How long will it take before the ball hits the ground? Recall: $h = -16t^2 + vt + h_1$.**

9. **Find the intercepts of the parabola $y = -3x^2 + 2x + 1$.**

10. **Graph the parabola: $y = 2x^2 + 4x - 5$.**

11. **Find the extreme *y* value for the function $y = \frac{1}{3}x^2 - 2x + 3$.**

Answers to Chapter 8 Quiz

1. **(5,0), (1,0), (0,5).** To find the x-intercepts, set y equal to 0 and solve for x.

   $0 = x^2 - 6x + 5 = (x-5)(x-1)$
   $x = 5$ or $x = 1$

   So the x-intercepts are (5,0) and (1,0).

   To find the y-intercept, set x equal to 0 and solve for y.

   $y = 0^2 - 6 \cdot 0 + 5 = 5$

   So the y-intercept is (0,5).

2. **Steep; opens downward.** The absolute value of –3 is greater than 1, so the parabola tends to be steep. The negative on that first coefficient makes the parabola open downward.

3. $\left(\frac{1}{3}, \frac{4}{3}\right)$. Using the formula for the x-intercept of the vertex, $x = \frac{-b}{2a}$, $x = \frac{-(2)}{2(-3)} = \frac{-2}{-6} = \frac{1}{3}$.

   Substitute the $\frac{1}{3}$ for x in the equation of the parabola: $y = -3\left(\frac{1}{3}\right)^2 + 2\left(\frac{1}{3}\right) + 1$
   $= -\frac{1}{3} + \frac{2}{3} + 1 = \frac{4}{3}$.

4. $x = 2$**.** The axis of symmetry is a vertical line going through the vertex. To find the x-value of the vertex, use $x = \frac{-b}{2a} = \frac{-(-12)}{2(3)} = \frac{12}{6} = 2$. So the equation of the axis of symmetry is $x = 2$.

5. **See the following figure.** The vertex is at $(-3,4)$, and the intercepts are: $(-5,0)$, $(-1,0)$, $(0,-5)$. The parabola opens downward and has an axis of symmetry of $x = -3$.

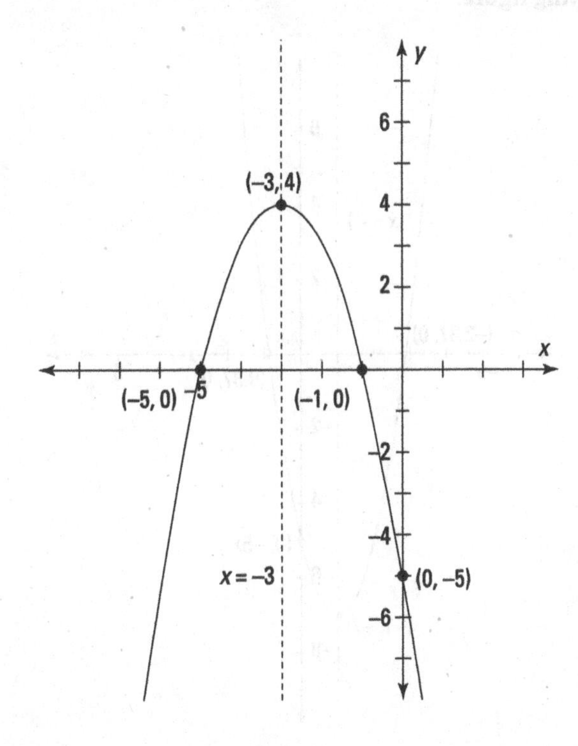

(6) **(3,−4).** Using the formula for the x-intercept of the vertex, $x=\frac{-b}{2a}$, $x=\frac{-(-6)}{2(1)}=\frac{6}{2}=3$.

Substitute the 3 for x in the equation of the parabola: $y=(3)^2-6(3)+5=9-18+5=-4$.

(7) **Maximum −5.** This is a parabola that opens downward, so you want the y-coordinate of the vertex, which will give you the maximum value. Solve for the x-coordinate:
$x=\frac{-b}{2a}=\frac{-(4)}{2(-2)}=\frac{-4}{-4}=1$. Solving for y: $y=-2(1)^2+4(1)-7=-2+4-7=-5$.

(8) **2.5 seconds.** The initial velocity, v, is 20 feet per second, and the initial height is 50 feet, or h_1. Let h = 0 and solve for t, the amount of time for the height to be 0 (hit the ground).

$$0=-16t^2+20t+50=(-4t+10)(4t+5)$$
$$t=2.5 \text{ or } -1.25$$

(9) $\left(-\frac{1}{3},0\right)$, **(1,0), (0,1).** To find the x-intercepts, set y equal to 0 and solve for x.

$$0=-3x^2+2x+1=(3x+1)(x-1)$$
$$x=-\frac{1}{3} \text{ or } x=1$$

So the x-intercepts are $\left(-\frac{1}{3},0\right)$ and (1,0).

To find the y-intercept, set x equal to 0 and solve for y.

$$y=-3(0)^2+2(0)+1=1$$

So the y-intercept is (0,1).

(10) **See the following figure.**

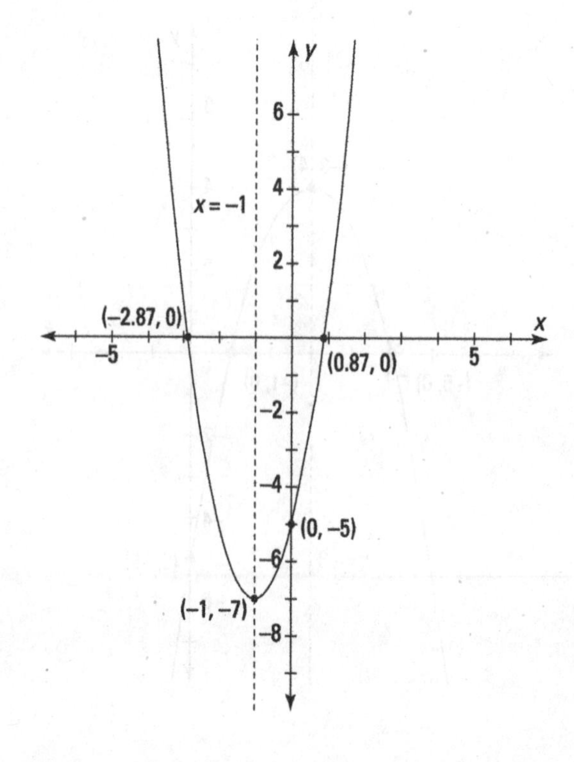

The vertex is at $(-1,-7)$, and the intercepts are: $(-2.87,0)$, $(0.87,0)$, $(0,-5)$. The parabola opens upward and has an axis of symmetry of $x=-1$. The x-intercepts are found using the quadratic formula: $x=\frac{-4\pm\sqrt{16-4(2)(-5)}}{2(2)}=\frac{-4\pm\sqrt{56}}{4}=\frac{-4\pm 2\sqrt{14}}{4}$

$=\frac{-2\pm\sqrt{14}}{2}\approx -2.87$ or 0.87.

11. **Minimum 0.** This is a parabola that opens upward, so you want the y-coordinate of the vertex, which will give you the minimum value. Solve for the x-coordinate:

$x=\frac{-b}{2a}=\frac{-(-2)}{2\left(\frac{1}{3}\right)}=\frac{2}{\frac{2}{3}}=2\cdot\frac{3}{2}=3$. Solving for y: $y=\frac{1}{3}(3)^2-2(3)+3=3-6+3=0$.

IN THIS CHAPTER

» Examining the standard polynomial form

» Graphing and finding polynomial intercepts

» Determining function signs on intervals

» Using the tools of algebra to dig up rational roots

» Taking on synthetic division in a natural fiber world

Chapter 9
Plugging In Polynomials

The word *polynomial* comes from *poly-*, meaning many, and *-nomial,* meaning name or designation. *Binomial* (two terms) and *trinomial* (three terms) are two of the many names or designations used for selected polynomials. The terms in a polynomial are made up of numbers and letters that get stuck together with multiplication.

Although the name may seem to imply complexity (much like Albert Einstein, Pablo Picasso, or Mary Jane Sterling), polynomials are some of the easier functions or equations to work with in algebra. The exponents used in polynomials are all whole numbers — no fractions or negatives. Polynomials get progressively more interesting as the exponents get larger; they can have more intercepts and turning points. This chapter outlines what you can do with polynomials: factor them, graph them, analyze them to pieces — everything but make a casserole with them. The graph of a polynomial looks like a Wisconsin landscape: smooth, rolling curves. Are you ready for this ride?

Getting into Polynomial Basics and Vocabulary

A polynomial function is a specific type of function that can be easily spotted in a crowd of other types of functions and equations. The exponents on the variable terms in a polynomial function are always whole numbers. And, by convention, you write the terms from the largest exponent to the smallest. Actually, the exponent 0 on the variable makes the variable factor equal to 1, so you don't see a variable there at all — just a constant, if there is one.

A traditional standard equation for the terms of a polynomial function is shown here. Don't let all the subscripts and superscripts throw you. The letter *a* is repeated with numbers, rather than giving the terms coefficients *a*, *b*, *c*, and so on, because a polynomial with a degree higher than 26 would run out of letters in the English alphabet.

REMEMBER

The general form for a polynomial function is

$$f(x) = a_n x^n + a_{n-1} x^{n-1} + a_{n-2} x^{n-2} + \ldots + a_1 x^1 + a_0$$

Here, the *a*'s are real numbers and the *n*'s are whole numbers. The last term is technically $a_0 x^0$, if you want to show the variable in every term.

Exploring polynomial intercepts and turning points

The intercepts of a polynomial are the points where the graph of the curve of the polynomial crosses the *x*- and *y*-axes. A polynomial function has exactly one *y*-intercept, but it can have any number of *x*-intercepts, depending on the degree of the polynomial (the powers of the variable). The higher the degree, the more *x*-intercepts you might have. And there may be none!

The *x*-intercepts of a polynomial are also called the *roots, zeros*, or *solutions*. You may think that mathematicians can't make up their minds about what to call these values, but they do have their reasons; depending on the application, the *x*-intercept has an appropriate name for what you're working on. The nice thing is that you use the same technique to solve for the intercepts, no matter what they're called. (Lest the *y*-intercept feel left out, it's frequently called the *initial value*.)

The *x*-intercepts are often where the graph of the polynomial goes from positive values (above the *x*-axis) to negative values (below the *x*-axis) or from negative values to positive values. Sometimes, though, the values on the graph don't change sign at an *x*-intercept: These graphs take on a touch-and-go appearance. The graphs approach the *x*-axis, seem to change their minds about crossing the axis, touch down at the intercepts, and then go back to the same side of the axis.

A turning point of a polynomial is where the graph of the curve changes direction. It can change from going upward to going downward, or vice versa. A turning point is where you find a relative maximum value of the polynomial, an absolute maximum value, a relative minimum value, or an absolute minimum value.

Interpreting relative value and absolute value

Many functions can have an absolute maximum or an absolute minimum value — the point at which the graph of the function has no higher or lower value, respectively. For example, a parabola opening downward has an absolute maximum; you see no point on the curve that's higher than the maximum. In other words, no value of the function is greater than that number. (Check out Chapter 8 for more on quadratic functions and their parabola graphs.) Some functions, however, also have *relative* maximum or minimum values:

- **Relative maximum:** A point on the graph — a value of the function — that's relatively large; the point is higher than anything around it, but you may be able to find a higher point somewhere else on the graph.

» **Relative minimum:** A point on the graph — a value of the function — that's lower than anything close to it; it's lower relative to all the points on the curve near it.

In Figure 9-1, you can see five turning points. Two are relative maximum values, which means they're higher than any points close to them. Three are minimum values, which means they're lower than any points around them. Two of the minimums are relative minimum values, and one is absolutely the lowest point on the curve. This function has no absolute maximum value because it keeps going up and up without end.

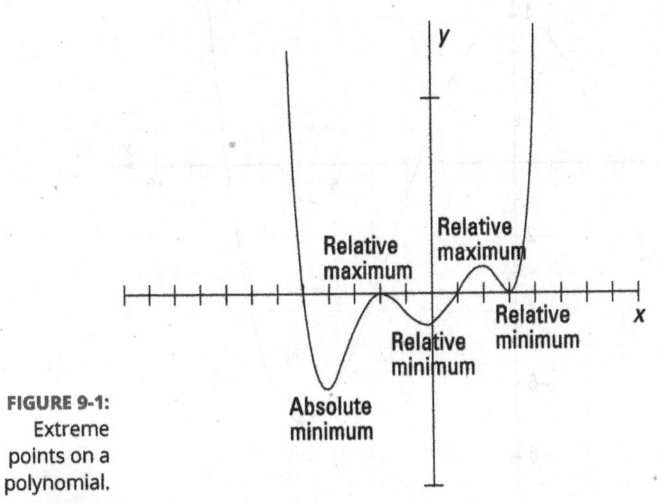

FIGURE 9-1: Extreme points on a polynomial.

EXAMPLE

Q. What are the coordinates of the intercepts in the graph?

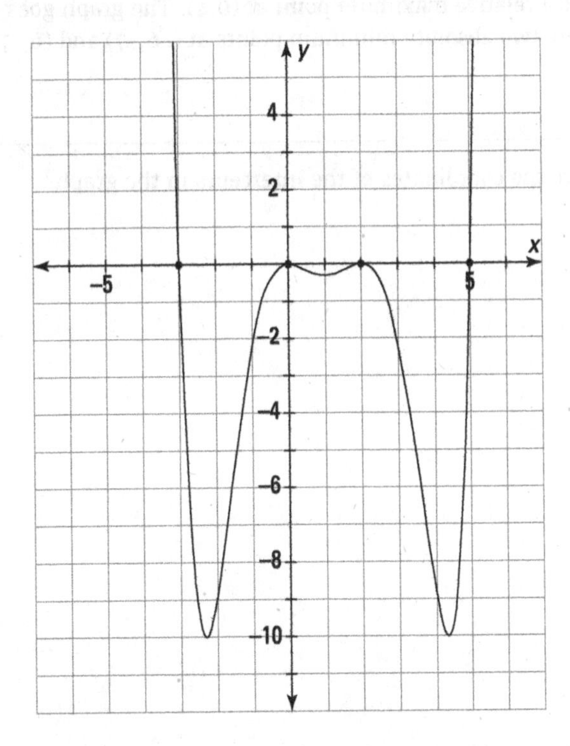

A. The intercepts occur at (–3,0), (0,0), (2,0), and (5,0).

Q. What are the coordinates of the relative and absolute maximum and minimum points in the graph?

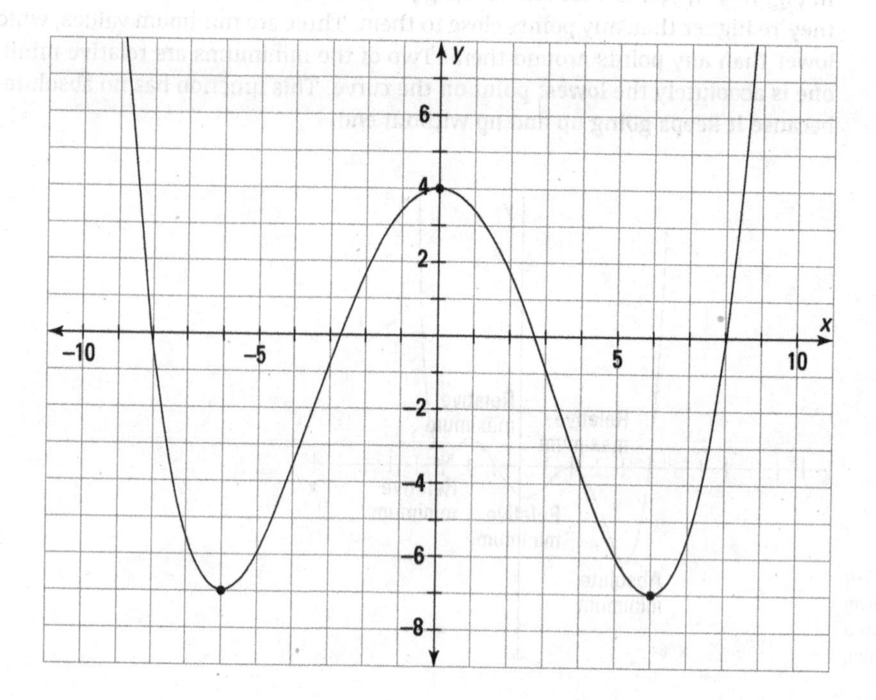

A. There is a relative maximum point at (0,4). The graph goes lower to the left and right. There are two absolute minimum points at (–6,–7) and (6,–7). The graph doesn't go any lower.

YOUR TURN

What are the coordinates of the intercepts in the graph?

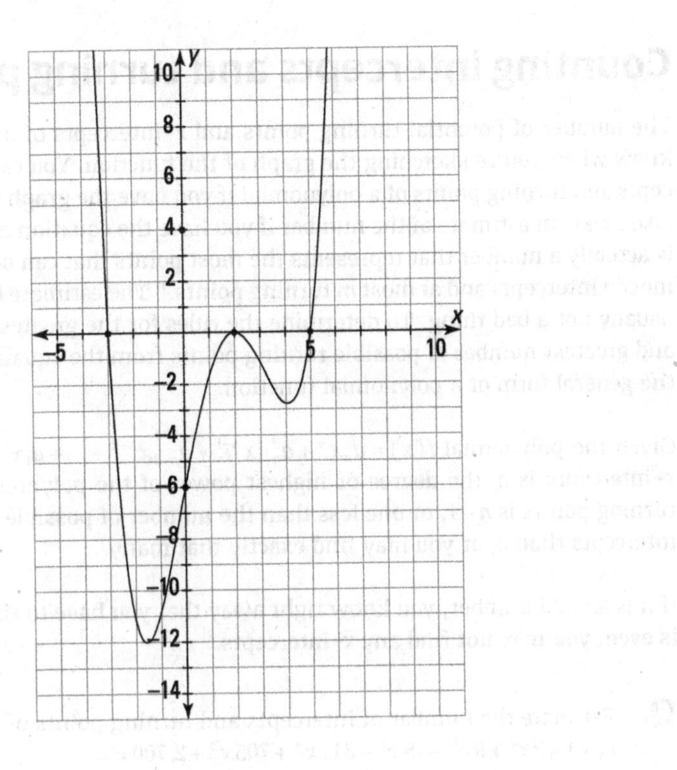

2 **What are the coordinates of the relative and absolute maximum and minimum points in the graph?**

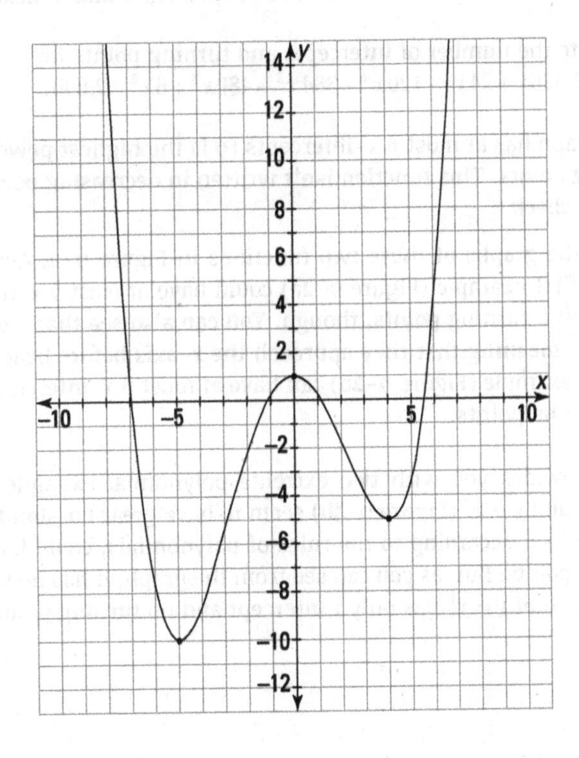

Counting intercepts and turning points

The number of potential turning points and x-intercepts of a polynomial function is good to know when you're sketching the graph of the function. You can count the number of x-intercepts and turning points of a polynomial if you have the graph of it in front of you, but you can also make an estimate of the number if you have the equation of the polynomial. Your estimate is actually a number that represents the most points that can occur. You can say, "There are at most n intercepts and at most m turning points." The estimate is the best you can do, but that's usually not a bad thing. To determine the rules for the greatest number of possible intercepts and greatest number of possible turning points from the equation of a polynomial, you look at the general form of a polynomial function.

Given the polynomial $f(x) = a_n x^n + a_{n-1}x^{n-1} + a_{n-2}x^{n-2} + \ldots + a_1 x^1 + a_0$, the maximum number of x-intercepts is n, the degree or highest power of the polynomial. The maximum number of turning points is $n-1$, or one less than the number of possible intercepts. You may find fewer intercepts than n, or you may find exactly that many.

TIP

If n is an odd number, you know right away that you have to find at least one x-intercept. If n is even, you may not find any x-intercepts.

EXAMPLE

Q. Estimate the number of intercepts and turning points of the graph of $f(x) = 2x^7 + 9x^6 - 75x^5 - 317x^4 + 705x^3 + 2{,}700x^2$.

A. To determine the possible number of intercepts and turning points for the functions, look for the values of n, the exponents that have the highest values. The highest power here is 7, so there are a maximum of 7 intercepts and a maximum of 6 turning points.

Q. Estimate the number of intercepts and turning points for $f(x) = 1{,}536x + 24x^5 - 120x^4 + 384x^2 - 480x^3 + 6x^6 - 2{,}000$.

A. This graph has at most 6 x-intercepts (6 is the highest power in the function) and 5 turning points. This function isn't written in decreasing powers of the exponent. Always check.

You can see the graphs of these two functions in Figure 9-2. According to its function, the graph of the first example (Figure 9-2a) could have at least 7 x-intercepts, but it has only 5; it does have all 6 turning points, though. You can also see that 2 of the intercepts are touch-and-go types, meaning that they approach the x-axis before heading away again. The graph of the second example (Figure 9-2b) can have at most 6 x-intercepts, but it has only 2; it does have all 5 turning points.

Figure 9-3 provides you with two extreme polynomial examples. The graphs of $y = x^8 + 1$ (Figure 9-3a) and $y = x^9$ (Figure 9-3b) seem to have great possibilities that don't pan out. The graph of $y = x^8 + 1$, according to the rules of polynomials, could have as many as 8 intercepts and 7 turning points. But, as you can see from the graph, it has no intercepts and just 1 turning point. The graph of $y = x^9$ has only 1 intercept and no turning points.

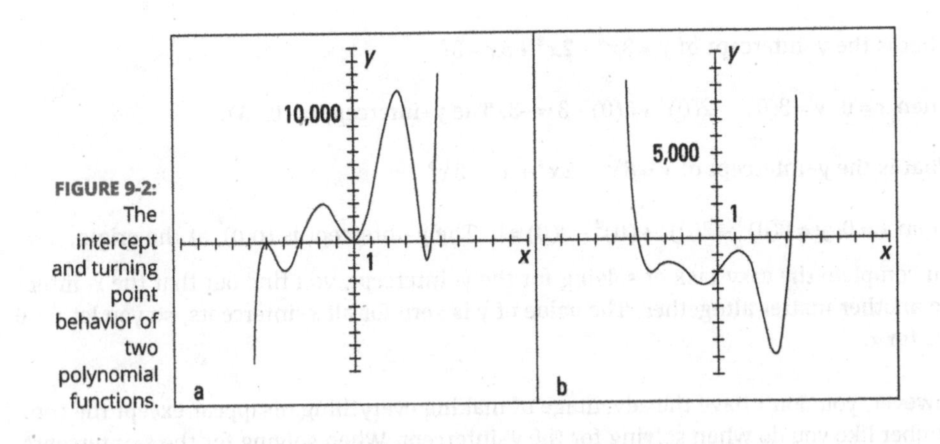

FIGURE 9-2: The intercept and turning point behavior of two polynomial functions.

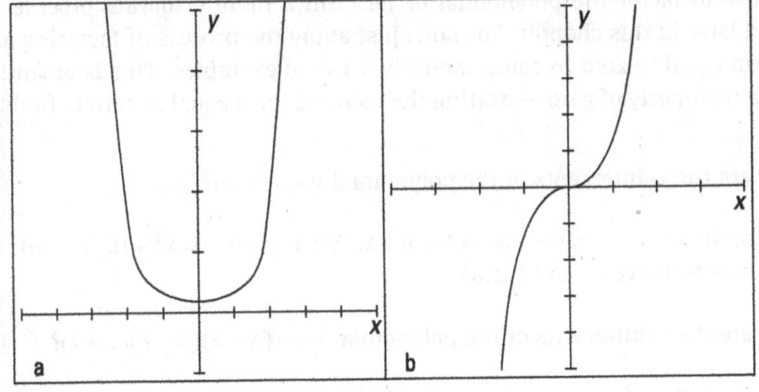

FIGURE 9-3: A polynomial's highest power provides information on the *most-possible* turning points and intercepts.

YOUR TURN

3 **Estimate the possible number of intercepts and turning points of the function $f(x) = 640 - x^4$.**

Solving for Polynomial Intercepts

You can easily solve for the y-intercept of a polynomial function, of which you can find only one. The y-intercept is where the curve of the graph crosses the y-axis, and that's when $x = 0$. So, to determine the y-intercept for any polynomial, simply replace all the x's with zeros and solve for y (that's the y part of the coordinate of that intercept).

Q. What is the y-intercept of $y=3x^4-2x^2+5x-3$?

EXAMPLE **A.** When $x=0$, $y=3(0)^4-2(0)^2+5(0)-3=-3$. The y-intercept is (0,–3).

Q. What is the y-intercept of $y=8x^5-2x^3+x^2-3x$?

A. When $x=0$, $y=8(0)^5-2(0)^3+(0)^2-3(0)=0$. The y-intercept is (0,0), at the origin.

After you complete the easy task of solving for the y-intercept, you find out that the x-intercepts are another matter altogether. The value of y is zero for all x-intercepts, so you let $y=0$ and solve for x.

Here, however, you don't have the advantage of making everything disappear except the constant number like you do when solving for the y-intercept. When solving for the x-intercepts, you may have to factor the polynomial or perform a more elaborate process — techniques you can find later in this chapter. For now, just apply the process of factoring and setting the factored form equal to zero to some carefully selected examples. This is essentially using the multiplication property of zero — setting the factored form equal to zero to find the intercepts.

Q. What are the x-intercepts of the polynomial $y=x^2-16$?

EXAMPLE **A.** Replace the y's with zeros and solve for x. When $y=0$, $0=x^2-16$, $x^2=16$, $x=\pm4$. So the x-intercepts are (4,0) and (–4,0).

Q. What are the x-intercepts of the polynomial $y=x(x-5)(x-2)(x+1)$?

A. When $y=0$, $0=x(x-5)(x-2)(x+1)$, $x=0,5,2,$ or -1. So the x-intercepts are (0,0), (5,0), (2,0), and (–1,0).

Q. What are the x-intercepts of the polynomial $y=x^4(x+3)^8$?

A. When $y=0$, $0=x^4(x+3)^8$, $x=0$ or -3. So the x-intercepts are (0,0) and (–3,0).

Both of the intercepts come from multiple roots (when a solution appears more than once). Another way of writing the factored form is

$$x\cdot x\cdot x\cdot x\cdot(x+3)\cdot(x+3)\cdot(x+3)\cdot(x+3)\cdot(x+3)\cdot(x+3)\cdot(x+3)\cdot(x+3).$$

You could list the answer as $0,0,0,0,-3,-3,-3,-3,-3,-3,-3,-3$. The number of times a root repeats is significant when you're graphing. A multiple root has a different kind of look or graph where it intersects the axis.

Q. What are the x-intercepts of the polynomial $y=x^4+1$?

A. This polynomial has no x-intercepts. When you set y = 0, you have $0=x^4+1$, which has no real solutions.

YOUR TURN

4 What are the x-intercepts of $y = x(x-7)^3$?

5 What are the x-intercepts of $y = (x-3)^4(x+6)^5$?

Finding the Roots of a Polynomial

Finding intercepts (or roots or zeros) of polynomials can be relatively easy or a little challenging, depending on the complexity of the function. Factored polynomials have roots that just stand up and shout at you, "Here I am!" Polynomials that factor easily are very desirable. Polynomials that don't factor at all, however, are relegated to computers or graphing calculators.

The polynomials that remain are those that factor — with a little planning and work. The planning process involves counting the number of possible positive and negative real roots and making a list of potential rational roots. The work is using synthetic division to test the list of choices to find the roots. Just keep in mind: A polynomial function of degree n (the highest power in the polynomial equation is n) can have as many as n roots.

Finding x-intercepts of polynomials isn't difficult as long as you have the polynomial in nicely factored form. You just set the y equal to zero and use the multiplication property of zero to pick the intercepts off like flies. But what if the polynomial isn't in factored form (and it should be)? What do you do? Well, you factor it, of course. This section deals with easily recognizable factors of polynomials — types that probably account for 70 percent of any polynomials you'll have to deal with. (I cover other, more challenging types in the following sections.)

Applying factoring patterns and groupings

Half the battle is recognizing the patterns in factorable polynomial functions. You want to take advantage of the patterns. If you don't see any patterns (or if none exist), you need to investigate further. The most easily recognizable factoring patterns used on polynomials are the following:

$a^2 - b^2 = (a+b)(a-b)$	Difference of squares
$ab \pm ac = a(b \pm c)$	Greatest common factor
$a^3 - b^3 = (a-b)(a^2 + ab + b^2)$	Difference of cubes
$a^3 + b^3 = (a+b)(a^2 - ab + b^2)$	Sum of cubes
$a^2 \pm 2ab + b^2 = (a \pm b)^2$	Perfect square trinomial
UnFOIL	Trinomial factorization
Grouping	Common factors in groups

EXAMPLE

Q. Factor the polynomial $y = x^3 - 9x$.

A. Use the greatest common factor and then the difference of squares: $y = x^3 - 9x = x(x^2 - 9) = x(x+3)(x-3)$

Q. Factor: $y = 27x^3 + 8$.

A. Use the sum of two perfect cubes: $y = 27x^3 + 8 = (3x+2)(9x^2 - 6x + 4)$

Q. Factor: $y = x^5 - x^3 - 125x^2 + 125$.

A. You initially factor the next polynomial by grouping. The first two terms have a common factor of x^3, and the second two terms have a common factor of -125. The terms in the new equation have a common factor of $x^2 - 1$. After performing the factorization, you see that the first factor is the difference of squares and the second is the difference of cubes:

$$\begin{aligned} y &= x^5 - x^3 - 125x^2 + 125 \\ &= x^3(x^2 - 1) - 125(x^2 - 1) = (x^2 - 1)(x^3 - 125) \\ &= (x+1)(x-1)(x-5)(x^2 + 5x + 25) \end{aligned}$$

Q. Factor: $y = x^6 - 12x^5 + 36x^4$.

A. First, use the greatest common factor and then factor with a perfect square trinomial:

$$y = x^6 - 12x^5 + 36x^4 = x^4(x^2 - 12x + 36) = x^4(x-6)^2$$

YOUR TURN

6 Factor the polynomial $y = x^4 - 6x^3 - 16x^2$ and write the corresponding x-intercepts.

7 Factor the polynomial $y = x^8 - 8x^5 - x^4 + 8x$ and write the corresponding x-intercepts.

Considering the unfactorable

Life would be wonderful if you would always find your paper on the doorstep in the morning and if all polynomials would factor easily. I won't get into the trials and tribulations of newspaper delivery, but polynomials that can't be factored do need to be discussed. You can't just give up and walk away. In some cases, polynomials can't be factored, but they do have intercepts that are decimal values that go on forever. Other polynomials both can't be factored and have no x-intercepts.

If a polynomial doesn't factor, you can attribute the roadblock to one of two things:

- **The polynomial doesn't have *x*-intercepts.** You can tell that from its graph using a graphing calculator. This situation happens with some even-degreed polynomials.
- **The polynomial has irrational roots or zeros.** These can be estimated with a graphing calculator or computer. All odd-powered polynomials have at least one x-intercept.

Irrational roots means that the x-intercepts are written with radicals or rounded-off decimals. Irrational numbers are those that you can't write as fractions; you usually find them as square roots of numbers that aren't perfect squares or cube roots of numbers that aren't perfect cubes. You can sometimes solve for irrational roots if one of the factors of the polynomial is a quadratic. You can apply the quadratic formula to find that solution (see Chapter 3).

TIP

A quicker way to find irrational roots is to use a graphing calculator. Some will find the zeros for you one at a time; others will list the irrational roots for you all at once. You can also save some time if you recognize the patterns in polynomials that don't factor (you won't try to factor them).

For instance, polynomials in these forms don't factor:

- $x^{2n}+c$ **where** $c>0$. A polynomial in this form has no real roots or solutions. See Chapter 17 for more on how to deal with imaginary numbers.
- x^2+ax+a^2. The quadratic formula will help you determine the imaginary roots.
- x^2-ax+a^2. This form requires the quadratic formula, too. You'll get imaginary roots.

The second and third examples are part of the factorization of the difference or sum of two cubes. In Chapter 3, you see how the difference or sum of cubes is factored and also that the resulting trinomial doesn't factor.

Saving Your Sanity: The Rational Root Theorem

What do you do if the factorization of a polynomial doesn't leap out at you? You have a feeling that the polynomial factors but the necessary numbers escape you. Never fear! Your help is in the form of the Rational Root Theorem. This theorem is really neat because it's so orderly and predictable, and it has an obvious end to it; you know when you're done so you can stop looking for solutions. But before you can put the theorem into play, you have to be able to recognize a rational root — or a rational number, for that matter.

A *rational number* is any real number that you can write as a fraction with one integer divided by another. (An integer is a positive or negative whole number or zero.) You often write a rational number in the form $\frac{p}{q}$, with the understanding that the denominator, q, can't equal zero. All whole numbers are rational numbers because you can write them as fractions, such as $4=\frac{12}{3}$. What distinguishes rational numbers from their opposites, irrational numbers, has to do with the decimal equivalents. The decimal associated with a fraction (rational number) will either terminate or repeat (have a pattern of numbers that occurs over and over again). The decimal equivalent of an irrational number never repeats and never ends; it just wanders on aimlessly.

Without further ado, here's the Rational Root Theorem: If the polynomial $f(x)=a_nx^n+a_{n-1}x^{n-1}+a_{n-2}x^{n-2}+\ldots+a_1x^1+a_0$ has any rational roots, they all meet the requirement that you can write them as a fraction equal to $\frac{\text{factor of } a_0}{\text{factor of } a_n}$.

In other words, according to the theorem, any rational root of a polynomial is formed by dividing a factor of the constant term by a factor of the lead coefficient.

EXAMPLE

Q. List all the possible rational roots of $y=x^4-3x^3+2x^2+12$.

A. All the factors of a_0 are: ±1, ±2, ±3, ±4, ±6, ±12. The factors of a_n are just ±1, so dividing the factors of a_0 by the factors of a_n will not change the listing.

Q. List all the possible rational roots of $y = 6x^5 - 4x^3 + 3x^2 - 5x + 4$.

A. All the factors of a_0 are: $\pm 1, \pm 2, \pm 4$, and the factors of a_n are $\pm 1, \pm 2, \pm 3, \pm 6$. Dividing the factors of a_0 by the factors of a_n and not listing any repeats, the total list of possible rational roots is: $\pm 1, \pm 2, \pm 4, \pm\frac{1}{2}, \pm\frac{1}{3}, \pm\frac{1}{6}, \pm\frac{2}{3}, \pm\frac{4}{3}$.

YOUR TURN

8 List all the possible rational roots of $y = 2x^6 - 9x^3 + 15$.

9 List all the possible rational roots of $y = x^4 - 6x^3 - 15x^2 + 8x$.

Putting the theorem to good use

The Rational Root Theorem gets most of its workout by letting you make a list of numbers that may be roots of a particular polynomial. After using the theorem to make your list of potential roots (and checking it twice), you plug the numbers into the polynomial to determine which roots, if any, work. You may run across an instance where none of the candidates work, which tells you that there are no rational roots. (And if a given rational number isn't on the list of possibilities that you come up with then it can't be a root of that polynomial.)

Before you start to plug and chug, however, check out the section, "Letting Descartes make a ruling on signs," later in this chapter; it helps you with your guesses. Also, you can refer to the section, "Synthesizing Root Findings," for a quicker method than plugging in.

Changing from roots to factors

When you have the factored form of a polynomial and set it equal to 0, you can solve for the solutions (or *x*-intercepts, if that's what you want). Just as important, if you have the solutions, you can go backward and write the factored form. Factored forms are needed when you have polynomials in the numerator and denominator of a fraction and you want to reduce the fraction. Factored forms are easier to compare with one another.

How can you use the Rational Root Theorem to factor a polynomial function? Why would you want to? The answer to the second question, first, is that you can reduce a factored form if it's in a fraction. Also, a factored form is more easily graphed. Now, for the first question: You use the Rational Root Theorem to find roots of a polynomial and then translate those roots into binomial factors whose product is the polynomial.

If $x = \frac{b}{a}$ is a root of the polynomial $f(x)$, the binomial $(ax - b)$ is a factor. This works because multiplying each side of $x = \frac{b}{a}$ by a gives you $ax = b$, and then subtracting each side of the equation by b gives you $ax - b = 0$.

The positive roots give factors of the form $x - c$, and the negative roots give factors of the form $x + c$, which comes from $x - (-c)$.

EXAMPLE

Q. Write an equation of a polynomial with the five roots $x = 1$, $x = -2$, $x = 3$, $x = \frac{3}{2}$, and $x = -\frac{1}{2}$.

A. Apply the previously stated rule to get $f(x) = k(x-1)(x+2)(x-3)\left(x-\frac{3}{2}\right)\left(x+\frac{1}{2}\right)$. The k indicates that a constant multiplier may exist so many polynomials actually have the same roots. To get rid of the fractions, you multiply the polynomial by 4 to get

$$4f(x) = 4k(x-1)(x+2)(x-3)\left(x-\frac{3}{2}\right)\left(x+\frac{1}{2}\right) = k(x-1)(x+2)(x-3)\cdot 2\left(x-\frac{3}{2}\right)\cdot 2\left(x+\frac{1}{2}\right)$$
$$= k(x-1)(x+2)(x-3)(2x-3)(2x+1)$$

Q. Write an equation of a polynomial with the multiple roots: $x = 0$, $x = 2$, $x = 2$, $x = -3$, $x = -3$, $x = -3$, $x = -3$, and $x = 4$.

A. To show multiple roots, or roots that occur more than once, use exponents on the factors. The corresponding polynomial is $f(x) = kx(x-2)^2(x+3)^4(x-4)$.

YOUR TURN

10 Write the equation of a polynomial with the roots: $x = 1$, $x = -3$, $x = \frac{1}{3}$.

11 Write the equation of a polynomial with the roots: $x = 0$, $x = \frac{1}{2}$, $x = \frac{1}{2}$, $x = \frac{1}{2}$.

Letting Descartes make a ruling on signs

René Descartes was a French philosopher and mathematician. One of his contributions to algebra is Descartes' Rule of Signs. This handy, dandy rule is a weapon in your arsenal for the fight to find roots of polynomial functions. If you pair this rule with the Rational Root Theorem from the previous section, you'll be well equipped to succeed.

The Rule of Signs tells you how many positive and negative *real* roots you may find in a polynomial. A real number is just about any number you can think of. It can be positive or negative, rational or irrational. A real number can be graphed on a number line. The only thing a real number can't be is imaginary. (I cover imaginary numbers in Chapter 17, if you want to know more about them.)

Counting up the positive roots

The first part of the Rule of Signs helps you identify how many of the real roots of a polynomial are positive.

Descartes' Rule of Signs (Part 1): The polynomial $f(x) = a_n x^n + a_{n-1}x^{n-1} + a_{n-2}x^{n-2} + \ldots a_1 x^1 + a_0$ has at most n real roots. Count the number of times the sign changes in f, and call that value p. The value of p is the maximum number of positive real roots of f. If the number of positive roots isn't p, it is $p-2$, $p-4$, or some number less by a multiple of two.

EXAMPLE

Q. Count the number of positive real roots in $f(x) = 2x^7 - 19x^6 + 66x^5 - 95x^4 + 22x^3 + 87x^2 - 90x + 27$.

A. Count the number of sign changes. The sign of the first term is positive, the second is a negative, and the third is positive, and then negative, positive, stays positive, negative, and then positive. Whew! In total, you count six sign changes. Therefore, you conclude that the polynomial has six positive roots, four positive roots, two positive roots, or none at all. Out of seven possible roots, it looks like at least one has to be negative. (By the way, this polynomial does have six positive roots; I built it that way! The only way you'd know that without being told is to go ahead and find the roots, with help from the Rule of Signs.)

Q. Count the number of positive real roots in $f(x) = 2x^8 + 18x^6 + 23x^5 + 11x^4 + 11x^4 + 22x^3 + 87x^2 + 90x + 2$.

A. Count the number of sign changes. All the signs are positive, so there are no sign changes. There are no possible positive real roots.

Changing the function to count negative roots

Along with the positive roots (see the previous section), Descartes' Rule of Signs deals with the possible number of negative roots of a polynomial. After you count the possible number of positive real roots, you combine that value with the number of possible negative real roots to make your guesses and solve the equation.

Descartes' Rule of Signs (Part 2): The polynomial $f(x)=a_nx^n+a_{n-1}x^{n-1}+a_{n-2}x^{n-2}+\dots a_1x^1+a_0$ has at most n real roots. Find $f(-x)$, and then count the number of times the sign changes in $f(-x)$ and call that value q. The value of q is the maximum number of negative roots of f. If the number of negative roots isn't q, the number is $q-2$, $q-4$, and so on, for as many multiples of two as necessary.

Q. Find the possible number of negative roots of $f(x)=2x^7-19x^6+66x^5-95x^4+22x^3+87x^2-90x+27$.

A. You first find $f(-x)$ by replacing each x with $-x$ and simplifying:

$$\begin{aligned}f(x)&=2(-x)^7-19(-x)^6+66(-x)^5-95(-x)^4+22(-x)^3+87(-x)^2-90(-x)+27\\&=-2x^7-19x^6-66x^5-95x^4-22x^3+87x^2+90x+27\end{aligned}$$

As you can see, the function has only one sign change, from negative to positive. Therefore, the function has exactly one negative root — no more, no less.

Q. Find the possible number of negative roots of $f(x)=x^7-29x^6-6x^5-5x^4-2x^3+7x^2-9x-7$.

A. You first find

$$\begin{aligned}f(-x)&=(-x)^7-29(-x)^6-6(-x)^5-5(-x)^4-2(-x)^3+7(-x)^2-9(-x)-7\\&=-x^7-29x^6+6x^5-5x^4+2x^3+7x^2+9x-7\end{aligned}$$

There are three sign changes, so there are possibly three or one negative real roots.

Knowing the potential number of positive and negative roots for a polynomial is very helpful when you want to pinpoint an exact number of roots. If a polynomial has only one negative real root, then that fact tells you to concentrate your guesses on positive roots; the odds are better that you'll find a positive root first. When you're using synthetic division (see the later section, "Synthesizing Root Findings") to find the roots, the steps get easier and easier as you find and eliminate roots. By picking off the roots that you have a better chance of finding first, you can save the harder-to-find roots for the end.

12 Find the number of possible positive and negative real roots of $f(x)=4x^{10}-x^9-2x^8-3x^5+4x^3-5x^2+1$.

13 **Find the number of possible positive and negative real roots of $f(x)=16-x^5$.**

Synthesizing Root Findings

Synthetic division can be used to test the list of possible roots for a polynomial that you come up with by using the Rational Root Theorem. Synthetic division is a method of dividing a polynomial by a binomial, using only the coefficients of the terms in the polynomial and the constant in the binomial. The method is quick, neat, and highly accurate — usually even more accurate than long division — and it uses most of the information from earlier sections in this chapter, putting it all together for the search for roots, zeros, and intercepts of polynomials. (You can find more information about, and practice problems for, long division and synthetic division in one of my other spine-tingling thrillers, *Algebra I Workbook For Dummies* [John Wiley & Sons, Inc.].)

You can interpret your results in three different ways, depending on what purpose you're using synthetic division for. I explain each way in the following sections.

Using synthetic division to test for roots

When you use synthetic division to test for roots in a polynomial, you're looking for and hoping for a zero. You want the last number on the bottom row of your synthetic division problem to be a zero. If that's the case, then the division had no remainder, and the number being tested is a root. The fact that there's no remainder means that the binomial represented by the number is dividing the polynomial evenly. The number is a root because the binomial is a factor.

So, how do you perform this process of synthetic division? Here are the steps:

1. **Write the polynomial being divided in order of decreasing powers of the exponents. Replace any missing powers with zero to represent the coefficient.**
2. **Write the coefficients of the polynomial in a row, including any zeros.**
3. **Put the number you want to divide by in front of the row of coefficients, separated by a half-box. When looking for roots, the root choice is the number.**
4. **Draw a horizontal line below the row of coefficients, leaving room for numbers under the coefficients.**
5. **Bring the first coefficient straight down below the line.**
6. **Multiply the number you bring below the line by the number that you're dividing into everything. Put the result under the second coefficient.**

7. **Add the second coefficient and the product, putting the result below the line.**

8. **Repeat the multiplication/addition from Steps 6 and 7 with the rest of the coefficients. The very last result is the remainder (which you're hoping is zero).**

EXAMPLE

Q. Find the roots of the polynomial $y = x^5 + 5x^4 - 2x^3 - 28x^2 - 8x + 32$.

A. Use synthetic division to test the possibilities. According to the Rule of Signs, there are 2 or 0 possible positive real roots and 3 or 1 negative real roots. And, using the Rational Root Theorem, your list of the potential rational roots is $\pm1, \pm2, \pm4, \pm8, \pm16, \pm32$. You choose one of these roots and apply synthetic division. When using synthetic division, you should generally go with the smaller numbers first, so use 1 and –1, 2 and –2, and so on.

USING THE PREVIOUS STEPS:

1. **Write the polynomial in order of decreasing powers of the exponents. Replace any missing powers with zero to represent the coefficient.**

   The polynomial is already in the correct order: $y = x^5 + 5x^4 - 2x^3 - 28x^2 - 8x + 32$.

2. **Write the coefficients in a row, including any zeros.**

   1 5 –2 –28 –8 32

3. **Put the number you want to divide by in front of the row of coefficients, separated by a half-box.**

   In this case, the guess is $x = 1$.

   $\underline{1|}$ 1 5 -2 -28 –8 32

4. **Draw a horizontal line below the row of coefficients, leaving room for numbers under the coefficients.**

   $\underline{1|}$ 1 5 -2 -28 -8 32

   ___

5. **Bring the first coefficient straight down below the line.**

   $\underline{1|}$ 1 5 -2 -28 -8 32

   ___

   1

6. **Multiply the number you bring below the line by the number that you're dividing into everything. Put the result under the second coefficient.**

   $\underline{1|}$ 1 5 -2 -28 -8 32

   1

   ___

   1

7. Add the second coefficient and the product, putting the result below the line.

$$\begin{array}{r|rrrrrr} \underline{1} & 1 & 5 & -2 & -28 & -8 & 32 \\ & & 1 & & & & \\ \hline & 1 & 6 & & & & \end{array}$$

8. Repeat the multiplication/addition from Steps 6 and 7 with the rest of the coefficients.

$$\begin{array}{r|rrrrrr} \underline{1} & 1 & 5 & -2 & -28 & -8 & 32 \\ & & 1 & 6 & 4 & -24 & -32 \\ \hline & 1 & 6 & 4 & -24 & -32 & 0 \end{array}$$

The last entry on the bottom is a zero, so you know $x = 1$ is a root. Now, you can do a modified synthetic division when testing for the next root; you just use the numbers across the bottom except the bottom-right zero entry. (These values are actually coefficients of the quotient, if you do long division; see the following section.)

If your next guess is to see if $x = -1$ is a root, the modified synthetic division appears as follows:

The last entry on the bottom row isn't zero, so -1 isn't a root.

If you're a really good guesser, then you'll decide to try $x = 2$, $x = -4$, $x = -2$, and $x = -2$ (a second time). These values represent the rest of the roots, and the synthetic division for all the guesses looks like this:

First, trying $x = 2$,

$$\begin{array}{r|rrrrr} \underline{2} & 1 & 6 & 4 & -24 & -32 \\ & & 2 & 16 & 40 & 32 \\ \hline & 1 & 8 & 20 & 16 & 0 \end{array}$$

The last number in the bottom row is 0. That's the remainder in the division. So now just look at all the numbers that come before the 0; they're the new coefficients to divide into. Notice that the last coefficient is now 16, so you can modify your list of possible roots to be just the factors of 16. Also, they're all positive with no sign changes. You don't have any positive roots left. Now, dividing by -4:

$$\begin{array}{r|rrrr} \underline{-4} & 1 & 8 & 20 & 16 \\ & & -4 & -16 & -16 \\ \hline & 1 & 4 & 4 & 0 \end{array}$$

You could keep dividing, using -2 twice. But, when you get down to 3 terms in that last row, it's usually more efficient to write the corresponding quadratic and factor it. The quadratic with those coefficients is $x^2 + 4x + 4$, which factors into $(x + 2)(x + 2)$, giving you the two solutions $x = -2$ and $x = -2$ when set equal to 0.

In this case, the roots are $x = 1, -1, 2, -2$, and -2. And the factorization is $y = (x - 1)(x + 1)(x - 2)(x + 2)^2$.

YOUR TURN

14 Find the roots of $f(x)=x^4-2x^3-28x^2+50x+75$ using synthetic division.

15 Find the roots of $f(x)=x^6-x^4-81x^2+81$ using synthetic division. (Yes, I know that this factors nicely by grouping, but just humor me and use synthetic division.)

16 Find the roots of $f(x)=2x^4-19x^3+37x^2+55x-75$ using synthetic division.

Synthetically dividing by a binomial

Finding the roots of a polynomial isn't the only excuse you need to use synthetic division. You can also use synthetic division to replace the long, drawn-out process of dividing a polynomial by a binomial. Divisions like this are found in lots of calculus problems where you need to make the expression more simplified so you can perform some wonderful operation.

The polynomial being divided can be any degree; the binomial has to be either $x+c$ or $x-c$, and the coefficient on the x is 1. This may seem rather restrictive, but a huge number of long divisions you'd have to perform fit in this category, so it helps to have a quick, efficient method to perform these basic division problems.

To use synthetic division to divide a polynomial by a binomial, you first write the polynomial in decreasing order of exponents, inserting a zero for any missing exponent. The number you put in front, or divide by, is the opposite of the number in the binomial. So, if you're dividing by $x+2$, you'll use –2. If you're dividing by $x-5$, you'll use 5.

EXAMPLE

Q. Divide the polynomial $2x^5+3x^4-8x^2-5x+2$ by the binomial $x+2$.

A. Use synthetic division, letting the divisor be –2. Be sure to put a 0 where there's a missing third-degree term.

$$\begin{array}{r|rrrrrr} -2 & 2 & 3 & 0 & -8 & -5 & 2 \\ & & -4 & 2 & -4 & 24 & -38 \\ \hline & 2 & -1 & 2 & -12 & 19 & -36 \end{array}$$

As you can see, the last entry on the bottom row isn't zero. If you're looking for roots of a polynomial equation, this fact tells you that –2 isn't a root. In this case, because you're working on a long division application (which you know because you need to divide to simplify the expression), the –36 is the remainder of the division — in other words, the division doesn't come out even.

You obtain the answer (quotient) of the division problem from the coefficients across the bottom of the synthetic division. You start with a power one value lower than the original polynomial's power, and you use all the coefficients, dropping the power by one with each successive coefficient. The last coefficient is the remainder, which you write over the divisor.

Here's the division problem and its solution. The original division problem is written first. After the problem, you see the coefficients from the synthetic division written in front of variables — starting with one degree lower than the original problem. The remainder of –36 is written in a fraction on top of the divisor, $x+2$.

$$(2x^5+3x^4-8x^2-5x+2)\div(x+2)=2x^4-x^3+2x^2-12x+19-\frac{36}{x+2}$$

Q. Divide the polynomial x^5-243 by the binomial $x-3$.

A. Use synthetic division, letting the divisor be 3. Be sure to put zeros for the missing terms.

$$\begin{array}{r|rrrrrr} 3 & 1 & 0 & 0 & 0 & 0 & -243 \\ & & 3 & 9 & 27 & 81 & 243 \\ \hline & 1 & 3 & 9 & 27 & 81 & 0 \end{array}$$

As you can see, the last entry on the bottom row is zero, so there's no remainder. Here's the division problem and its solution. The original division problem is written first. After the problem, you see the coefficients from the synthetic division written in front of variables — starting with one degree lower than the original problem.

$$(x^5-243)\div(x-3)=x^4+3x^3+9x^2+27x+81$$

YOUR TURN

17 **Use synthetic division to do the following: $\left(x^5-4x^4-5x^3-8x^2+32x+40\right)\div(x+1)$.**

18 **Use synthetic division to do the following: $\left(2x^6-3x^4+2x^3-4x+1\right)\div(x-1)$.**

Wringing out the Remainder (Theorem)

In the two previous sections, you use synthetic division to test for roots of a polynomial equation and then to do a long division problem. You use the same synthetic division process but you read and use the results differently. In this section, I present yet another use of synthetic division involving the Remainder Theorem. When you're looking for roots or solutions of a polynomial equation, you always want the remainder from the synthetic division to be zero. In this section, you get to see how to make use of all those pesky remainders that weren't zeros.

The Remainder Theorem: When the polynomial $f(x)=a_nx^n+a_{n-1}x^{n-1}+a_{n-2}x^{n-2}+\ldots a_1x^1+a_0$ is divided by the binomial $x-c$, the remainder of the division is equal to $f(c)$. In other words: this is a quick and easy way of evaluating the function — instead of raising to powers, multiplying, and so on.

EXAMPLE

Q. Use the Remainder Theorem on $f(x)=2x^5+3x^4-8x^2-5x+2$ to find $f(-3)$.

A. With synthetic division,

$$\begin{array}{r|rrrrrr} -3 & 2 & 3 & 0 & -8 & -5 & 2 \\ & & -6 & 9 & -27 & 105 & -300 \\ \hline & 2 & -3 & 9 & -35 & 100 & -298 \end{array}$$

So $f(-3)=-298$.

Do you doubt this? Do you need reassurance? Okay, I'll check.

Using the old method,

$$f(-3) = 2(-3)^5 + 3(-3)^4 - 8(-3)^2 - 5(-3) + 2 = 2(-243) + 3(81) - 8(9) + 15 + 2$$
$$= -486 + 243 - 72 + 17 = -243 - 55 = -298$$

Q. Use the Remainder Theorem to find $f(3)$ when $f(x) = x^8 - 3x^7 + 2x^5 - 14x^3 + x^2 - 15x + 11$.

A. Applying the Remainder Theorem,

3	1	−3	0	2	0	−14	1	−15	11
		3	0	0	6	18	12	39	72
	1	0	0	2	6	4	13	24	83

$f(3) = 83$

Compare the process you use here with substituting the 3 into the function: $f(3) = (3)^8 - 3(3)^7 + 2(3)^5 - 14(3)^3 + (3)^2 - 15(3) + 11$. These numbers get really large. For instance, $3^8 = 6{,}561$. The numbers are much more manageable when you use synthetic division and the Remainder Theorem.

YOUR TURN

19 Given $f(x) = x^5 - 3x^4 + 4x^3 - 6x^2 + 5x - 1$, find $f(2)$.

20 Given $f(x) = 6x^{10} - 16x^6 + 150x^5 - 80x^2$, find $f(-2)$.

Determining Positive and Negative Intervals

When a polynomial has positive y-values for some interval — between two x-values — its graph lies above the x-axis. When a polynomial has negative values, its graph lies below the x-axis in that interval. The only way to change from positive to negative values or vice versa is to go through zero — in the case of a polynomial, at an x-intercept. Polynomials can't jump from one side of the x-axis to the other because their domains are all real numbers; nothing is skipped to allow such a jump. The fact that x-intercepts work this way is good news for you because x-intercepts play a large role in the big picture of solving polynomial equations and determining the positive and negative natures of polynomials.

REMEMBER

The positive versus negative values of polynomials are important in various applications in the real world, especially where money is involved. If you use a polynomial function to model the profit in your business or the depth of water (above or below flood stage) near your house, you should be interested in positive versus negative values and in what intervals they occur. The technique you use to find the positive and negative intervals also plays a big role in calculus, so you get an added bonus by using it here first.

If you're a visual person like me, you'll appreciate the interval method I present in this section. Using a sign-line and marking the intervals between x-values allows you to determine where a polynomial is positive or negative, and it appeals to your artistic bent!

The process of using a sign-line involves these steps:

1. **Draw a number line, and place the values of the x-intercepts in their correct positions on the line.**
2. **Choose random values to the right and left of the intercepts to test whether the function is positive or negative in those positions.**
3. **Place a + or – symbol in each interval to show the function sign.**

EXAMPLE

Q. Determine where the graph of the function $f(x)=x(x-2)(x-7)(x+3)$ is above or below the x-axis.

A. Determine the signs of the function values between the intercepts. The intercepts are at $x=0, 2, 7,$ and -3. Draw a number line and place the values of the x-intercepts in their correct positions on the line.

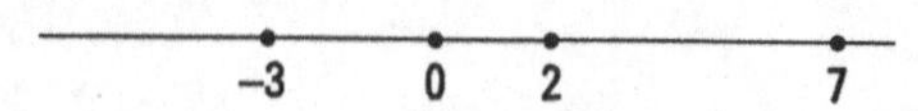

Choose random values to the right and left of the intercepts to test whether the function is positive or negative in those positions.

Some possible random number choices are $x=-4, -1, 1, 3,$ and 8. These values represent numbers in each interval determined by the intercepts. (*Note:* These aren't the only possibilities; you can pick your favorites.)

You don't need the actual number value, just the sign of the result. For example, $f(-4)=(-)(-)(-)(-)=+$.

$f(-1)=(-)(-)(-)(+)=-$ $\quad f(1)=(+)(-)(-)(+)=+$

$f(3)=(+)(+)(-)(+)=-$ $\quad f(8)=(+)(+)(+)(+)=+$

You need to check only one point in each interval; the function values all have the same sign within that interval.

Place a + or – symbol in each interval to show the function sign.

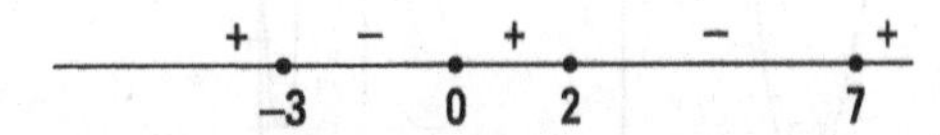

The graph of this function is positive, or above the x-axis, whenever x is smaller than –3, between 0 and 2, and bigger than 7. You write that $f(x)>0$ when: $x<-3$ or $0<x<2$ or $x>7$. In interval notation, that's $(-\infty,-3)\cup(0,2)\cup(7,\infty)$.

Q. Determine where the graph of the function $f(x)=(x-1)^2(x-3)^5(x+2)^4$ is above or below the x-axis.

A. The function doesn't change at each intercept. The intercepts are where $x=1, 3,$ and -2.

Draw the number line and insert the intercepts.

–2 1 3

Test values to the left and right of each intercept. Some possible random choices are to let $x=-3, 0, 2, 4$.

When you can, you should always use 0, because it combines so nicely.

$f(-3)=(-)^2(-)^5(-)^4=(+)(-)(+)=-$ $\quad f(0)=(-)^2(-)^5(+)^4=(+)(-)(+)=-$

$f(2)=(+)^2(-)^5(+)^4=(+)(-)(+)=-$ $\quad f(4)=(+)^2(+)^5(+)^4=(+)(+)(+)=+$

Mark the signs in the appropriate places on the number line.

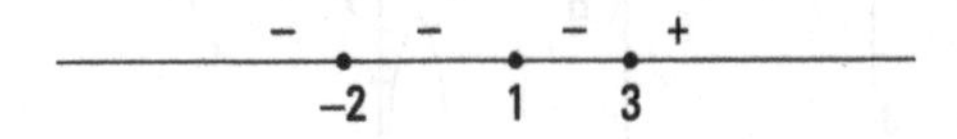

You probably noticed that the factors raised to an even power were always positive. The factor raised to an odd power is only positive when the result in the parentheses is positive.

Look back at the two polynomial examples in the previous section. Did you notice that in the first example the sign changed every time, and in the second, the signs were sort of stuck on the same sign for a while? When the signs of functions don't change, the graphs of the polynomials don't cross the x-axis at any intercepts, and you see touch-and-go graphs. Why do you suppose this is? First, look at graphs of two functions from the previous section in Figures 9-4a and 9-4b, $y = x(x-2)(x-7)(x+3)$ and $y = (x-1)^2(x-3)^5(x+2)^4$.

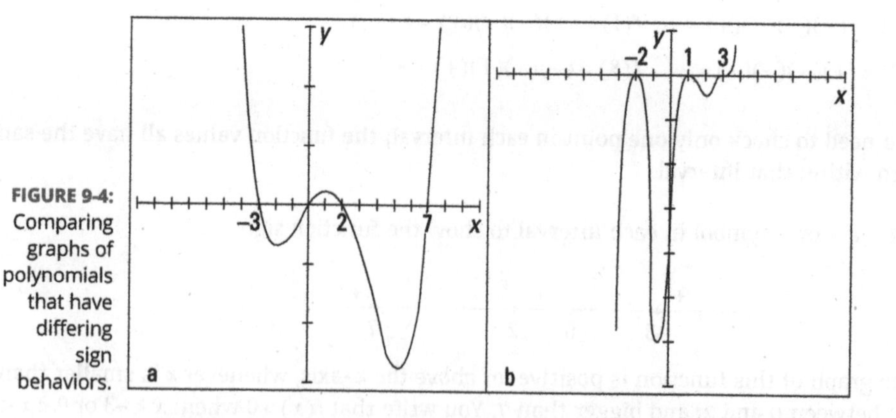

FIGURE 9-4: Comparing graphs of polynomials that have differing sign behaviors.

The rule for whether a function displays sign changes or not at the intercepts is based on the exponent on the factor that provides you with a particular intercept.

If a polynomial function is factored in the form $y = (x-a_1)^{n_1}(x-a_2)^{n_2}\ldots$, you see a sign change whenever n_1 is an odd number (meaning it crosses the x-axis), and you see no sign change whenever n_1 is even (meaning the graph of the function is touch and go).

EXAMPLE

Q. Discuss how the sign changes or lack of sign changes affect the graph of $y = x^4(x-3)^3(x+2)^8(x+5)^2$, shown in Figure 9-5a.

A. You see a sign change at $x = 3$, where the exponent is odd. There is no sign change at $x = 0$, -2, or -5. The curve crosses the x-axis at $x = 3$ only.

Q. Discuss how the sign changes or lack of sign changes affects the graph of $y = (2-x)^2(4-x)^2(6-x)^2(2+x)^2$, shown in Figure 9-5b.

A. You never see a sign change. All the exponents are even. The curve never crosses the x-axis.

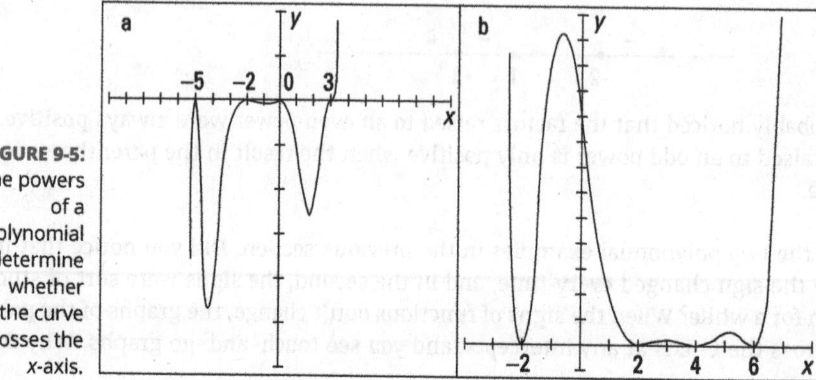

FIGURE 9-5: The powers of a polynomial determine whether the curve crosses the x-axis.

YOUR TURN

21 Determine when the graph of $f(x) = x^2(x-3)^4(x+1)^5(x-11)$ is above the x-axis, below the x-axis, and when it crosses the x-axis.

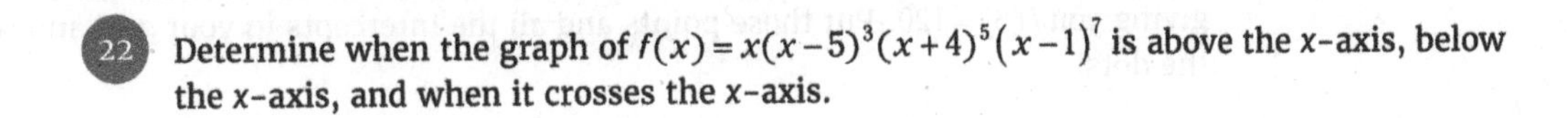

22 Determine when the graph of $f(x) = x(x-5)^3(x+4)^5(x-1)^7$ is above the x-axis, below the x-axis, and when it crosses the x-axis.

Graphing Polynomials

In the previous sections of this chapter, you find out all you could ever want to know about a polynomial function: intercepts, above or below the x-axis, crossing or touching the axis, and so on. Now all these processes can be put to use when graphing a polynomial function. You know that you're going to have a curvy figure that goes from negative infinity to positive infinity, crossing the y-axis just once. It may never cross the x-axis, but it could also cross it any number of times. Or it may just do a touch-and-go at one or more of the intercepts.

REMEMBER

Make use of all the tools discussed earlier in this chapter when you take on the task of constructing a graph. Your steps are

1. **Determine the y-intercept and x-intercepts. Graph them in your grid.**
2. **Determine where the curve is above and below the x-axis.**
3. **Pick a few points, if necessary, to help you draw in the curve; cross the x-axis when indicated and just touch it in the other instances.**

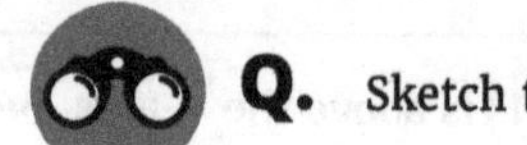

Q. Sketch the graph of $f(x) = -x^4 - 4x^3 + 25x^2 + 28x$.

EXAMPLE

A. Factoring the polynomial, $y = -x^4 - 4x^3 + 25x^2 + 28x = -x(x-4)(x+7)(x+1)$, giving you x-intercepts of 0, 4, −7, and −1. Letting $x = 0$ to solve for the y-intercept, you get $y = 0$. This is one of the x-intercepts. Checking on where the curve is using a sign-line, you have:

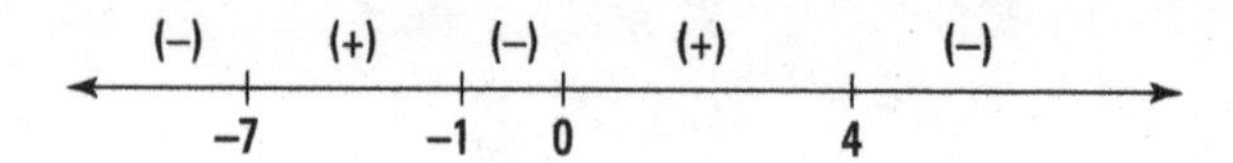

The curve crosses at each intercept. It comes up from negative infinity on the left and departs to negative infinity on the right. Just check out a few points to help with the graphing. Between −7 and −1, try out $x = -5$, giving you $f(-5) = 360$. Also, try when $x = 3$, giving you $f(3) = 120$. Put those points and all the intercepts in your grid and connect the dots!

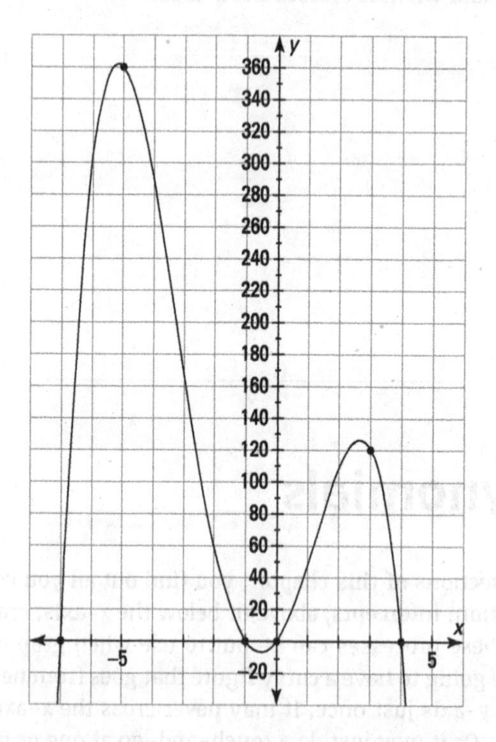

Q. Sketch the graph of $f(x) = (x-1)^2(x+2)^2(x-3)$.

A. The intercepts are (1,0), (−2,0), (3,0), and (0,−12). With the even exponents on two of the factors, you know that the curve will just touch at those two intercepts and cross at (3,0). Select some points to help with the graph and determine when the graph is above or below the x-axis. Choosing $x = -1$, you have $f(-1) = -16$, and when $x = 2$, $f(2) = -16$. Plot these points and sketch the graph.

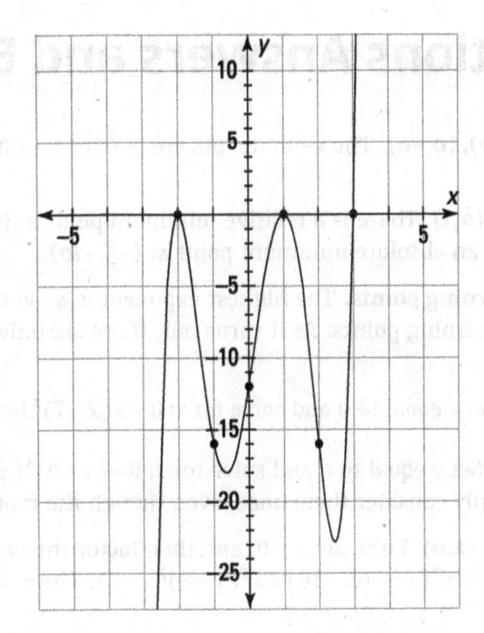

23 **Sketch the graph of $f(x) = x^2(x-3)(x+2)$.**

YOUR TURN

24 **Sketch the graph of $f(x) = x^4 - 5x^3 + x^2 + 21x - 18$.**

Practice Questions Answers and Explanations

1. **(-3,0), (2,0), (5,0), (0,-6).** The x-intercepts are at (–3,0), (2,0), and (5,0). The y-intercept is at (0,–6).

2. **(-5,-10), (4,-5), (0,1).** There is a relative minimum point at (4,–5), a relative maximum point at (0,1) and an absolute minimum point at (–5,–10).

3. **4 intercepts, 3 turning points.** The highest exponent is 4, so there are a maximum of 4 exponents and 3 turning points. As it turns out, there are only 2 intercepts and 1 turning point.

4. **(0,0) and (7,0).** Set y equal to 0 and solve for x: $0 = x(x-7)^3$ has two solutions, x = 0 and x = 7.

5. **(3,0) and (-6,0).** Set y equal to 0 and solve for x: $0 = (x-3)^4(x+6)^5$ has two solutions, x = 3 and x = –6. You only consider them once, even though the roots are repeated.

6. **(0,0), (8,0), and (-2,0).** First, set $y = 0$, and then factor the polynomial: $0 = x^4 - 6x^3 - 16x^2 = x^2(x^2 - 6x - 16) = x^2(x-8)(x+2)$. The x-intercepts are (0,0), (8,0), and (–2,0).

7. **(0,0), (2,0), (1,0), (-1,0).** First, factor out the common factor of x, and then use grouping.

   $$y = x^8 - 8x^5 - x^4 + 8x = x(x^7 - 8x^4 - x^3 + 8) = x(x^4(x^3-8) - 1(x^3-8)) = x(x^3-8)(x^4-1)$$

   Now factor the difference of cubes and the difference of squares — which will be done twice.

   $$\begin{aligned} &= x((x-2)(x^2+2x+4))(x^4-1) \\ &= x((x-2)(x^2+2x+4))((x^2-1)(x^2+1)) \\ &= x((x-2)(x^2+2x+4))((x-1)(x+1)(x^2+1)) \end{aligned}$$

   Only 4 of the factors provide solutions that can be used in the intercepts.

8. $\mathbf{\pm1, \pm3, \pm5, \pm15, \pm\frac{1}{2}, \pm\frac{3}{2}, \pm\frac{5}{2}, \pm\frac{15}{2}}$. All the possible factors of 15 are $\pm1, \pm3, \pm5, \pm15$, and the factors of 2 are $\pm1, \pm2$. Dividing all the factors of 15 by the factors of 2 (and ignoring repeats), you have $\pm1, \pm3, \pm5, \pm15, \pm\frac{1}{2}, \pm\frac{3}{2}, \pm\frac{5}{2}, \pm\frac{15}{2}$.

9. $\mathbf{\pm1, \pm2, \pm4, \pm8}$. You first have to factor out an x: $y = x^4 - 6x^3 - 15x^2 + 8x = x(x^3 - 6x^2 - 15x + 8)$. Now the Rational Root Theorem can be applied to the polynomial in the parentheses. All the factors of 8 are $\pm1, \pm2, \pm4, \pm8$. The only factors of the first term are ±1, so the list will not change.

10. $\mathbf{k(x-1)(x+3)(3x-1)}$. Starting with $f(x) = k(x-1)(x+3)\left(x - \frac{1}{3}\right)$, multiply each side of the equation by 3: $3f(x) = 3k(x-1)(x+3)\left(x-\frac{1}{3}\right) = k(x-1)(x+3)\cdot 3\left(x - \frac{1}{3}\right)$ $= k(x-1)(x+3)(3x-1)$.

(11) $kx(2x-1)(2x-1)(2x-1)$. Starting with $f(x)=kx\left(x-\frac{1}{2}\right)\left(x-\frac{1}{2}\right)\left(x-\frac{1}{2}\right)$, multiply each side of the equation by 8: $8f(x)=8kx\left(x-\frac{1}{2}\right)\left(x-\frac{1}{2}\right)\left(x-\frac{1}{2}\right)=kx\cdot 2\left(x-\frac{1}{2}\right)\cdot 2\left(x-\frac{1}{2}\right)\cdot 2\left(x-\frac{1}{2}\right)$ $=kx(2x-1)(2x-1)(2x-1)$.

(12) **Positive: 4, 2, 0; negative: 4, 2, 0.** First, consider $f(x)$ and count the number of sign changes. The signs change 4 times, so there are 4, 2, or 0 possible positive real roots. Then looking at $f(-x)$, you have $f(-x)=4x^{10}+x^9-2x^8+3x^5-4x^3-5x^2+1$, where there are also 4 changes in sign, so there are 4, 2, or 0 possible negative real roots.

(13) **Positive: 1; negative: none.** First, consider $f(x)$ and count the number of sign changes. The sign changes 1 time, so there is 1 positive real root. Then looking at $f(-x)$, you have $f(-x)=16+x^5$, where there is no change in sign, so there is no possible negative real root.

(14) **3, 5, -1, -5.** Using the sign rule, you see that there are 2 or 0 possible positive real roots and 2 or 0 possible negative real roots. The list of possible rational roots is: $\pm 1, \pm 3, \pm 5, \pm 15, \pm 25, \pm 75$. Performing synthetic division:

3	1	−2	−28	50	75
		3	3	−75	−75
5	1	1	−25	−25	
		5	30	25	
	1	6	5		

The bottom row now represents $x^2+6x+5=(x+1)(x+5)$, giving you the roots of $x=-1$ and $x=-5$.

So the roots are $x=3$, $x=5$, $x=-1$, and $x=-5$.

(15) **1, -1, 3, -3.** Using the sign rule, you see that there are 2 or 0 possible positive real roots and 2 or 0 possible negative real roots. The list of possible rational roots is: $\pm 1, \pm 3, \pm 9, \pm 81$. Performing synthetic division, you have to place zeros for the missing terms:

1	1	0	−1	0	−81	0	81
		1	1	0	0	−81	−81
−1	1	1	0	0	−81	−81	
		−1	0	0	0	81	
3	1	0	0	0	−81		
		3	9	27	81		
−3	1	3	9	27			
		−3	0	−27			
	1	0	9				

The last row gives a quadratic equation of $x^2+9=0$. This has no real solution, so the roots are $x=\pm 1$ and $x=\pm 3$.

(16) **1, 5, 5, $-\frac{3}{2}$.** Using the sign rule, you see that there are 3 or 1 possible positive real roots and 1 possible negative real root. The list of possible rational roots is: $\pm1, \pm3, \pm5, \pm15, \pm75, \pm\frac{1}{2}, \pm\frac{3}{2}, \pm\frac{5}{2}, \pm\frac{15}{2}, \pm\frac{75}{2}$. Performing synthetic division:

1	2	−19	37	55	−75
		2	−17	20	75
5	2	−17	20	75	
		10	−35	−75	
	2	−7	−15		

The last row gives a quadratic equation of $2x^2-7x-15=0$, which factors into $(2x+3)(x-5)=0$, giving you the two solutions $x=-\frac{3}{2}$ and $x=5$. The solutions are $x=1$, $x=5$, $x=-\frac{3}{2}$, $x=5$. There is that double root at 5.

(17) $x^4-5x^3-8x+40$. Change the constant in the divisor to −1. Then perform the synthetic division:

−1	1	−4	−5	−8	32	40
		−1	5	0	8	−40
	1	−5	0	−8	40	

There is no remainder, so the result is written: $\left(x^5-4x^4-5x^3+32x+40\right)\div(x+1)=x^4-5x^3-8x+40$.

(18) $2x^5+2x^4-x^3+x^2+x-3-\frac{2}{x-1}$. Change the constant to +1. Then perform the synthetic division. Be sure to place zeros where there are missing terms.

1	2	0	−3	2	0	−4	1
		2	2	−1	1	1	−3
	2	2	−1	1	1	−3	−2

The remainder is −2. The result is written:

$$\left(2x^6-3x^4+2x^3-4x+1\right)\div(x-1)=2x^5+2x^4-x^3+x^2+x-3-\frac{2}{x-1}.$$

(19) **1.** Using the Remainder Theorem and synthetic division,

2	1	−3	4	−6	5	−1
		2	−2	4	−4	2
	1	−1	2	−2	1	1

(20) **0.** Using the Remainder Theorem and synthetic division,

−2	6	0	0	0	−16	150	0	0	−80	0	0
		−12	24	−48	96	−160	20	−40	80	0	0
	6	−12	24	−48	80	−10	20	−40	0	0	0

(21) **Above at $x < -1$ or $x > 11$; below at $-1 < x < 0$ or $0 < x < 3$ or $3 < x < 11$; cross at $x = -1$, $x = 11$.** Place the intercepts on a sign-line and choose a number for each interval of the line determined by the intercepts.

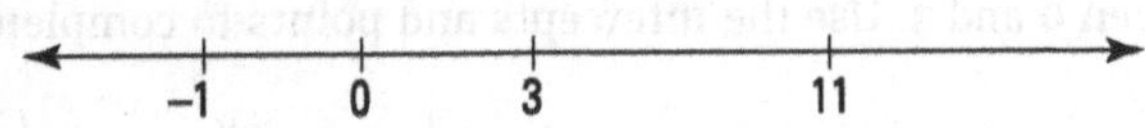

Test the numbers and determine the sign of the product of the factors.

$$
\begin{aligned}
x < -1,&\quad f(x) = (+)(+)(-)(-) = + \\
-1 < x < 0,&\quad f(x) = (+)(+)(+)(-) = - \\
0 < x < 3,&\quad f(x) = (+)(+)(+)(-) = - \\
3 < x < 11,&\quad f(x) = (+)(+)(+)(-) = - \\
x > 11,&\quad f(x) = (+)(+)(+)(+) = +
\end{aligned}
$$

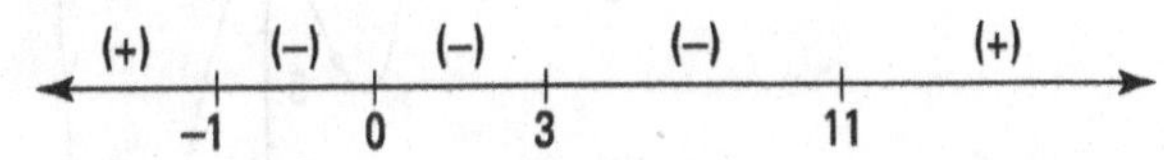

The graph is above the axis when $x < -1$ or $x > 11$, below the axis when $-1 < x < 0$ or $0 < x < 3$ or $3 < x < 11$, and only crosses the axis at –1 and 11. At the two intercepts (0,0) and (3,0) the curve just touches the x-axis and doesn't cross.

(22) **Above at $x < -4$ or $0 < x < 11$ or $x > 5$; below at $-4 < x < 0$ or $1 < x < 5$; cross at $x = -4, 0, 1, 5$.** Place the intercepts on a sign-line and choose a number for each interval of the line determined by the intercepts.

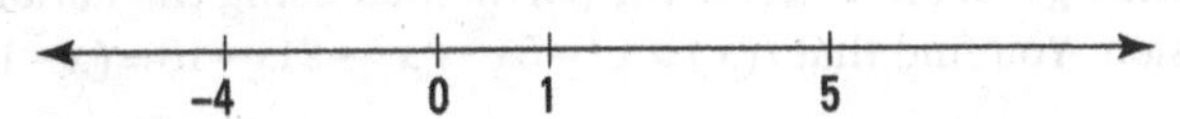

Test the numbers and determine the sign of the product of the factors.

$$
\begin{aligned}
x < -4,&\quad f(x) = (-)(-)(-)(-) = + \\
-4 < x < 0,&\quad f(x) = (-)(-)(+)(-) = - \\
0 < x < 1,&\quad f(x) = (+)(+)(+)(+) = + \\
1 < x < 5,&\quad f(x) = (+)(-)(+)(+) = - \\
x > 5,&\quad f(x) = (+)(+)(+)(+) = +
\end{aligned}
$$

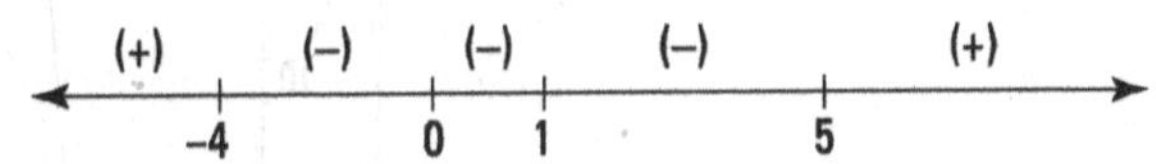

The graph is above the axis when $x < -4$ or $0 < x < 1$ or $x > 5$, below the axis when $-4 < x < 0$ or $1 < x < 5$, and crosses the axis at $x = -4, 0, 1, 5$, each of the intercepts.

23. **See the figure.** The x-intercepts are (0,0) — which is also the y-intercept — (3,0) and (−2,0). The curve touches the axis at 0 and crosses at 3 and −2. Some points on the graph are (−1,−4) and (2,−16), telling you that the curve is below the axis between −2 and 0 and below the axis between 0 and 3. Use the intercepts and points to complete the graph.

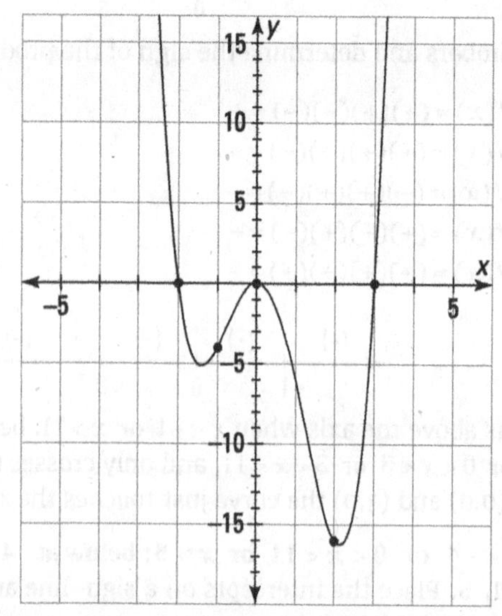

24. **See the figure.** First, factor the polynomial using the Rational Root Theorem and synthetic division. You find that $f(x) = x^4 - 5x^3 + x^2 + 21x - 18 = (x-1)(x+2)(x-3)^2$.

There are x-intercepts at (1,0), (−2,0), and (3,0). The y-intercept is (0,−18). The curve crosses the x-axis at 1 and −2, and it touches the axis when $x = 3$. Finding the points (−1,−32) and (2,4) tells you that the curve is below the x-axis between −2 and 0 and it's above the axis between 1 and 3. Use the intercepts and points to complete the graph.

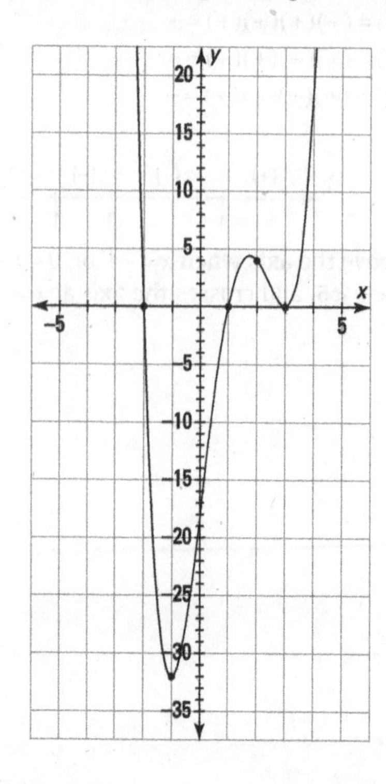

Whaddya Know? Chapter 9 Quiz

Quiz time! Complete each problem to test your knowledge on the various topics covered in this chapter. You can then find the solutions and explanations in the next section.

1. Given the graph of a function, determine any relative maximum or minimum points and any absolute maximum or minimum points.

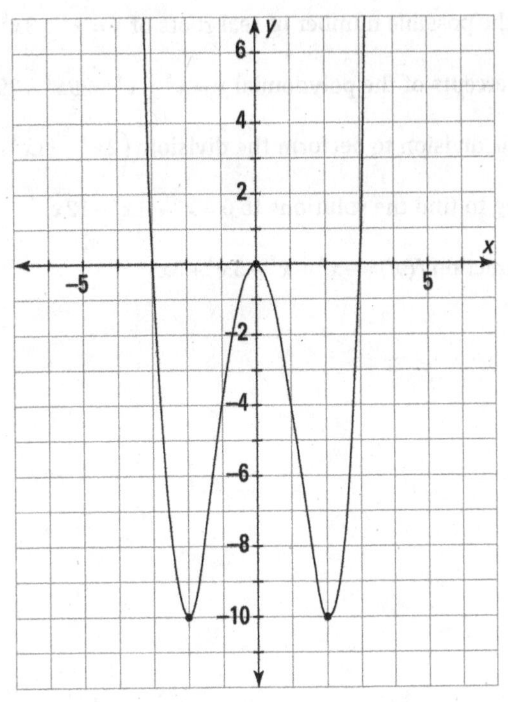

2. Given $f(x) = 3x^5 - 9x^3 - 8x^2 + 13$, use the Remainder Theorem to find $f(-2)$.

3. Determine the intercepts of the polynomial: $y = -3x(x-4)(x+7)(x^2-9)$.

4. Given $f(x) = -3x(x-4)(x+7)(x^2-9)$, determine where the graph is above and below the x-axis.

5. Find the x-intercepts of $f(x) = x^3 + x^2 - 4x - 4$.

6. Use the Rational Root Theorem to find the solutions to $0 = x^5 - 49x^3 - 8x^2 + 392$.

7. Determine when the graph of the function $y = x(x+3)(x-2)(x^2+5)$ is above the x-axis.

8. Determine the intercepts of the polynomial: $y = x^3 - 27$.

9. Determine all the possible real roots of the function $y = x^3 + 6x^2 + 3x - 10$ using the Rational Root Theorem and Descartes' Rule of Signs. Then find the roots using synthetic division.

10. Determine when the graph of the function $y = x^5 + 3x^4 - 9x^3 - 31x^2 + 36$ is above the x-axis and when it crosses the axis.

11. Determine all the possible real roots of the function $y = 2x^5 + 5x^4 - 96x^3 - 149x^2 + 1150x + 600$ using the Rational Root Theorem and Descartes' Rule of Signs. Then find the roots using synthetic division.

12. Use synthetic division to perform the division: $\left(x^3 - 3x^2 - 10x + 24\right) \div (x - 2)$.

13. Determine the possible number of real roots of $y = x^4 - 3x^2 + 2x + 12$.

14. Find the intercepts of the polynomial $y = x^5 + x^4 - 29x^3 - 29x^2 + 100x + 100$.

15. Use synthetic division to perform the division: $\left(2x^4 + 11x^3 - 3x^2 - 44x - 11\right) \div (x + 5)$.

16. Use factoring to find the solutions to $0 = x^3 + 4x^2 - 12x$.

17. Graph the function $f(x) = -x^4 + x^3 + 5x^2 + 3x$.

Answers to Chapter 9 Quiz

1. **Absolute minima: (−2,−10) and (2,−10). Relative maximum: (0,0).** The graph continues on up to positive infinity to the left of $x=-2$ and to the right of $x=2$.

2. **−43.** Applying the Remainder Theorem by using synthetic division:

−2	3	0	−9	−8	0	13	
		−6	12	−6	28	−56	$f(-2)=-43$
	3	−6	3	−14	28	−43	

3. **(0,0), (4,0), (−7,0), (3,0), (−3,0).** To find the *x*-intercepts, set *y* equal to 0 and solve for *x*. Because the equation is already factored, set each factor equal to 0.

$$-3x(x-4)(x+7)(x^2-9)=0$$

$-3x=0,\ x=0$

$x-4=0,\ x=4$

$x+7=0,\ x=-7$

$x^2-9=0,\ x=\pm 3$

To find the *y*-intercept, let *x* equal 0 and solve for *y*.

$$y=-3\cdot 0\cdot(0-4)(0+7)(0^2-9)=0$$

This is also an *x*-intercept.

4. **Above: $x<-7$ or $-3<x<0$ or $3<x<4$. Below: $-7<x<-3$ or $0<x<3$ or $x>4$.** The *x*-intercepts are at 0, 4, −7, 3, and −3. Place these points on a number line and determine the signs of the product in each interval.

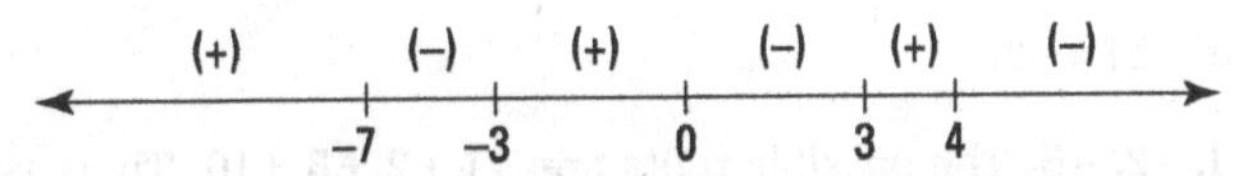

The curve is above the axis when the product is positive and below the axis when the product is negative.

5. **(−1,0), (2,0), (−2,0).** First, use grouping.

$$0=x^3+x^2-4x-4=x^2(x+1)-4(x+1)=(x+1)(x^2-4)$$

Now factor the difference of squares.

$$=(x+1)(x-2)(x+2)$$

Set each factor equal to zero to get: $x=-1$, $x=2$, $x=-2$. These are the *x*-coordinates of the intercepts.

6. **2, 7, −7.** The possible rational roots are: $\pm1, \pm2, \pm4, \pm7, \pm8, \pm14, \pm28, \pm49, \pm56, \pm98, \pm196, \pm392$. There are 2 sign changes, so you have 2 or 0 positive real roots possible. When replacing x with $-x$, there are 3 sign changes, so you have 3 or 1 possible negative real roots. Using synthetic division,

2	1	0	−49	−8	0	392
		2	4	−90	−196	−392
7	1	2	−45	−98	−196	
		9	63	126	196	
−7	1	9	18	28		
		−7	−14	−28		
	1	2	4			

The resulting coefficients create a polynomial that has no real roots. The solutions are $x = 2, 7, -7$. These are all the possible real solutions, because $x^2 + 2x + 4$ is not factorable.

As it turns out, this polynomial can also be factored using grouping.

7. $-3 < x < 0$ **and** $x > 2$. Indicate the x-intercepts on the number line and determine whether the product is positive or negative in the regions determined. Note that the factor $(x^2 + 5)$ is always positive.

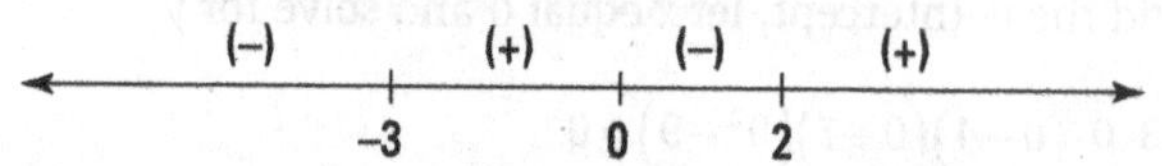

8. $(3,0)$, $(0,-27)$. To find the x-intercepts, set y equal to 0 and solve for x.

$x^3 - 27 = 0 \rightarrow x^3 = 27 \rightarrow x = 3$

To find the y-intercept, let x equal 0 and solve for y.

$y = 0^3 - 27 = -27$

9. $x = 1, -2, -5$. The possible roots are: $\pm1, \pm2, \pm5, \pm10$. There is one change of sign in $x^3 + 6x^2 + 3x - 10$, so there is only 1 possible positive root; there are two changes of sign in $-x^3 + 6x^2 - 3x - 10$, so there are 2 or 0 negative real roots. Using synthetic division:

1	1	6	3	−10
		1	7	10
−2	1	7	10	
		−2	−10	
−5	1	5		
		−5		
	1			

Of course, you can stop after the first division and factor the quadratic formed from the 1, 7, 10: $x^2 + 7x + 10 = (x+2)(x+5)$.

(10) **Above: $-3 < x < -2$, $-2 < x < 1$, and $x > 3$; crosses at $-3, 1$, and 3.** Use synthetic division. The possible solutions are $\pm 1, \pm 2, \pm 3, \pm 4, \pm 6, \pm 9, \pm 12, \pm 18, \pm 36$.

-3	1	3	-9	-31	0	36
		-3	0	27	12	-36
-2	1	0	-9	-4	12	0
		-2	4	10	-12	
3	1	-2	-5	6	0	
		3	3	-6		
1	1	1	-2	0		
		1	2			
-2	1	2	0			
		-2				
	1	0				

The solutions are $x = -2, 1, -3$, or 3. Writing the factorization of the polynomial: $y = (x+2)^2(x-1)(x+3)(x-3)$. Indicate the x-intercepts on the number line and determine whether the product is positive or negative in the regions determined.

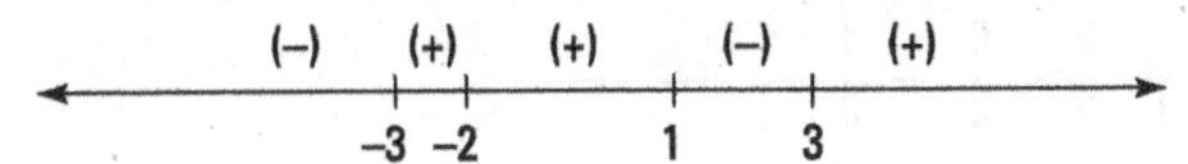

The signs change at -3, 1, and 3; this is where the curve crosses the x-axis. With the double root at -2, it just touches the axis at $(-2,0)$.

(11) **$x = 4, 5, -5, -6, -\frac{1}{2}$.** The possible roots are:

$\pm 1, \pm 2, \pm 3, \pm 4, \pm 5, \pm 6, \pm 8, \pm 10, \pm 12, \pm 15, \pm 20, \pm 24,$
$\pm 25, \pm 30, \pm 50, \pm 60, \pm 75, \pm 100, \pm 150, \pm 300, \pm 600$

There are two changes of sign in $2x^5 + 5x^4 - 96x^3 - 149x^2 + 1150x + 600$ and three changes of sign in $-2x^5 + 5x^4 + 96x^3 - 149x^2 - 1150x + 600$. So there are either 2 or 0 positive roots or 3 or 1 negative roots.

Using synthetic division:

4	2	5	-96	-149	1150	600
		8	52	-176	-1300	-600
5	2	13	-44	-325	-150	0
		10	115	355	150	
-5	2	23	71	30	0	
		-10	-65	-30		
	2	13	6	0		

You find the roots 4, 5, and -5.

Then factor the quadratic formed from the last quotient: $2x^2 + 13x + 6 = (2x+1)(x+6)$. The two roots are $-\frac{1}{2}$ or -6.

(12) $\boldsymbol{x^2 - x - 12}$. **Write the coefficients of the cubic, and use the opposite of the divisor's number.**

2	1	−3	−10	24
		2	−2	−24
	1	−1	−12	0

There is no remainder. Fill in the variables starting with one power less than the highest in the polynomial being divided: $\left(x^3 - 3x^2 - 10x + 24\right) \div \left(x - 2\right) = x^2 - x - 12$.

(13) **2 or 0 positive, 2 or 0 negative.** There are 2 sign changes in $x^4 - 3x^2 + 2x + 12$ and 2 sign changes in $x^4 - 3x^2 - 2x + 12$.

(14) **(5,0), (−5,0), (2,0), (−2,0), −1,0), (0,100).** There are 2 or 0 possible positive roots (x-intercepts) and 3 or 1 possible negative roots. Using synthetic division:

5	1	1	−29	−29	100	100
		5	30	5	−120	−100
−5	1	6	1	−24	−20	0
		−5	−5	20	20	
2	1	1	−4	−4	0	
		2	6	4		
−2	1	3	2	0		
		−2	−2			
−1	1	1	0			
		−1				
	1	0				

Writing the factorization of the polynomial, $y = x^5 + x^4 - 29x^3 - 29x^2 + 100x + 100$ $= (x-5)(x+5)(x-2)(x+2)(x+1)$, **giving you x-intercepts at 5, −5, 2, −2, and −1. To find the y-intercept, set x equal to 0 and solve for y:** $y = 0^5 + 0^4 - 29 \cdot 0^3 - 29 \cdot 0^2 + 100 \cdot 0 + 100 = 100$.

(15) $\boldsymbol{2x^3 + x^2 - 8x - 4 + \frac{9}{x+5}}$. **Write the coefficients of the polynomial, and use the opposite of the divisor's number.**

−5	2	11	−3	−44	−11
		−10	−5	40	20
	2	1	−8	−4	9

The remainder is 9. Fill in the variables of the answer starting with one power less than the highest in the polynomial being divided: $\left(2x^4 + 11x^3 - 3x^2 - 44x - 11\right) \div \left(x + 5\right) =$ $2x^3 + x^2 - 8x - 4 + \frac{9}{x+5}$

(16) $\boldsymbol{x = 0, x = -6, x = 2}$. **First, factor x from each term. Then factor the quadratic trinomial:**

$$0 = x^3 + 4x^2 - 12x = x\left(x^2 + 4x - 12\right) = x\left(x+6\right)\left(x-2\right)$$

Set each factor equal to zero to get: $x = 0,\ x = -6,\ x = 2$.

17 **See the figure.** First factor the polynomial by dividing out $-x$. $f(x) = -x^4 + x^3 + 5x^2 + 3x = -x\left(x^3 - x^2 - 5x - 3\right)$. Then, using synthetic division on the polynomial in the parentheses:

$$\begin{array}{r|rrrr} -1 & 1 & -1 & -5 & -3 \\ & & -1 & 2 & 3 \\ \hline -1 & 1 & -2 & -3 & 0 \\ & & -1 & 3 & \\ \hline 3 & 1 & -3 & 0 & \\ & & 3 & & \\ \hline & 1 & 0 & & \end{array}$$

Writing the factorization of the polynomial, $f(x) = -x^4 + x^3 + 5x^2 + 3x = -x(x+1)^2(x-3)$. The x-intercepts are $(0,0)$, $(-1,0)$, and $(3,0)$; $(0,0)$ is also the y-intercept. The curve touches the axis when $x = -1$ and crosses at the other two intercepts. The curve dives to negative infinity to the left of -1 and to the right of 3. Use the point $(2,18)$ to help you graph the function.

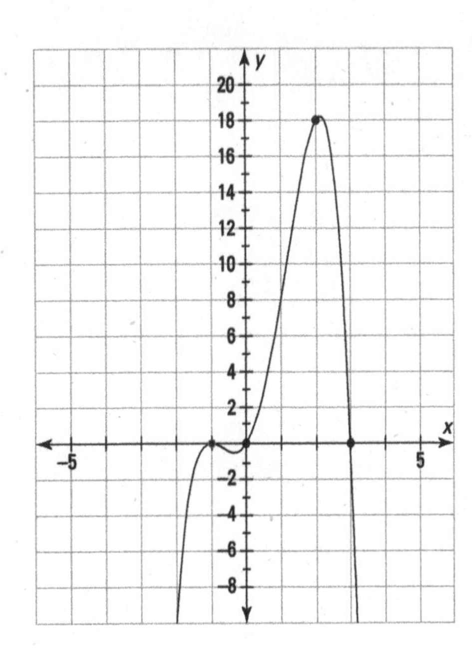

IN THIS CHAPTER

- » Covering the basics of rational functions
- » Identifying vertical, horizontal, and oblique asymptotes
- » Spotting removable discontinuities in rational graphs
- » Edging up to rational limits
- » Taking rational cues to sketch graphs

Chapter 10
Acting Rationally with Functions

The word *rational* has many interpretations. You say that rational people act reasonably and predictably. You can also say that rational numbers are reasonable and predictable — their decimals either repeat (have a distinctive pattern as they go on forever and ever) or terminate (come to an abrupt end). This chapter gives your rational repertoire another boost — it deals with rational functions.

A rational function may not appear to be reasonable but it's definitely predictable. In this chapter, you refer to the intercepts, the asymptotes, any removable discontinuities, and the limits of rational functions to tell where the function values have been, what they're doing for particular values of the domain, and what they'll be doing for large values of *x*. You also need all this information to discuss or graph a rational function. A nice feature of rational functions is that you can use the intercepts, asymptotes, and removable discontinuities to help you sketch the graphs of the functions. And, by the way, you can throw in a few limits to help you finish the whole thing off with a bow on top.

Whether you're graphing rational functions by hand (yes, of course, it's holding a pencil) or with a graphing calculator, you need to be able to recognize their various characteristics (domain, intercepts, asymptotes, and so on). If you don't know what these characteristics are and how to find them, your calculator is no better than a paperweight to you.

Exploring the Domain and Intercepts of Rational Functions

You see rational functions written, in general, in the form of a fraction: $y = \frac{f(x)}{g(x)}$, where f and g are polynomials (expressions with whole-number exponents; see Chapter 9) and g cannot be equal to 0.

Rational functions (and more specifically their graphs) are distinctive because of what they do and don't have. The graphs of rational functions do have helpers called *asymptotes* (lines drawn in to help with the shape and direction of the curve), and the graphs often don't have all the real numbers in their domains. Polynomials and exponential functions (which I cover in Chapters 9 and 11, respectively) make use of all the real numbers — their domains aren't restricted.

Sizing up domain

As I explain in Chapter 7, the domain of a function consists of all the real numbers that you can use in the function equation. Values in the domain have to work in the equation and avoid producing imaginary or nonexistent answers.

You write the equations of rational functions as fractions — and fractions have denominators. The denominator of a fraction can't equal zero, so you exclude anything that makes the denominator of a rational function equal to zero from the domain of the function.

EXAMPLE

Q. What is the domain of $y = \frac{x-1}{x-2}$?

A. $(-\infty, 2) \cup (2, \infty)$. The domain of $y = \frac{x-1}{x-2}$ is all real numbers except 2, because x cannot be 2 or the denominator would be equal to 0. In interval notation, you write the domain as $(-\infty, 2) \cup (2, \infty)$. (The symbol ∞ signifies that the numbers increase without end, and the $-\infty$ signifies that they decrease without end; the $\cup$ between the two parts of the answer means *or*.)

Q. What is the domain of $y = \frac{x+1}{x(x+4)}$?

A. $(-\infty, -4) \cup (-4, 0) \cup (0, \infty)$. The domain of $y = \frac{x+1}{x(x+4)}$ is all real numbers except 0 and −4. In interval notation, you write the domain as $(-\infty, -4) \cup (-4, 0) \cup (0, \infty)$.

Q. What is the domain of $y = \frac{x}{x^2+3}$?

A. $(-\infty, \infty)$. The domain of $y = \frac{x}{x^2+3}$ is all real numbers; no number makes the denominator equal to zero. You can write the domain as $(-\infty, \infty)$.

Q. What is the domain of $y=\frac{x(x-1)}{x^2-1}$?

A. **All real numbers except 1 and –1.** Even though you can factor out the binomial $(x-1)$ from both the numerator and denominator, you still have to eliminate the number 1 from the domain. You'll see more on this in the section, "Removing Discontinuities."

YOUR TURN

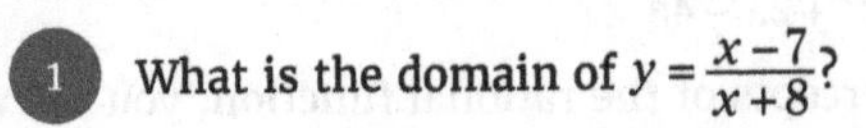

1 What is the domain of $y=\frac{x-7}{x+8}$?

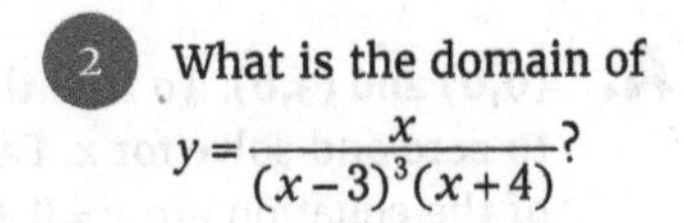

2 What is the domain of $y=\frac{x}{(x-3)^3(x+4)}$?

Introducing intercepts

Functions in algebra can have intercepts. A rational function may have some x-intercepts and/or a y-intercept, but it doesn't have to have either. You can determine whether a given rational function has intercepts by looking at its equation.

Using zero to find y-intercepts

The coordinate $(0,b)$ represents the y-intercept of a rational function. To find the value of b, you substitute a zero for x and solve for y.

EXAMPLE

Q. What is the y-intercept of the function $y=\frac{x+6}{x-3}$?

A. $(0,-2)$. Replace each x with zero to get $y=\frac{0+6}{0-3}=\frac{6}{-3}=-2$. The y-intercept is $(0,-2)$.

Q. What is the y-intercept of the function $y=\frac{x+1}{x^2}$?

A. **No y-intercept.** You can't replace x with 0, because you end up dividing by 0. This function has no y-intercept.

If zero is in the domain of a rational function, you can be sure that the function has at least a y-intercept. A rational function doesn't have a y-intercept if its denominator equals zero when you substitute zero in the equation for x.

X marks the spot

The coordinate (a,0) represents an x-intercept of a rational function. To find the value(s) of a, you let y equal zero and solve for x. (Basically, you just set the numerator of the fraction equal to zero — after you completely reduce the fraction.) You could also multiply each side of the equation by the denominator to get the same equation — it just depends on how you look at it.

EXAMPLE

Q. What are the x-intercepts of $y=\dfrac{x^2-3x}{x^2+2x-48}$?

A. **(0,0) and (3,0). To find the x-intercepts of the rational function, you set x^2-3x equal to zero and solve for x. Factoring the numerator, you get $x(x-3)=0$. The two solutions of the equation are $x=0$ and $x=3$. The two intercepts, therefore, are (0,0) and (3,0). Neither $x=0$ nor $x=3$ creates a zero in the denominator.**

Q. What are the x-intercepts of $y=\dfrac{2}{x^2-3x+11}$?

A. **The numerator is a constant and is never equal to zero, so there are no x-intercepts.**

YOUR TURN

3 Find the intercepts of the rational function $y=\dfrac{x^2-3x-4}{x^2+2x+1}$.

4 Find the intercepts of the rational function $y=\dfrac{x^2+5x+11}{x^2+3x+11}$.

5 Find the intercepts of the rational function $y=\dfrac{x}{x-3}$.

Adding Asymptotes to the Rational Pot

The graphs of rational functions take on some distinctive shapes because of asymptotes. An asymptote is a sort of ghost line. Asymptotes are drawn into the graph of a rational function to show the shape and direction of the function. The asymptotes aren't really part of the graphs, though, because they aren't made up of function values. Rather, they indicate where the function isn't (usually). You lightly sketch in the asymptotes when you're graphing to help you with the final product. The types of asymptotes that you usually find in a rational function include the following:

- Vertical asymptotes
- Horizontal asymptotes
- Oblique (slant) asymptotes

In this section, I explain how you crunch the numbers of rational equations to identify asymptotes and graph them.

Determining the equations of vertical asymptotes

The equations of vertical asymptotes appear in the form $x = h$. This equation of a line has only the x variable — no y variable — and the number h. A vertical asymptote occurs in the rational function $y = \frac{f(x)}{g(x)}$ if f(x) and g(x) have no common factors, and it appears at whatever values the denominator equals zero — $g(x) = 0$. (In other words, vertical asymptotes occur at values that don't fall in the domain of the rational function.)

A vertical asymptote is a *discontinuity*. A discontinuity is a place where a rational function doesn't exist; you find a break in the flow of the numbers being used in the function equation (the domain). A discontinuity is indicated by a numerical value that tells you where the function isn't defined; this number isn't in the domain of the function. You know that a function is discontinuous wherever a vertical asymptote appears in the graph because vertical asymptotes indicate breaks or gaps in the domain.

EXAMPLE

Q. Find the vertical asymptotes of $y = \frac{x}{x^2 - 4x + 3}$.

A. $x = 1$, $x = 3$. First, note that there's no common factor in the numerator and denominator. Then set the denominator equal to zero. Factoring $x^2 - 4x + 3 = 0$, you get $(x-1)(x-3) = 0$. The solutions are $x = 1$ and $x = 3$, which are the equations of the vertical asymptotes.

Q. Find the vertical asymptote(s) of $y = \frac{x+3}{x}$.

A. **x = 0.** Set the denominator equal to zero. This occurs when x = 0, which is the equation of the vertical asymptote.

Heading after horizontal asymptotes

The horizontal asymptote of a rational function has an equation that appears in the form $y = k$. This linear equation has only the variable y — no x — and the k is some number. A rational function $y = \frac{f(x)}{g(x)}$ has only one horizontal asymptote, if it has one at all. (Some rational functions have no horizontal asymptotes, others have one, and none of them have more than one.) A rational function has a horizontal asymptote when the degree (highest power) of $f(x)$, the polynomial in the numerator, is less than or equal to the degree of $g(x)$, the polynomial in the denominator.

REMEMBER

Here's a rule for determining the equation of a horizontal asymptote. The horizontal asymptote of $y = \frac{f(x)}{g(x)} = \frac{a_n x^n + a_{n-1}x^{n-1} + \ldots + a_0}{b_m x^m + b_{m-1}x^{m-1} + \ldots + b_0}$ is $y = \frac{a_n}{b_m}$ when $n = m$, meaning that the highest degrees of the polynomials in the numerator and denominator are the same. The fraction here is made up of the lead coefficients of the two polynomials. When $n < m$, meaning that the degree in the numerator is less than the degree in the denominator, the horizontal asymptote is $y = 0$.

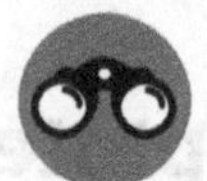

EXAMPLE

Q. Find the horizontal asymptote of $y = \frac{3x^4 - 2x^3 + 7}{x^5 - 3x^2 - 5}$.

A. $y = 0$. You use the previously stated rules. Because $4 < 5$, the horizontal asymptote is $y = 0$.

Q. Find the horizontal asymptote of $y = \frac{3x^4 - 2x^3 + 7}{x^4 - 3x^2 - 5}$.

A. $y = 3$. Because the highest degree is the same in both the numerator and denominator, the horizontal asymptote is $y = 3$: $\frac{a_n}{b_m}$. The fraction formed by the lead coefficients is $y = \frac{3}{1} = 3$.

YOUR TURN

6 Find the equations of the asymptotes of the function $y = \frac{5x^2 - x}{x^2 - 4}$.

7 Find the equations of the asymptotes of the function $y = \frac{x^3 - 27}{x^4 - 5x^2 + 4}$.

Graphing vertical and horizontal asymptotes

When a rational function has one vertical asymptote and one horizontal asymptote, its graph usually looks like two flattened-out, C-shaped curves that appear diagonally opposite to one another from the intersection of the asymptotes. Occasionally, the curves appear side by side, but that's the exception rather than the rule. Figure 10-1 shows you two examples of the more frequently found graphs in the one horizontal and one vertical classification. *Note:* The asymptotes are graphed with dashes rather than solid lines to emphasize that they are not really a part of the graph of the function.

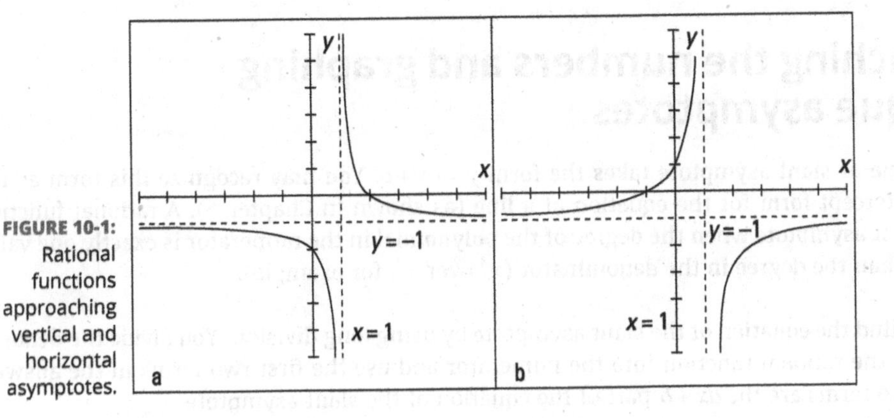

FIGURE 10-1: Rational functions approaching vertical and horizontal asymptotes.

In both graphs, the vertical asymptotes are at $x = 1$, and the horizontal asymptotes are at $y = -1$. In Figure 10-1a, the intercepts are $(0,-2)$ and $(2,0)$. Figure 10-1b has intercepts of $(-1,0)$ and $(0,1)$.

You can have only one horizontal asymptote in a rational function but you can have more than one vertical asymptote. Typically, the curve to the right of the right-most vertical asymptote and to the left of the left-most vertical asymptote are like flattened-out or slowly turning C's. They nestle in the corner and follow along the asymptotes. Between the vertical asymptotes is where some graphs get more interesting. Some graphs between vertical asymptotes can be U-shaped, going upward or downward (see Figure 10-2a), or they can cross in the middle, clinging to the vertical asymptotes on one side or the other (see Figure 10-2b). You find out which case you have by calculating a few points — intercepts and a couple more — to give you clues as to the shape. The graphs in Figure 10-2 show you some of the possibilities.

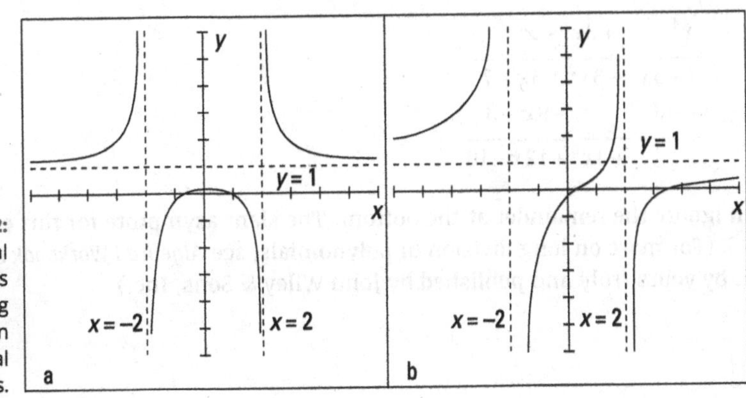

FIGURE 10-2: Rational functions curving between vertical asymptotes.

In Figure 10-2, the vertical asymptotes are at $x = 2$ and $x = -2$. The horizontal asymptotes are at $y = 1$. Figure 10-2a has two x-intercepts lying between the two vertical asymptotes; the y-intercept is there, too. In Figure 10-2b, the y-intercept and one x-intercept lie between the two vertical asymptotes; another x-intercept is to the right of the right-most vertical asymptote.

When graphing rational functions, find the intercepts and the asymptotes, and then find a few points to help you sketch the curve. The graph of a rational function can cross a horizontal asymptote but it never crosses a vertical asymptote. Horizontal asymptotes show what happens for very large or very small values of x.

Crunching the numbers and graphing oblique asymptotes

An oblique or slant asymptote takes the form $y = ax + b$. You may recognize this form as the slope-intercept form for the equation of a line (as shown in Chapter 6). A rational function has a slant asymptote when the degree of the polynomial in the numerator is exactly one value greater than the degree in the denominator (x^4 over x^3, for example).

You can find the equation of the slant asymptote by using long division. You divide the denominator of the rational function into the numerator and use the first two terms in the answer. Those two terms are the $ax + b$ part of the equation of the slant asymptote.

EXAMPLE

Q. Find the slant asymptote of $y = \dfrac{x^3 - 3}{x^2}$.

A. $y = x$. This division is simple. Just put each term in the numerator over the denominator.

$$\frac{x^3 - 3}{x^2} = x - \frac{3}{x^2}$$

The slant asymptote is $y = x$.

Q. Find the slant asymptote of $y = \dfrac{x^4 - 3x^3 + 2x - 7}{x^3 + 3x - 1}$.

A. $y = x - 3$. Do the long division:

$$\begin{array}{r} x - 3 \\ x^3 + 3x - 1 \overline{\big) x^4 - 3x^3 + 2x - 7} \\ \underline{x^4 + 3x^2 - x } \\ -3x^3 - 3x^2 + 3x - 7 \\ \underline{-3x^3 - 9x + 3} \\ + 3x^2 + 12x - 10 \end{array}$$

You can ignore the remainder at the bottom. The slant asymptote for this example is $y = x - 3$. (For more on long division of polynomials, see *Algebra I Workbook For Dummies*, written by yours truly and published by John Wiley & Sons, Inc.)

An oblique (or slant) asymptote creates two new possibilities for the graph of a rational function. If a function has an oblique asymptote, its curve tends to be a very-flat C on opposite sides of the intersection of the slant asymptote and a vertical asymptote (see Figure 10-3a), or the curve has U-shapes between the asymptotes (see Figure 10-3b).

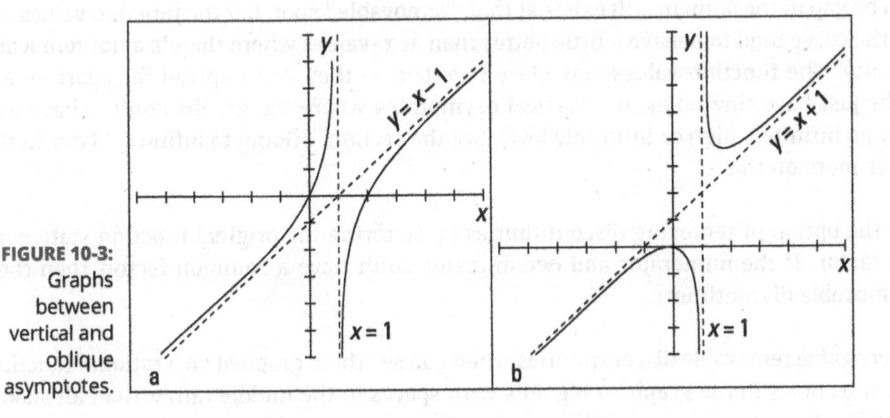

FIGURE 10-3: Graphs between vertical and oblique asymptotes.

Figure 10-3a has a vertical asymptote at $x = 1$ and a slant asymptote at $y = x - 1$; its intercepts are at (0,0) and (2,0). Figure 10-3b has a vertical asymptote at $x = 1$ and a slant asymptote at $y = x + 1$; its only intercept is at (0,0).

YOUR TURN

8 Find the equation of the slant asymptote for $y = \dfrac{x^2 + 3x - 2}{x - 1}$.

9 Find the equation of the slant asymptote for $y = \dfrac{x^3 + 2x^2 - 4x - 6}{x^2 - 3x + 1}$.

10 Find the intercepts and equations of all the asymptotes of $y = \dfrac{2x^2 + 5x - 3}{x - 4}$.

11 Find the intercepts and equations of all the asymptotes of $y = \dfrac{x^3 - 9x}{x^2 - 3x + 2}$.

Removing Discontinuities

Discontinuities at vertical asymptotes can't be removed. But rational functions sometimes have *removable* discontinuities in other places. The removable designation is, however, a bit misleading. The gap in the domain still exists at that "removable" spot, but the function values and graph of the curve tend to behave a little better than at x-values where there's a nonremovable discontinuity. The function values stay close together — they don't spread far apart — and the graphs just have tiny holes, not vertical asymptotes where the graphs don't behave very well (they go infinitely high or infinitely low). See the section, "Going to infinity," later in this chapter for more on this.

You have the option of removing discontinuities by factoring the original function statement, if it does factor. If the numerator and denominator don't have a common factor, then there isn't a removable discontinuity.

You can recognize removable discontinuities when you see them graphed on a rational function; they appear as holes in the graph — big dots with spaces in the middle rather than all shaded in. Removable discontinuities aren't big, obvious discontinuities like vertical asymptotes; you have to look carefully for them. If you just can't wait to see what these things look like, skip ahead to the section, "Showing removable discontinuities on a graph," later in the chapter.

Removal by factoring

Discontinuities are removed when they no longer have an effect on the rational function equation. You know this is the case when you find a factor that's common to both the numerator and the denominator. You accomplish the removal process by factoring the polynomials in the numerator and denominator of the rational function and then reducing the fraction.

EXAMPLE

Q. Remove the discontinuity in $y = \frac{x^2 - 4}{x^2 - 5x - 14}$.

A. To remove the discontinuity in the rational function, you first factor the numerator and denominator of the fraction $y = \frac{(x-2)(x+2)}{(x-7)(x+2)}$.

Now you reduce the fraction to the new function statement: $y = \frac{(x-2)\cancel{(x+2)}}{(x-7)\cancel{(x+2)}}$, giving you $y = \frac{(x-2)}{(x-7)}$.

Q. Remove the discontinuity in $y = \frac{x^3 - 8}{x^2 + 7x - 18}$.

A. Factor the numerator and denominator and then reduce the fraction.

$$y = \frac{x^3 - 8}{x^2 + 7x - 18} = \frac{\cancel{(x-2)}(x^2 + 2x + 4)}{\cancel{(x-2)}(x+9)} = \frac{x^2 + 2x + 4}{x + 9}$$

By getting rid of the removable discontinuity, you simplify the equation that you're graphing. It's easier to graph a line with a little hole in it than to deal with an equation that has a fraction — with all the computations involved.

Evaluating the removal restrictions

The first function that you see as an example the previous section starts out with a quadratic in the denominator. You factor the denominator, and when you set it equal to zero, you find that the solutions $x = -2$ and $x = 7$ don't appear in the domain of the function. Now what?

Numbers excluded from the domain stay excluded even after you remove the discontinuity. The function still isn't defined for the two values you find. Therefore, you can conclude that the function behaves differently at each of the discontinuities. When $x = -2$, the graph of the function has a hole; the curve approaches the value, skips it, and goes on. It behaves in a reasonable fashion: The function values skip over the discontinuity, but the values get really close to it. When $x = 7$, however, a vertical asymptote appears; the discontinuity doesn't go away. The function values go haywire at that x value and don't settle down at all.

In the case of the second example, when x = 2 you have a hole, and the vertical asymptote is at $x = -9$.

YOUR TURN

12 Find any removable discontinuities in the graph of $y = \frac{x^2 - 4x}{x^2 - 16}$.

13 Find any removable discontinuities in the graph of $y = \frac{x^2 + 4x - 21}{3x^2 - 7x - 6}$.

Showing removable discontinuities on a graph

A vertical asymptote of a rational function indicates a discontinuity, or a place on its graph where the function isn't defined. On either side of a vertical asymptote, the graph rises toward positive infinity or falls toward negative infinity. The rational function has no limit wherever you see a vertical asymptote (see the section, "Going the Limit: Limits at a Number and Infinity," later in the chapter). Some rational functions may have discontinuities at which limits exist. When a function has a removable discontinuity, a limit exists, and its graph shows this by putting a hollow circle in place of a piece of the graph.

VULGAR UNCLES VERSUS VULGAR FRACTIONS

You should try to avoid vulgar people because they can't speak without introducing some unpleasant words or thoughts — like an overbearing uncle at Christmas trying to induce some laughs. A vulgar fraction, however, should elicit the opposite reaction on your part; you should in no way avoid it. Unlike the uncle, vulgar fractions are perfectly proper and G-rated.

A fraction is vulgar if it has one integer divided by another integer, as long as the integer that's doing the dividing isn't equal to zero. For instance, the fraction "two divided by three," or $\frac{2}{3}$, is considered vulgar. The value 2 is in the numerator of the fraction, and the value 3 is in the denominator. The word *numerator* comes from enumerate, which means to announce an amount. So, the numerator announces how many of the fractional parts you're talking about at that time. In the case of $\frac{2}{3}$, you have 2 of the 3 pieces of the whole picture. The word *denominator* means to name something or to tell what kind of item you're talking about. The 3 in $\frac{2}{3}$ tells what kind of divisions of the whole have been made.

Figure 10-4 shows a rational function with a vertical asymptote at $x = -2$ and a removable discontinuity at $x = 3$. The horizontal asymptote is the x-axis (written $y = 0$). Unfortunately, graphing calculators don't show the little hollow circles indicating removable discontinuities. Oh, sure, they leave a gap there, but the gap is only one pixel wide, so you can't see it with the naked eye. You just have to know that the discontinuity is there. We're still better than the calculators!

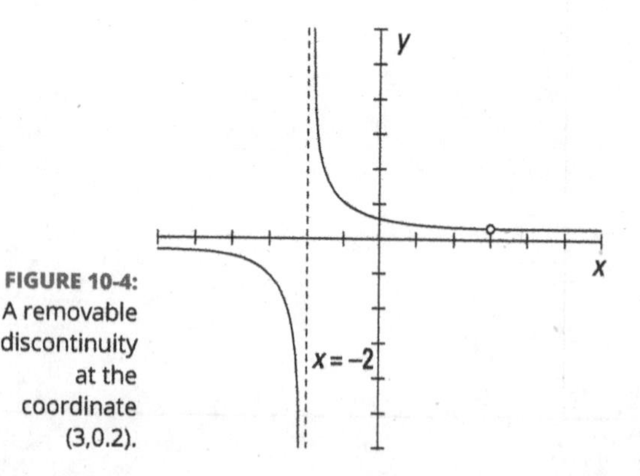

FIGURE 10-4: A removable discontinuity at the coordinate (3,0.2).

Going the Limit: Limits at a Number and Infinity

The limit of a rational function is something like the speed limit on a road. The speed limit tells you how fast you can go (legally). As you approach the speed limit, you adjust the pressure you put on the gas pedal accordingly, trying to keep close to the limit. Most drivers want to stay at least slightly above or slightly below the limit.

The limit of a rational function acts this way, too — homing in on a specific number, either slightly greater than or slightly less than that number. If a function has a limit at a particular number, as you approach the designated number from the left or from the right (from below or above the value, respectively), you approach the same place or function value. The function doesn't have to be defined at the number you're approaching (sometimes they are and sometimes not); there could be a discontinuity. But, if a limit rests at the number, the values of the function have to be really close together — but not touching.

The special notation for limits is as follows:

$$\lim_{x \to a} f(x) = L$$

You read the notation as, "The limit of the function, $f(x)$, as x approaches the number a, is equal to L." The number a doesn't have to be in the domain of the function. You can talk about a limit of a function whether a is in the domain or not. And you can approach a, as long as you don't actually reach it.

Allow me to relate the notation back to following the speed limit. The value a is the exact pressure you need to put on the pedal to achieve the exact speed limit — often impossible to attain.

Look at the function $f(x) = x^2 + 2$. Suppose that you want to see what happens on either side of the value $x = 1$. In other words, you want to see what's happening to the function values as you get close to 1 coming from the left and then coming from the right. Table 10-1 shows you some selected values.

TABLE 10-1 Approaching $x = 1$ from Both Sides in $f(x) = x^2 + 2$

x Approaching 1 from the Left	Corresponding Behavior in $x^2 + 2$	x Approaching 1 from the Right	Corresponding Behavior in $x^2 + 2$
0.0	2.0	2.0	6.0
0.5	2.25	1.5	4.25
0.9	2.81	1.1	3.21
0.999	2.998001	1.001	3.002001
0.99999	2.9999800001	1.00001	3.0000200001

As you approach $x = 1$ from the left or the right, the value of the function approaches the number 3. The number 3 is the limit. You may wonder why I didn't just plug the number 1 into the function equation: $f(1) = 1^2 + 2^2 = 3$. My answer is, in this case, you can. I just used the table to illustrate how the concept of a limit works.

Evaluating limits at discontinuities

The beauty of a limit is that it can also work when a rational function isn't defined at a particular number. The function $y = \frac{x-2}{x^2 - 2x}$, for example, is discontinuous at $x = 0$ and at $x = 2$. You find these numbers by factoring the denominator; setting it equal to zero, $x(x-2) = 0$; and solving for x. This function has no limit when x approaches zero, but it has a limit when x approaches 2. Sometimes it's helpful to actually see the numbers — see what you get from evaluating a function at different values — so I've included Table 10-2. It shows what happens as x approaches zero from the left and right, and it illustrates that the function has no limit at that value.

TABLE 10-2 Approaching $x = 0$ from Both Sides in $y = \frac{x-2}{x^2-2x}$

x Approaching 0 from the Left	Corresponding Behavior of $y = \frac{x-2}{x^2-2x}$	x Approaching 0 from the Right	Corresponding Behavior of $y = \frac{x-2}{x^2-2x}$
–1.0	–1	1.0	1
–0.5	–2	0.5	2
–0.1	–10	0.1	10
–0.001	–1,000	0.001	1,000
–0.00001	–100,000	0.00001	100,000

Table 10-2 shows you that $\lim_{x \to 0} \frac{x-2}{x^2-2x}$ doesn't exist. As x approaches from less than zero, the values of the function drop down lower and lower toward negative infinity. Coming from greater than zero, the values of the function rise higher and higher toward positive infinity. The sides will never come to an agreement; no limit exists.

Table 10-3 shows you how a function can have a limit even when the function isn't defined at a particular number. Sticking with the previous example function, you find a limit as x approaches 2.

TABLE 10-3 Approaching $x = 2$ from Both Sides in $y = \frac{x-2}{x^2-2x}$

x Approaching 2 from the Left	Corresponding Behavior of $y = \frac{x-2}{x^2-2x}$	x Approaching 2 from the Right	Corresponding Behavior of $y = \frac{x-2}{x^2-2x}$
1.0	1.0	3.0	0.3333
1.5	0.6666	2.5	0.4
1.9	0.526316	2.1	0.476190
1.99	0.502512	2.001	0.499750
1.999	0.500250	2.00001	0.4999975

Table 10-3 shows $\lim_{x \to 2} \frac{x-2}{x^2-2x} = 0.5$. The numbers get closer and closer to 0.5 as x gets closer and closer to 2 from both directions. You find a limit at $x = 2$, even though the function isn't defined there.

Determining an existent limit without tables

If you've examined the two tables from the previous section, you may think that the process of finding limits is exhausting. Allow me to tell you that algebra offers a much easier way to find limits — if they exist.

Functions with removable discontinuities have limits at the values where the discontinuities exist. To determine the values of these limits, you follow these steps:

1. Factor the rational function equation.
2. Reduce the function equation.
3. Evaluate the new, revised function equation at the value of x in question.

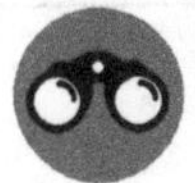

EXAMPLE

Q. Solve for the limit of $y = \frac{x-2}{x^2-2x}$ when x = 2.

A. You first factor and then reduce the fraction:

$$y = \frac{x-2}{x^2-2x} = \frac{\cancel{x-2}}{x\cancel{(x-2)}} = \frac{1}{x}$$

Now you replace the x with 2 and get $y = 0.5$, the limit when $x = 2$. Wow! How simple! In general, if a rational function factors, then you'll find a limit at the number excluded from the domain if the factoring makes that exclusion seem to disappear.

Q. Solve for the limit of $y = \frac{x^2-9}{x^2+8x+15}$ at the point of the removable discontinuity.

A. You first factor and then reduce the fraction:

$$y = \frac{x^2-9}{x^2+8x+15} = \frac{(x-3)\cancel{(x+3)}}{\cancel{(x+3)}(x+5)} = \frac{x-3}{x+5}$$

The removable discontinuity occurs when x = –3. Replace the x in the reduced fraction with –3, and you have $y = \frac{-3-3}{-3+5} = \frac{-6}{2} = -3$. The limit when x = –3 is –3: $\lim_{x \to -3} \frac{x^2-9}{x^2+8x+15} = -3$.

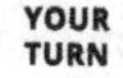

YOUR TURN

14 Find the limit of the function $y = \frac{x^3 - 16x}{x^2 - 5x + 4}$ at the point of the removable discontinuity.

15 Find the limit of the function $y = \frac{x+3}{x^3 + 27}$ at the point of the removable discontinuity.

Determining which functions have limits

Some rational functions have limits at discontinuities and some don't. You can determine whether to look for a removable discontinuity in a particular function by first trying the x value in the function. Replace all the x's in the function with the number in the limit (what x is approaching). The result of that substitution tells you if you have a limit or not. You use the following rules of thumb:

» If $\lim_{x \to a} \frac{f(x)}{g(x)} = \frac{\text{some number}}{0}$, the function has no limit at *a*.

» If $\lim_{x \to a} \frac{f(x)}{g(x)} = \frac{0}{0}$, the function has a limit at *a*. You reduce the fraction and evaluate the newly formed function equation at *a* (as I explain how to do in the preceding section).

A fraction has no value when a zero sits in the denominator, but a zero divided by a zero does have a form — called an *indeterminate form*. Take this form as a signal that you can look for a value for the limit.

EXAMPLE

Q. Determine whether the function $y = \frac{x^2 - 4x - 5}{x^2 - 1}$ has a limit when $x = 1$.

A. You're looking at what x is approaching in the limit statement, so you're only concerned about the 1.

$$\lim_{x \to 1} \frac{x^2 - 4x - 5}{x^2 - 1} = \frac{1 - 4 - 5}{1 - 1} = \frac{-8}{0}$$

The function has no limit at 1 because the substitution creates a number over zero.

Q. Determine whether the function $y = \frac{x^2 - 4x - 5}{x^2 - 1}$ has a limit when $x = -1$.

A. If you try this function at $x = -1$, you see that the function has a removable discontinuity:

$$\lim_{x \to 1} \frac{x^2 - 4x - 5}{x^2 - 1} = \frac{1 - 4(-1) - 5}{1 - 1} = \frac{0}{0}$$

The function has a limit at -1 because the substitution gives you zero over zero. When you factor $(x+1)$ out of the numerator and denominator and then evaluate the new fraction for $x = -1$, you get a limit of 3, so you can say that $\lim_{x \to 1} \frac{x^2 - 4x - 5}{x^2 - 1} = 3$.

Going to infinity

When a rational function doesn't have a limit at a particular value, the function values and graph have to go somewhere. A particular function may not have the number 3 in its domain, and its graph may have a vertical asymptote when $x = 3$. Even though the function has no limit, you can still say something about what's happening to the function as it approaches 3 from the left and the right. The graph has no numerical limit at that point, but you can still tell something about the behavior of the function. The behavior is attributed to one-sided limits.

REMEMBER

A one-sided limit tells you what a function is doing at an x-value as the function approaches from one side or the other. One-sided limits are more restrictive; they work only from the left or from the right.

The notation for indicating one-sided limits from the left or right is shown here:

- The limit as *x* approaches the value *a* from the left is $\lim_{x \to a^-} f(x)$.
- The limit as *x* approaches the value *a* from the right is $\lim_{x \to a^+} f(x)$.

Do you see the little positive or negative sign after the *a*? You can think of "from the left" as coming from the same direction as all the negative numbers on the number line and "from the right" as coming from the same direction as all the positive numbers.

Table 10-4 shows some values of the function $y = \frac{1}{x-3}$, which has a vertical asymptote at $x = 3$.

TABLE 10-4 Approaching $x = 3$ from Both Sides in $y = \frac{1}{x-3}$

x Approaching 3 from the Left	Corresponding Behavior of $y = \frac{1}{x-3}$	*x* Approaching 3 from the Right	Corresponding Behavior of $y = \frac{1}{x-3}$
2.0	–1	4.0	1
2.5	–2	3.5	2
2.9	–10	3.1	10
2.999	–1,000	3.001	1,000
2.99999	–100,000	3.00001	100,000

You express the one-sided limits for the function from Table 10-4 as follows:

$$\lim_{x\to 3^-}\frac{1}{x-3}=-\infty \text{ and } \lim_{x\to 3^+}\frac{1}{x-3}=+\infty$$

The function goes down to negative infinity as it approaches 3 from less than the value, and up to positive infinity as it approaches 3 from greater than the value. "And nary the twain shall meet."

Catching rational limits at infinity

The previous section describes how function values can go to positive or negative infinity as x approaches some specific number. This section also talks about infinity, but it focuses on what rational functions do as their x-values become very large or very small (approaching infinity themselves).

A function such as the parabola $y=x^2+1$ opens upward. If you let x be some really big number, y gets very big, too. Also, when x is very small (a "big" negative number), you square the value, making it positive, so that y is very big for the small x. In function notation, you describe what's happening to this function as the x-values approach infinity with $\lim_{x\to\infty}(x^2+1)=+\infty$.

You can indicate that a function approaches positive infinity going in one direction and negative infinity going in the other direction with the same kind of notation you use for one-sided limits (see the previous section).

EXAMPLE

Q. Describe the movement of the function $y=-x^3+6$ as it approaches negative and positive infinity.

A. The function $y=-x^3+6$ approaches negative infinity as x gets very large — think about what $x=1{,}000$ does to the y value (you'd get $-1{,}000{,}000{,}000+6$). On the other hand, when $x=-1{,}000$, the y value gets very large, because you have $y=-(-1{,}000{,}000{,}000)+6$, so the function approaches positive infinity. The two limit statements you use to describe this are: $\lim_{x\to-\infty}(-x^3+6)=+\infty$ and $\lim_{x\to+\infty}(-x^3+6)=-\infty$.

Q. Describe $y=\dfrac{4x^2+3}{2x^2-3x-7}$ with respect to its horizontal asymptote.

A. When looking for the horizontal asymptote of the function $y=\dfrac{4x^2+3}{2x^2-3x-7}$, you can use the rules in the section, "Heading after horizontal asymptotes" to determine that the horizontal asymptote of the function is $y=2$. Using limit notation, you can write the solution as $\lim_{x\to\infty}\dfrac{4x^2+3}{2x^2-3x-7}=2$.

In the case of rational functions, the limits at infinity — as x gets very large or very small — may be specific, finite, describable numbers. When a rational function has a horizontal asymptote, its limit at infinity is the same value as the number in the equation of the asymptote.

The proper algebraic method for evaluating limits at infinity is to divide every term in the rational function by the highest power of x in the fraction and then look at each term. Here's an important property to use: As x approaches infinity, any term with $\frac{1}{x}$ or any positive power of x, $\frac{1}{x^n}$, in it approaches zero — in other words, gets very small — so you can replace those terms with zero and simplify.

Q. Evaluate the limit at infinity of the function $y=\dfrac{4x^2+3}{2x^2-3x-7}$.

EXAMPLE

A. The highest power of the variable in the fraction is x^2, so every term is divided by x^2:

$$\lim_{x\to\infty}\frac{\frac{4x^2}{x^2}+\frac{3}{x^2}}{\frac{2x^2}{x^2}-\frac{3x}{x^2}-\frac{7}{x^2}}=\lim_{x\to\infty}\frac{4+\frac{3}{x^2}}{2-\frac{3}{x}-\frac{7}{x^2}}=\frac{4+0}{2-0-0}=\frac{4}{2}=2$$

The limit as x approaches infinity is 2. As predicted, the number 2 is the number in the equation of the horizontal asymptote.

Q. Evaluate the limit at infinity of the function $y=\dfrac{x+5}{4x^2-x+6}$.

A. The highest power of the variable in the fraction is x^2, so every term is divided by x^2:

$$\lim_{x\to\infty}\frac{\frac{x}{x^2}+\frac{5}{x^2}}{\frac{4x^2}{x^2}-\frac{x}{x^2}+\frac{6}{x^2}}=\lim_{x\to\infty}\frac{\frac{1}{x}+\frac{3}{x^2}}{4-\frac{1}{x}+\frac{6}{x^2}}=\frac{0+0}{4-0-0}=\frac{0}{4}=0$$

The limit as x approaches infinity is 0. The equation x = 0, which is the x-axis, is the horizontal asymptote.

This quick method for determining horizontal asymptotes is an easier way to find limits at infinity, and this procedure is also the correct mathematical way of doing it — and it shows why the other rule (the quick method) works. You also need this quicker method for more involved limit problems found in calculus and other higher mathematics.

YOUR TURN

16 Find the limit of $y=\dfrac{x^2-5x-6}{2x^2-x-3}$ when $x=-1$.

17 Determine whether the function $y=\dfrac{x^2+4x}{x^2+7x+12}$ has a limit when $x=-4$.

18 Evaluate the limit at infinity of the function $y = \frac{6x^2 - 2x + 3}{2x^2 + 5x + 1}$.

19 Evaluate $\lim_{x\to\infty} \frac{4x^4 - 4x^3 + 9}{36 + 36x^2 - x^4}$.

Graphing Rational Functions

The graphs of rational functions can include intercepts, asymptotes, and removable discontinuities. In fact, some graphs include all three. Sketching the graph of a rational function is fairly simple if you prepare carefully. Make use of any and all information you can glean from the function equation, sketch in any intercepts and asymptotes, and then plot a few points to determine the general shape of the curve.

EXAMPLE

Q. Sketch the graph of the function $y = \frac{x^2 - 2x - 3}{x^2 - x - 2}$.

A. First, look at the powers of the numerator and denominator. The degrees, or highest powers, are the same, so you find the horizontal asymptote by making a fraction of the lead coefficients. Both coefficients are 1, and 1 divided by 1 is still 1, so the equation of the horizontal asymptote is $y = 1$.

The rest of the necessary information for graphing is more forthcoming if you factor the numerator and denominator:

$$y = \frac{(x-3)(x+1)}{(x-2)(x+1)} = \frac{(x-3)\cancel{(x+1)}}{(x-2)\cancel{(x+1)}} = \frac{x-3}{x-2}$$

You factor out a common factor of $x + 1$ from the numerator and denominator. This action tells you two things. First, because $x = -1$ makes the denominator equal to zero, you know that -1 isn't in the domain of the function. Furthermore, the fact that -1 is removed by the factoring signals a removable discontinuity when $x = -1$. You can plug -1 into the new equation to find out where to graph the hole or open circle:

$$y = \frac{x-3}{x-2} = \frac{-1-3}{-1-2} = \frac{-4}{-3} = \frac{4}{3}$$

The hole is at $\left(-1, \frac{4}{3}\right)$. The remaining terms in the denominator tell you that the function has a vertical asymptote at $x = 2$. You find the y-intercept by letting $x = 0$: $y = \frac{0^2 - 2(0) - 3}{0^2 - 0 - 2} = \frac{-3}{-2} = \frac{3}{2}$, so the y-intercept is $\left(0, \frac{3}{2}\right)$. You find the x-intercepts by setting the new numerator equal to zero and solving for x. When $x - 3 = 0$, $x = 3$, so the x-intercept is (3,0). You place all this information on a graph, which is shown in Figure 10-5. Figure 10-5a shows how you sketch in the asymptotes, intercepts, and removable discontinuity.

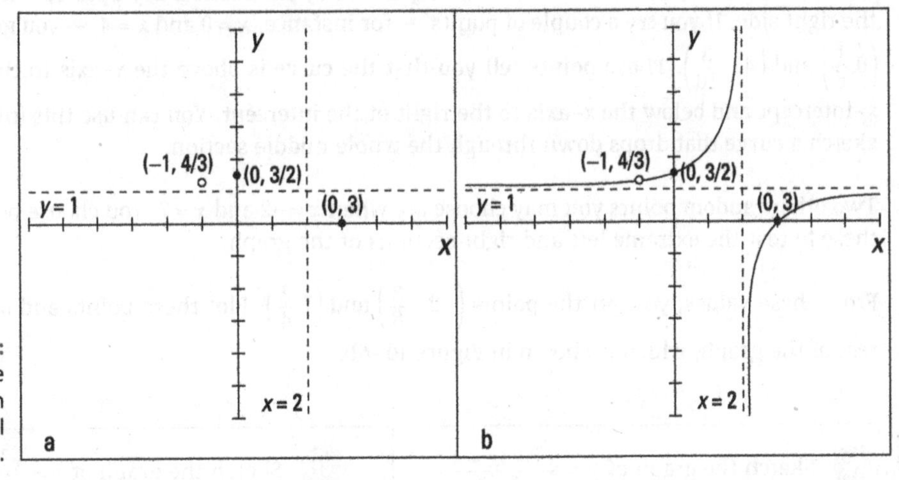

FIGURE 10-5: Following the steps to graph a rational function.

Figure 10-5a seems to indicate that the curve will have soft-C shapes in the upper-left and lower-right parts of the graph, opposite one another through the asymptotes. If you plot a few points to confirm this, you see that the graph approaches positive infinity as it approaches $x = 2$ from the left and goes to negative infinity from the right. You can see the completed graph in Figure 10-5b.

Q. Sketch the graph of the function $y = \frac{x-1}{x^2 - 5x - 6}$.

A. First, look at the powers of the numerator and denominator. When you divide each term in the fraction by x^2 and evaluate it as x approaches infinity, you have 0 divided by 1, so the equation of the horizontal asymptote is $y = 0$, which is the x-axis.

By factoring the denominator and setting it equal to zero, you get $(x+1)(x-6) = 0$; the vertical asymptotes are at $x = -1$ and $x = 6$. The only x-intercept is (1,0), and a y-intercept is located at $\left(0, \frac{1}{6}\right)$. Figure 10-6a shows the asymptotes and intercepts on a graph.

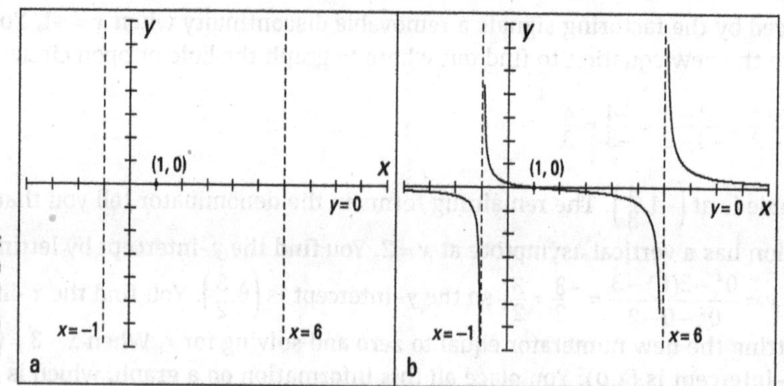

FIGURE 10-6: Graphing a rational function with two vertical asymptotes.

The only place the graph of the function crosses the x-axis is at (1,0), so the curve must come from the left side of that middle section separated by the vertical asymptotes and continue to the right side. If you try a couple of points — for instance, $x = 0$ and $x = 4$ — you get the points $\left(0, \frac{1}{6}\right)$ and $\left(4, -\frac{3}{10}\right)$. These points tell you that the curve is above the x-axis to the left of the x-intercept and below the x-axis to the right of the intercept. You can use this information to sketch a curve that drops down through the whole middle section.

Two other random points you may choose are when $x = -2$ and $x = 7$. You choose points such as these to test the extreme left and right sections of the graph.

From these values, you get the points $\left(-2, -\frac{3}{8}\right)$ and $\left(7, \frac{3}{4}\right)$. Plot these points and sketch in the rest of the graph, which is shown in Figure 10-6b.

YOUR TURN

20 Sketch the graph of $y = \frac{x^2 - 3x - 4}{x^2 - 9}$.

21 Sketch the graph of $y = \frac{x^2 + 2x - 15}{x - 2}$.

22 Sketch the graph of $y = \frac{x^2 + x - 30}{x^2 + 7x + 6}$.

Practice Questions Answers and Explanations

(1) **$(-\infty,-8)\cup(-8,\infty)$.** The domain of $y=\frac{x-7}{x+8}$ is all real numbers except −8, because x cannot be −8 or the denominator would be equal to 0. In interval notation, you write the domain as $(-\infty,-8)\cup(-8,\infty)$.

(2) **$(-\infty,-4)\cup(-4,3)\cup(3,\infty)$.** The domain of $y=\frac{x}{(x-3)^3(x+4)}$ is all real numbers except 3 and −4. In interval notation, you write the domain as $(-\infty,-4)\cup(-4,3)\cup(3,\infty)$.

(3) **(0,−4), (4,0), (−1,0).** To find the y-intercept, set each x equal to 0 and solve for y: $y=\frac{0^2-3\cdot 0-4}{0^2+2\cdot 0+1}=\frac{-4}{1}=-4$. To find the x-intercepts, set the numerator equal to zero, factor the quadratic, and solve for x: $x^2-3x-4=(x-4)(x+1)=0$ when $x=4$ or $x=-1$.

(4) **(0,1).** To find the y-intercept, set each x equal to 0 and solve for y: $y=\frac{0^2+5\cdot 0+11}{0^2+3\cdot 0+11}=\frac{11}{11}=1$. To find the x-intercepts, set the numerator equal to zero, and solve for x. The numerator is never equal to 0, so there are no x-intercepts.

(5) **(0,0).** To find the y-intercept, set each x equal to 0 and solve for y: $y=\frac{0}{0-3}=\frac{0}{-3}=0$. To find the x-intercepts, set the numerator equal to zero, and solve for x. This only happens when x = 0. The x-intercept and y-intercept are the same point.

(6) **x = 2, x = −2, y = 5.** Find the equations of the vertical asymptotes, set the denominator equal to 0, and solve for x: $x^2-4=(x-2)(x+2)=0$ when $x=2$ or $x=-2$. These are the equations of the asymptotes. The horizontal asymptote is equal to the fraction formed by the coefficients of the highest powers — which are the same: $y=\frac{5}{1}=5$.

(7) **x = 2, x = −2, x = 1, x = −1, y = 0.** Find the equations of the vertical asymptotes, set the denominator equal to 0, and solve for x: $x^4-5x^2+4=\left(x^2-4\right)\left(x^2-1\right)=(x-2)(x+2)(x-1)(x+1)=0$. The vertical asymptotes are: $x=2$, $x=-2$, $x=1$, $x=-1$. The horizontal asymptote is y = 0, because the highest power in the numerator is smaller than the highest power in the denominator.

(8) **$y=x+4$.** Divide the numerator by the denominator.

$$\begin{array}{r} x+4 \\ x-1\overline{)x^2+3x-2} \\ \underline{x^2-x} \\ 4x-2 \\ \underline{4x-4} \\ 2 \end{array}$$

The equation of the slant asymptote is $y=x+4$.

(9) **$y=x+5$.** Divide the numerator by the denominator.

$$\begin{array}{r} x+5 \\ x^2-3x+1\overline{)x^3+2x^2-4x-6} \\ \underline{x^3-3x^2+x} \\ 5x^2-5x-6 \\ \underline{5x^2-15x+5} \\ 10x-11 \end{array}$$

The equation of the slant asymptote is $y=x+5$.

10 $\left(0,\frac{3}{4}\right), \left(\frac{1}{2},0\right), (-3,0), x=4, y=2x+13.$

For the y-intercept, set $x=0$: $y=\frac{-3}{-4}=\frac{3}{4}$. The y-intercept is $\left(0,\frac{3}{4}\right)$.

For the x-intercepts, solve $2x^2+5x-3=(2x-1)(x+3)=0$. The x-intercepts are $\left(\frac{1}{2},0\right)$ and $(-3,0)$.

The vertical asymptote is found by setting the denominator equal to 0: $x=4$.

There is no horizontal asymptote.

The slant asymptote is found by dividing the numerator by the denominator: $y=2x+13$.

$$\begin{array}{r} 2x+13 \\ x-4\overline{)2x^2+5x-3} \\ \underline{2x^2-8x} \\ 13x-3 \\ \underline{13x-52} \\ 49 \end{array}$$

11 $(0,0), (3,0), (-3,0), x=1, x=2, y=x+3.$ For the y-intercept, set $x=0$: $y=\frac{0}{2}=0$. The y-intercept is $(0,0)$.

For the x-intercepts, solve $x^3-9x=x\left(x^2-9\right)=x(x-3)(x+3)=0$. The x-intercepts are $(0,0), (3,0), (-3,0)$.

The vertical asymptotes are found by setting the denominator equal to 0: $x^2-3x+2=(x-1)(x-2)=0$. The vertical asymptotes are $x=1, x=2$.

There is no horizontal asymptote.

The slant asymptote is found by dividing the numerator by the denominator: $y=x+3$.

$$\begin{array}{r} x+3 \\ x^2-3x+2\overline{)x^3\qquad\quad -9x} \\ \underline{x^3-3x^2+2x} \\ 3x^2-11x \\ \underline{3x^2-9x+6} \\ -2x-6 \end{array}$$

12 $x=4$. Factor the numerator and denominator. Then reduce the fraction:

$$y=\frac{x^2-4x}{x^2-16}=\frac{x\cancel{(x-4)}}{\cancel{(x-4)}(x+4)}=\frac{x}{x+4}.$$

The removable discontinuity is at $x=4$.

13 $x=3$. Factor the numerator and denominator. Then reduce the fraction:

$$y=\frac{x^2+4x-21}{3x^2-7x-6}=\frac{(x+7)\cancel{(x-3)}}{(3x+2)\cancel{(x-3)}}=\frac{x+7}{3x+2}.$$

The removable discontinuity is at $x=3$.

14. $\frac{32}{3}$**.** First, factor and reduce the function equation:

$y=\frac{x^3-16x}{x^2-5x+4}=\frac{x\left(x^2-16\right)}{(x-4)(x-1)}=\frac{x\cancel{(x-4)}(x+4)}{\cancel{(x-4)}(x-1)}=\frac{x(x+4)}{(x-1)}$. The removable discontinuity occurs when x = 4. Evaluating the factored expression using 4, $y=\frac{4(4+4)}{(4-1)}=\frac{32}{3}$. The limit as x approaches 4 is $\frac{32}{3}$. This is written $\lim_{x\to4}\frac{x^3-16x}{x^2-5x+4}=\frac{32}{3}$.

15. $\frac{1}{27}$**.** First, factor and reduce the function equation

$y=\frac{x+3}{x^3+27}=\frac{\cancel{x+3}}{\cancel{(x+3)}\left(x^2-3x+9\right)}=\frac{1}{x^2-3x+9}$. The removable discontinuity occurs at x = −3. Evaluating the factored expression using −3, $y=\frac{1}{(-3)^2-3(-3)+9}=\frac{1}{9+9+9}=\frac{1}{27}$. The limit as x approaches −3 is $\frac{1}{27}$. This is written $\lim_{x\to-3}\frac{x+3}{x^3+27}=\frac{1}{27}$.

16. $\frac{7}{5}$**.** When replacing each x with −1, you have $y=\frac{(-1)^2-5(-1)-6}{2(-1)^2-(-1)-3}=\frac{1+5-6}{2+1-3}=\frac{0}{0}$. This tells you that there is a removable discontinuity. Factor the expression and remove the discontinuity:

$y=\frac{x^2-5x-6}{2x^2-x-3}=\frac{(x-6)\cancel{(x+1)}}{(2x-3)\cancel{(x+1)}}=\frac{x-6}{2x-3}$. Now, replacing each x with −1, you have

$y=\frac{-1-6}{2(-1)-3}=\frac{-7}{-5}=\frac{7}{5}$.

17. $\lim_{x\to-4}\frac{x^2+4x}{x^2+7x+12}=4$**.** When replacing each x with −4, you have

$y=\frac{(-4)^2+4(-4)}{(-4)^2+7(-4)+12}=\frac{16-16}{16-28+12}=\frac{0}{0}$. This tells you that there is a removable discontinuity. Factor the expression and remove the discontinuity: $y=\frac{x^2+4x}{x^2+7x+12}=\frac{x\cancel{(x+4)}}{\cancel{(x+4)}(x+3)}=\frac{x}{x+3}$.

Now, replacing each x with −4, you have $y=\frac{-4}{-4+3}=\frac{-4}{-1}=4$.

18. **3.** Divide each term in the fraction by x^2: $\lim_{x\to\infty}\frac{\frac{6x^2}{x^2}-\frac{2x}{x^2}+\frac{3}{x^2}}{\frac{2x^2}{x^2}+\frac{5x}{x^2}+\frac{1}{x^2}}=\lim_{x\to\infty}\frac{6-\frac{2}{x}+\frac{3}{x^2}}{2+\frac{5}{x}+\frac{1}{x^2}}=\frac{6-0+0}{2+0+0}=3$.

19. **−4.** Divide each term in the fraction by x^4:

$$\lim_{x\to\infty}\frac{\frac{4x^4}{x^4}-\frac{4x^3}{x^4}+\frac{9}{x^4}}{\frac{36}{x^4}+\frac{36x^2}{x^4}-\frac{x^4}{x^4}}=\lim_{x\to\infty}\frac{4-\frac{4}{x}+\frac{9}{x^4}}{\frac{36}{x^4}+\frac{36}{x^2}-1}=\frac{4-0+0}{0+0-1}=-4.$$

20. **See the figure.** Gathering information: $\lim_{x\to\infty}\frac{x^2-3x-4}{x^2-9}=1$; the horizontal asymptote is y = 1.

Factoring: $y=\frac{x^2-3x-4}{x^2-9}=\frac{(x-4)(x+1)}{(x-3)(x+3)}$; the vertical asymptotes are $x=3$ and $x=-3$.

When y = 0, $(x-4)(x+1)=0$; the x-intercepts are (4,0) and (−1,0).

When x = 0, $y=\frac{(-4)(1)}{(-3)(3)}=\frac{4}{9}$; the y-intercept is $\left(0,\frac{4}{9}\right)$.

Some points on the graph are: $\left(1,\frac{3}{4}\right)$, $\left(2,\frac{6}{5}\right)$, $\left(-4,\frac{24}{7}\right)$

Place the information in the graph and complete the curves.

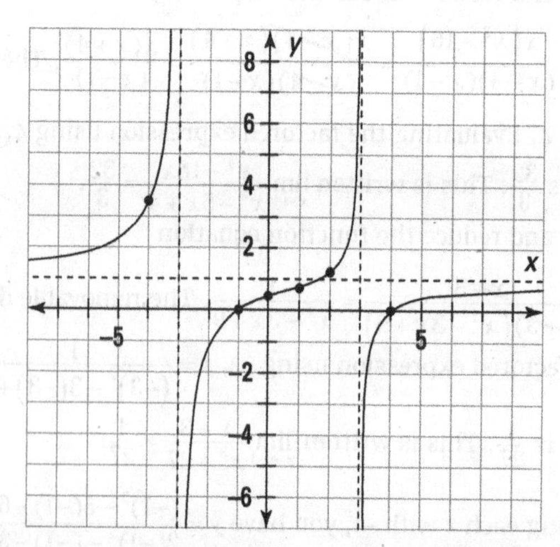

(21) **See the figure.** Gathering information: There is no horizontal asymptote. Find the equation of the slant asymptote by dividing the numerator by the denominator: $\frac{x^2+2x-15}{x-2} = x+4-\frac{7}{x-2}$; the slant asymptote is $y = x+4$.

$$\begin{array}{r} x+4 \\ x-2\overline{)x^2+2x-15} \\ \underline{x^2-2x} \\ 4x-15 \\ \underline{4x-8} \\ -7 \end{array}$$

Factoring: $y = \frac{x^2+2x-15}{x-2} = \frac{(x+5)(x-3)}{x-2}$; the vertical asymptote is $x = 2$.

When $y = 0$, $(x+5)(x-3) = 0$; the x-intercepts are $(-5,0)$ and $(3,0)$.

When $x = 0$, $y = \frac{(5)(-3)}{-2} = \frac{-15}{-2} = \frac{15}{2}$; the y-intercept is $\left(0,\frac{15}{2}\right)$.

Some points on the graph are: $\left(4,\frac{9}{2}\right)$, $\left(-4,\frac{7}{6}\right)$.

Place the information in the graph and complete the curves.

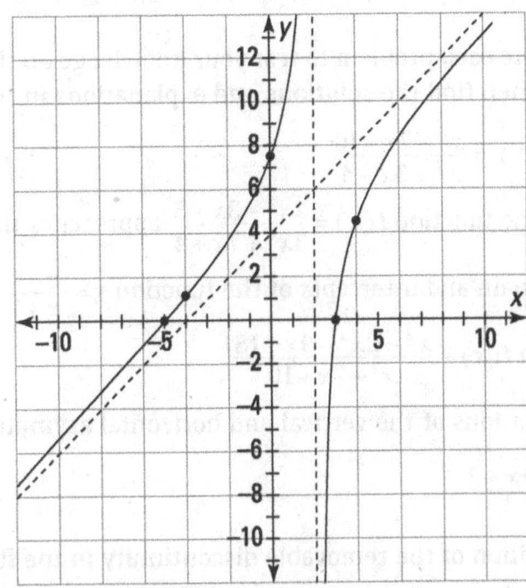

(22) **See the figure.** Gathering information: $\lim_{x\to\infty}\frac{x^2+x-30}{x^2+7x+6}=1$; the horizontal asymptote is $y=1$.

Factoring: $y=\frac{x^2+x-30}{x^2+7x+6}=\frac{\cancel{(x+6)}(x-5)}{\cancel{(x+6)}(x+1)}=\frac{x-5}{x+1}$; there is a removable discontinuity when $x=-6$. The vertical asymptote is $x=-1$, and there is a hole at $\left(-6,\frac{11}{5}\right)$.

When $y=0$, $x-5=0$; the x-intercept is $(5,0)$.

When $x=0$, $y=\frac{-5}{1}=-5$; the y-intercept is $(0,-5)$.

Some points on the graph are: $(-7,2)$, $(-3,4)$, $\left(8,\frac{1}{3}\right)$

Place the information in the graph and complete the curves.

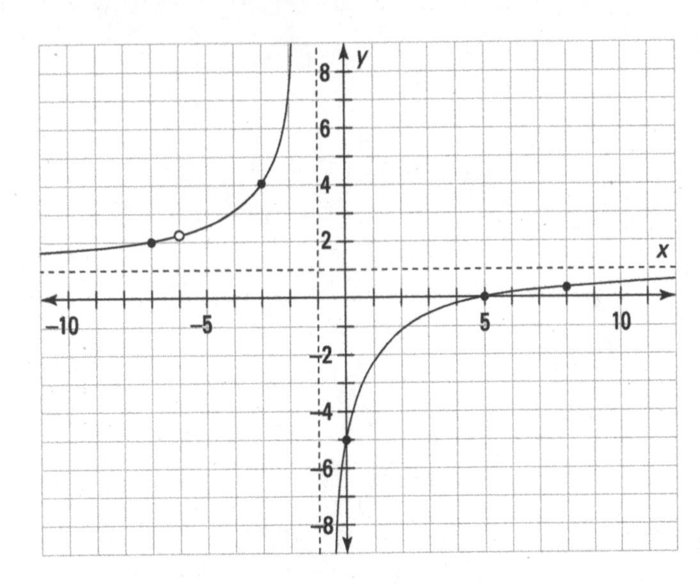

Whaddya Know? Chapter 10 Quiz

Quiz time! Complete each problem to test your knowledge on the various topics covered in this chapter. You can then find the solutions and explanations in the next section.

1. Graph the function $y=\frac{x^2-3x-10}{x^2+3x-4}$.
2. Find the limit as the function $f(x)=\frac{2x^2-3x-5}{3x^2+7x+4}$ approaches the removable discontinuity.
3. Determine the domain and intercepts of the function $y=\frac{x-4}{x+1}$.
4. Graph the function $f(x)=\frac{x^3-2x^2-9x+18}{x^2-7x+10}$.
5. Determine the equations of the vertical and horizontal asymptotes of $y=\frac{2x+7}{x^2-4}$.
6. Evaluate: $\lim_{x\to 3}\frac{x^2-4x+2}{x+1}$
7. Determine the position of the removable discontinuity in the function $y=\frac{x^2+x-6}{x^2-9}$.
8. Graph the function $y=\frac{x-3}{x+2}$.
9. Evaluate: $\lim_{x\to\infty}\frac{x^3-3x^2+6x+5}{x^3+4x^2-11}$
10. Determine the domain and intercepts of the function $y=\frac{x^2-4x-5}{x^2+x-2}$.
11. Graph the function $y=\frac{x^2-4}{x^2+4x-12}$.
12. Find the equation of the oblique asymptote in the function $y=\frac{x^2-6x-7}{x+3}$.

Answers to Chapter 10 Quiz

1. First, factor the numerator and denominator of the function: $y = \frac{x^2 - 3x - 10}{x^2 + 3x - 4} = \frac{(x-5)(x+2)}{(x+4)(x-1)}$.
The intercepts are $(-2,0)$, $(5,0)$, and $\left(0, \frac{5}{2}\right)$. The two vertical asymptotes are $x = -4$ and $x = 1$. The horizontal asymptote is $y = 1$.

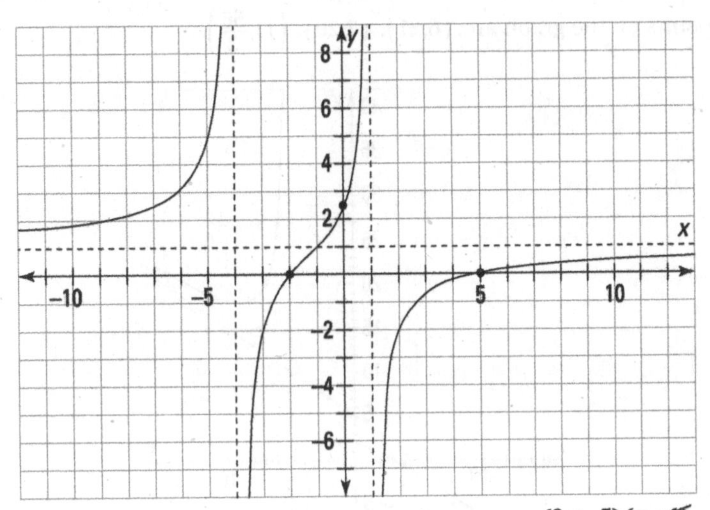

2. **-7.** Factor the fraction and reduce: $f(x) = \frac{2x^2 - 3x - 5}{3x^2 + 7x + 4} = \frac{(2x-5)\cancel{(x+1)}}{(3x+4)\cancel{(x+1)}} = \frac{2x-5}{3x+4}$. The removable discontinuity occurs when $x = -1$. Evaluating the reduced fraction at –1,
$f(-1) = \frac{2(-1) - 5}{3(-1) + 4} = \frac{-7}{1} = -7$.

3. **Domain: $x \neq -1$, intercepts: (4,0), (0,–4).** The domain consists of all the real numbers that can be used. Because the denominator is equal to 0 if x is –1, the domain consists of all real numbers except –1, written $(-\infty, -1) \cup (-1, \infty)$ in interval notation. To find the x intercept, set y equal to 0 and solve for x; to find the y-intercept, let the x's be 0 and solve for y.

$0 = \frac{x-4}{x+1} \rightarrow 0 = x - 4 \rightarrow 4 = x$ $\quad$ $y = \frac{0-4}{0+1} \rightarrow y = \frac{-4}{1} \rightarrow y = -4$

Intercept: (4,0) $\quad$ Intercept: (0,-4)

4. **See the figure.** There is no horizontal asymptote. Find the equation of the slant asymptote by dividing the numerator by the denominator: $\frac{x^3 - 2x^2 - 9x + 18}{x^2 - 7x + 10} = x + 5 + \frac{16x - 32}{x^2 - 7x + 10}$; the slant asymptote is $y = x + 5$.

Factoring:

$$y = \frac{x^3 - 2x^2 - 9x + 18}{x^2 - 7x + 10} = \frac{x^2(x-2) - 9(x-2)}{(x-2)(x-5)} = \frac{(x-2)(x^2-9)}{(x-2)(x-5)} = \frac{\cancel{(x-2)}(x-3)(x+3)}{\cancel{(x-2)}(x-5)}$$
$$= \frac{(x-3)(x+3)}{x-5}$$

The vertical asymptote is $x = 5$.

A removable discontinuity occurs when $x = 2$.

When $y = 0$, $(x-3)(x+3) = 0$; the x-intercepts are (3,0) and (–3,0).

When $x = 0$, $y = \frac{(-3)(3)}{-5} = \frac{-9}{-5} = \frac{9}{5}$; the y-intercept is $\left(0, \frac{9}{5}\right)$.

When $x = 2$, the removable discontinuity occurs at $\left(2, \frac{5}{3}\right)$. When graphing, place a hollow circle at that point.

Some points on the graph are: $(6,27)$, $(7,20)$, $\left(11, \frac{56}{3}\right)$.

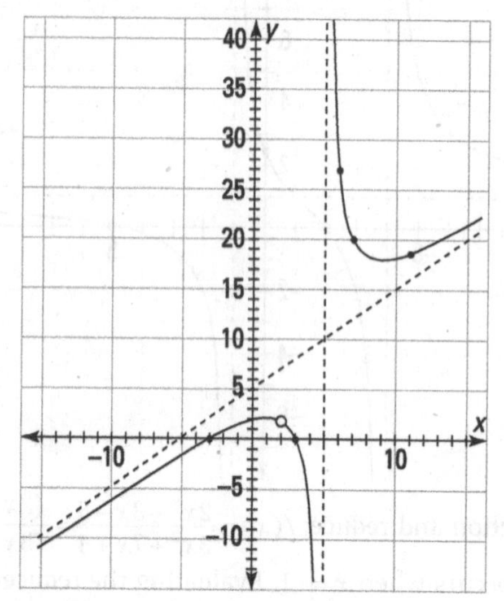

5 $\boldsymbol{x = 2, x = -2, y = 0}$**.** For the vertical asymptotes, set the denominator equal to 0 and solve for x.

$$x^2 - 4 = 0 \rightarrow (x-2)(x+2) = 0 \rightarrow x = 2 \text{ or } x = -2$$

For the horizontal asymptote, because the highest power in the numerator is smaller than the denominator, the asymptote is $y = 0$.

6 $\boldsymbol{-\frac{1}{4}}$**.** $\lim_{x \to 3} \frac{x^2 - 4x + 2}{x + 1} = \frac{3^2 - 4 \cdot 3 + 2}{3 + 1} = \frac{9 - 12 + 2}{4} = \frac{-1}{4} = -\frac{1}{4}$

7 $\boldsymbol{x = -3}$**.** Factor the numerator and denominator to find the common factor.

$$y = \frac{x^2 + x - 6}{x^2 - 9} = \frac{\cancel{(x+3)}(x-2)}{\cancel{(x+3)}(x-3)}$$

The removable discontinuity occurs when $x + 3 = 0$ or when $x = -3$.

(8) **See the figure.** The intercepts are $(3,0)$ and $\left(0,-\frac{3}{2}\right)$. The vertical asymptote is $x=-2$, and the horizontal asymptote is $y=1$.

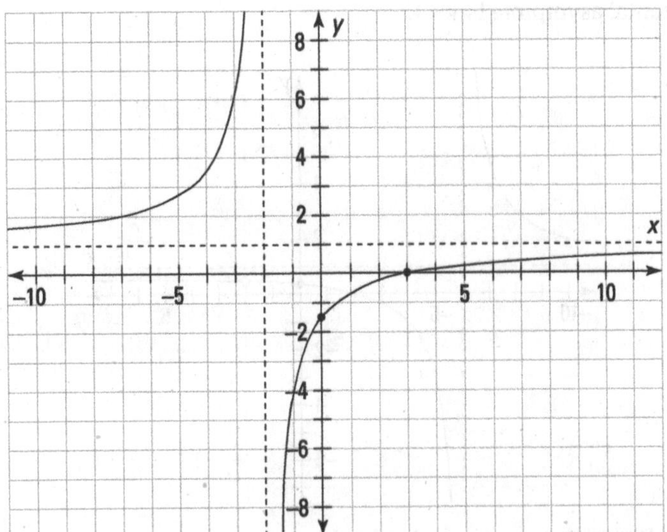

(9) **1.** First, divide each term by the highest power of the variable, and then evaluate at infinity.

$$\lim_{x\to\infty}\frac{x^3-3x^2+6x+5}{x^3+4x^2-11}=\lim_{x\to\infty}\frac{\frac{x^3}{x^3}-\frac{3x^2}{x^3}+\frac{6x}{x^3}+\frac{5}{x^3}}{\frac{x^3}{x^3}+\frac{4x^2}{x^3}-\frac{11}{x^3}}=\lim_{x\to\infty}\frac{1-\frac{3}{x}+\frac{6}{x^2}+\frac{5}{x^3}}{1+\frac{4}{x}-\frac{11}{x^3}}=\frac{1-0+0+0}{1+0-1}=1$$

(10) **Domain: x not equal to −2, 1; intercepts: (5,0), (−1,0), $\left(0,\frac{5}{2}\right)$**

The domain consists of all the real numbers that can be used. Because the denominator is equal to 0 if x is −2 or 1, the domain consists of all real numbers except −2 or 1, written $(-\infty,-2)\cup(-2,1)\cup(1,\infty)$ in interval notation. To find the x intercept, set y equal to 0 and solve for x; to find the y-intercept, let the x's be 0 and solve for y.

$$0=\frac{x^2-4x-5}{x^2+x-2}\to 0=x^2-4x-5\to$$
$$0=(x-5)(x+1)$$

Zeros: $5,-1\to$ Intercepts: $(5,0)$, $(-1,0)$

$$y=\frac{0^2-4\cdot 0-5}{0^2+0-2}\to y=\frac{-5}{-2}$$
$$y=\frac{5}{2}\to \text{Intercept: }\left(0,\frac{5}{2}\right)$$

(11) **See the figure.** Factor the numerator and denominator of the function:

$$y=\frac{x^2-4}{x^2+4x-12}=\frac{\cancel{(x-2)}(x+2)}{\cancel{(x-2)}(x+6)}.$$

You see a removable discontinuity when $x = 2$. You place a hollow circle at $\left(2, \frac{1}{2}\right)$ to indicate the discontinuity. The intercepts are $(-2, 0)$ and $\left(0, \frac{1}{3}\right)$. The vertical asymptote is $x = -6$, and the horizontal asymptote is $y = 1$.

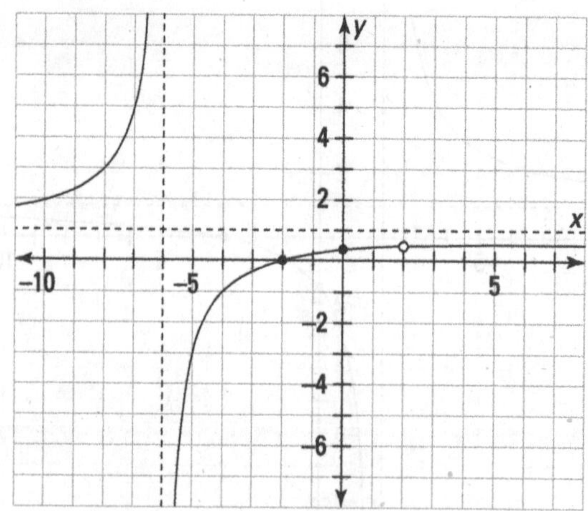

12 $y = x - 9$. Divide the numerator by the denominator.

$$
\begin{array}{r}
x - 9 \\
x + 3\,\overline{)\,x^2 - 6x - 7} \\
\underline{x^2 + 3x } \\
-9x - 7 \\
\underline{-9x - 27} \\
+20
\end{array}
$$

Just ignore the remainder. The equation of the oblique asymptote is $y = x - 9$.

IN THIS CHAPTER

- **Getting comfortable with exponential expressions**
- **Functioning with base *b* and base *e***
- **Taming exponential equations and compound interest**
- **Working with logarithmic functions**
- **Rolling over on log equations**
- **Picturing exponentials and logs**

Chapter 11 Exploring Exponential and Logarithmic Functions

Exponential growth and decay are natural phenomena. They happen all around us. And being the thorough, worldly people they are, mathematicians have come up with ways of describing, formulating, and graphing these phenomena. Thanks to their hard work, you can express the patterns observed when exponential growth and decay occur mathematically using exponential and logarithmic functions.

Why else are these functions important? Certain algebraic functions, such as polynomials and rational functions, share certain characteristics that exponential functions lack. For example, algebraic functions all show their variables as bases raised to some power, such as x^2 or x^8. Exponential functions, on the other hand, show their variables up in the powers of the expressions and lower some numbers as the bases, such as 2^x or e^x. In this chapter, I discuss the properties, uses, and graphs of exponential and logarithmic functions in detail.

Evaluating *e*-Expressions and Powers of *e*

An exponential function is unique because its variable appears in the exponential position and its constant appears in the base position. You write an exponent, or power, as a superscript just after the base. In the expression 3^x, for example, the variable x is the exponent, and the constant 3 is the base. The general form for an exponential function is $f(x)=a\cdot b^{k\cdot x}$, where

- The base b is any positive number.
- The coefficients a and k are any real numbers.
- The exponent x is any real number.

The base of an exponential function can't be zero or negative, the domain is all real numbers, and the range is all positive numbers when a is positive. When you enter a number into an exponential function, you evaluate it by using the order of operations (along with other rules for working with exponents; I discuss these topics in Chapters 1 and 4). The order of operations dictates that you evaluate the function in the following order:

1. **Perform all operations inside grouping symbols (parentheses, brackets, radicals, and so on).**
2. **Evaluate powers and roots.**
3. **Perform multiplication and division.**
4. **Deal with addition and subtraction.**

EXAMPLE

Q. Evaluate $f(x)=3^x+1$ for $x=2$.

A. **10.** You replace the x with the number 2. So, $f(2)=3^2+1=9+1=10$.

Q. Evaluate the exponential function $g(x)=4\left(\frac{1}{2}\right)^{2x}-3$ for $x=-2$.

A. **61.** You raise to the power first, after flipping the fraction to get rid of the negative sign, and then multiply by 4, and finally subtract 3.

$$g(x)=4\left(\frac{1}{2}\right)^{2(-2)}-3=4\left(\frac{1}{2}\right)^{-4}-3=4(2)^4-3=4(16)-3=64-3=61$$

YOUR TURN

1 Evaluate $f(x)=3(2)^x$ for $x=3$.

2 Evaluate $f(x)=5^x-11$ for x = 2.

---

3 Evaluate $f(x)=7-3^{x-2}$ for x = −1.

---

4 Evaluate $f(x)=4^x+2\cdot 4^{2x+1}$ for x = 0.

Exponential functions: it's all about the base, baby

The base of an exponential function can be any positive number. The bigger the number, the bigger the function value becomes as you raise it to higher and higher powers. (It's sort of like "the more money you have, the more money you make.") The bases can get downright small, too. In fact, when the base is some number between zero and 1, you don't have a function that grows; instead, you have a function that falls.

Observing the trends in bases

The base of an exponential function tells you a lot about the nature and character of the function, making it one of the first things you should look at and classify. The main way to classify the bases of exponential functions is to determine whether they're larger or smaller than 1. After you make that designation, you look at how much larger or smaller they are. The exponents themselves affect the expressions that contain them in somewhat predictable ways, making them prime places to look when grouping.

Because the domain of exponential functions is all real numbers, and the base is always positive, the result of b^{kx} is always a positive number. Even when you raise a positive base to a negative power, you get a positive answer. Exponential functions can end up with negative values if you subtract from the power or multiply by a negative number, but the power itself is always positive.

Grouping exponential functions according to their bases

Algebra offers three classifications for the base of an exponential function, due to the fact that the numbers used as bases appear to react in distinctive ways when raised to positive powers:

- When $b > 1$, the values of b^x grow larger as x gets larger.
- When $b = 1$, the values of b^x show no movement. Raising the number 1 to higher powers always results in the number 1. You see no exponential growth or decay. In fact, some mathematicians leave the number 1 out of the listing of possible bases for exponentials. Others leave it in as a bridge between functions that increase in value and those that decrease in value. It's just a matter of personal taste.
- When $0 < b < 1$, the value of b^x grows smaller as x gets larger. The base b has to be positive, and the numbers $0 < b < 1$ are all proper fractions (fractions with the numerators smaller than the denominators).

EXAMPLE

Q. Evaluate: 2^2, 2^4, 2^5, 2^7, and 2^{10}.

A. **1,024.** $2^2 = 4$, $2^4 = 16$, $2^5 = 32$, $2^7 = 128$, and $2^{10} = 1024$. **As the exponent gets larger, the power of 2 gets larger.**

Q. Evaluate: 1^2, 1^2, 1^2, and 1^{112}.

A. **1. The number 1 raised to any power is equal to 1:** $1^2 = 1$, $1^5 = 1$, $1^7 = 1$, and $1^{112} = 1$.

Q. Evaluate: $\left(\frac{1}{3}\right)^2$, $\left(\frac{1}{3}\right)^5$, and $\left(\frac{1}{3}\right)^8$.

A. $\left(\frac{1}{3}\right)^2 = \frac{1}{9}$, $\left(\frac{1}{3}\right)^5 = \frac{1}{243}$, and $\left(\frac{1}{3}\right)^8 = \frac{1}{6,561}$. **The numbers get smaller and smaller as the powers get bigger.**

Q. Evaluate: 0.01^2, 0.01^3, 0.01^5, and 0.01^{10}.

A. $0.01^2 = 0.0001$, $0.01^3 = 0.000001$, $0.01^5 = 0.0000000001$, and $0.01^{10} = 0.00000000000000000001$. **You get the point.**

Adding a coefficient of k to the exponent (making it kx) can change the behavior of the function, depending on what sign k is. After you take into account the effect of multiplying by k, though, the basic properties remain the same.

Grouping exponential functions according to their exponents

An exponent placed with a number can affect the expression that contains the number in somewhat predictable ways. The exponent makes the result take on different qualities, depending on whether the exponent is greater than zero, equal to zero, or smaller than zero:

- When the base $b > 1$ and the exponent $kx > 0$ (meaning that k is a positive number), the values of b^{kx} get bigger and bigger as x gets larger. You say that the values grow exponentially.

» When the base $b > 1$ and the exponent $x = 0$, the only value of b^{kx} you get is 1. The rule is that $b^0 = 1$ for any number except $b = 0$. So, an exponent of zero really flattens things out.

» When the base $b > 1$ and the exponent $kx < 0$ (meaning k is a negative number), the values of b^{kx} get smaller and smaller as the exponents get further and further from zero.

5 Compare the values of 4^{3k} when $k = -2$ to when $k = -1$.

6 Compare the values of 2^{-5k} when $k = -2$ to when $k = -1$.

Meeting the most frequently used bases: 10 and *e*

Exponential functions feature bases represented by numbers greater than zero. The two most frequently used bases are 10 and *e*, where $b = 10$ and $b = e$.

It isn't too hard to understand why mathematicians like to use base 10 — to understand their motivation, just hold all your fingers in front of your face! All the powers of 10 are made up of ones and zeros — for instance, $10^2 = 100$, $10^9 = 1{,}000{,}000{,}000$, and $10^{-5} = 0.00001$. How much more simple can it get? Our number system, the decimal system, is based on tens.

Like the value 10, base *e* occurs naturally. Members of the scientific community prefer base *e* because powers and multiples of *e* keep creeping up in models of natural occurrences. Including *e*'s in computations also simplifies things for financial professionals, mathematicians, and engineers.

If you use a scientific calculator to get the value of *e*, you see only some of *e*. The numbers you see only estimate what *e* is; most calculators give you seven or eight decimal places. Here are the first nine decimal places in the value of *e*, in rounded form: $e \approx 2.718281828$. The decimal value of *e* actually goes on forever without repeating in a pattern. The hint of a pattern that you see in the nine digits doesn't hold for long.

The expression $\lim_{x\to\infty}\left(1+\frac{1}{x}\right)^x$ represents the exact value of *e*. The larger and larger *x* gets, the more correct decimal places you get. Most algebra courses call for you to memorize only the first four digits of *e*: 2.718. Now for the bad news: As wonderful as *e* is for taming function equations and scientific formulas, its powers aren't particularly easy to deal with. The base *e* is approximately 2.718, and when you take a number with a decimal value that never ends and raise it to a power, the decimal values get even more unwieldy. The common practice in mathematics is to leave your answers as multiples and powers of *e* instead of switching to decimals, unless you need an approximation in some application. Scientific calculators change your final answer, in terms of *e*, to an answer that is correct to as many decimal places as needed.

You follow the same rules when simplifying an expression that has a factor of *e* that you use on a variable base (see Chapter 3). The following list presents the rules involved in simplifying expressions with a base of *e*.

- When you multiply expressions with the same base, you add the exponents: $e^x \cdot e^y = e^{x+y}$.
- When you divide expressions with the same base, you subtract the exponents: $\frac{e^x}{e^y} = e^{x-y}$.
- Use the order of operations — powers first, followed by multiplication, addition, or subtraction.

EXAMPLE

Q. Simplify: $e^{2x} \cdot e^{3x}$.

A. $e^{2x} \cdot e^{3x} = e^{5x}$.

Q. Simplify: $\frac{e^{15x^2}}{e^{3x}}$.

A. $\frac{e^{15x^2}}{e^{3x}} = e^{15x^2-3x}$. You can't simplify the exponent any further, unless you want to factor the equation.

Q. Simplify: $e^2 \cdot e^4 + 2(3e^3)^2$.

A. $19e^6$. Using the order of operations: first, raise the power, then multiply that result by 2 and add the exponents in the first term. Finally, add the two results.

$e^2 \cdot e^4 + 2(3e^3)^2 = e^6 + 2(9e^6) = e^6 + 18e^6 = 19e^6$

Q. Simplify: $\sqrt{e} \cdot \sqrt[3]{e} =$.

A. $e^{5/6}$. First, change the radicals to fractional exponents. Then perform the multiplication by adding the exponents: $\sqrt{e} \cdot \sqrt[3]{e} = e^{1/2} \cdot e^{1/3} = e^{5/6}$.

YOUR TURN

7 Simplify: $e^{2x} \cdot e^{4x} + e^x(e^5)^x$.

8 Simplify: $\dfrac{e^{-3x} \cdot e^{8x}}{2e^{x}}$.

9 Simplify: $\sqrt[3]{5e^{6x} + 3e^{6x}}$.

Solving Exponential Equations

To solve an algebraic equation, you work toward finding the numbers that replace any variables and make a true statement. The process of solving exponential equations incorporates many of the same techniques you use in algebraic equations — adding to or subtracting from each side, multiplying or dividing each side by the same number, factoring, squaring both sides, and so on.

Solving exponential equations requires some additional techniques, however. That's what makes them so much fun! Some techniques you use when solving exponential equations involve changing the original exponential equations into new equations that have matching bases. Other techniques involve putting the exponential equations into more recognizable forms — such as quadratic equations — and then using the appropriate formulas. (If you can't change the bases to matching bases or put the equations into quadratic or linear forms, you have to switch to logarithms or use a change-of-base formula — neither of which is within the scope of this book.)

Taking advantage of matching bases

If you have an equation written in the form $b^x = b^y$, where the same number represents the bases b, the following rule holds:

$$b^x = b^y \leftrightarrow x = y$$

You read the rule as follows: "If b raised to the xth power is equal to b raised to the yth power, that implies that $x = y$." The double-pointed arrow indicates that the rule is true in the opposite direction, too.

EXAMPLE

Q. Solve for x in: $2^{3+x} = 2^{4x-9}$.

A. 4. Using the base rule to solve the equation $2^{3+x} = 2^{4x-9}$, you see that the bases (the 2s) are the same, so the exponents must also be the same. You just pull the exponents down and solve the linear equation $3 + x = 4x - 9$ for the value of x: $3 + x = 4x - 9 \rightarrow 12 = 3x$ or $x = 4$. You then put the 4 back into the original equation to check your answer: $2^{3+4} = 2^{4 \cdot 4-9} \rightarrow 2^7 = 2^7$.

Q. Solve for x in: $3^{x^2} = 3^{2x+3}$.

A. 3, −1. The bases are the same, so you can write the equation $x^2 = 2x + 3$. This is a quadratic. Solve for x: $x^2 = 2x + 3 \rightarrow x^2 - 2x - 3 = 0 \rightarrow (x-3)(x+1) = 0$. The roots are 3 and −1. Do they both work? When $x = 3$, the original equation reads: $3^{3^2} = 3^{2 \cdot 3+3} \rightarrow 3^9 = 3^9$. So far, so good. Now, when $x = -1$, you have $3^{(-1)^2} = 3^{2(-1)+3} \rightarrow 3^1 = 3^1$. Both work!

YOUR TURN

10 Solve for x in: $5^{2x+7} = 5^{4x-9}$.

11 Solve for x in: $e^{6+x^2} = e^{5x}$.

What do you do if the bases aren't the same? Many times, bases are related to one another by being powers of the same number. When that happens, you change the bases to powers of the same number and proceed as usual.

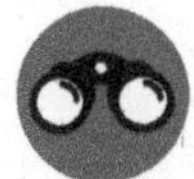

EXAMPLE

Q. Solve the equation $4^{x+3}=8^{x-1}$ for x.

A. **16,777,216.** You need to write both the bases as powers of 2 and then apply the rules of exponents. First, change the 4 and 8 to powers of 2.

$$4^{x+3}=8^{x-1}\to\left(2^2\right)^{x+3}=\left(2^3\right)^{x-1}$$

Now raise the powers to powers.

$$2^{2(x+3)}=2^{3(x-1)}\to 2^{2x+6}=2^{3x-3}$$

Now that the bases are the same, set the exponents equal to one another and solve for x.

$$2x+6=3x-3\to 9=x$$

Check your answer in the original equation.

$$4^{9+3}=8^{9-1}\to 4^{12}=8^8\to 16{,}777{,}216=16{,}777{,}216$$

Q. Solve the equation $3^{x+3}=9^{x^2}$ for x.

A. $x=\frac{3}{2}$ **and** -1. Write the base 9 as base 3 and simplify: $3^{x+3}=9^{x^2}\to 3^{x+3}=\left(3^2\right)^{x^2}\to 3^{x+3}=3^{2x^2}$. Set the exponents equal to one another and solve for x: $x+3=2x^2\to 0=2x^2-x-3\to 0=(2x-3)(x+1)$. This gives you two solutions: $x=\frac{3}{2}$ and $x=-1$. Checking in the original equation, when $x=\frac{3}{2}$, $3^{\frac{3}{2}+3}=9^{\left(\frac{3}{2}\right)^2}\to 3^{\frac{9}{2}}=9^{\frac{9}{4}}\to 3^{\frac{9}{2}}=\left(3^2\right)^{\frac{9}{4}}=3^{\frac{9}{2}}$. And, checking in the original equation, when $x=-1$, $3^{-1+3}=9^{(-1)^2}\to 3^2=9^1\to 9=9$. They both work.

YOUR TURN

12 Solve for x in: $8^{x^2/3}=4^x$.

13 Solve for x in: $25^{x-2}=125^{x+1}$.

14 Solve for x in: $\frac{27^{x+1}}{\sqrt{3}} = 9^{2x-3}$.

Recognizing and using quadratic patterns

When exponential terms appear in equations with two or three terms, you may be able to treat the equations as you do quadratic equations (see Chapter 3) to solve them with familiar methods. Using the methods for solving quadratic equations offers a big advantage because you can factor the exponential equations or you can resort to the quadratic formula.

You factor quadratics by dividing every term by a common factor or, with trinomials, by using unFOIL to determine the two binomials whose product is the trinomial. (See Chapters 1 and 3 if you need a refresher on these types of factoring.)

You can make use of just about any equation pattern that you see when solving exponential functions. If you can simplify the exponential to the form of a quadratic or cubic and then factor, find perfect squares, find sums and differences of squares, and so on, then you've made life easier by changing the equation into something recognizable and doable. In the sections that follow, I provide examples of the two most common types of problems you're likely to run up against: those involving common factors and unFOIL.

Taking out a greatest common factor

When you solve a quadratic equation by factoring out a greatest common factor (GCF), you write that greatest common factor outside the parentheses and show all the results of dividing by it inside the parentheses.

EXAMPLE

Q. Solve for x: $3^{2x} - 9 \cdot 3^x = 0$.

A. **x = 2.** Factor 3^x from each term and get $3^x(3^x - 9) = 0$. After factoring, you use the multiplication property of zero by setting each of the separate factors equal to zero. (If the product of two numbers is zero, at least one of the numbers must be zero; see Chapter 1.) You set the factors equal to zero to find out what value of x satisfies the equation: $3^x = 0$ has no solution; 3 raised to a power can't be equal to 0. Setting $3^x - 9 = 0$, you change the 9 to a power of 3 to get $3^x - 3^2 = 0$. Move the 3^2 to the right, and the equation becomes $3^x = 3^2$. Setting the exponents equal to one another, $x = 2$. You find only one solution to the entire equation.

Q. Solve for x: $2^{4x} - 2^x = 0$.

A. **x = 0.** Factor 2^x from each term to get $2^x(2^{3x} - 1) = 0$. The equation $2^x = 0$ has no solution, but when $2^{3x} - 1 = 0$, you write $2^{3x} = 1$, and this has a solution when $x = 0$.

Factoring like a quadratic trinomial

A quadratic trinomial has a term with the variable squared, a term with the variable raised to the first power, and a constant term. This is the pattern you're looking for if you want to solve an exponential equation by treating it like a quadratic.

EXAMPLE

Q. Solve for x: $5^{2x} - 26 \cdot 5^x + 25 = 0$.

A. **x = 0 and 2.** One option is to consider the quadratic $y^2 - 26y + 25 = 0$, which would look something like the exponential equation if you replaced each 5^x with a y. The quadratic in y's factors into $(y-1)(y-25) = 0$. Using the same pattern on the exponential version, you get the factorization $(5^x - 1)(5^x - 25) = 0$. Setting each factor equal to zero, when $5^x - 1 = 0$, $5^x = 1$. This equation holds true when $x = 0$, making that one of the solutions. Now, when $5^x - 25 = 0$, you say that $5^x = 25$, or $5^x = 5^2$. In other words, $x = 2$. You find two solutions to this equation: $x = 0$ and $x = 2$.

Q. Solve for x: $3^{2x} - 10 \cdot 3^x + 9 = 0$.

A. **x = 9 and 1.** Replace each 3^x with y and write the quadratic $y^2 - 10y + 9 = 0$, which factors into $(y-9)(y-1) = 0$. The two solutions are 9 and 1. So, write the two solutions using the exponential and you have $3^x = 9 \rightarrow 3^x = 3^2$ or **x = 2**. The other solution is written $3^x = 1 \rightarrow 3^x = 3^0$ or **x = 0**.

YOUR TURN

15 Solve for x: $5^{2x} - 5 \cdot 5^x = 0$.

16 Solve for x: $2^{2x} - 5 \cdot 2^x + 4 = 0$.

17 Solve for x: $7^{4x} - 50 \cdot 7^{2x} + 49 = 0$.

Making Cents: Applying Compound Interest

Professionals (you included, although you may not know it) use exponential functions in many financial applications. If you have a mortgage on your home, an annuity for your retirement, or a credit-card balance, you should be interested in interest — and in the exponential functions that drive it.

Applying the compound interest formula

When you deposit your money in a savings account, an individual retirement account (IRA), or other investment vehicle, you get paid for the money you invest; this payment comes from the proceeds of compound interest — interest that earns interest. For instance, if you invest $100 and earn $2 in interest, the two amounts come together, and you now earn interest on $102. The interest compounds. This, of course, is a wonderful thing.

Here's the formula you can use to determine the total amount of money you have (*A*) after you deposit the principal (*P*) that earns interest at the rate of *r* percent (written as a decimal), compounding *n* times each year for *t* years:

$$A = P\left(1 + \frac{r}{n}\right)^{nt}$$

EXAMPLE

Q. You receive a windfall of $20,000 for an unexpected inheritance, and you want to sock it away for 10 years. You invest the cash in a fund at 4.5 percent interest, compounded monthly. How much money will you have at the end of 10 years if you can manage to keep your hands off it?

A. **31,339.86.** Apply the formula as follows:

$$A = 20{,}000\left(1 + \frac{0.045}{12}\right)^{(12)(10)} = 20{,}000(1.00375)^{120} = 20{,}000(1.566993) = 31{,}339.86$$

You'd have over $31,300. This growth in your money shows you the power of compounding and exponents.

Q. You want to have $100,000 available 18 years from now, when your baby will be starting college. How much do you have to deposit in an account that earns 5 percent interest, compounded monthly?

A. **$40,733.06.** To find out, you take the compound interest formula and work backward. This time, you'll be solving for the principal (*P*):

$$100{,}000 = P\left(1+\frac{0.05}{12}\right)^{(12)(18)} \rightarrow 100{,}000 = P(1.0041667)^{216}$$

You solve for *P* in the equation by dividing each side by the value in the parentheses raised to the power 216. According to the order of operations (see Chapter 1), you raise to a power before you multiply or divide. So, after you set up the division to solve for *P*, you raise what's in the parentheses to the 216th power and divide the result into 100,000:

$$100{,}000 = P(1.0041667)^{216} \rightarrow \frac{100{,}000}{(1.0041667)^{216}} = P \rightarrow \frac{100{,}000}{2.455008} = P \rightarrow 40{,}733.06 = P$$

A deposit of almost $41,000 will result in enough money to pay for college in 18 years (not taking into account the escalation of tuition fees). You may want to start talking to your baby about scholarships now!

Measuring the actual compound: effective rates

When you go into a bank or credit union, you see all sorts of interest rates posted. You may have noticed the terms "nominal rate" and "effective rate" in previous visits. The *nominal rate* is the named rate, or the value entered into the compounding formula. The named rate may be 4 percent or 7.5 percent, but that value isn't indicative of what's really happening because of the compounding. The *effective rate* represents what's really happening to your money after it compounds. A nominal rate of 4 percent translates into an effective rate of 4.074 percent when compounded monthly. This may not seem like much of a difference — the effective rate is about 0.07 higher — but it makes a big difference if you're talking about fairly large sums of money or long time periods.

You compute the effective rate by using the middle portion of the compound interest formula: $\left(1+\frac{r}{n}\right)^n$.

Q. Determine the effective rate of 1.3 percent compounded quarterly.

A. $\left(1+\frac{0.013}{4}\right)^4 = 1.013063512$

Q. Determine the effective rate of 4 percent compounded monthly.

A. $\left(1+\frac{.04}{12}\right)^{12} = 1.04071543$

The 1 before the decimal point in the answer indicates the original amount. You subtract that 1, and the rest of the decimals are the percentage values used for the effective rate.

Table 11-1 shows what happens to a nominal rate of 4 percent when you compound it different numbers of times per year.

TABLE 11-1 Compounding a Nominal 4 Percent Interest Rate

Times Compounded	Computation	Effective Rate
Annually	$(1+0.04/1)^1 = 1.04$	4.00%
Biannually	$(1+0.04/2)^2 = 1.0404$	4.04%
Quarterly	$(1+0.04/4)^4 = 1.04060401$	4.06%
Monthly	$(1+0.04/12)^{12} = 1.04074154292$	4.07%
Daily	$(1+0.04/365)^{365} = 1.04080849313$	4.08%
Hourly	$(1+0.04/8{,}760)^{8{,}760} = 1.04081067873$	4.08%
Every second	$(1+0.04/31{,}536{,}000)^{31{,}536{,}000} = 1.04081104727$	4.08%

Looking at continuous compounding

Typical compounding of interest occurs annually, quarterly, monthly, or perhaps even daily. Continuous compounding occurs immeasurably quickly or often. To accomplish continuous compounding, you use a different formula than you use for other compounding problems.

Here's the formula you use to determine a total amount (*A*) when the initial value or principal is *P* and when the amount grows continuously at the rate of *r* percent (written as a decimal) for *t* years:

$$A = Pe^{rt}$$

The *e* represents a constant number (the *e* base; see the section, "Meeting the most frequently used bases: 10 and *e*," earlier in this chapter) — approximately 2.71828.

EXAMPLE

Q. Determine how much money you'd have after 10 years of investment when the interest rate is 4.5 percent and you deposit $20,000.

A. Using the formula:

$$A = 20{,}000e^{(.045)(10)} = 20{,}000(1.568312) = 31{,}366.24$$

Q. Which account will earn you more interest on an investment of $10,000 deposited for 20 years: 4% compounded monthly or 3.5% compounded continuously?

A. For 4% compounded monthly, use $A = P\left(1+\frac{r}{n}\right)^{nt}$: $A = 10{,}000\left(1+\frac{0.04}{12}\right)^{12\cdot 20} = 22{,}225.82$.

And for 3.5% compounded continuously, use $A = Pe^{rt}$: $A = 10{,}000e^{0.035\cdot 20} = 20{,}137.53$.

You do better with the monthly compounding when the time is 20 years.

You should use the continuous compounding formula as an approximation in appropriate situations — when you're not actually paying out the money. The compound interest formula is much easier to deal with and gives a good estimate of total value.

Using the continuous compounding formula to approximate the effective rate of 4 percent compounded continuously, you get $e^{0.04} = 1.0408$ (an effective rate of 4.08 percent). Compare this with the value in Table 11-1 in the previous section.

YOUR TURN

18 You invest $12,500 in an account earning 3% interest compounded monthly. How much will you have in 10 years?

19 What is the effective rate of that account that pays 3% interest compounded monthly?

20 You're considering investing that $12,500 in an account earning 3% interest compounded continuously. How much will you have in 10 years?

21 You want to save up $40,000 for a down-payment on a yacht. How much do you have to invest right now in an account earning 4.8% interest compounded monthly if you want this to be your only investment and you need the down payment in 5 years?

Logging On to Logarithmic Functions

A logarithm is the exponent of a number. Logarithmic (log) functions are the inverses of exponential functions. They answer the question, "What power gave me that answer?" The log function associated with the exponential function $f(x)=b^x$, for example, is $f^{-1}(x)=\log_b x$. The superscript –1 after the function name *f* indicates that you're looking at the inverse of the function *f*. So, $\log_2 8$, for example, asks, "What power of 2 gave me 8?"

A logarithmic function has a base and an argument. The logarithmic function $f(x)=\log_b x$ has a base *b* and an argument *x*. The base must always be a positive number and not equal to 1. The argument must always be positive.

You can see how a function and its inverse work as exponential and log functions by evaluating the exponential function for a particular value and then seeing how you get that value back after applying the inverse function to the answer.

EXAMPLE

Q. What is the inverse of the function $f(x)=2^x$?

A. $f^{-1}(x)=\log_2 x$

Q. What is the inverse of the function $h(x)=\log_3 x$?

A. $h^{-1}(x)=3^x$

Using some actual number examples, first let $x=3$ in $f(x)=2^x$; you get $f(3)=2^3=8$. You put the answer, 8, into the inverse function $f^{-1}(x)=\log_2 x$, and you get $f^{-1}(8)=\log_2 8=3$. The answer comes from the definition of how logarithms work; the 2 raised to the power of 3 equals 8. You have the answer to the fundamental logarithmic question, "What power of 2 gave me 8?"

Meeting the properties of logarithms

Logarithmic functions share similar properties with their exponential counterparts. When necessary, the properties of logarithms allow you to manipulate log expressions so you can solve equations or simplify terms. As with exponential functions, the base *b* of a log function has to be positive. I show the properties of logarithms in Table 11-2.

TABLE 11-2 Properties of Logarithms

Property Name	Property Rule	Example
Equivalence	$y=\log_b x \leftrightarrow b^y=x$	$y=\log_9 3 \leftrightarrow 9^y=3$
Log of a product	$\log_b x\cdot y=\log_b x+\log_b y$	$\log_2 8z=\log_2 8+\log_2 z$
Log of a quotient	$\log_b \frac{x}{y}=\log_b x-\log_b y$	$\log_2 \frac{8}{5}=\log_2 8-\log_2 5$
Log of a power	$\log_b x^n=n\cdot\log_b x$	$\log_3 8^{10}=10\cdot\log_3 8$
Log of 1	$\log_b 1=0$	$\log_4 1=0$
Log of the base	$\log_b b=1$	$\log_4 4=1$

Exponential terms that have a base e (see the earlier section, "Meeting the most frequently used bases: 10 and e") have special logarithms just for the e's (the ease?). Instead of writing the log base e as $\log_e x$, you insert a special symbol, ln, for the log. The symbol ln is called the natural logarithm, and it designates that the base is e. The equivalences for base e and the properties of natural logarithms are the same, but they look just a bit different. Table 11-3 shows them.

TABLE 11-3 Properties of Natural Logarithms

Property Name	Property Rule	Example
Equivalence	$y = \ln x \leftrightarrow e^y = x$	$6 = \ln x \leftrightarrow e^6 = x$
Natural log of a product	$\ln x \cdot y = \ln x + \ln y$	$\ln 4z = \ln 4 + \ln z$
Natural log of a quotient	$\ln \frac{x}{y} = \ln x - \ln y$	$\ln \frac{4}{z} = \ln 4 - \ln z$
Natural log of a power	$\ln x^n = n \cdot \ln x$	$\ln x^5 = 5 \cdot \ln x$
Natural log of 1	$\ln 1 = 0$	$2 \cdot \ln 1 = 2 \cdot 0 = 0$
Natural log of e	$\ln e = 1$	$2 \cdot \ln e = 2 \cdot 1 = 2$

As you can see in Table 11-3, the natural logs are much easier to write — you have no subscripts. Professionals use natural logs extensively in mathematical, scientific, and engineering applications.

Putting your logs to work

You can use the basic exponential/logarithmic equivalence $y = \log_b x \leftrightarrow b^y = x$ to simplify equations that involve logarithms. Applying the equivalence makes the equation much nicer to work with.

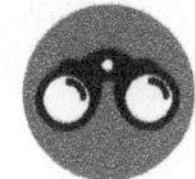

EXAMPLE

Q. Evaluate $\log_9 3$.

A. $\frac{1}{2}$. First, write the expression as an equation, $\log_9 3 = x$. Then use the equivalence to change this equation to $9^x = 3$. Now you have it in a form that you can solve for x (the x that you get is the answer or value of the original expression). You solve by changing the 9 to a power of 3 and then finding x in the new, more familiar form:

$$\left(3^2\right)^x = 3 \rightarrow 3^{2x} = 3^1 \rightarrow 2x = 1 \rightarrow x = \frac{1}{2}$$

The result tells you that $\log_9 3 = \frac{1}{2}$.

Q. Evaluate $\log_4 \frac{1}{4}$.

A. **−1.** Write the expression as an equation, and then use the equivalence to change the format.

$$\log_4 \frac{1}{4} = x \rightarrow 4^x = \frac{1}{4}$$

Write the fraction as a power of 4 and solve for x.

$4^x = 4^{-1} \rightarrow x = -1$

This could also be evaluated using the laws of logarithms.

Use the law $\log_b \frac{x}{y} = \log_b x - \log_b y$ to rewrite the expression as $\log_4 1 - \log_4 4$.

Then use the two laws $\log_b 1 = 0$ and $\log_b b = 1$ to say $\log_4 1 - \log_4 4 = 0 - 1 = -1$.

Q. Evaluate $10\log_3 27$.

A. **30.** Write the expression as an equation, and then divide each side by 10. This will allow you to use the equivalence to change the format.

$10\log_3 27 = x \rightarrow \frac{\cancel{10}\log_3 27}{\cancel{10}} = \frac{x}{10} \rightarrow \log_3 27 = \frac{x}{10}$ can be written as $3^{x/10} = 27$.

Write the 27 as a power of 3: $3^{x/10} = 3^3$.

Now set the two exponents equal to one another and solve for x.

$\frac{x}{10} = 3 \rightarrow x = 30$

Q. Evaluate $\log_5\left(\frac{1}{25}\right)$ using the laws of logarithms.

A. **−2.** Using the rules for the log of 1, the log of the base, the log of a power, and the log of a quotient (see Table 11-2), you can do the following:

$$\begin{aligned}
\log_5\left(\frac{1}{25}\right) &= \log_5 1 - \log_5 25 \text{ (using log of a quotient)} \\
&= \log_5 1 - \log_5 5^2 \text{ (rewriting 25 as a power of 5)} \\
&= \log_5 1 - 2\log_5 5 \text{ (using log of a power)} \\
&= 0 - 2(1) \text{ (using log of 1 and log of the base)} \\
&= -2 \text{ (simplifying)}
\end{aligned}$$

YOUR TURN

22 Evaluate $\log_6 \frac{1}{36}$.

23 Evaluate $4\log_2 16$.

24 Evaluate $3\log_4 \frac{1}{64}$.

Expanding and contracting expressions with log notation

You write logarithmic expressions and create logarithmic functions by combining all the usual algebraic operations of addition, subtraction, multiplication, division, powers, and roots. Expressions with two or more of these operations can get pretty complicated. A big advantage of logs, though, is their properties. Because of the special features of log properties, you can change multiplication to addition and powers to products. Put all the log properties together and you can change a single complicated expression into several simpler terms.

Also, results of computations in science and mathematics can involve sums and differences of logarithms. When this happens, experts usually prefer to have the answers written all in one term, which is where the properties of logarithms come in. You apply the properties in just the opposite way that you break down expressions for greater simplicity (see the previous section). Instead of spreading the work out, you want to create one compact, complicated expression.

EXAMPLE

Q. Simplify $\log_3 \frac{x^3\sqrt{x^2+1}}{(x-2)^7}$ by using the properties of logarithms.

A. You first use the property for the log of a quotient and then use the property for the log of a product on the first term you get (refer to Table 11-2 to review these properties):

$$\log_3 \frac{x^3\sqrt{x^2+1}}{(x-2)^7} = \log_3 x^3\sqrt{x^2+1} - \log_3 (x-2)^7 = \log_3 x^3 + \log_3 \sqrt{x^2+1} - \log_3 (x-2)^7$$

The last step is to use the log of a power on each term, changing the radical to a fractional exponent first:

$$\log_3 x^3 + \log_3 \left(x^2+1\right)^{1/2} - \log_3 (x-2)^7 = 3\log_3 x + \frac{1}{2}\log_3 \left(x^2+1\right) - 7\log_3 (x-2)$$

The three new terms you create are each much simpler than the original expression.

Q. Write the expression as a single term: $4\ln(x+2)-8\ln(x^2-7)-\frac{1}{2}\ln(x+1)$.

A. You first apply the property involving the natural log (ln) of a power to all three terms. You then factor out –1 from the last two terms and write them in a bracket:

$$\ln(x+2)^4-\ln(x^2-7^8)-\ln(x+1)^{1/2}=\ln(x+2)^4-\left[\ln(x^2-7)^8+\ln(x+1)^{1/2}\right]$$

You now use the property involving the ln of a product on the terms in the bracket, change the $\frac{1}{2}$ exponent to a radical, and use the property for the ln of a quotient to write everything as the ln of one big fraction:

$$\ln(x+2)^4-\left[\ln(x^2-7)^8+\ln(x+1)^{1/2}\right]=\ln(x+2)^4-\left[\ln(x^2-7)^8(x+1)^{1/2}\right]$$

$$=\ln(x+2)^4-\ln(x^2-7)^8\sqrt{x+1}=\ln\frac{(x+2)^4}{(x^2-7)^8\sqrt{x+1}}$$

The expression is messy and complicated, but it sure is compact.

25 Simplify $\log_4\sqrt{\frac{x+1}{(3x+4)^5}}$ by using the properties of logarithms.

26 Write the expression as a single term: $\log(x-3)+2\log(x+6)-3\log(x^2+7)$.

Solving Logarithmic Equations

Logarithmic equations can have one or more solutions, just like many other types of algebraic equations. What makes solving log equations a bit different is that you get rid of the log part as quickly as possible, leaving you to solve either a polynomial or an exponential equation in its place. Polynomial and exponential equations are easier and more familiar, and you may already know how to solve them (if not, see Chapter 8 and the section, "Solving Exponential Equations," earlier in this chapter).

The only caution I present before you begin solving logarithmic equations is that you need to check the answers you get from the new, revised forms. You may get answers to the polynomial or exponential equations, but they may not work in the original logarithmic equation. Switching to another type of equation introduces the possibility of extraneous roots — answers that fit the new, revised equation that you choose but sometimes don't fit in with the original equation.

When no log base is shown, you assume that the log's base is 10. Base 10 logarithms are *common* logarithms. So when you see $\log x = 8$, you know that the log base is 10.

Setting log equal to log

One type of log equation features each term carrying a logarithm in it (all the logarithms have to have the same base). You need to have exactly one log term on each side, so if an equation has more, you apply any properties of logarithms that form the equation to fit the rule (refer to Table 11-2 for these properties). After you do, you can apply the following rule:

If $\log_b x = \log_b y$, then $x = y$.

EXAMPLE

Q. Solve for x in $\log_4 x^2 = \log_4 (x+6)$.

A. Apply the rule so that you can write and solve the equation $x^2 = x+6$ by first setting it equal to 0 and then factoring:

$x^2 - x - 6 = 0 \rightarrow (x-3)(x+2) = 0$

$x-3=0$ gives you $x=3$; $x+2=0$ gives you $x=-2$.

The $x=3$ and $x=-2$ that you find are solutions of the *quadratic* equation, but you must check to see if they work in the original logarithmic equation.

If $x=3$:

$\log_4 3^2 = \log_4 (3+6) \rightarrow \log_4 9 = \log_4 9$

So, 3 is a solution.

If $x=-2$:

$\log_4 (-2)^2 = \log_4 (-2+6) \rightarrow \log_4 4 = \log_4 4$

You have another winner.

Q. Solve for x: $\log(x-8) + \log x = \log 9$.

A. First, apply the property involving the log of a product to get just one log term on the left: $\log(x-8)x = \log 9$. Next, you use the property that allows you to drop the logs and get the equation $(x-8)x = 9$. This is a quadratic equation that you can solve by multiplying, setting it equal to 0, and then factoring:

$$x^2-8x=9\to x^2-8x-9=0\to(x-9)(x+1)=0$$

$x-9=0$ gives you $x=9$; $x+1=0$ gives you $x=-1$.

Checking the answers, you see that the solution 9 works just fine:

$$\log(9-8)+\log 9=\log 9\to\log 1+\log 9=\log 9\to 0+\log 9=\log 9$$

However, the solution –1 doesn't work:

$$\log(-1-8)+\log(-1)\neq\log 9$$

You can stop right there. Both of the logs on the left have negative arguments. The argument in a logarithm has to be positive, so letting $x=-1$ doesn't work in the log equation (even though it was just fine in the quadratic equation). You determine that –1 is an extraneous solution.

Rewriting log equations as exponentials

When a log equation has log terms as well as a term that doesn't have a logarithm in it, you need to use algebra techniques and log properties (refer to Table 11-2) to put the equation in the form $y=\log_b x$. After you create the right form, you can apply the equivalence to change it to a purely exponential equation.

EXAMPLE

Q. Solve $\log_3(x+8)-2=\log_3 x$.

A. First, subtract $\log_3 x$ from each side and add 2 to each side to get $\log_3(x+8)-\log_3 x=2$.

Now you apply the property involving the log of a quotient: $\log_3\frac{x+8}{x}=2$. You then rewrite the equation by using the equivalence and solve for x.

$$3^2=\frac{x+8}{x}\to 9x=x+8\to 8x=8\to x=1$$

The only solution is $x=1$, which works in the original logarithmic equation:

$$\log_3(x+8)-2=\log_3 x\to\log_3(1+8)-2=\log_3 1\to\log_3 9-2=0\to$$
$$\log_3 9=2\to 3^2=9$$

Q. Solve $\log(3x^2+1)-\log(x+3)=\log(3x-2)$.

A. First, write the terms on the left as the quotient of two logs. Then apply the property to write a rational equation.

$$\log\frac{3x^2+1}{x+3}=\log(3x-2)\to\frac{3x^2+1}{x+3}=3x-2$$

Multiply each side of the equation by the denominator and solve the equation for x.

$$\frac{\cancel{x+3}}{1}\cdot\frac{3x^2+1}{\cancel{x+3}}=(3x-2)(x+3)\to 3x^2+1=3x^2+7x-6$$
$$1=7x-6\to 7=7x\to 1=x$$

The solution checks.

YOUR TURN

27 Solve for x: $\log_3(x^2-6)=\log_3 x$.

28 Solve for x: $\log_5 x+\log_5(x+3)=\log_5(2x+12)$.

29 Solve for x: $\log_2(x+3)-\log_2(x-4)=3$.

Graphing Exponential and Logarithmic Functions

Exponential and logarithmic functions have rather distinctive graphs because they're so plain and simple. The graphs are lazy C's that can slope upward or downward. The main trick when graphing them is to determine any intercepts, which way the graphs move as you go from left to right, and how steep the curves are.

Expounding on the exponential

Exponential functions have curves that usually look like the graphs you see in Figure 11-1a and 11-1b.

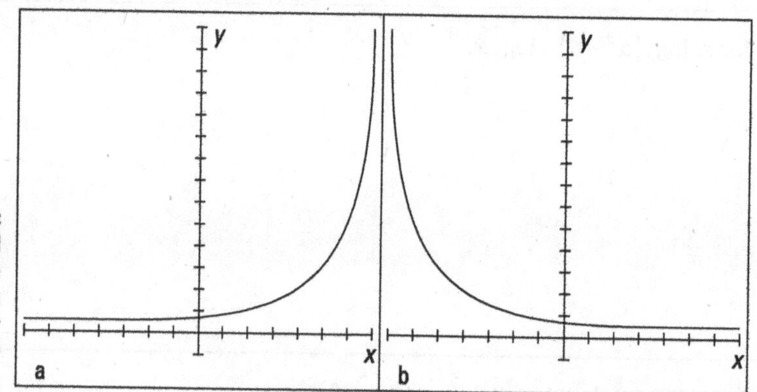

FIGURE 11-1: Exponential graphs rise away from the x-axis or fall toward the x-axis.

The graph in Figure 11-1a depicts exponential growth when the values of the function are increasing. Figure 11-1b depicts exponential decay when the values of the function are decreasing. Both graphs intersect the y-axis but not the x-axis, and they both have a horizontal asymptote: the x-axis.

Identifying a rise or fall

You can tell whether a graph will feature exponential growth or decay by looking at the equation of the function. In order to determine whether a function represents exponential growth or decay, you look at its base:

- If the exponential function $y = b^x$ has a base $b > 1$, the graph of the function rises as you read from left to right, meaning that you observe exponential growth.
- If the exponential function $y = b^x$ has a base $0 < b < 1$, the graph falls as you read from left to right, meaning that you observe exponential decay.

EXAMPLE

Q. Determine whether the graph of $g(x) = 4e^{3x}$ rises or falls.

A. **The graph rises, because e is greater than 1.**

Q. Determine whether the graph of $g(x) = 4(0.9)^{3x}$ rises or falls.

A. **The graph falls, because 0.9 is between 0 and 1.**

Sketching exponential graphs

In general, exponential functions have no x-intercepts, but they do have single y-intercepts. The exception to this rule is when you change the function equation by subtracting a number from the exponential term; this action drops the curve down below the x-axis. And the coefficient of the exponent x affects the degree of the rise (or fall) of the curve. Acting like a slope, a coefficient greater than 1 makes the curve steeper. A coefficient between 0 and 1 flattens the curve.

EXAMPLE

Q. Find the y-intercept of $y=(0.5)^{3x}-2$. Describe the movement of the curve.

A. To find the y-intercept of an exponential function, you set $x=0$ and solve for y: $y=(0.5)^{3(0)}-2=(0.5)^0-2=1-2=-1$. The y-intercept is (0,–1). The multiplier of 3 on the x in the exponent makes the graph rise more steeply.

Q. Find the y-intercept of $y=3(2)^{0.4x}$. Describe the movement of the curve.

A. To find the y-intercept of an exponential function, you set $x=0$ and solve for y: $y=3(2)^{0.4(0)}=3(2)^0=3(1)=3$. So, the y-intercept is (0,3). This function rises from left to right because the base is greater than 1. The multiplier 0.4 on the x in the exponent acts like the slope of a line — in this case, making the graph rise more slowly or gently.

Before you try to graph an equation, you should find another point or two for help with the shape. For instance, if $x=5$ in the previous example, $y=3(2)^{0.4(5)}=3(2)^2=3(4)=12$. So, the point (5,12) falls on the curve. Also, if $x=-5$, $y=3(2)^{0.4(-5)}=3(2)^{-2}=3(0.25)=0.75$ (see Chapter 5 for info on dealing with negative exponents). The horizontal asymptote is $y=0$. The point $(-5, 0.75)$ is also on the curve. Figure 11-2 shows the graph of this example curve with the intercept and points drawn in.

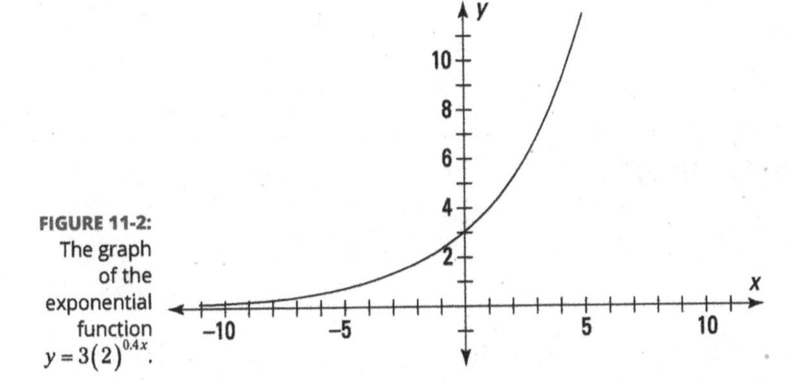

FIGURE 11-2: The graph of the exponential function $y=3(2)^{0.4x}$.

To graph the function $y=10(0.9)^x$, you first find the y-intercept. When $x=0$, $y=10(0.9)^0=10(1)=10$. So, the y-intercept is (0,10). Two other points you may decide to use are (1,9) and $(-4, 15.24)$. The graph of this function falls as you read from left to right because the base is smaller than 1. And the horizontal asymptote is $y=0$. Figure 11-3 shows the graph of this example curve and the random points of reference.

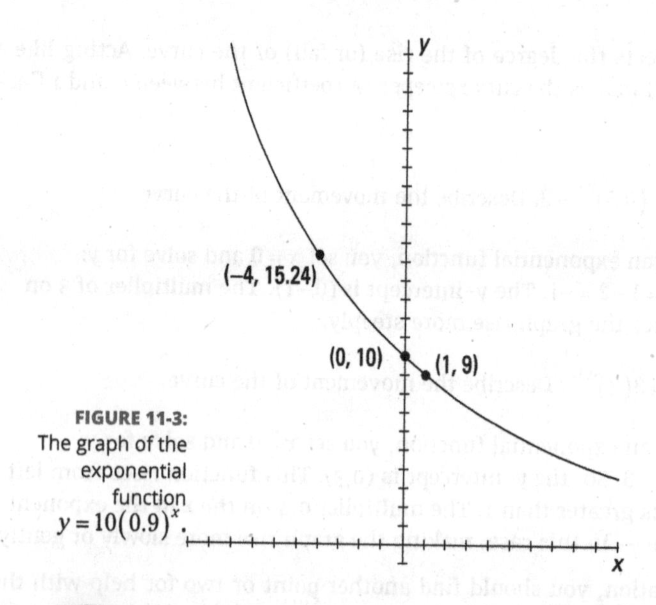

FIGURE 11-3: The graph of the exponential function $y = 10(0.9)^x$.

YOUR TURN

30 Sketch the graph of $y = 2(3)^x + 1$.

31 Sketch the graph of $y = 4(0.5)^{0.2x}$.

Graphing log functions using intercepts

The graphs of logarithmic functions either rise or fall, and they usually look like one of the sketches in Figure 11-4. The graphs have a single vertical asymptote: the y-axis. Having the y-axis as an asymptote is the opposite of an exponential function, whose asymptote is the x-axis (see the previous section). Log functions are also different from exponential functions in that they have an x-intercept but not (usually) a y-intercept.

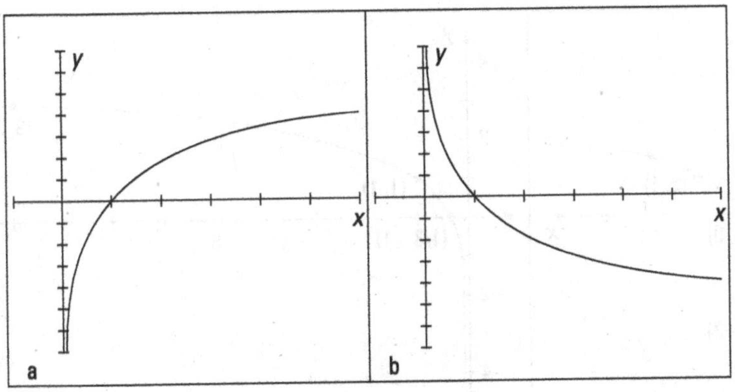

FIGURE 11-4: Logarithmic functions rise or fall, breaking away from the asymptote: the *y*-axis.

When you graph a log function, you look at its x-intercept and the base of the function:

- If the base is a number greater than 1, the graph rises from left to right.
- If the base is between zero and 1, the graph falls as you go from left to right.

EXAMPLE

Q. Sketch the graph of $y = \log_2 x$.

A. The graph has an x-intercept of (1,0) and rises from left to right. You get the intercept by letting $y = 0$ and solving the equation $0 = \log_2 x$ for x. You should choose a couple other points on the curve to help with shaping the graph. The graph of $y = \log_2 x$ contains the points (2,1) and $\left(\frac{1}{8},-3\right)$. You compute those points by substituting the chosen x value into the function equation and solving for y. Figure 11-5a shows you the graph of the function (the points are indicated).

Q. Sketch the graph of $y = 1 + \log_3 x$.

A. The graph has an x-intercept of $\left(\frac{1}{3},0\right)$ and rises from left to right. You get the intercept by letting $y = 0$ and solving the equation $0 = 1 + \log_3 x$ for x. To do this, subtract 1 from each side, and then apply the inverse function rule to create an exponential equation: $-1 = \log_3 x \rightarrow 3^{-1} = x \rightarrow x = \frac{1}{3}$. You should choose a couple of other points on the curve to help with shaping the graph. The graph of $y = 1 + \log_3 x$ contains the points (1,1) and (9,3). Figure 11-5a shows you the graph of the function (the points are indicated).

Reflecting on inverses of exponential functions

Exponential and logarithmic functions are inverses of one another. You may have noticed in previous sections that the flattened, C-shaped curves of log functions look vaguely familiar. In fact, they're mirror images of the graphs of exponential functions.

An exponential function, $y_1 = b^x$, and its logarithmic inverse, $y_2 = \log_b x$, have graphs that mirror one another over the line $y = x$. The symmetry of exponential and log functions about the diagonal line $y = x$ is very helpful when graphing the functions.

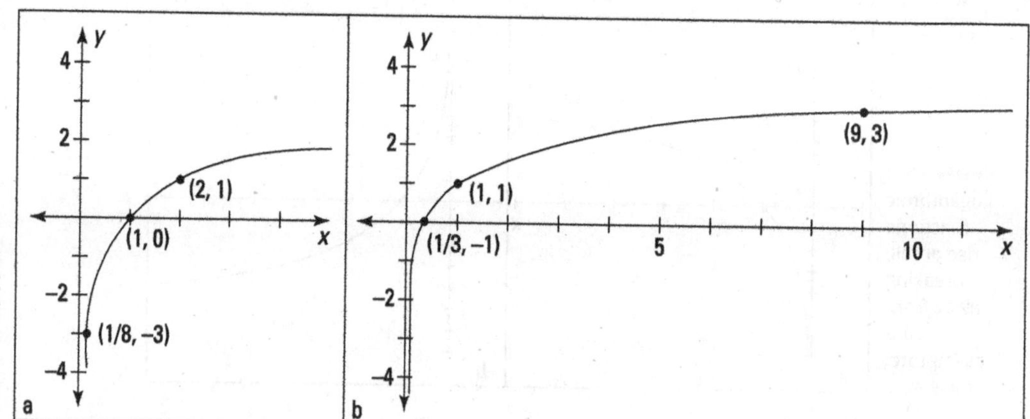

FIGURE 11-5: With a log base of 2 or 3, the curve of the function rises.

Q. Sketch the graphs of $y = 3^x$ and $y = \log_3 x$ on the same graph.

EXAMPLE

A. The exponential function $y = 3^x$ has a y-intercept of $(0,1)$, the x-axis as its horizontal asymptote, and the plotted points $(1,3)$ and $\left(-2, \frac{1}{9}\right)$. You can compare that function to its inverse function, $y = \log_3 x$, which has an x-intercept at $(1,0)$, the y-axis as its vertical asymptote, and the plotted points $(3,1)$ and $\left(\frac{1}{9}, -2\right)$. Figure 11-6 shows you both of the graphs and some of the points.

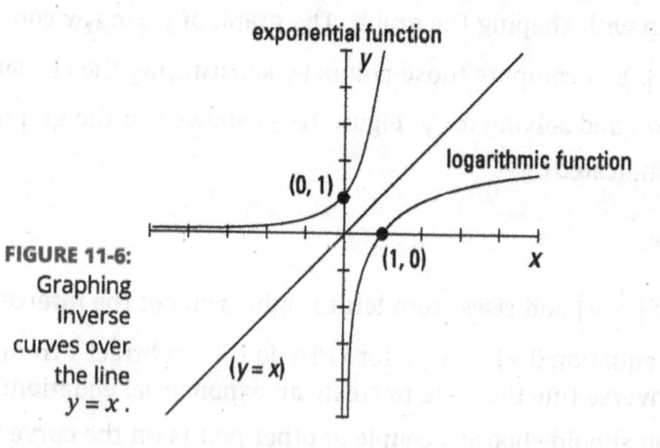

FIGURE 11-6: Graphing inverse curves over the line $y = x$.

Q. Sketch the graphs of $y = \log_{1/4} x$ and $y = \left(\frac{1}{4}\right)^x$ on the same graph.

A. If you want to graph $y = \log_{1/4} x$ and don't want to mess with its fractional base, you can graph $y = \left(\frac{1}{4}\right)^x$ instead and flip the graph over the diagonal line to get the graph of the log function. The graph of $y = \left(\frac{1}{4}\right)^x$ contains the points $(0,1)$, $\left(1, \frac{1}{4}\right)$, and $(-2, 16)$. These points are easier to compute than the log values. You just reverse the coordinates of those points to $(1,0)$, $\left(\frac{1}{4}, 1\right)$, and $(16, -2)$; you now have points on the graph of the log function. Figure 11-7 illustrates this process.

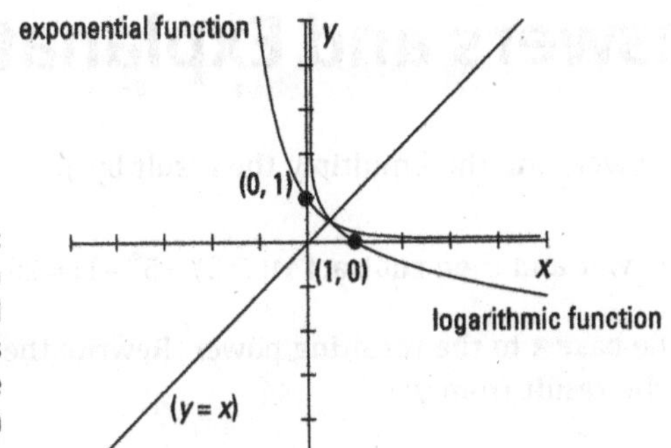

FIGURE 11-7: Using an exponential function as an inverse to graph a log function.

Notice that the exponential function and its inverse log function cross the line $y = x$ at the same point. This is true of all functions and their inverses — they cross the line $y = x$ in the same place or places. Keep this in mind when you're graphing.

YOUR TURN

32 Sketch the graphs of $y = \log_2 x$ and $y = 2^x$ on the same graph.

33 Sketch the graphs of $y = \log_{1/2} x$ and $y = \left(\frac{1}{2}\right)^x$ on the same graph.

Practice Questions Answers and Explanations

1. **24.** Replace the x with 3. Raise the power, and then multiply the result by 3: $f(3)=3(2)^3=3(8)=24$.

2. **14.** Replace the x with 2. Raise the power and then subtract 11: $f(2)=5^2-11=25-11=14$.

3. $6\frac{26}{27}$. Replace the x with −1. Raise the base 3 to the resulting power. Rewrite the negative power as a fraction. Then subtract the result from 7:

$$f(-1)=7-3^{-1-2}=7-3^{-3}=7-\frac{1}{3^3}=7-\frac{1}{27}=6\frac{26}{27}.$$

4. **9.** Replace both x's with 0. Raise the powers. Then multiply the second term by 2. Finally, add the results: $f(0)=4^0+2\cdot 4^{2\cdot 0+1}=1+2\cdot 4^1=1+2\cdot 4=1+8=9$.

5. $\frac{1}{4096}<\frac{1}{64}$. When the exponent is negative, the values get smaller as they get further from 0. The value, then, when $k=-2$ will be smaller than the value when $k=-1$:

$$4^{3(-2)}<4^{3(-1)}\rightarrow 4^{-6}<4^{-3}\rightarrow \frac{1}{4096}<\frac{1}{64}.$$

6. $2^{10}>2^5$. When the exponent is positive, the values get larger as they get further from 0. The value of 2^{-5k} when $k=-2$ is 2^{10}, and the value of 2^{-5k} when $k=-1$ is 2^5.

7. $2e^{6x}$. Raise the power. Then perform the two multiplications. The resulting products can be added.

$$e^{2x}\cdot e^{4x}+e^x\left(e^5\right)^x=e^{2x}\cdot e^{4x}+e^x\cdot e^{5x}=e^{2x+4x}+e^{x+5x}=e^{6x}+e^{6x}=2e^{6x}$$

8. $\frac{e^{4x}}{2}$. Multiply the terms in the numerator; then divide: $\frac{e^{-3x}\cdot e^{8x}}{2e^x}=\frac{e^{-3x+8x}}{2e^x}=\frac{e^{5x}}{2e^x}=\frac{e^{5x-x}}{2}=\frac{e^{4x}}{2}$.

9. $2\cdot 3^{2x}$. Combine the terms in the radical. Change the radical to a fractional exponent. Then simplify by raising the factors to the power: $\sqrt[3]{5e^{6x}+3e^{6x}}=\sqrt[3]{8e^{6x}}=\sqrt[3]{8}\cdot\sqrt[3]{e^{6x}}$ $=8^{1/3}\left(e^{6x}\right)^{1/3}=\left(2^3\right)^{1/3}\left(e^{6x}\right)^{1/3}=2e^{2x}$.

10. **8.** The bases are the same, so set the exponents equal to one another and solve for x: $2x+7=4x-9\rightarrow 16=2x\rightarrow 8=x$.

11. **2, 3.** The bases are the same, so set the exponents equal to one another and solve for x: $6+x^2=5x\rightarrow x^2-5x+6=0\rightarrow (x-2)(x-3)=0$. The two solutions are $x=2$ and $x=3$. When checking $x=2$, you have $e^{6+2^2}=e^{5\cdot 2}\rightarrow e^{10}=e^{10}$. And when checking $x=3$, you have $e^{6+3^2}=e^{5\cdot 3}\rightarrow e^{15}=e^{15}$. Both check.

12. **0, 2.** Change both bases to powers of 2 and simplify the terms:
$8^{x^2/3}=4^x\rightarrow\left(2^3\right)^{x^2/3}=\left(2^2\right)^x\rightarrow 2^{x^2}=2^{2x}$. Set the two exponents equal to one another and solve for x: $x^2=2x\rightarrow x^2-2x=0\rightarrow x(x-2)=0$. There are two solutions: $x=0$ and $x=2$. Checking $x=0$: $8^{x^2/3}=4^x\rightarrow 8^{0^2/3}=4^0\rightarrow 1=1$. And checking $x=2$:
$8^{2^2/3}=4^2\rightarrow 8^{4/3}=4^2\rightarrow\left(2^3\right)^{4/3}=4^2\rightarrow 2^4=4^2\rightarrow 16=16$. They both check.

13. **−7.** Change both bases to powers of 5 and simplify the terms: $25^{x-2} = 125^{x+1} \rightarrow \left(5^2\right)^{x-2} = \left(5^3\right)^{x+1} \rightarrow 5^{2x-4} = 5^{3x+3}$. Set the exponents equal and solve for x: $2x-4 = 3x+3 \rightarrow -7 = x$. Checking: $25^{-7-2} = 125^{-7+1} \rightarrow 25^{-9} = 125^{-6} \rightarrow \left(5^2\right)^{-9} = \left(5^3\right)^{-6} \rightarrow 5^{-18} = 5^{-18}$.

14. $\mathbf{\frac{13}{2}}$. Change each of the bases to a power of 3 and then apply the rules of exponents. You can change the bases 9 and 27 to powers of 3, replace the radical with a fractional exponent, and then raise the powers to other powers by multiplying the exponents.

$$\frac{27^{x+1}}{\sqrt{3}} = 9^{2x-3} \rightarrow \frac{\left(3^3\right)^{x+1}}{3^{1/2}} = \left(3^2\right)^{2x-3} \rightarrow \frac{3^{3x+3}}{3^{1/2}} = 3^{4x-6}$$

When you divide two numbers with the same base, you subtract the exponents. After you have a single power of 3 on each side, you can equate the exponents and solve for x:

$$3^{3x+3-1/2} = 3^{4x-6} \rightarrow 3^{3x+5/2} = 3^{4x-6} \rightarrow 3x + \frac{5}{2} = 4x - 6 \rightarrow 6 + \frac{5}{2} = x \rightarrow \frac{17}{2} = x$$

15. **1.** Factor out 5^x: $5^x\left(5^x - 5\right) = 0$. The first factor, 5^x, cannot equal 0. But when $5^x - 5 = 0$, you have $5^x = 5$, and $x = 1$.

16. **2, 0.** Replace 2^x with y and solve the resulting quadratic equation: $y^2 - 5y + 4 = (y-4)(y-1) = 0$. Now substitute the 2^x back in and set the two factors equal to 0: $2^{2x} - 5\cdot 2^x + 4 = (2^x - 4)(2^x - 1) = 0$. When $2^x = 4$, $x = 2$, and when $2^x = 1$, $x = 0$.

17. **0, 1.** Replace 7^x with y and solve the resulting quadratic-like equation: $y^4 - 50y^2 + 49 = \left(y^2 - 1\right)\left(y^2 - 49\right) = (y-1)(y+1)(y-7)(y+7) = 0$. This provides four different solutions to investigate, after replacing the y variables with 7^x. When $7^x = 1$, $x = 0$. When $7^x = -1$, there is no solution. When $7^x = 7$, $x = 1$. And when $7^x = -7$, again, there is no solution.

18. **$16,866.92.** Using the formula, $A = 12{,}500\left(1 + \frac{0.03}{12}\right)^{12\cdot 10} = 16{,}866.92$.

19. **3.04%.** Using the formula, $\left(1 + \frac{0.03}{12}\right)^{12} = 0.0304$.

20. **$16,873.24.** Using the formula, $A = 12{,}500e^{0.03\cdot 10} = 16{,}873.24$.

21. **$31,480.18.** Using the formula, you solve for the amount P: $40{,}000 = P\left(1 + \frac{0.048}{12}\right)^{12\cdot 5}$.

First, simplify what's in the parentheses and raise that amount to the 60th power:

$$40{,}000 = P\left(1 + \frac{0.048}{12}\right)^{12\cdot 5} \rightarrow 40{,}000 = P(1.004)^{60} \rightarrow 40{,}000 = P(1.270640719)$$

Now divide each side by that multiplier of P:

$$\frac{40{,}000}{1.270640719} = \frac{P(1.270640719)}{1.270640719} \rightarrow 31{,}480.18$$

Looks like it'll take about $31,500 to have enough for the down payment. Maybe there could be some monthly additions after an initial deposit?

22. **−2.** There are two ways of approaching this problem.

Method 1: Write an equation and change the equation to the inverse.

$$\log_6 \frac{1}{36} = x \rightarrow 6^x = \frac{1}{36}$$

Now write the fraction as a power of 6 and solve for x.

$$6^x = \frac{1}{6^2} = 6^{-2} \rightarrow x = -2$$

Method 2: Use the laws of logs to write the fraction as a difference.

$$\log_6 \frac{1}{36} = \log_6 1 - \log_6 36$$

Change the 36 to a power of 6, and then write the log of a power as a product.

$$\log_6 1 - \log_6 36 = \log_6 1 - \log_6 6^2 = \log_6 1 - 2\log 6$$

Now use the laws of logs involving the log of 1 and the log of the base.

$$\log_6 1 - 2\log 6 = 0 - 2(1) = -2$$

(23) **16.** Change the 16 to a power of 2, and then write the log of a power as a product.

$$4\log_2 16 = 4\log_2 2^4 = 4(4\log_2 2) = 16\log_2 2$$

Now use the law of logs involving the log of the base.

$$16\log_2 2 = 16(1) = 16$$

(24) **−9.** Use the laws of logs to write the fraction as a difference.

$$3\log_4 \frac{1}{64} = 3(\log_4 1 - \log_4 64)$$

Change the 64 to a power of 4, and then write the log of a power as a product.

$$3(\log_4 1 - \log_4 64) = 3(\log_4 1 - \log_4 4^3) = 3(\log_4 1 - 3\log_4 4)$$

Now use the laws of logs involving the log of 1 and the log of the base.

$$3(\log_4 1 - 3\log_4 4) = 3(0 - 3 \cdot 1) = 3(-3) = -9$$

(25) $\frac{1}{2}\log(x+1) - \frac{5}{2}\log(3x+4)$. Using the laws of logarithms:

Write the radical as a fractional exponent and apply the law of logarithms that makes the exponent a multiplier.

$$\log_4 \sqrt{\frac{x+1}{(3x+4)^5}} = \log_4 \left(\frac{x+1}{(3x+4)^5}\right)^{1/2} = 1/2 \log \frac{x+1}{(3x+4)^5}$$

Write the fraction as the difference of logs. Distribute the $\frac{1}{2}$.

$$\frac{1}{2}\log \frac{x+1}{(3x+4)^5} = \frac{1}{2}\left(\log(x+1) - \log(3x+4)^5\right) = \frac{1}{2}\log(x+1) - \frac{1}{2}\log(3x+4)^5$$

Use the law of logarithms involving an exponent on the last term.

$\frac{1}{2}\log(x+1)-\frac{1}{2}\log(3x+4)^5=\frac{1}{2}\log(x+1)-\frac{5}{2}\log(3x+4)$

(26) $\log\frac{(x-3)(x+6)^2}{(x^2+7)^3}$. Using the laws of logarithms:

First, write the coefficients of the second and third terms as exponents.

$\log(x-3)+2\log(x+6)-3\log(x^2+7)=\log(x-3)+\log(x+6)^2-\log(x^2+7)^3$

The first two terms are multiplied together, so write them as the log of a product.

$\log(x-3)(x+6)^2-\log(x^2+7)^3$

The two terms are the result of division, so write them as the log of a quotient.

$$\log(x-3)(x+6)^2-\log(x^2+7)^3=\log\frac{(x-3)(x+6)^2}{(x^2+7)^3}$$

(27) **3.** First, write the quadratic equation and then solve for x:
$x^2-6=x\rightarrow x^2-x-6=0\rightarrow(x-3)(x+2)=0$. This has two solutions: $x=3$ and $x=-2$. The 3 works, but the -2 is extraneous.

(28) **3.** First, write the two terms on the left as the log of a product. Then create the equation.

$$\log_5 x+\log_5(x+3)=\log_5(2x+12)$$
$$\log_5 x(x+3)=\log_5(2x+12)$$
$$x(x+3)=2x+12$$

Solve for x: $x^2+3x=2x+12\rightarrow x^2+x-12=0\rightarrow(x+4)(x-3)=0$.

This has two solutions: $x=-4$ and $x=3$. The 3 works, but the -4 is extraneous.

(29) **5.** First, write the two terms on the left as the log of a quotient. Then apply the inverse rule to write the equation.

$\log_2(x+3)-\log_2(x-4)=3\rightarrow\log_2\frac{x+3}{x-4}=3\rightarrow 2^3=\frac{x+3}{x-4}$

Solving for x, $8=\frac{x+3}{x-4}\rightarrow 8(x-4)=x+3\rightarrow 8x-32=x+3\rightarrow 7x=35\rightarrow x=5$.

This checks.

(30) **See the figure.** The y-intercept is at $y = 2(3)^0 + 1 = 2(1) + 1 = 2 + 1 = 3$, or at (0,3). The horizontal asymptote is $y = 1$. Some points that can be included in the graph are $\left(-1, \frac{5}{3}\right)$ and (2,19).

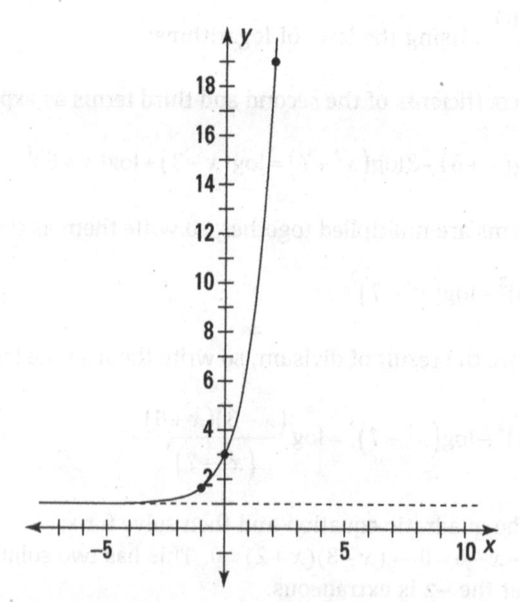

(31) **See the figure.** The y-intercept is at $y = 4(0.5)^{0.2(0)} = 4(0.5)^0 = 4(1) = 4$, or at (0,4). The horizontal asymptote is the x-axis. The 0.2 multiplier on the x of the exponent will flatten out the graph. And the base of 0.5 will cause the graph to fall as you move from left to right. Some points that can be included in the graph are (−5,8) and (5,2).

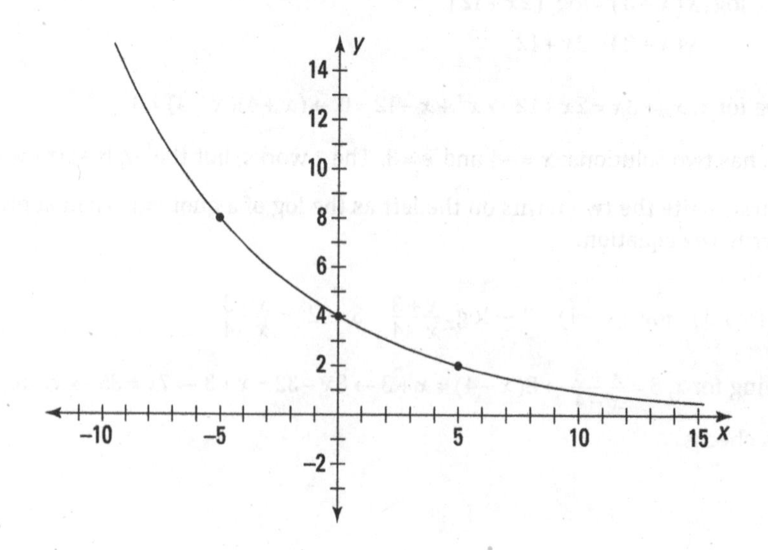

32 **See the figure.** The graph of $y = 2^x$ contains the points $(0,1)$, $(1,2)$, $\left(-2, \frac{1}{4}\right)$, and $(3,8)$. Reverse the coordinates of those points to create points on the log graph: $(1,0)$, $(2,1)$, $\left(\frac{1}{4}, -2\right)$, and $(8,3)$; you now have points on the graph of the log function.

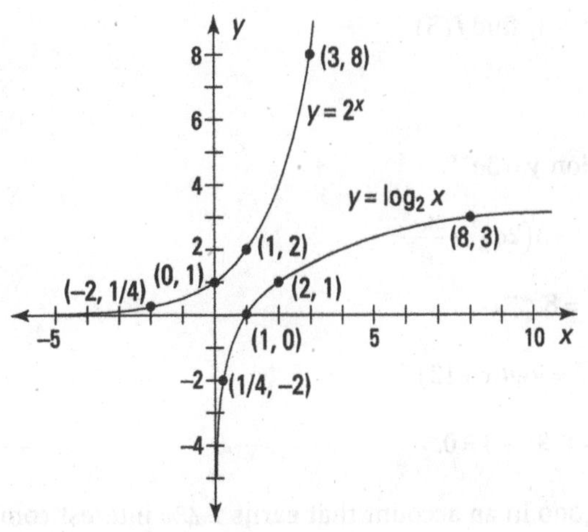

33 **See the figure.** The graph of $y = \left(\frac{1}{2}\right)^x$ contains the points $(0,1)$, $\left(1, \frac{1}{2}\right)$, $(-2,4)$, and $(-3,8)$. Reverse the coordinates of those points to create points on the log graph: $(1,0)$, $\left(\frac{1}{2}, 1\right)$, $(4,-2)$, and $(8,-3)$; you now have points on the graph of the log function.

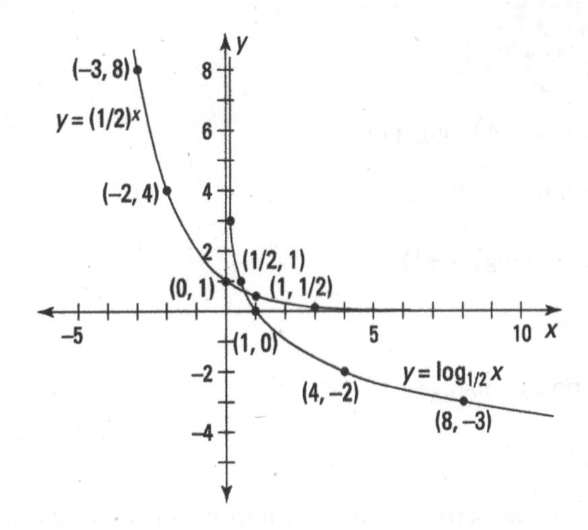

Whaddya Know? Chapter 11 Quiz

Quiz time! Complete each problem to test your knowledge on the various topics covered in this chapter. You can then find the solutions and explanations in the next section.

1. Given $f(x) = 4^{x-2} + 3$, find $f(5)$.
2. Simplify: $2e^x \cdot e^{3x}$.
3. Graph the function $y = 3e^{-x}$.
4. Simplify: $4e^4 \cdot e^{-1} + 3(2e)^5 - \frac{3e^6}{e^3}$.
5. Solve for x: $4^{2x+1} = 8^{3x+4}$.
6. Solve for x: $\log x^2 = \log(x+12)$.
7. Solve for x: $3^{2x} - 4 \cdot 3^x + 3 = 0$.
8. You deposit $10,000 in an account that earns 2.4% interest compounded quarterly. You leave the investment in the account for 15 years. What is your new total at the end of that time?
9. Write the expression as a single log term with no coefficients: $2\log x + \log 3 - \log(x+4)$.
10. Expand: $\log_4 \frac{5(x-6)^3}{\sqrt{x^2+7}}$.
11. Solve for x: $\log_3 \frac{3x+2}{x^2} = 2$.
12. Solve for x: $\log_2(2x^2 - 4) = \log_2(7x)$.
13. Graph the function $y = 2^{x+2} + 1$.
14. Solve for x: $\log_5 3x = \log_5(x+1)$.
15. Evaluate $\log_4 16 =$.
16. Graph the function $y = \log_{1/3} x$.
17. Evaluate $\log_3 \frac{1}{9} =$.
18. Which account has the better result: 3% interest compounded continuously for ten years or 4% compounded daily for ten years? (Use 365 for the number of days.)

Answers to Chapter 11 Quiz

1. **67.** $f(5) = 4^{5-2} + 3 = 4^3 + 3 = 64 + 3 = 67$

2. **$2e^{4x}$.** $2e^x \cdot e^{3x} = 2e^{x+3x} = 2e^{4x}$

3. **See the figure.** The y-intercept is (0,3), and the horizontal asymptote is $y = 0$, the x-axis.

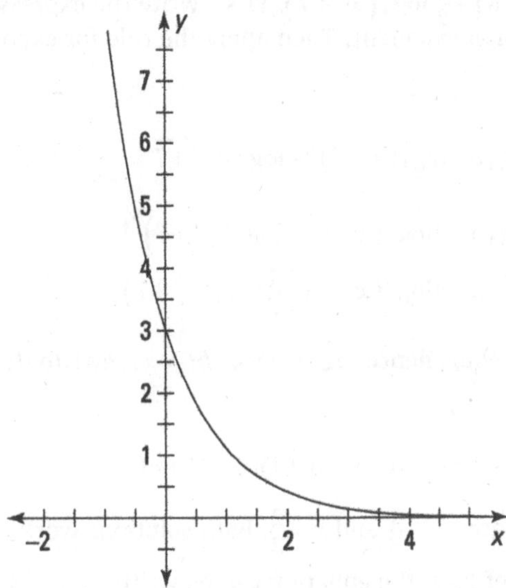

4. **$e^3 + 96e^5$.** First, simplify each term. Then combine like terms.

$$4e^4 \cdot e^{-1} + 3(2e)^5 - \frac{3e^6}{e^3} = 4e^{4+(-1)} + 3\left(32e^5\right) - 3e^{6-3} = 4e^3 + 96e^5 - 3e^3 = e^3 + 96e^5$$

5. **−2.** First, change the two bases to powers of 2 and simplify the terms.

$$\left(2^2\right)^{2x+1} = \left(2^3\right)^{3x+4} \rightarrow 2^{2(2x+1)} = 2^{3(3x+4)} \rightarrow 2^{4x+2} = 2^{9x+12}$$

Set the exponents equal to one another and solve for x.

$$4x + 2 = 9x + 12 \rightarrow -10 = 5x \rightarrow -2 = x$$

6. **4 or −3.** Using the laws of logs, the equation can be written $x^2 = x + 12$. Solve for x.

$$\log x^2 = \log(x+12) \rightarrow x^2 = x + 12 \rightarrow x^2 - x - 12 = 0 \rightarrow (x-4)(x+3) = 0$$
$$x = 4 \text{ or } x = -3$$

Checking the solutions, both work.

7. **0,1.** Factor the quadratic. Then set each factor equal to 0 and solve for x.

$$3^{2x} - 4 \cdot 3^x + 3 = 0 \rightarrow \left(3^x - 1\right)\left(3^x - 3\right) = 0$$
$$3^x - 1 = 0,\ 3^x = 1,\ x = 0$$
$$3^x - 3 = 0,\ 3^x = 3,\ x = 1$$

8. **$14,317.88.** Using $A = P\left(1+\frac{r}{n}\right)^{nt}$, you have

$$A = 10{,}000\left(1+\frac{0.024}{4}\right)^{4\cdot 15} = 10{,}000(1.006)^{60} = 14{,}317.88$$

9. $\log\frac{3x^2}{x+4}$. Change the coefficient of 2 on the first term to be the exponent of x. Multiply 3 times that result. Then divide the product by $x+4$.

10. $\log_4 5 + 3\log_4(x-6) - \frac{1}{2}\log_4(x^2+7)$. First, write the expression as three terms. Change the radical to a fractional exponent. Then apply the rule for exponents on the second and third terms.

$$\begin{aligned}\log_4\frac{5(x-6)^3}{\sqrt{x^2+7}} &= \log_4 5 + \log_4(x-6)^3 - \log_4\sqrt{x^2+7}\\ &= \log_4 5 + 3\log_4(x-6) - \log_4(x^2+7)^{1/2}\\ &= \log_4 5 + 3\log_4(x-6) - \frac{1}{2}\log_4(x^2+7)\end{aligned}$$

11. $-\frac{1}{3}, \frac{2}{3}$. Using the equivalence $\log_b x = y \Leftrightarrow b^y = x$, rewrite the equation as $3^2 = \frac{3x+2}{x^2}$. Solve for x.

$$9x^2 = 3x+2 \rightarrow 9x^2-3x-2=0 \rightarrow (3x+1)(3x-2)=0$$

The two solutions are $x = -\frac{1}{3}$ and $x = \frac{2}{3}$. Both solutions work.

12. **4.** Using the laws of logs, the equation can be written $2x^2-4=7x$. Solve for x.

$$2x^2-4=7x \rightarrow 2x^2-7x-4=0 \rightarrow (2x+1)(x-4)=0$$

$$x = -\frac{1}{2} \text{ or } x = 4$$

The solution $x = -\frac{1}{2}$ of the quadratic equation does not work in the log equation. It creates the log of a negative number. This solution is extraneous.

13. **See the figure.** The y-intercept is (0,5) and the horizontal asymptote is $y = 1$.

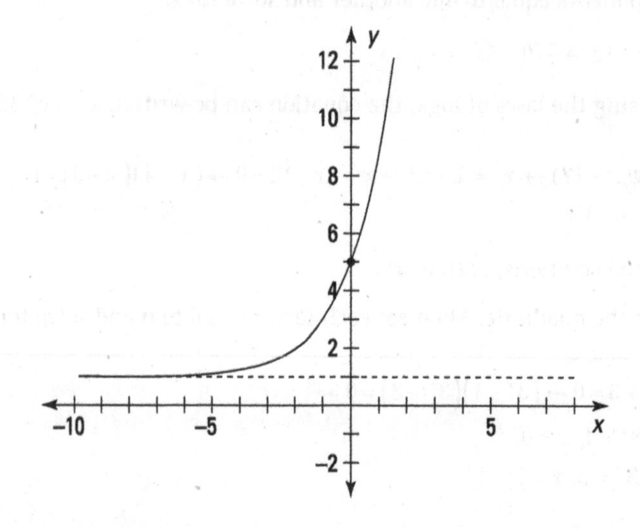

(14) $\frac{1}{2}$. Using the laws of logs, the equation can be written $3x = x + 1$. Solve for x.

$$3x = x + 1 \rightarrow 2x = 1 \rightarrow x = \frac{1}{2}$$

(15) **2.** Using the equivalence $\log_b x = y \Leftrightarrow b^y = x$, change $\log_4 16 = y$ to $4^y = 16$. Solving for y,

$$4^y = 16 \rightarrow 4^y = 4^2 \rightarrow y = 2.$$

(16) **See the figure.** The x-intercept is (1,0). Some other points on the curve are $\left(\frac{1}{3},1\right)$, $(3,-1)$, and $(9,-2)$.

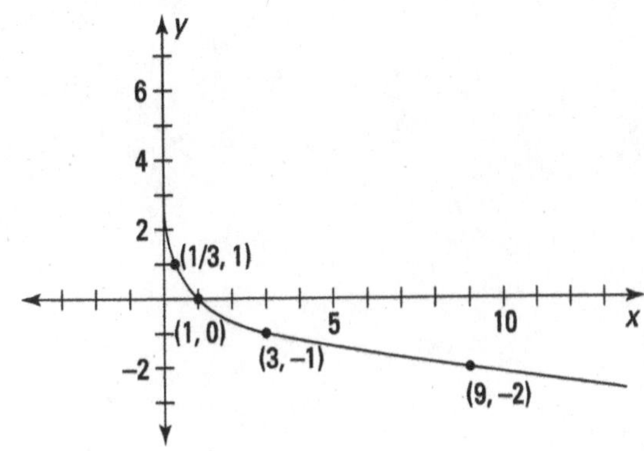

(17) **−2.** Using the equivalence $\log_b x = y \Leftrightarrow b^y = x$, change $\log_3 \frac{1}{9} =$ to $3^y = \frac{1}{9}$. Solving for y,

$$3^y = \frac{1}{9} \rightarrow 3^y = \frac{1}{3^2} \rightarrow 3^y = 3^{-2}$$

$$y = -2$$

(18) **4% compounded daily.**

$$A = Pe^{0.03(10)} \approx P(1.34986) \rightarrow A = P\left(1 + \frac{0.04}{365}\right)^{365(10)} \approx P(1.49179)$$

IN THIS CHAPTER

» Describing function transformations

» Applying one or more transformations

» Working with various function types

» Identifying properties of functions

Chapter 12 Transforming and Critiquing Functions

A *function transformation* consists of one or more changes in a function that doesn't really change the initial structure. Oh, yes, a transformation can move the function values around, change the signs, flip, twist, turn, and stretch or flatten it, but you still recognize the original. Function transformations are very helpful in situations where you just want to raise the base rate or change the due date or create the opposite situation. There are applications all over the place.

Running through the Transformations

Three basic transformation types are covered here: translations, reflections, and scaling. You start with the basic graph of the function $f(x)$ and then perform the particular transformation.

1. A *translation* moves the basic graph up or down or right or left by h units. The shape does not change.

   a. Translating up: $f(x)+h$

   b. Translating down: $f(x)-h$

   c. Translating right: $f(x-h)$

   d. Translating left: $f(x+h)$

2. A *reflection* flips the basic graph either horizontally or vertically over an axis or line.

   a. Reflecting over the x-axis: $-f(x)$

   b. Reflecting over the y-axis: $f(-x)$

3. *Scaling* makes the curve steeper or flatter (stretches or compresses)

   a. Stretching (steepening): $a \cdot f(x)$ when $a > 1$

   b. Compressing (flattening): $a \cdot f(x)$ when $0 < a < 1$

And now you get to see how all of these transformations act when performed on different types of functions: linear, absolute value, quadratic, other polynomial, and radical.

Translating and understanding the language

When a function is translated from one place to another on a graph, it really involves nothing more than sliding all the points to the left, right, up, or down. You don't change the shape of the graph; you merely adjust its position. Translations upward and downward are accomplished by adding or subtracting a constant number from the function rule. This adjusts the y-values. To translate to the left or right, you have to add or subtract a constant to the input. This way, the function operations work on the new, adjusted value.

EXAMPLE

Q. Determine the kind of translations that occur when you add or subtract 1 from the function $f(x) = x$.

A. The graph of $y = x$ is a diagonal line that divides the first and third quadrants evenly and goes through the origin (see Figure 12-1a). When you add 1 to the function equation and get $y = x + 1$, every point on the graph moves up one unit higher (see Figure 12-1b). Note how the point (1,1) in the first graph has moved up to (1,2) in the second graph. And when you subtract 1 from the base equation, you get $y = x - 1$, which is shown in Figure 12-1c. The y-coordinates are all 1 less than in the base equation.

Q. Determine the kind of translations that occur when you add or subtract 2 from the input into the function $f(x) = |x|$.

A. The graph of $y = |x|$ has a V-shape. The V can point upward or downward, depending on the multiplier of the operation (see Figure 12-2a). When you add 2 to the function input value and get $y = |x + 2|$, every point on the graph moves 2 units to the left (see Figure 12-2b). Note how the point (0,0) in Figure 12-2a has moved over to (−2,0) in Figure 12-2b. And when you subtract 2 from the base equation, you get $y = |x - 2|$, which moves every point 2 units to the right. You see in Figure 12-2c that the point (−2,2) has moved to (0,2). The y-coordinates stay the same; it's the x-coordinates that change.

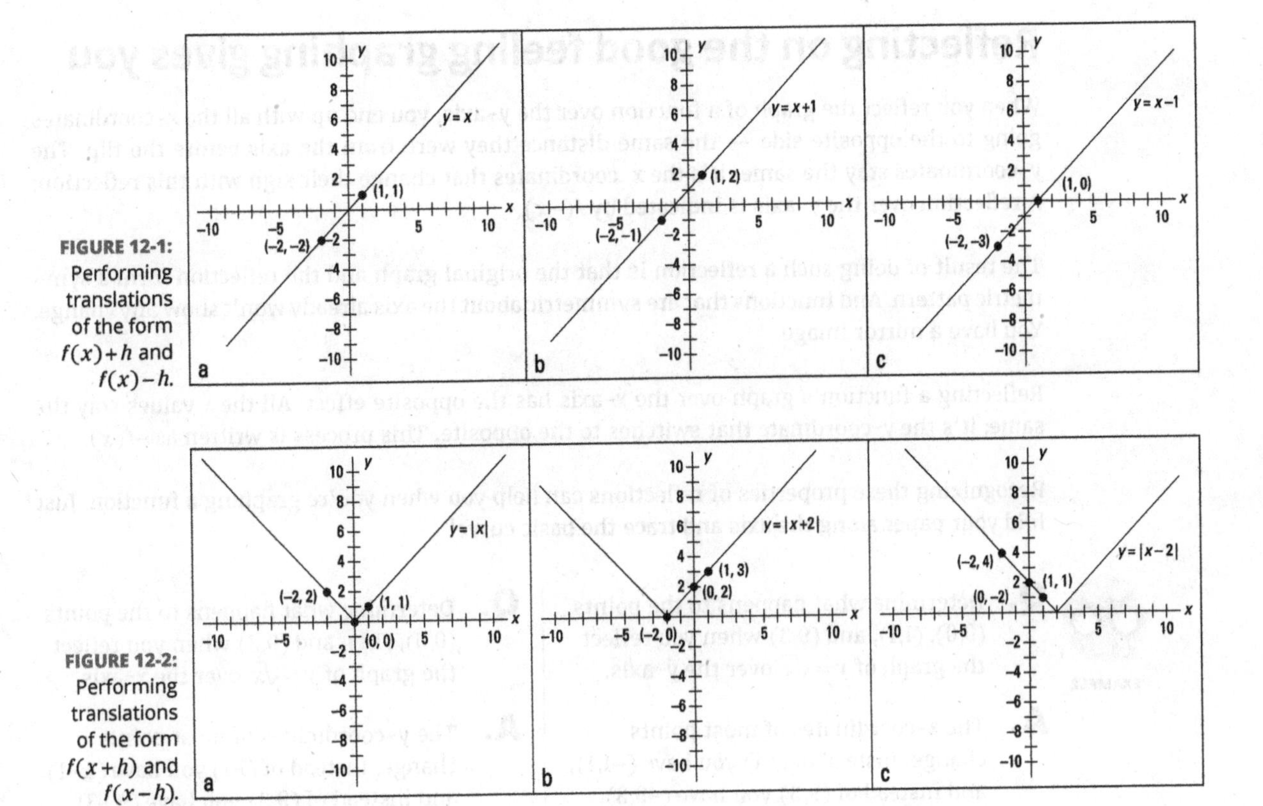

FIGURE 12-1: Performing translations of the form $f(x)+h$ and $f(x)-h$.

FIGURE 12-2: Performing translations of the form $f(x+h)$ and $f(x-h)$.

1 **Sketch the graphs of $y = x^2 - 1$ and $y = (x-1)^2$ using translations.**

2 **Sketch the graphs of $y = \sqrt{x+2}$ and $y = \sqrt{x} - 2$ using translations.**

Reflecting on the good feeling graphing gives you

When you reflect the graph of a function over the y-axis, you end up with all the x-coordinates going to the opposite side — the same distance they were from the axis before the flip. The y-coordinates stay the same; it's the x-coordinates that change their sign with this reflection. A reflection over the y-axis is indicated by $f(-x)$.

The result of doing such a reflection is that the original graph and the reflection form a symmetric pattern. And functions that are symmetric about the axis already won't show any change. You have a mirror image.

Reflecting a function's graph over the x-axis has the opposite effect. All the x values stay the same; it's the y-coordinate that switches to the opposite. This process is written as $-f(x)$.

Recognizing these properties of reflections can help you when you're graphing a function. Just fold your paper along the axis and trace the basic curve!

EXAMPLE

Q. Determine what happens to the points (0,0), (1,1), and (9,3) when you reflect the graph of $y=\sqrt{x}$ over the y-axis.

A. The x-coordinates of most points change. Instead of (1,1) you have (−1,1), and instead of (9,3) you have (−9,3). The point (0,0) is the only one that doesn't change. See Figure 12-3a, which has the original graph. Then Figure 12-3b shows the refection over the y-axis.

Q. Determine what happens to the points (0,0), (1,1), and (9,3) when you reflect the graph of $y=\sqrt{x}$ over the x-axis.

A. The y-coordinates of most points change. Instead of (1,1) you have (1,−1), and instead of (9,3) you have (9,−3). The point (0,0) is the only one that doesn't change. See Figure 12-3c, which shows the refection over the x-axis.

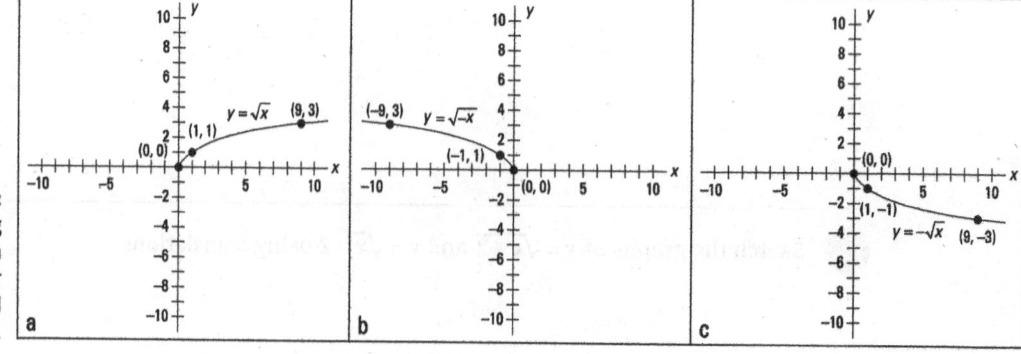

FIGURE 12-3: Reflecting the function $y=\sqrt{x}$ over the y- and x-axes.

YOUR TURN

3 **Sketch the graph of the reflection of $y = x^3$ over the y-axis.**

4 **Sketch the graph of the reflection of $y = x^2$ over the x-axis.**

5 **Reflect the graph of $y = |x|$ over both the x- and y-axes.**

Scaling the Cliff of Numerical Possibilities

The word *scaling* in this context refers to changing the value of a numerical multiplier. You can reduce the size of the graph by making the absolute values of the y-values smaller. The same goes for increasing the y-values. When a multiplier is used and it's a number whose absolute value is greater than 1, the graph of the original function gets steeper or stretches. And when the multiplier is between 0 and 1, the graph gets flatter; it shrinks.

EXAMPLE

Q. Sketch the graph of $y = 4 \cdot |x|$.

A. All the y-values of the original function, $y = |x|$, are 4 times as great. The graph is much steeper. See Figure 12-4a for the original graph of $y = |x|$ and see how much steeper the graph is in Figure 12-4b.

Q. Sketch the graph of $y = \frac{1}{4} \cdot |x|$.

A. All the y-values of the original function, $y = |x|$, are $\frac{1}{4}$ the value of the original. The graph is much flatter. See Figure 12-4c for the graph of this function.

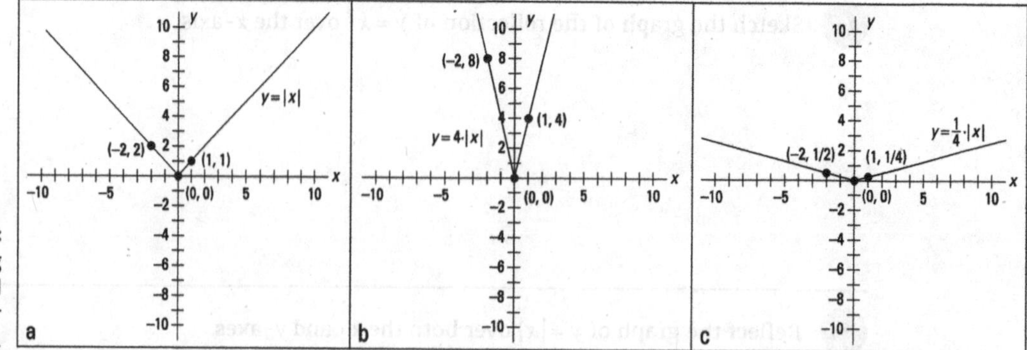

FIGURE 12-4: Creating steeper and flatter graphs.

YOUR TURN

6 Sketch the graph of $y = 3\sqrt{x}$.

7 Sketch the graph of $y = \frac{1}{5}x^2$.

Multiplying the Opportunities for Transforming

There are multiple ways of transforming a function. And you aren't restricted to using just one. You can use two or more transformations on a single function. You can make a graph steeper and shove it over 2 units to the right. You can flip it over and make it flatter and move it downward. All it takes is to write the function equation in a format that allows others to see what you want to do.

EXAMPLE

Q. Sketch the graph of $f(x) = 3x^2 - 2$.

A. Starting with the basic quadratic function $f(x) = x^2$, this function is steeper because of the multiplier of 3 and it drops down 2 units.

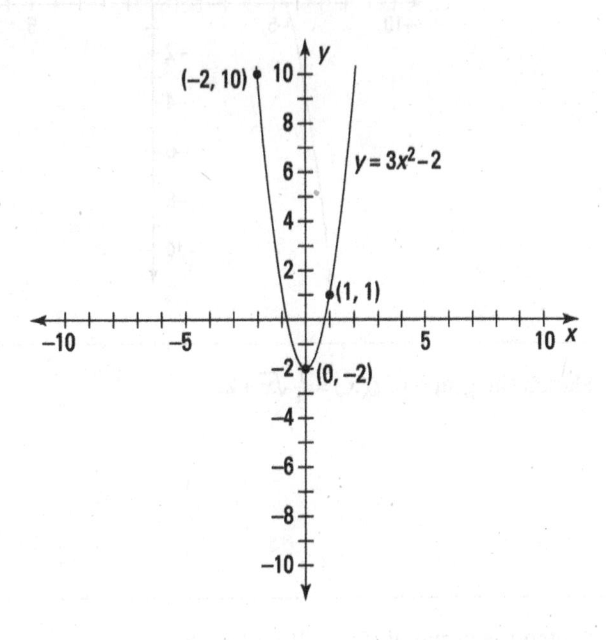

Q. Sketch the graph of $f(x) = \frac{1}{3}(x+4)^3 + 1$.

A. Starting with the basic graph of $f(x) = x^3$, you have a translation 4 units to the left, and you make it flatter by multiplying by $\frac{1}{3}$. And, finally, there is a translation of 1 unit upward.

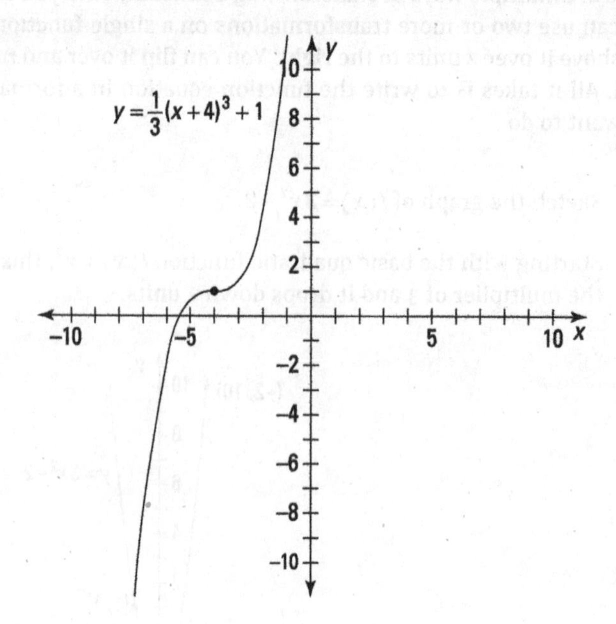

8 Sketch the graph of $f(x) = \frac{1}{3}\sqrt{x} + 2$.

9 Sketch the graph of $f(x) = 3(x+1)^2 - 5$.

10 Sketch the graph of $f(x) = -2|x-3| + 1$.

Noting the Ups, Downs, Highs, and Lows

One reason to study the aspects of a function and its graph is to see if it is really modelling a situation. Does it represent how the income of a company is progressing? Does it model the current weather pattern? Does it model your physical strength based on your exercising?

Many types of functions are chosen to model many different situations. The algebraic functions you find here are a good starting point. Throw in the trigonometric functions and it's looking even better. This section covers how to determine when a curve is increasing or decreasing and if it's at a high point or at a low point. You find a discussion of relative maximum and minimum points and absolute maximum and minimum points in Chapter 9, where you work with polynomials. This discussion just broadens your horizons!

Determining increasing and decreasing intervals

A function is increasing when its y-values get larger and larger as you move from left to right. And, conversely, a function is decreasing when its y-values get smaller as you move from left to right. These properties can be determined by looking at the graph of the function and observing the starting and ending points of the property. You will find an even nicer way to find these properties in a calculus course. (You will learn to love finding the derivative and what it tells you.)

EXAMPLE

Q. Determine where the graph of the function shown in Figure 12-5a is increasing and where it is decreasing.

A. **See Figure 12-5a.** This is a cubic polynomial with two points indicated. The function is increasing in two different intervals. It is increasing from negative infinity until it reaches the point $(-4,7)$, and it increases again from the point $(3,-4.3)$ through positive infinity. Using interval notation, you say that the function is increasing $(-\infty, -4) \cup (3, \infty)$. The function is decreasing between -4 and 3. This is written $(-4,3)$. Be sure that you recognize the context here. This isn't the point $(-4,3)$. This is the interval starting at $x=-4$ and ending at $x=3$. Note that it's the x-coordinates that are referenced when you talk about where things are happening in a function.

Q. Determine where the graph of the function $f(x) = -\frac{1}{x^2}$ is increasing and where it is decreasing.

A. **See Figure 12-5b.** This rational function has a vertical asymptote of $x = 0$. The function is decreasing from negative infinity to 0, and it increases from 0 through positive infinity. Using interval notation, you say that the function is decreasing $(-\infty, 0)$ and increasing $(0, \infty)$. The x-value of 0 is not included. The solutions begin with any numbers greater than 0.

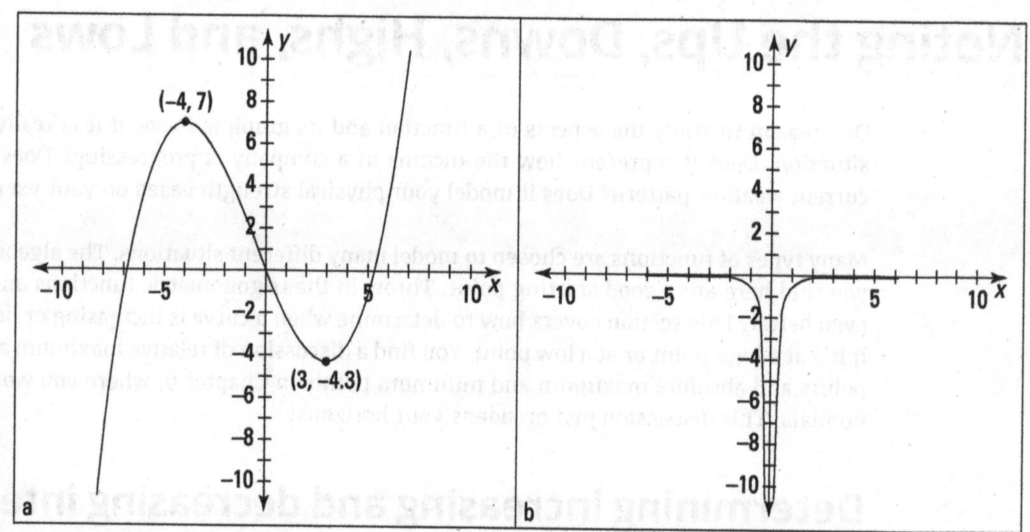

FIGURE 12-5: Increasing and decreasing intervals in a graph.

Charting the highs and lows

Another property of functions and their graphs that is of great interest is when they reach a high point and when they reach a low point. These points can also be "qualified" as to whether they are just the highest or lowest in the near vicinity (calling them "relative") or whether they are the "absolute" maximum or minimum. Again, these points are found very handily using calculus, but using the graphs here will work fine.

EXAMPLE

Q. Determine where the graph of the function shown in Figure 12-5a has any maximum or minimum points. Then classify them as relative or absolute.

A. Referring to Figure 12-5a, the point (−4,7) is a relative maximum. The graph increases coming from the left and decreases going to the right. Relative to all the points around it, this point is the highest. But the graph does go higher as you move to the right of $x = 7$. Also, you see a relative minimum point at (3,−4.3). There are lower points as you move to the left of $x = -8$.

Q. Determine where the graph of the function $f(x) = -\frac{1}{x^2}$ in Figure 12-5b has any maximum or minimum points. Then classify them as relative or absolute.

A. There are no maximum or minimum values — relative or absolute. The highest point does approach 0 going both to the left and right, but it never gets there!

YOUR TURN

The following exercises all refer to Figure 12-6.

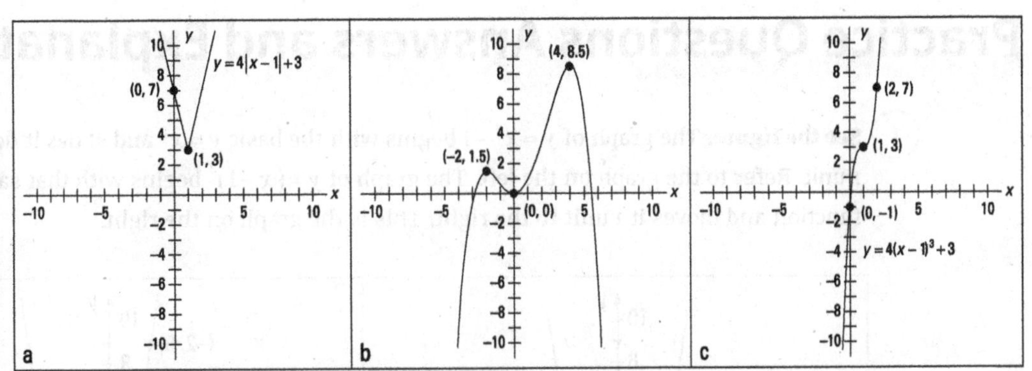

FIGURE 12-6: Increasing and decreasing around maximum and minimum points.

11 **Determine where the function $f(x) = 4|x - 1| + 3$ is increasing or decreasing and where there are any maximum or minimum values. Refer to Figure 12-6a.**

12 **Determine where the function shown in Figure 12-6b is increasing or decreasing and where there are any maximum or minimum values.**

13 **Determine where the function $f(x) = 4(x - 1)^3 + 3$ is increasing or decreasing and where there are any maximum or minimum values. Refer to Figure 12-6c.**

Practice Questions Answers and Explanations

1. **See the figure.** The graph of $y = x^2 - 1$ begins with the basic $y = x^2$ and slides it downward 1 unit. Refer to the graph on the left. The graph of $y = (x-1)^2$ begins with that same basic function and moves it 1 unit to the right. This is the graph on the right.

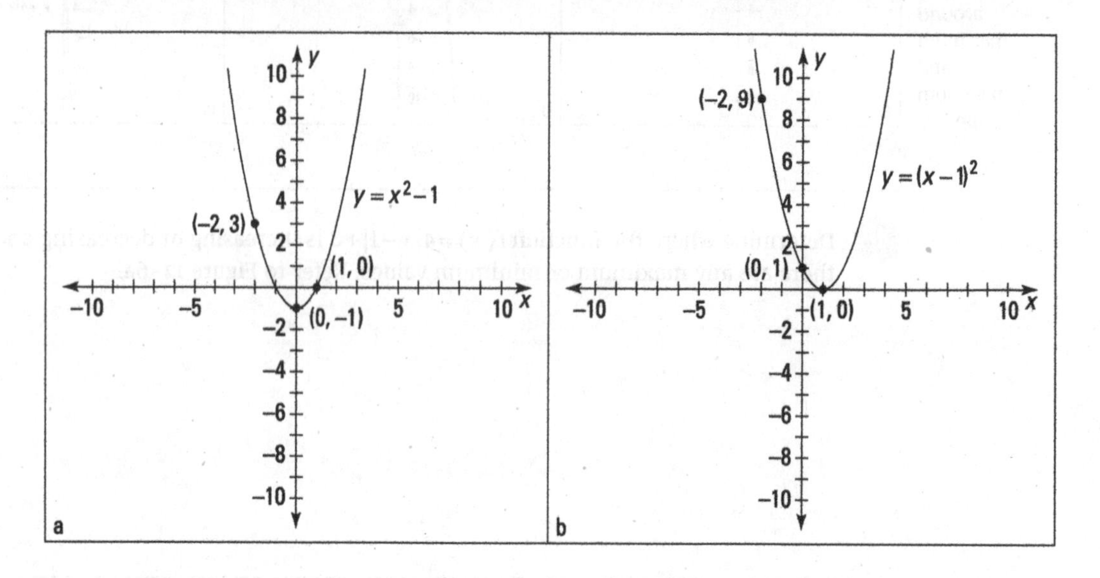

2. **See the figure.** The graph of $y = \sqrt{x+2}$ begins with the basic $y = \sqrt{x}$ and slides it 2 units to the left. You find it in the graph on the left. And $y = \sqrt{x} - 2$ begins with that same basic graph and moves it 2 units downward. Look at the graph on the right.

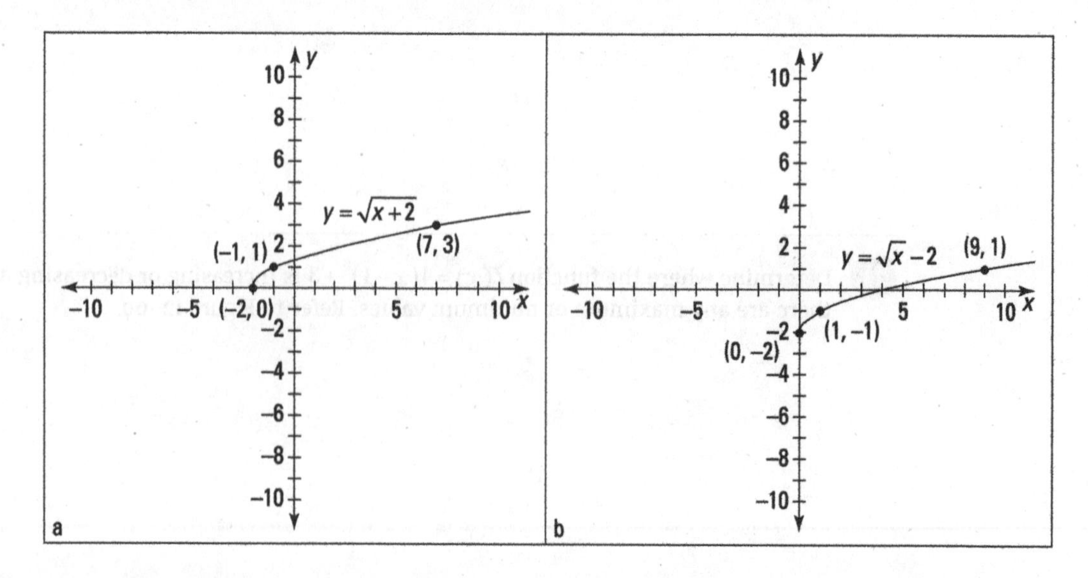

3) **See the figure.** The graph of $y = x^3$ is shown with dashes, and the reflection is shown with a solid curve. With this reflection, every x of the new curve is the opposite of the original x.

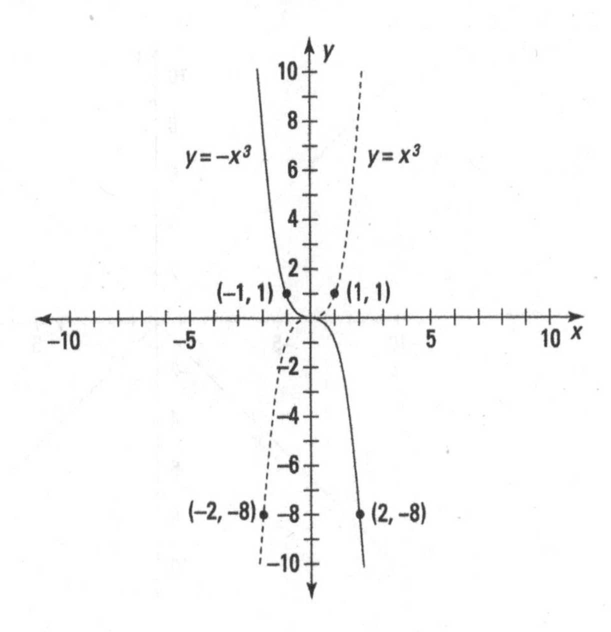

4) **See the figure.** The graph of $y = x^2$ is shown with dashes, and the reflection is shown with a solid curve. With this reflection, every y of the new curve is the opposite of the original y.

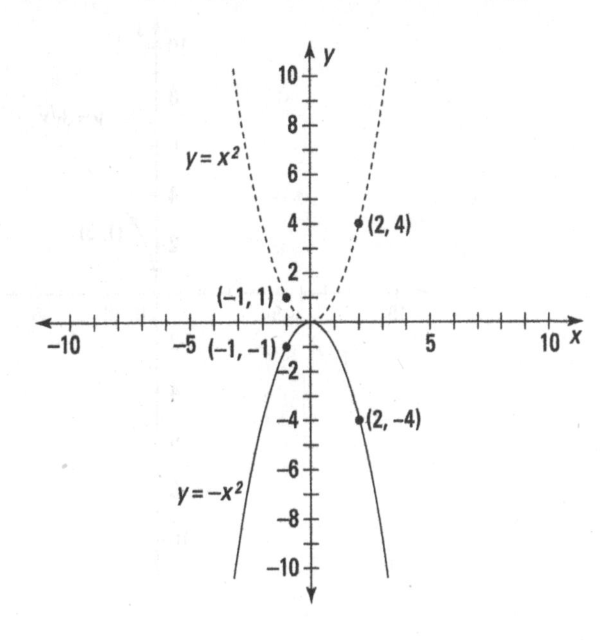

5) **See the figure.** The graph of $y = |x|$ and its reflection over the y-axis are the same graph. The graph of the reflection over the x-axis opens upward.

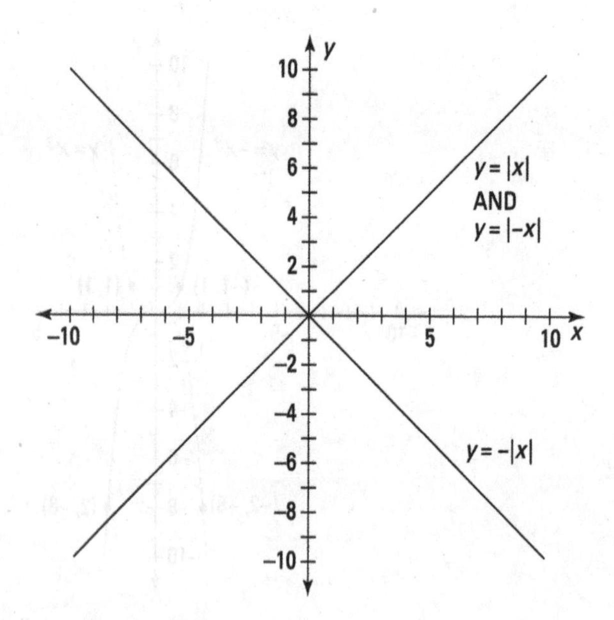

6) **See the figure.** The graph of $y = 3\sqrt{x}$ is three times as steep as the graph of $y = \sqrt{x}$.

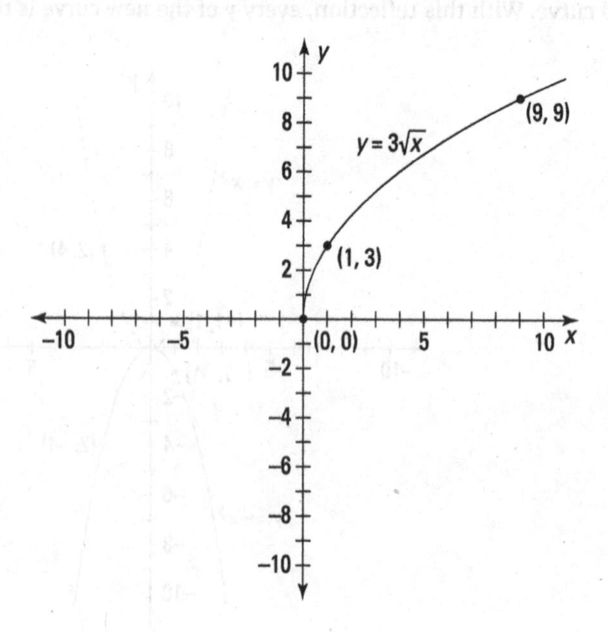

7. **See the figure.** The graph of $y = \frac{1}{5}x^2$ is much flatter than the graph of the base function. The vertex is still at (0,0).

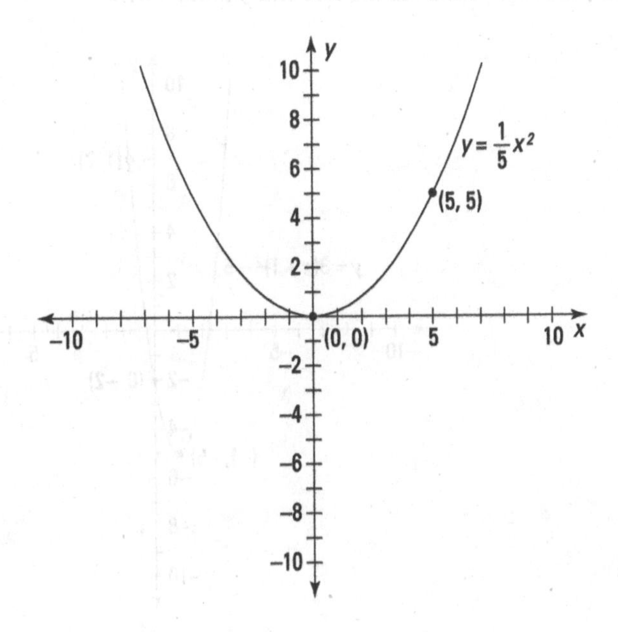

8. **See the figure.** The graph of $f(x) = \frac{1}{3}\sqrt{x} + 2$ is much flatter than the base graph. The y-intercept is (0,2).

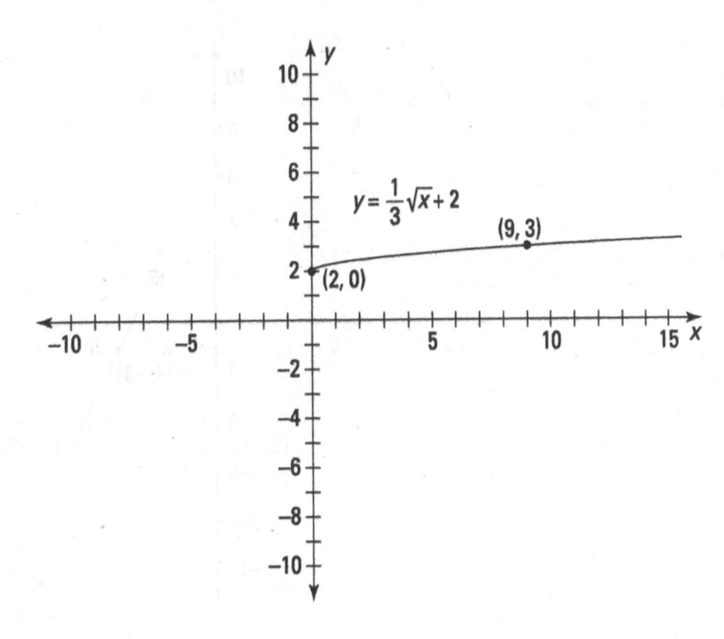

9 **See the figure.** The graph of $f(x)=3(x+1)^2-5$ is much steeper than the base graph. The vertex is moved 1 unit to the left and 5 units down.

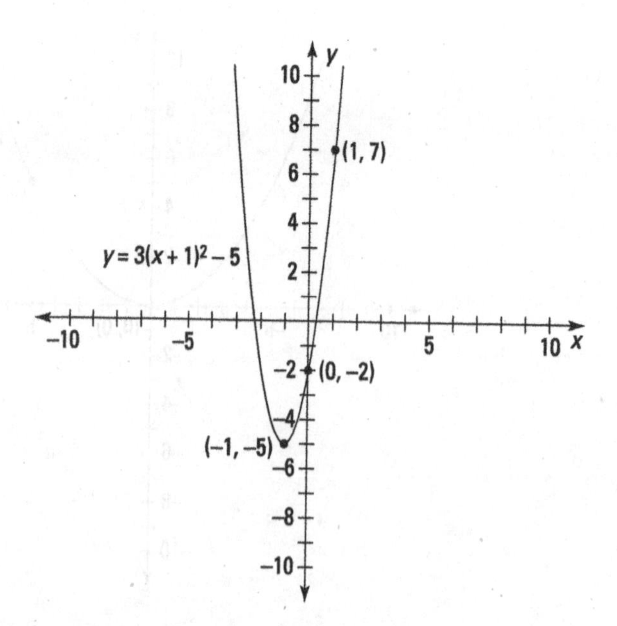

10 **See the figure.** The graph of $f(x)=-2|x-3|+1$ points upward instead of downward. The highest point is at (3,1). The two lines are steeper than those of the base graph.

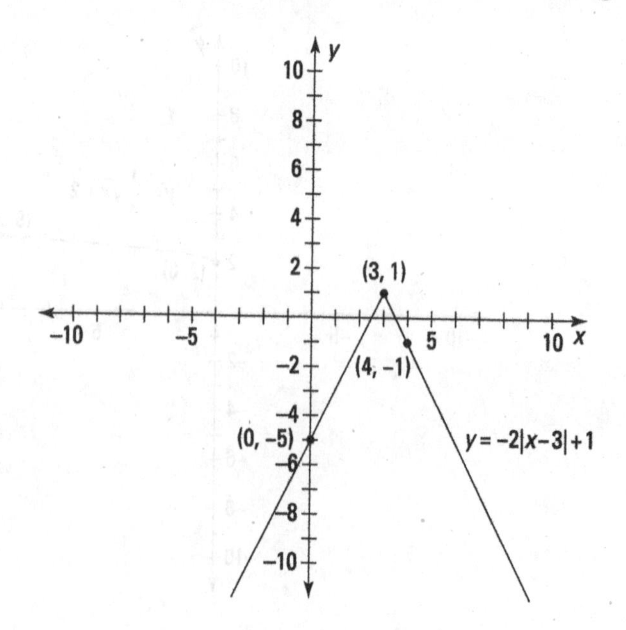

11 **Increases: $(1,\ \infty)$; decreases: $(-\infty,\ 1)$; abs. min.: $(1,3)$.** The function neither increases nor decreases at the minimum.

12 **Increases: $(-\infty,\ -2)\cup(0,\ 4)$; decreases: $(-2,\ 0)\cup(4,\ \infty)$; rel. max.: $(-2,1.5)$; abs. max.: $(4,8.5)$; rel. min.: $(0,0)$.**

13 **Increases: $(-\infty,\ 1)\cup(1,\ \infty)$.** It never decreases. There are no maximum or minimum values.

Whaddya Know? Chapter 12 Quiz

Quiz time! Complete each problem to test your knowledge on the various topics covered in this chapter. You can then find the solutions and explanations in the next section.

1. What kind of transformation changes $y=\sqrt{x}$ to $y=\sqrt{x+2}$?

2. Sketch the graph of $y=4\sqrt[3]{x-1}$.

3. Describe the transformations involved in changing $y=x$ to $y=-2x-3$.

4. Determine any maximum or minimum points on the graph of the polynomial.

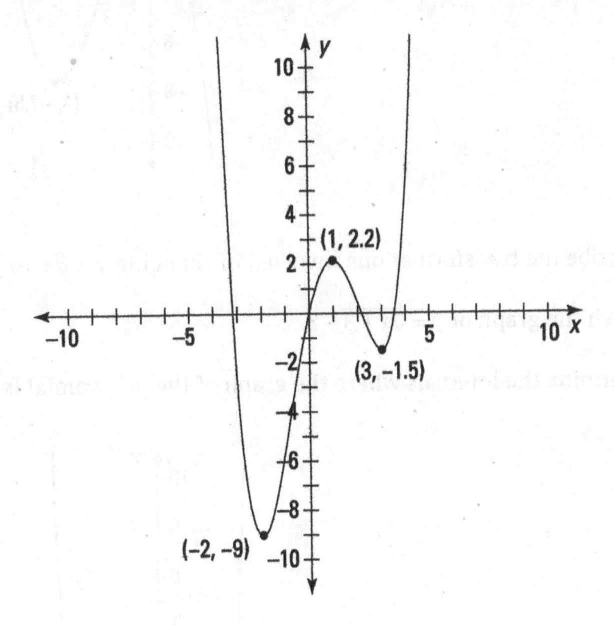

5. Sketch the graph of $y=-2(x+1)^2+4$.

6. What kind of transformation changes $y=x^3$ to $y=x^3-4$?

7 Determine any maximum or minimum points on the graph of the polynomial.

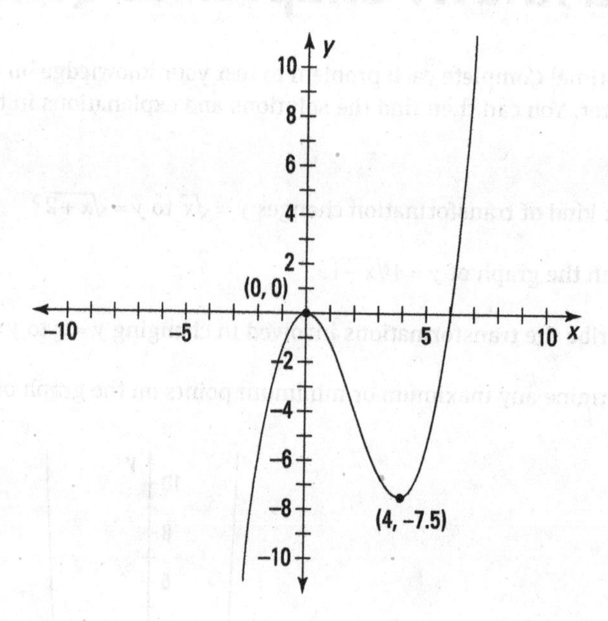

8 Describe the transformations involved in changing $y = \sqrt[3]{x}$ to $y = 2\sqrt[3]{x} + 5$.

9 Sketch the graph of $y = \sqrt{x+3} - 3$.

10 Determine the intervals where the graph of the polynomial is increasing or decreasing.

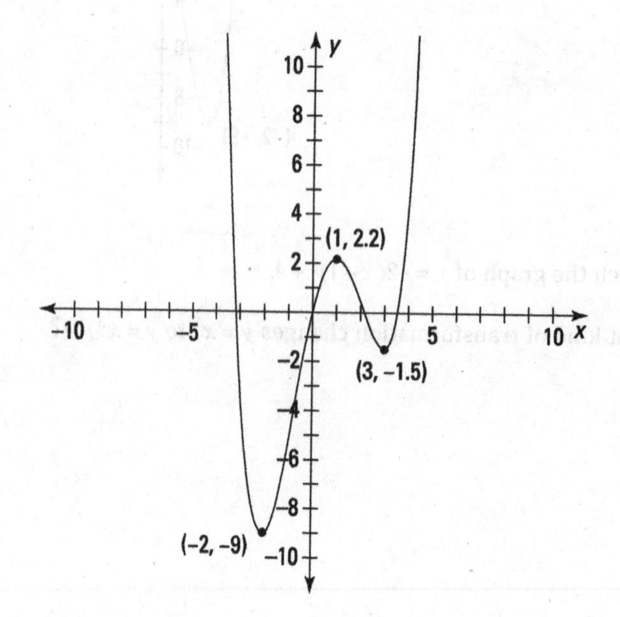

11. What kind of transformation changes $y = |x|$ to $y = 3|x|$?

12. Sketch the graph of $y = 2 - |x + 2|$.

13. What kind of transformation changes $y = x^2$ to $y = (x-1)^2$?

14. Determine the intervals where the graph of the polynomial is increasing or decreasing.

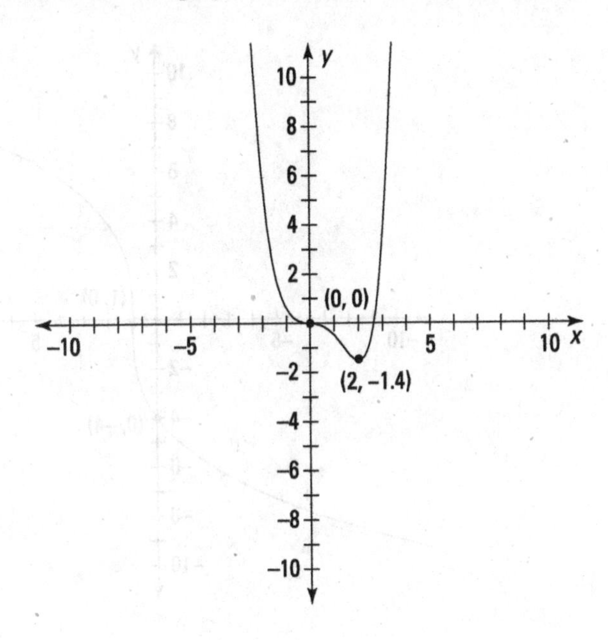

Answers to Chapter 12 Quiz

1. **Translation.** The graph of the function slides 2 units to the left.

2. **See the figure.** The basic graph $y = \sqrt[3]{x}$ has been translated 1 unit to the right. The multiplier of 4 makes the graph much steeper.

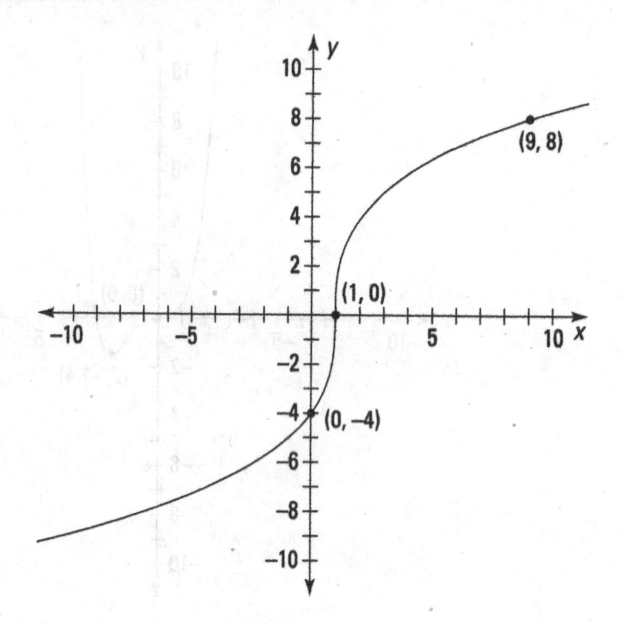

3. **Translation and reflection.** The graph of the function slides 3 units downward and is reflected over a horizontal axis. The multiplier also makes the line twice as steep.

4. **Abs. min.: (−2,−9); rel. min.: (3,−1.5); rel. max.: (1,2.2).** The absolute minimum occurs when $x = -2$. There is no absolute maximum, because the curve rises to positive infinity to the left and to the right.

(5) **See the figure.** The basic graph $y = x^2$ has been translated 1 unit to the left and 4 units up. The -2 multiplier reflects the graph over horizontally and makes it steeper.

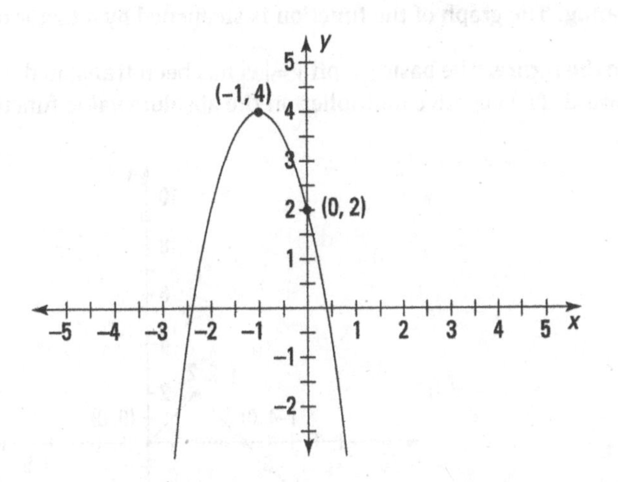

(6) **Translation.** The graph of the function slides 4 units downward.

(7) **Rel. min.: (4,−7.5); rel. max.: (0,0).** There is no absolute maximum or absolute minimum, because the curve rises to positive infinity to the left and to the right.

(8) **Translation and scaling.** The graph of the function slides 5 units upward and is steepened by a factor of 2.

(9) **See the figure.** The basic graph $y = \sqrt{x}$ has been translated 3 units to the left and 3 units downward.

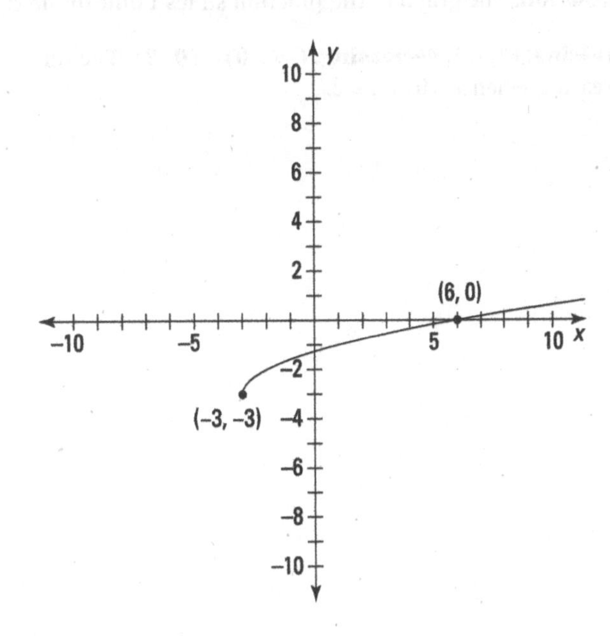

10. **Increasing: $(-2, 1)\cup(3, \infty)$; decreasing: $(-\infty, -2)\cup(1, 3)$.** The function doesn't increase or decrease at the three turning points.

11. **Scaling.** The graph of the function is steepened by a factor of 3.

12. **See the figure.** The basic graph $y=|x|$ has been translated 2 units to the left and 2 units upward. The negative multiplier on the absolute value function flips the graph downward.

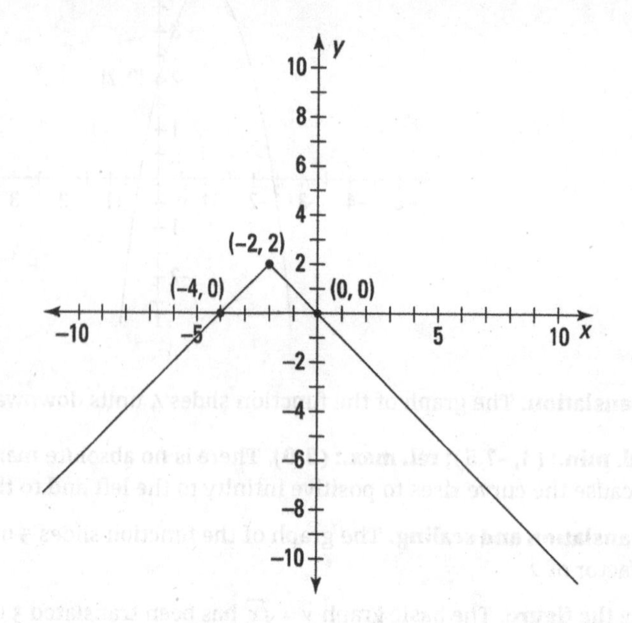

13. **Translation.** The graph of the function slides 1 unit to the right.

14. **Increasing: $(2, \infty)$; decreasing: $(-\infty, 0)\cup(0, 2)$.** The function is neither increasing nor decreasing when $x=0$ or $x=2$.

3

Using Conics and Systems of Equations

In This Unit . . .

IN THIS CHAPTER

» Grasping the layout of a sliced cone

» Investigating the standard equations and graphs of the four conic sections

» Properly identifying conics with nonstandard equations

Chapter 13 Slicing the Way You Like It: Conic Sections

Conic is the name given to a special group of curves. What they have in common is how they're constructed — points lying relative to an anchored point or points with respect to a line. But that sounds a bit stuffy, doesn't it? Maybe it works better to think of conic sections in terms of how you can best describe the curves visually. Picture the Rainbow Bridge across the Niagara River. Imagine the earth's path swinging around the sun. Reflect on the curved reflection plate in a car's headlights. All these pictures circling your mind are related to curves called conics.

If you take a cone — imagine one of those yummy sugar cones that you put ice cream in — and slice it through in a particular fashion, the resulting edge you create will trace one of the four conic sections: a parabola, circle, ellipse, or hyperbola. (You can see a sketch of a cone in the following section. Well, actually, it's two cones, lined up point-to-point.)

Each conic section has a specific equation, and I cover each thoroughly in this chapter. You can glean a good deal of valuable information from a conic section's equation, such as where it's centered in a graph, how wide it opens, and its general shape. I also discuss the techniques that work best for you when you're called on to graph conics. Grab a pizza and an ice-cream cone for visual motivation and read on!

Carefully Cutting Circular Cones

A *conic section* is a curve formed by the intersection of a cone or two cones and a plane (a *cone* is a shape whose base is a circle and whose sides taper up to a point). The curve that is formed depends on where the cone is sliced:

- If you slice the cone straight across, you create a circle along the edge, just like the top.
- If you slice a side piece off at an angle, you form a U-shaped parabola along the edge.
- If you slice the cone at a slant, you have an ellipse, or an oval shape, along the edge.
- If you picture two cones, tip to tip, and a slice going straight down through both, you picture a hyperbola. You have two wide, U-shaped edges that sort of come nose to nose. A hyperbola takes some special gymnastics!

Figure 13-1 shows each of the four conic sections sliced and diced. Each conic section has a specific equation that you use to graph the conic or to use it in an application (such as when the conic equation represents the curvature of a tunnel). You can head to the sections in this chapter that deal with the different conics to find out more about these topics.

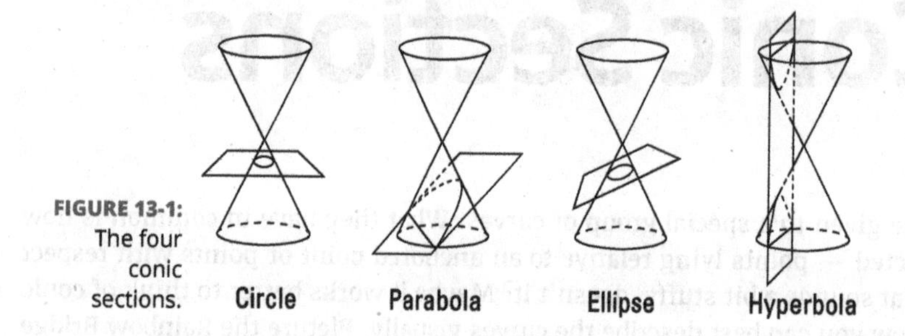

FIGURE 13-1: The four conic sections.

Opening Every Which Way with Parabolas

A parabola, a U-shaped conic that I first introduce in Chapter 8 (the parabola is the only conic section that can fit the definition of a polynomial), is defined as all the points that fall the same distance from some fixed point, called its *focus*, and a fixed line, called its *directrix*. The focus is denoted by F, and the directrix by $y = d$. Figure 13-2 shows you some of the points on a parabola and how they are the same distance from the parabola's focus as they are from the directrix.

A parabola has a couple other defining features. The *axis of symmetry* of a parabola is a line that runs through the focus and is perpendicular to the directrix (imagine a line running through F in Figure 13-2). The axis of symmetry does just what its name suggests: It shows off how symmetric a parabola is. A parabola is a mirror image on either side of its axis. Another feature is the parabola's vertex. The *vertex* is the curve's extreme point — the lowest or highest point, or the point on the curve farthest right or farthest left. The vertex is also the point where the axis of symmetry crosses the curve (you can create this point by putting a pencil on the curve in Figure 13-2 where the imaginary axis crosses the curve after spearing through F).

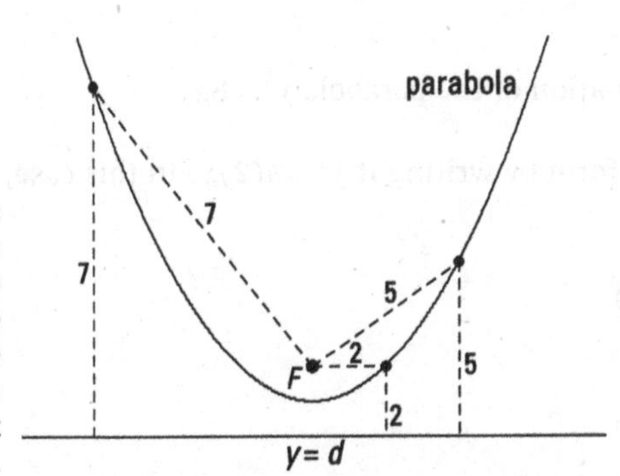

FIGURE 13-2: Points on a parabola are the same distance away from a fixed point and a line.

Looking at parabolas with vertices at the origin

Parabolas can have graphs that go every which way and have vertices at any point in the coordinate system. When possible, though, you like to deal with parabolas that have their vertices at the origin. The equations are easier to deal with, and the applications are easier to solve. Therefore, I put the cart before the horse with this section to cover these specialized parabolas. (In the section, "Observing the general form of parabolic equations," later in the chapter, I deal with all the other parabolas you may run across.)

Opening to the right or left

Parabolas with their vertices (the plural form of vertex is *vertices* — just a little Latin for you) at the origin and opening to the right or left have a standard equation $y^2 = 4ax$ and are known as *relations*; you see a relationship between the variables. A relation is different from a function in that you have more than one y-value for each x-value (a no-no for functions). The standard form comes packed with information about the focus, directrix, vertex, axis of symmetry, and direction of a parabola. The equation also gives you a hint as to whether the parabola is narrow or opens wide.

The general form of a parabola with the equation $y^2 = 4ax$ gives you the following info:

Focus: $(a,0)$

Directrix: $x = -a$

Vertex: $(0,0)$

Axis of symmetry: $y = 0$

Opening: To the right if a is positive; to the left if a is negative.

Shape: Narrow if $|4a|$ is less than 1; wide if $|4a|$ is greater than 1.

I use the absolute value operation, $|\ |$, instead of saying that $4a$ has to be between 0 and 1 or between −1 and 0. I think it's just a neater way of dealing with the rule.

EXAMPLE

Q. Extract information from the equation of the parabola $y^2 = 8x$.

A. Put the equation in the $y^2 = 4ax$ form by writing it $y^2 = 4(2)x$. In this case, you extract the following info:

The value of a is 2 (from $4 \cdot 2 = 8$).

The focus is at (2,0).

The directrix is the line $x = -2$.

The vertex is at (0,0).

The axis of symmetry is $y = 0$.

The parabola opens to the right.

The parabola is wide because $|4(2)|$ is greater than 1.

Figure 13-3 shows the graph of the parabola $y^2 = 8x$ with all the key information listed in the sketch.

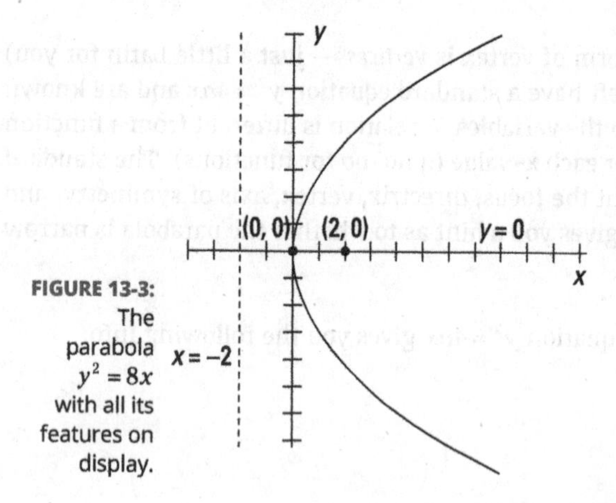

FIGURE 13-3: The parabola $y^2 = 8x$ with all its features on display.

Of course, extracting information is always more convenient when the equation of the parabola has an even-number coefficient, such as the 8 in $y^2 = 8x$, but the process works for odd numbers, too. You just have to deal with fractions.

Opening upward or downward

Parabolas that open left or right are relations, but parabolas that have curves opening upward or downward are a bit more special. The parabolas that open upward or downward are functions — they have only one y value for every x value. (For more on functions, see Chapter 7.) Parabolas of this variety with their vertex at the origin have the following standard equation: $x^2 = 4ay$.

You can tell that you have a relation (opening left or right) rather than a function (opening up or down) if the y variable is squared. The quadratic polynomials (from Chapter 8) and the functions all have the x variable squared.

The information you glean from the standard equation tells much the same story as the equation of parabolas that open sideways; however, many of the rules are reversed.

From the general form for the parabola $x^2 = 4ay$, you extract the following information:

Focus: $(0,a)$

Directrix: $y = -a$

Vertex: $(0,0)$

Axis of symmetry: $x = 0$

Opening: Upward if a is positive; downward if a is negative.

Shape: Narrow if $|4a|$ is less than 1; wide if $|4a|$ is greater than 1.

EXAMPLE

Q. Extract information from the equation of the parabola $x^2 = -\frac{1}{2}y$.

A. Convert the form $x^2 = -\frac{1}{2}y$ to the form $x^2 = 4ay$. To perform the conversion, just divide the coefficient by 4, and then write the coefficient as 4 times the result of the division. You haven't changed the value of the coefficient; you've just changed how it looks. You create the equation: $x^2 = 4\left(-\frac{1}{8}\right)y$.

In this case, you extract the following info:

The value of a is $-\frac{1}{8}$. The focus falls on the point $\left(0,-\frac{1}{8}\right)$. The directrix forms a line at $y = \frac{1}{8}$. The vertex is at (0,0). The axis of symmetry is $x = 0$. The graph opens downward. The parabola is wide because $|4(a)| = \left|4\left(-\frac{1}{8}\right)\right| = \left|-\frac{1}{2}\right| < 1$.

Figure 13-4 shows the graph of the parabola $x^2 = -\frac{1}{2}y$ with all the elements illustrated in the sketch.

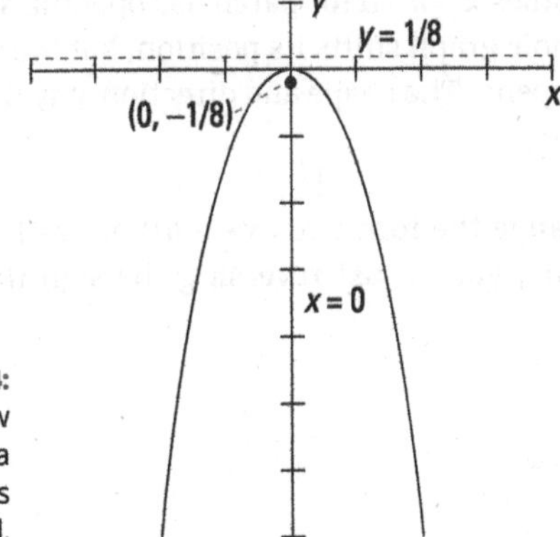

FIGURE 13-4: A narrow parabola that opens downward.

YOUR TURN

Extract the points of information about the parabola, given the equation: $x^2 = -8y$.

Extract the points of information about the parabola, given the equation: $y^2 = \frac{2}{3}x$.

Observing the general form of parabolic equations

The curves of parabolas can open upward, downward, to the left, or to the right, but the curves don't always have to have their vertices at the origin. Parabolas can wander all around a graph. So, how do you track the curves down to pin them on a graph? You look to their equations, which give you all the information you need to find out where they are and what they're doing.

The standard forms used for parabolas with their vertices at the origin (which I discuss in the previous sections) are special cases of these more general parabolas. If you replace the h and k coordinates with zeros, you have the special parabolas anchored at the origin.

A move in the position of the vertex (away from the origin, for example) changes the focus, directrix, and axis of symmetry of a parabola. In general, a move of the vertex just adds the value of h or k to the basic form. For instance, when the vertex is at (h,k), the focus is at $(h+a,k)$ for parabolas opening to the right and at $(h,k+a)$ for parabolas opening upward. The directrix is also affected by h or k in its equation; it becomes $x = h - a$ for parabolas opening sideways and $y = k - a$ for those opening up or down. The whole graph shifts its position, but the shift doesn't affect which direction it opens or how wide it opens. The shape and direction stay the same. The following summarizes the information for you.

If the equation contains $(x+h)$ or $(y+k)$, change the forms to $(x-[-h])$ or $(y-[-k])$, respectively, to determine the correct signs. Actually, you're just reversing the sign that's already there.

Opening left or right: $(y-k)^2 = 4a(x-h)$. When the *y* variable is squared, the parabola opens left or right.

- The vertex is at (h,k).
- The focus is at $(h+a,k)$.
- The axis of symmetry is $y = k$.
- The directrix is $x = h - a$.
- If $4a$ is positive, the curve opens right; if $4a$ is negative, the curve opens left.
- If $|4a| > 1$, the parabola is relatively wide; if $|4a| < 1$, the parabola is relatively narrow.

Opening up or down: $(x-h)^2 = 4a(y-k)$. When the *x* variable is squared, the parabola opens up or down.

- The vertex is at (h,k).
- The focus is at $(h,k+a)$.
- The axis of symmetry is $x = h$.
- The directrix is $y = k - a$.
- If $4a$ is positive, the curve opens up; if $4a$ is negative, the curve opens down.
- If $|4a| > 1$, the parabola is relatively wide; if $|4a| < 1$, the parabola is relatively narrow.

EXAMPLE

Q. Given the equation of the parabola $(x+4)^2 = 12(y-2)$, determine the following: vertex, focus, axis of symmetry, directrix, opening direction, steepness or flatness.

A. Referring to the general form $(x-h)^2 = 4a(y-k)$, you determine that $h=-4$, $k=2$, and $a=3$. From this, the vertex, $(h,k)=(-4,2)$. The focus, $(h,k+a)=(-4,5)$. The axis of symmetry, $x = h \rightarrow x = -4$. The directrix, $y = k - a \rightarrow y = -1$. Because $4a$ is positive, it opens upward. And because $3 > 1$, it will be relatively wide, making it flattened rather than steep.

Q. Given the equation of the parabola $(y+3)^2 = -2(x+1)$, determine the following: vertex, focus, axis of symmetry, directrix, opening direction, steepness or flatness.

A. Referring to the general form $(y-k)^2 = 4a(x-h)$, you determine that $h=-1$, $k=-3$, and $a=-\frac{1}{2}$. From this, the vertex, $(h,k)=(-1,-3)$. The focus, $(h+a,k)=\left(-\frac{3}{2},-3\right)$. The axis of symmetry, $y = k \rightarrow y = -3$. The directrix, $h = h - a \rightarrow x = -\frac{1}{2}$. Because $4a$ is negative, it opens left. And because $\frac{1}{2} < 1$, it will be relatively steep.

YOUR TURN

Find the vertex, focus, axis of symmetry, directrix, opening direction, and steepness or flatness of the parabola $(x-1)^2 = -8(y+7)$.

Find the vertex, focus, axis of symmetry, directrix, opening direction, and steepness or flatness of the parabola $(y+3)^2 = 8(x-5)$.

Converting parabolic equations to the standard form

When the equation of a parabola appears in standard form, you have all the information you need to graph it or to determine some of its characteristics, such as direction or size. Not all equations come packaged that way, though. You may have to do some work on the equation first to be able to identify anything about the parabola.

The standard form of a parabola is $(x-h)^2 = a(y-k)$ or $(y-k)^2 = a(x-h)$, where (h,k) is the vertex.

The methods used here to rewrite the equation of a parabola into its standard form also apply when rewriting equations of circles, ellipses, and hyperbolas. (See the later section, "Identifying Conics from Their Equations," for a more generalized view of changing the forms of conic equations.) The standard forms for conic sections are factored forms that allow you to immediately identify needed information. Different algebra situations call for different standard forms — the form just depends on what you need from the equation.

If you want to convert the equation of a parabola into the standard form, you perform the following steps, which contain a method called *completing the square* (a method you use to solve quadratic equations; see Chapter 3 for a review of completing the square):

1. **Rewrite the equation with the x^2 and x terms (and/or the y^2 and y terms) on one side of the equation and the rest of the terms on the other side.**
2. **Factor the left side, if necessary.**
3. **Add a number to each side to make the side with the squared term into a perfect square trinomial (thus completing the square; see Chapter 3 for more on trinomials).**
4. **Rewrite the perfect square trinomial in factored form, and factor the terms on the other side by the coefficient of the variable.**

EXAMPLE

Q. Write the equation of the parabola $x^2+10x-2y+23=0$ in the standard form.

A. First, rewrite the equation with just the x-terms on the left: $x^2+10x=2y-23$.

Add 25 to each side of the equation to create a perfect square trinomial on the left: $x^2+10x+25=2y-23+25$.

Write the left side as a binomial squared and the right side as a binomial factored: $(x+5)^2=2(y+1)$.

Q. Write the equation of the parabola $3y^2-12y-x+9=0$ in the standard form.

A. First, rewrite the equation with just the y-terms on the left: $3y^2-12y=x-9$.

Factor the terms on the left: $3\left(y^2-4y\right)=x-9$.

Add 4 to the terms in the parentheses to complete the square. This adds 12 to the left side, so also add 12 to the right side: $3\left(y^2-4y+4\right)=x-9+12$.

Write the left side as a binomial squared multiplied by 3 and the right side as a binomial: $3\left(y-2\right)^2=x+3$.

Divide each side by 3: $\left(y-2\right)^2=\frac{1}{3}(x+3)$.

YOUR TURN

 5 Write the equation of the parabola $5x^2-10x-8y+21=0$ in the standard form.

6 Write the equation of the parabola $2y^2+4y+x+5=0$ in the standard form.

Sketching the graphs of parabolas

Parabolas have distinctive U-shaped graphs, and with just a little information, you can make a relatively accurate sketch of the graph of a particular parabola. The first step is to think of all parabolas as being in one of the general forms I list in the previous two sections.

Taking the necessary graphing steps

Here's the full list of steps to follow when sketching the graph of a parabola — either $(x-h)^2 = 4a(y-k)$ or $(y-k)^2 = 4a(x-h)$:

1. **Determine the coordinates of the vertex, (h,k), and plot it.**
2. **Determine the direction the parabola opens, and decide if it's wide or narrow by looking at the $4a$ portion of the general parabola equation.**
3. **Lightly sketch in the axis of symmetry that goes through the vertex ($x = h$ when the parabola opens upward or downward and $y = k$ when it opens sideways).**
4. **Choose a couple of other points on the parabola and find each of their partners on the other side of the axis of symmetry to help you with the sketch.**

EXAMPLE

Q. Graph the parabola $(y+2)^2 = 8(x-1)$.

A. First, note that this parabola has its vertex at the point (1,–2) and opens to the right, because the y is squared (if the x had been squared, it would open up or down) and a (2) is positive. The graph is relatively wide about the axis of symmetry, $y=-2$, because $a=2$, which makes $|4a|$ greater than 1. To find a random point on the parabola, try letting $y=6$ and solve for x: $(6+2)^2 = 8(x-1) \rightarrow 8^2 = 8(x-1) \rightarrow 8^{\not 2} = \not 8(x-1) \rightarrow 8 = x-1 \rightarrow 9 = x$; so the point on the parabola is (9,6). Its partner point, on the other side of the axis of symmetry, is (9,–10). The 6 is 8 units above the axis of symmetry and the –10 is 8 units below the axis. Another example point, which you find by using the same process, is (5.5,4). The point (5.5,4) is 6 units above the axis of symmetry, so its partner is the point (5.5,–8). Figure 13-5 shows the vertex, axis of symmetry, and the two points placed in a sketch.

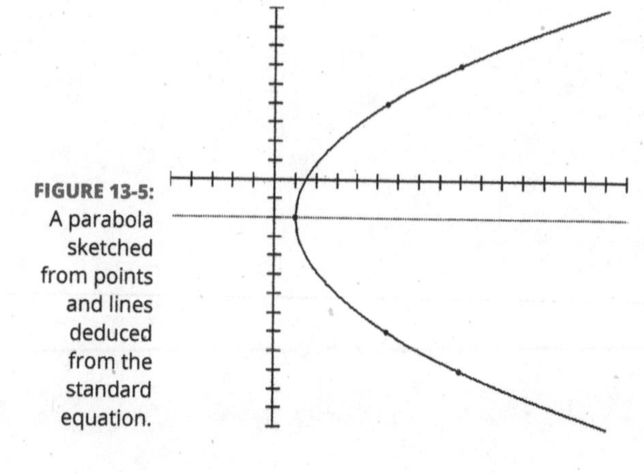

FIGURE 13-5: A parabola sketched from points and lines deduced from the standard equation.

YOUR TURN

7 Sketch a graph of the parabola $(y+2)^2 = -8(x+4)$.

8 Sketch a graph of the parabola $(x-1)^2 = \frac{1}{4}(y+3)$.

9 Sketch a graph of the parabola $x^2 - 4x - 12y + 40 = 0$.

Applying suspense to the parabola

Sketching parabolas helps you visualize how they're used in an application, so you want to be able to sketch quickly and accurately if need be. A real-world occurrence of a curve close to the parabola involves the cables that hang between the towers of a suspension bridge. These cables form a curve called a *catenary*, which is usually very close to a parabolic curve. Consider the following situation: An electrician wants to put a decorative light on the cable of a suspension bridge at a point 100 feet (horizontally) from where the cable touches the roadway at the middle of the bridge. You can see the electrician's blueprint in Figure 13-6.

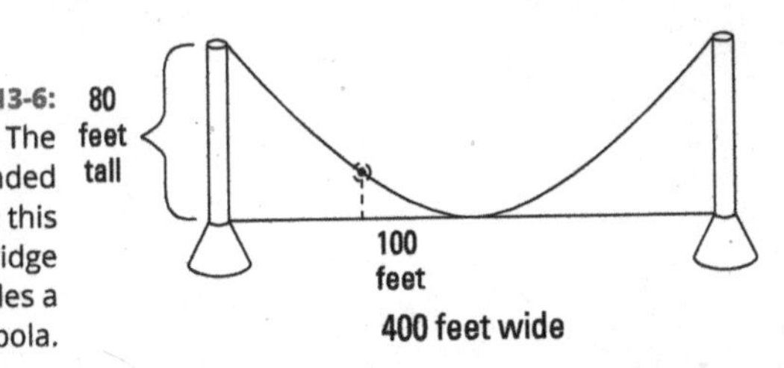

FIGURE 13-6: The suspended cable on this bridge resembles a parabola.

The electrician needs to know how high the cable is at a point 100 feet from the center of the bridge so that they can plan their lighting experiment. The towers holding the cable are 80 feet high, and the total length of the bridge is 400 feet.

You can help the electrician solve this problem by writing the equation of the parabola that fits all these parameters. The easiest way to handle the problem is to let the bridge roadbed be the x-axis and the center of the bridge be the origin (0,0). The origin, therefore, is the vertex of the parabola. The parabola opens upward, so you use the equation of a parabola that opens upward with its vertex at the origin, which is $x^2 = 4ay$ (see the earlier section, "Looking at parabolas with vertices at the origin").

To solve for a, you put the coordinates of the point (200,80) in the equation. Where do you get these seemingly random numbers? Half the total of 400 feet of bridge is 200 feet. You move 200 feet to the right of the middle of the bridge and 80 feet up to get to the top of the right tower. Replacing the x in the equation with 200 and the y with 80, you get $40{,}000 = 4a(80)$. Dividing each side of the equation by 80, you find that $4a$ is 500, so the equation of the parabola that represents the cable is $x^2 = 500y$.

So, how high is the cable at a point 100 feet from the center? In Figure 13-6, the point 100 feet from the center is to the left, so -100 represents x. A parabola is symmetric about its vertex, so it doesn't matter whether you use positive or negative 100 to solve this problem. But, sticking to the figure, let $x = -100$ in the equation; you get $(-100)^2 = 500y$, which becomes $10{,}000 = 500y$. Dividing each side by 500, you get $y = 20$. The cable is 20 feet high at the point where the electrician wants to put the light. They need a ladder!

Going Round and Round in Circles

A circle, probably the most recognizable of the conic sections, is defined as all the points plotted at the same distance from a fixed point — the circle's center, C. The fixed distance is the radius, r, of the circle.

The standard form for the equation of a circle with radius r and with its center at the point (h,k) is $(x-h)^2 + (y-k)^2 = r^2$. Figure 13-7 shows the sketch of a circle.

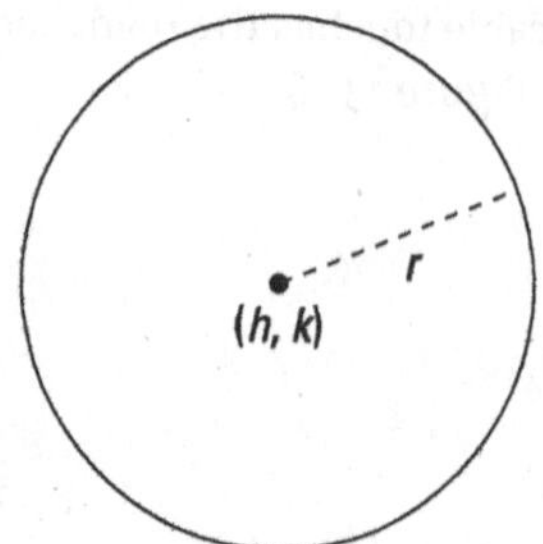

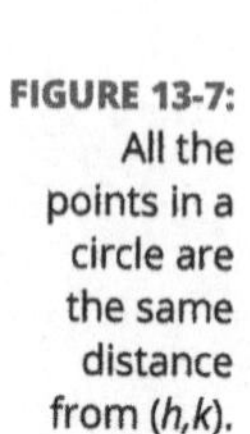
FIGURE 13-7: All the points in a circle are the same distance from (h,k).

Standardizing the circle

When the equation of a circle appears in the standard form, it provides you with all you need to know about the circle: its center and radius. With these two bits of information, you can sketch the graph of the circle. When the circle you're given isn't in the standard form, you can perform the mathematics (very similar to what is used to standardize a parabola's equation) needed to put it in that form.

EXAMPLE

Q. Write the equation $x^2 + y^2 + 6x - 4y - 3 = 0$ in the standard form of a circle.

A. You can change this equation to the standard form by completing the square for each of the variables (see Chapter 4 if you need a review of this process). Just follow these steps:

1. **Change the order of the terms so that the x's and y's are grouped together and the constant appears on the other side of the equal sign. Leave a space after the groupings for the numbers that you need to add:**

   $x^2 + 6x \quad + y^2 - 4y \quad = 3$

2. **Complete the square for each variable, adding the number that creates perfect square trinomials.** In the case of the x's, you add 9, and with the y's, you add 4. Don't forget to also add 9 and 4 to the right: $x^2 + 6x + 9 + y^2 - 4y + 4 = 3 + 9 + 4$. When it's simplified, you have $x^2 + 6x + 9 + y^2 - 4y + 4 = 16$.

3. **Factor each perfect square trinomial.** The standard form for the equation of this circle is $(x+3)^2 + (y-2)^2 = 16$.

   The circle has its center at the point $(-3, 2)$ and has a radius of 4 (the square root of 16). To sketch this circle, you locate the point $(-3, 2)$ and then count 4 units up, down, left, and right; sketch in a circle that includes those points. Figure 13-8 shows you the way.

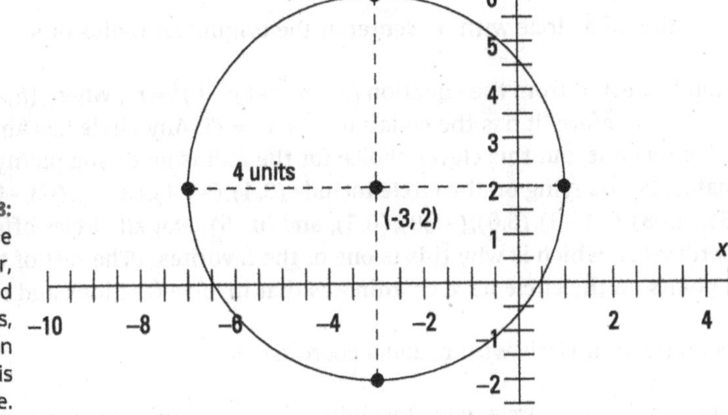

FIGURE 13-8: With the center, radius, and a compass, you too can sketch this circle.

YOUR TURN

10 Sketch a graph of the circle $(x+2)^2 + (y-3)^2 = 9$.

11 Sketch a graph of the circle $x^2+y^2-6x+4y-12=0$.

Specializing in circles

Two circles you should consider special are the circle with its *center at the origin*, and the *unit circle.*

A circle with its center at the origin has, of course, a center at (0,0), so its standard equation becomes $x^2+y^2=r^2$. The equation of the center origin is simple and easy to work with, so you should take advantage of its simplicity and try to manipulate any application you're working with that uses a circle into one with its center at the origin.

The unit circle also has its center at the origin but it always has a radius of 1. The equation of the unit circle is $x^2+y^2=1$. This circle is also convenient and nice to work with. You use it to define trigonometric functions, and you find it in analytic geometry and calculus applications.

EXAMPLE

Q. Write the equation of a circle with its center at the origin and radius of 5.

A. This equation is created from the equation $(x-h)^2+(y-k)^2=r^2$, where (h,k) is (0,0) and $r^2=5^2=25$; therefore, it has the equation $x^2+y^2=25$. Any circle has an infinite number of points on it, but this clever choice for the radius gives you plenty of integers for coordinates. Points lying on the circle include $(3,4),(-3,4),(3,-4),(-3,-4),$ $(4,3),(-4,3),(4,-3),(-4,-3),(5,0),(-5,0),(0,5)$, and $(0,-5)$. Not all circles offer this many integral coordinates, which is why this is one of the favorites. (The rest of the infinite number of points on the circle have coordinates that involve fractions and radicals.)

Q. Find points on the unit circle with rational coordinates.

A. To find points on the unit circle, you start with $x^2+y^2=1$ and solve for y: $y=\pm\sqrt{1-x^2}$. You can pick points for x and then solve for the two possible y-coordinates. The value of x has to be less than 1, and, if you want nice, rational results, it is best to pick some cooperative numbers to insert into the equation. The best choices for x are fractions whose numbers come from Pythagorean triples: $3-4-5$, $5-12-13$, $7-24-25$, and so on. You let the largest number be the denominator and either of the other numbers the numerator. Choosing $x=\frac{3}{5}$, the value of y is found: $y=\pm\sqrt{1-\left(\frac{3}{5}\right)^2}=\pm\sqrt{1-\frac{9}{25}}=\pm\sqrt{\frac{16}{25}}=\pm\frac{4}{5}$. The points $\left(\frac{3}{5},\frac{4}{5}\right)$ and $\left(\frac{3}{5},-\frac{4}{5}\right)$ are both points on the unit circle.

YOUR TURN

12 Write the equation of a circle with its center at the origin and radius of 16.

13 Find points on the unit circle with rational coordinates involving the Pythagorean triple 5–12–13.

Preparing for Solar Ellipses

The ellipse is considered the most aesthetically pleasing of all the conic sections. It has a nice oval shape often used for mirrors, windows, and art forms. Our solar system seems to agree: All the planets take an elliptical path around the sun.

The definition of an ellipse is all the points on a curve where the sum of the distances from any of those points to two fixed points is a constant. The two fixed points are the foci (plural of focus), with each denoted by the letter F. You see that Figure 13-9 illustrates this definition. You can pick a point on the ellipse, and the two distances from that point to the two foci sum to a number equal to any other distance sum from other points on the ellipse. In Figure 13-9, the distances from point A to the two foci are 3.2 and 6.8, which add to 10. The distances from point B to the two foci are 5 and 5, which also add to 10.

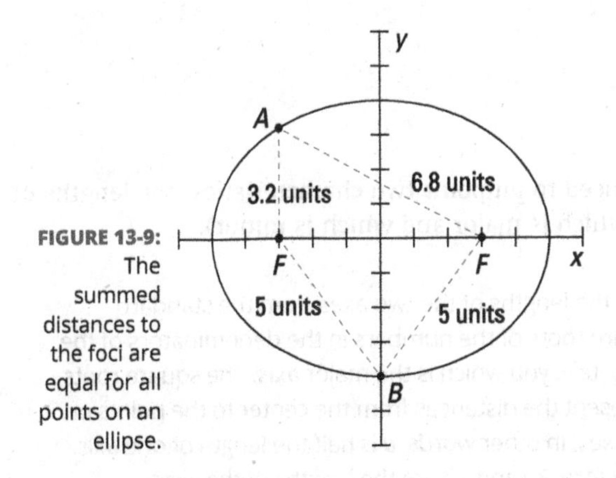

FIGURE 13-9: The summed distances to the foci are equal for all points on an ellipse.

Raising the standards of an ellipse

You can think of the ellipse as a sort of squished circle. Of course, there's much more to ellipses than that, but the label sticks because the standard equation of an ellipse has a vague resemblance to the equation for a circle (see the previous section).

The standard equation for an ellipse with its center at the point (h,k) is $\frac{(x-h)^2}{a^2}+\frac{(y-k)^2}{b^2}=1$, where

- x and y are points on the ellipse.
- a is half the length of the ellipse from left to right at its widest point.
- b is half the distance up or down the ellipse at its tallest point.

To be successful at solving elliptical problems, you need to find out more from the standard equation than just the center. You want to know if the ellipse is long and narrow or tall and slim. How long is it across, and how far up and down does it go? You may even want to know the coordinates of the foci. You can determine all these elements from the equation.

Determining the shape

An ellipse is criss-crossed by a major axis and a minor axis. Each axis divides the ellipse into two equal halves, with the major axis being the longer of the segments (such as the x-axis in Figure 13-9; if no axis is longer, you have a circle). The two axes intersect at the center of the ellipse. At the ends of the major axis, you find the vertices of the ellipse. Figure 13-10 shows two ellipses with their axes and vertices identified.

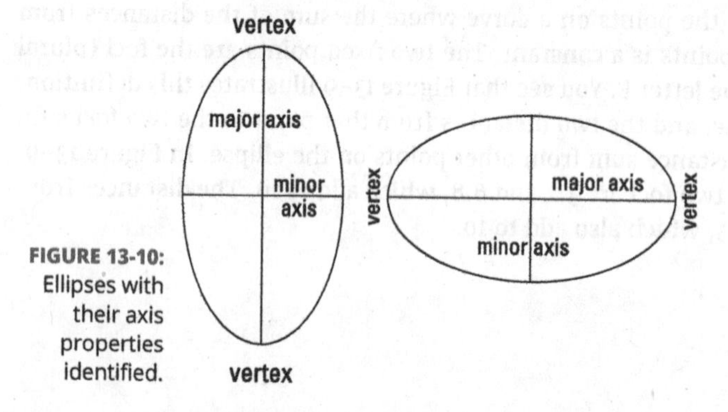

FIGURE 13-10: Ellipses with their axis properties identified.

To determine the shape of an ellipse, you need to pinpoint two characteristics: the lengths of the axes and the assignment of the axes (which is major and which is minor):

- **Lengths of the axes:** You can determine the lengths of the two axes from the standard equation of the ellipse. You take the square roots of the numbers in the denominators of the fractions. Whichever value is larger, a or b, tells you which is the major axis. The square roots of the numbers in the denominator represent the distances from the center to the points on the ellipse at the end of their respective axes. In other words, a is half the length of one axis, and b is half the length of the other. Therefore, $2a$ and $2b$ are the lengths of the axes.

» **Assignment of the axes:** The positioning of the axes is significant. The denominator that falls under the *x* signifies the axis that runs parallel to the *x*-axis; it is horizontal. The denominator that falls under the *y* factor is the axis that runs parallel to the *y*-axis; it is vertical. The denominator that is larger indicates that is the major axis.

EXAMPLE

Q. Determine the lengths of the major and minor axes of the ellipse $\frac{(x-4)^2}{25}+\frac{(y+1)^2}{49}=1$.

A. **14, 10.** To find the lengths of the major and minor axes of the ellipse, you take the square roots of 25 and 49. The square root of the larger number, 49, is 7. Twice 7 is 14, so the major axis, the vertical axis, is 14 units long. The square root of 25 is 5, and twice 5 is 10. Thus, the minor axis, the horizontal axis, is 10 units long.

Q. Determine the lengths of the major and minor axes of the ellipse $\frac{(x+7)^2}{4}+\frac{(y-3)^2}{1}=1$.

A. **4, 2.** Take the square roots of 4 and 1. The square root of the larger number, 4, is 2. Twice 2 is 4, so the major axis, the horizontal axis, is 4 units long. The square root of 1 is 1, and twice 1 is 2. Thus, the minor axis, the vertical axis, is 2 units long.

Finding the foci

You can find the two foci of an ellipse by using information from the standard equation. The foci, for starters, always lie on the major axis. They lie *c* units from the center. To find the value of *c*, you use parts of the ellipse equation to form the equation $c^2=a^2-b^2$ or $c^2=b^2-a^2$, depending on which is larger, a^2 or b^2. The value of c^2 has to be positive.

EXAMPLE

Q. Determine the center, foci, and vertices of the ellipse $\frac{(x+1)^2}{625}+\frac{(y-3)^2}{49}=1$.

A. The ellipse has its center at (–1,3). The major axis is horizontal, running parallel to the *x*-axis. You find the foci by solving $c^2=a^2-b^2$, which, in this case, is $c^2=25^2-7^2=625-49=576$. The value of *c* is either 24 (the root of 576) or –24 from the center, so the foci are (23,3) and (–25,3). The major axis is $2(25)=50$ units, and the minor axis is $2(7)=14$ units. The 25 and 7 come from the square roots of 625 and 49, respectively. And the vertices, the endpoints of the major axis, are at (24,3) and (–26,3).

Q. Determine the center, foci, and vertices of the ellipse $\frac{x^2}{25}+\frac{y^2}{9}=1$.

A. The major axis runs across the ellipse, parallel to the *x*-axis (see the previous section to find out why). Actually, the major axis is on the *x*-axis, because the center of this ellipse is the origin. You know this because the *h* and *k* are missing from the equation (actually, they're both equal to zero). You find the foci of this ellipse by solving the foci equation: $c^2=a^2-b^2$. Substituting 25 for a^2 and 9 for b^2, you have $c^2=25-9=16$. Taking the square root, $c=\pm4$.

So, the foci are 4 units on either side of the center of the ellipse. In this case, the coordinates of the foci are (−4,0) and (4,0). Figure 13-11 shows the graph of the ellipse with the foci identified. Also, for this example ellipse, the major axis is 10 units long, running from (−5,0) to (5,0). These two points are the vertices. The minor axis is 6 units long, running from (0,3) down to (0,−3).

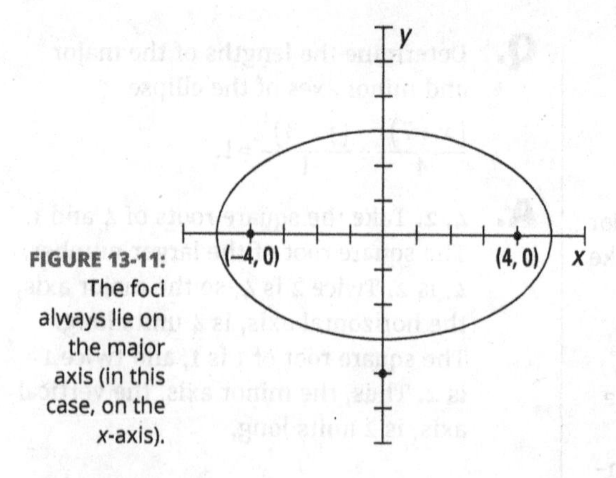

FIGURE 13-11: The foci always lie on the major axis (in this case, on the *x*-axis).

YOUR TURN

14 Determine the center and the lengths of the two axes of the ellipse $\frac{(x-7)^2}{25}+\frac{(y-1)^2}{16}=1$.

15 Determine the center and the lengths of the two axes of the ellipse $\frac{x^2}{49}+\frac{(y+3)^2}{100}=1$.

16 Determine the center, foci, and vertices of the ellipse $\frac{x^2}{100}+\frac{y^2}{64}=1$.

17 Determine the center, foci, and vertices of the ellipse $\frac{(x+9)^2}{16}+\frac{(y-2)^2}{25}=1$.

Sketching an elliptical path

Have you ever been in a whispering gallery? I'm talking about a room or auditorium where you can stand at a spot and whisper a message, and a person standing a great distance away from you can hear your message. This phenomenon was much more impressive before the days of hidden microphones, which tend to make us more skeptical of how this works. Anyway, here's the algebraic principle behind a whispering gallery. You're standing at one focus of an ellipse, and the other person is standing at the other focus. The sound waves from one focus reflect off the surface or ceiling of the gallery and move over to the other focus.

Suppose you run across a problem on a test that asks you to sketch the ellipse associated with a whispering gallery that has foci 240 feet apart and a major axis (length of the room) of 260 feet. Your first task is to construct the equation of the ellipse.

EXAMPLE

Q. Sketch the graph of an ellipse whose foci are 240 feet apart and whose major axis is 260 feet.

A. The foci are 240 feet apart, so they each stand 120 feet from the center of the ellipse. The major axis is 260 feet long, so the vertices are each 130 feet from the center. Using the equation $c^2 = a^2 - b^2$ — which gives the relationship between c, the distance of a focus from the center; *a*, the distance from the center to the end of the major axis; and *b*, the distance from the center to the end of the minor axis — you get $120^2 = 130^2 - b^2$, or $b^2 = 50^2$. Armed with the values of a^2 and b^2, you can write the equation of the ellipse representing the curvature of the ceiling: $\frac{x^2}{130^2} + \frac{y^2}{50^2} = 1$.

To sketch the graph of this ellipse, you first locate the center at (0,0). You count 130 units to the right and left of the center and mark the vertices, and then you count 50 units up and down from the center for the endpoints of the minor axis. You can sketch the ellipse by using these endpoints. The sketch in Figure 13-12 shows the points described and the ellipse. It also shows the foci — where the two people would stand in the whispering gallery.

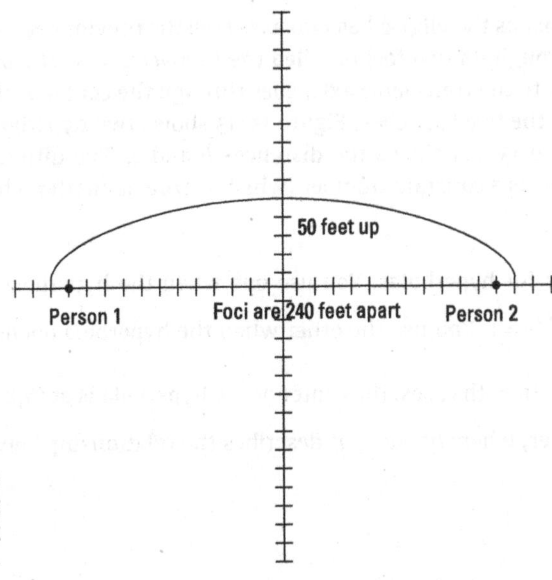

FIGURE 13-12: A whispering gallery is long and narrow.

18 Sketch the graph of the ellipse $\frac{(x-1)^2}{25}+\frac{(y+2)^2}{9}=1.$

19 Sketch the graph of the ellipse $\frac{(x+3)^2}{4}+\frac{y^2}{16}=1.$

Getting Hyper Excited about Hyperbolas

The hyperbola is a conic section that seems to be in combat with itself. It features two completely disjoint curves, or branches, that face away from one another but are mirror images across a line that runs halfway between them.

A *hyperbola* is defined as all the points such that the difference of the distances between two fixed points (called foci) is a constant value. In other words, you pick a value, such as the number 6; you find two distances whose difference is 6, such as 10 and 4; and then you find a point that rests 10 units from the one point and 4 units from the other point.

The hyperbola has two axes, just as the ellipse has two axes (see the previous section). The axis of the hyperbola that goes through its two foci is called the *transverse axis*. The other axis, the *conjugate axis*, is perpendicular to the transverse axis, goes through the center of the hyperbola, and acts as the mirror line for the two branches. Figure 13-13 shows two hyperbolas with their axes and foci identified. Figure 13-13a shows the distances *P* and *q*. The difference between these two distances and the foci is a constant number (which is true no matter what points you pick on the hyperbola).

There are two basic equations for hyperbolas. You use one when the hyperbola opens to the left and right: $\frac{(x-h)^2}{a^2}-\frac{(y-k)^2}{b^2}=1$. You use the other when the hyperbola opens upward and downward: $\frac{(y-k)^2}{a^2}-\frac{(x-h)^2}{b^2}$. In both cases, the center of the hyperbola is at (h,k), and the foci are c units away from the center, where $b^2=c^2-a^2$ describes the relationship between the different parts of the equation.

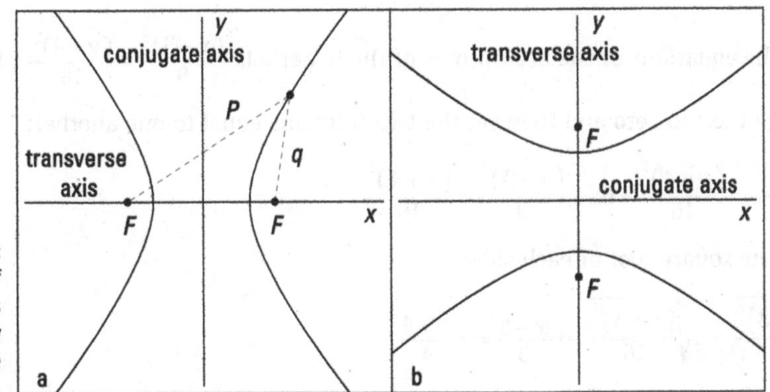

FIGURE 13-13: The curves of hyperbolas face away from one another.

EXAMPLE

Q. Identify the center and foci of the hyperbola $\frac{(x-3)^2}{9}-\frac{(y+4)^2}{16}=1$.

A. **The center is at (3,–4).** The hyperbola opens left and right, so the foci are to the left and right of the center. Using $b^2=c^2-a^2$, $16=c^2-9\rightarrow 25=c^2\rightarrow \pm 5=c$. So the foci are at (–2,–4) and (8,–4), which you obtain by adding and subtracting 5 from the x-coordinate.

Q. Identify the center and foci of the hyperbola $\frac{(y-2)^2}{25}-\frac{(x+1)^2}{4}=1$.

A. **The center is at (–1,2).** The hyperbola opens upward and downward, so the foci are above and below the center. Using $b^2=c^2-a^2$, $4=c^2-25\rightarrow 29=c^2\rightarrow \pm\sqrt{29}=c$. The coordinates of the foci are: $\left(-1,2+\sqrt{29}\right)$ and $\left(-1,2-\sqrt{29}\right)$.

Including the asymptotes and vertices

Two very helpful techniques you can use to sketch hyperbolas are to first lightly sketch in the two diagonal asymptotes of the hyperbola and then find the vertices.

Asymptotes aren't actual parts of the graph; they just help you determine the shape and direction of the curves. The asymptotes of a hyperbola intersect at the center of the hyperbola. You find the equations of the asymptotes by replacing the 1 in the equation of the hyperbola with a zero and simplifying the resulting equation into the equations of two lines.

The vertices are points where the curves change direction. They are found using $(h\pm a,k)$ for hyperbolas opening horizontally (left and right) and $(h,k\pm a)$ for those opening vertically (up and down).

Q. Find the equations of the asymptotes of the hyperbola $\frac{(x-3)^2}{9}-\frac{(y+4)^2}{16}=1$.

EXAMPLE **A.** Change the 1 to zero and then set the two fractions equal to one another:

$$\frac{(x-3)^2}{9}-\frac{(y+4)^2}{16}=0 \rightarrow \frac{(x-3)^2}{9}=\frac{(y+4)^2}{16}.$$

Take the square root of each side:

$$\sqrt{\frac{(x-3)^2}{9}}=\pm\sqrt{\frac{(y+4)^2}{16}} \rightarrow \frac{x-3}{3}=\pm\frac{y+4}{4}$$

Now you multiply each side by 4 to get the equations of the asymptotes in better form:

$$4\cdot\left(\frac{x-3}{3}\right)=\pm\left(\frac{y+4}{\cancel{4}}\right)\cdot\cancel{4} \rightarrow \frac{4}{3}(x-3)=\pm(y+4)$$

Consider the two cases — one using the positive sign, and the other using the negative sign. When using the positive sign, $\frac{4}{3}(x-3)=+(y+4)$, giving you $\frac{4}{3}x-4=y+4$ or $\frac{4}{3}x-8=y$.

And, with the negative sign, $\frac{4}{3}(x-3)=-(y+4)$ becomes $\frac{4}{3}x-4=-y-4$. Adding 4 to each side, you get $\frac{4}{3}x=-y$ or $-\frac{4}{3}x=y$.

The two asymptotes you find are $y=\frac{4}{3}x-8$ and $y=-\frac{4}{3}x$, and you can see them in Figure 13-14. Notice that the slopes of the lines are the opposites of one another. (For a refresher on graphing lines, see Chapter 2.)

Q. Find the vertices of the hyperbola $\frac{(x-3)^2}{9}-\frac{(y+4)^2}{16}=1$.

A. The hyperbola opens left and right, so use $(h\pm a,k)$. The two vertices are: $(3\pm3,-4)\rightarrow(6,-4)$ and $(0,-4)$. You find them in Figure 13-14.

Graphing hyperbolas

Hyperbolas are relatively easy to sketch if you pick up the necessary information from the equations. To graph a hyperbola, use the following steps as guidelines:

1. **Determine if the hyperbola opens to the sides or up and down by noting whether the *x* term is first or second. The *x* term first means it opens to the sides; if second, it opens up and down.**
2. **Find the center of the hyperbola by looking at the values of *h* and *k*.**
3. **Lightly sketch in a rectangle twice as wide as the square root of the denominator under the *x* value and twice as high as the square root of the denominator under the *y* value. The rectangle's center is the center of the hyperbola.**
4. **Lightly sketch in the asymptotes through the vertices of the rectangle.**
5. **Draw in the hyperbola, making sure it touches the midpoints of the sides of the rectangle. These are the vertices.**

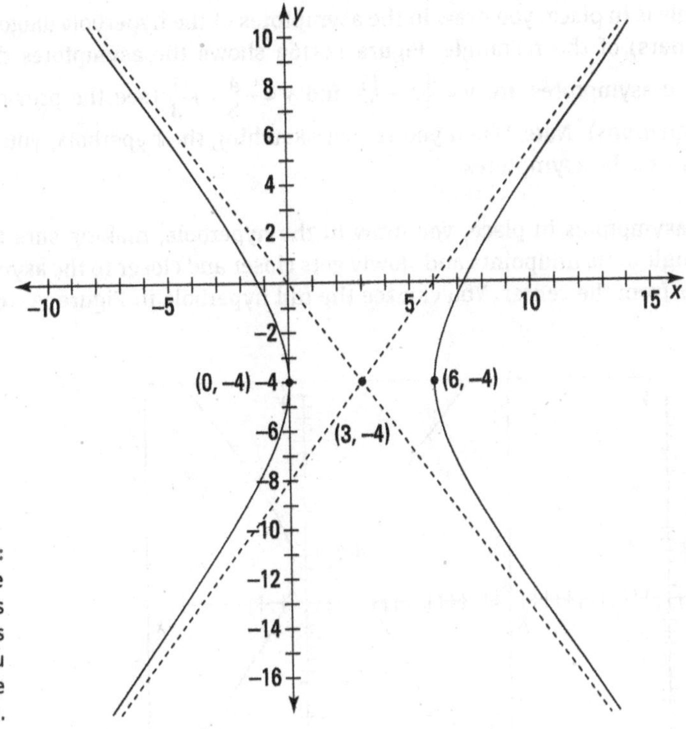

FIGURE 13-14: The asymptotes and vertices help you sketch the hyperbola.

EXAMPLE

Q. Graph the hyperbola $\frac{(x+2)^2}{9} - \frac{(y-3)^2}{16} = 1$.

A. First, note that this equation opens to the left and right because the x value comes first in the equation. The center of the hyperbola is at (–2,3).

Now comes the mysterious rectangle. In Figure 13-15a, you see the center placed on the graph at (–2,3). You count 3 units to the right and left of center (totaling 6), because twice the square root of 9 is 6. Now you count 4 units up and down from center, because twice the square root of 16 is 8. A rectangle 6 units wide and 8 units high is shown in Figure 13-15b.

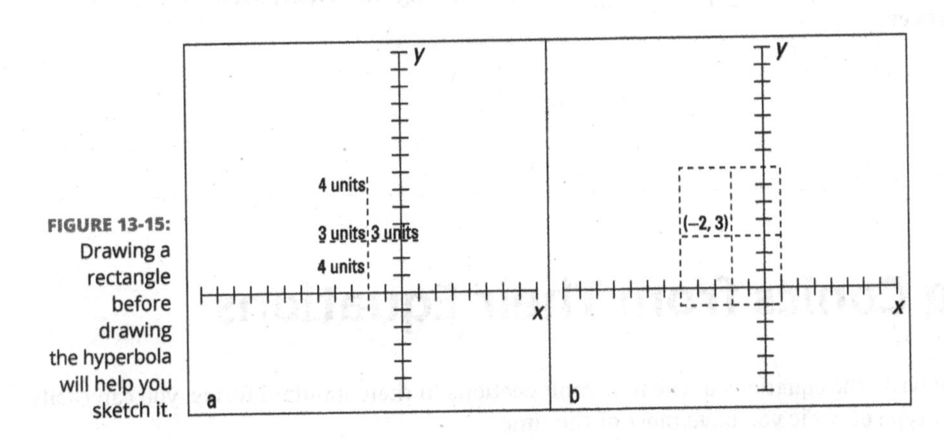

FIGURE 13-15: Drawing a rectangle before drawing the hyperbola will help you sketch it.

When the rectangle is in place, you draw in the asymptotes of the hyperbola diagonally through the vertices (corners) of the rectangle. Figure 13-16a shows the asymptotes drawn in. The equations of those asymptotes are $y=\frac{4}{3}x+\frac{17}{3}$ and $y=-\frac{4}{3}x+\frac{1}{3}$ (see the previous section to calculate these equations). ***Note:*** When you're just sketching the hyperbola, you usually don't need the equations of the asymptotes.

Lastly, with the asymptotes in place, you draw in the hyperbola, making sure it touches the sides of the rectangle at its midpoints and slowly gets closer and closer to the asymptotes as the curves get farther from the center. You can see the full hyperbola in Figure 13-16b.

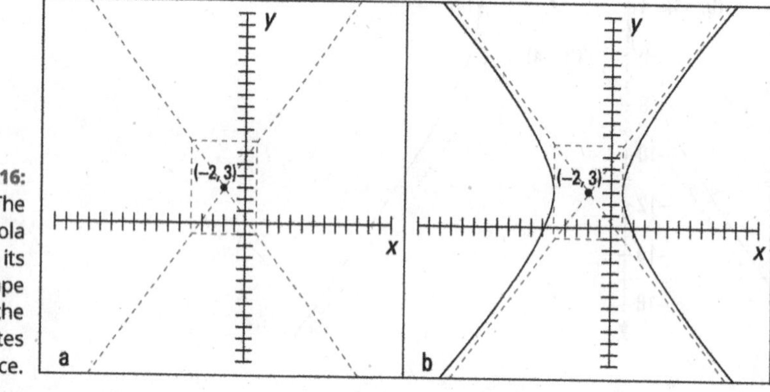

FIGURE 13-16: The hyperbola takes its shape with the asymptotes in place.

YOUR TURN

20 Sketch the graph of $\frac{(x+1)^2}{4}-\frac{(y-2)^2}{9}=1$, identifying the center, asymptotes, foci, and vertices.

21 Sketch the graph of $\frac{(y+3)^2}{1}-\frac{(x+2)^2}{16}=1$, identifying the center, asymptotes, foci, and vertices.

Identifying Conics from Their Equations

When you write the equations of the four conic sections in their standard forms, you can easily tell which type of conic you have most of the time.

- A parabola has only one squared variable:

  $(y-h)^2 = 4a(x-h)$ or $(x-h)^2 = 4a(y-k)$

- A circle can be written without the variables in fractions:

  $(x-h)^2 + (y-k)^2 = r^2$

  - An ellipse shows a *sum* of the two variable terms (at least one as a fraction):

    $$\frac{(x-h)^2}{a^2} + \frac{(y-k)^2}{b^2} = 1$$

- A hyperbola shows a *difference* of the two variable terms:

  $$\frac{(x-h)^2}{a^2} - \frac{(y-k)^2}{b^2} = 1 \text{ or } \frac{(y-k)^2}{a^2} - \frac{(x-h)^2}{b^2} = 1$$

Sometimes, however, the equation you're given isn't in standard form. It has yet to be changed, by completing the square (see Chapter 3) or whatever method it takes. So, in these situations, how do you tell which type of conic you have by just looking at the equation (without having to rewrite the form)?

Consider all the equations of the form $Ax^2 + By^2 + Cx + Dy + E = 0$. You have squared terms and first-degree terms of the two variables *x* and *y*. By observing what *A*, *B*, *C*, and *D* are, you can determine which type of conic you have.

When you have an equation in the form $Ax^2 + By^2 + Cx + Dy + E = 0$, follow these rules:

- If $A = B$, you have a circle (as long as *A* and *B* don't equal zero).
- If $A \neq B$, and *A* and *B* are the same sign, you have an ellipse.
- If *A* and *B* are different signs, you have a hyperbola.
- If *A* or *B* is zero, you have a parabola (they can't both be zero).

EXAMPLE

Q. Determine which conic you have and write its equation in the standard form: $9x^2 + 4y^2 - 72x - 24y + 144 = 0$

A. This is an ellipse, because $A \neq B$, and *A* and *B* are both positive. You find the standard form for this equation by rearranging the terms and completing the squares.

$$9x^2 + 4y^2 - 72x - 24y + 144 = 0 \rightarrow 9x^2 - 72x + 4y^2 - 24y = -144 \rightarrow$$
$$9\left(x^2 - 8x\right) + 4\left(y^2 - 6y\right) = -144 \rightarrow 9\left(x^2 - 8x + 16\right) + 4\left(y^2 - 6y + 9\right) = -144 + 144 + 36 \rightarrow$$
$$9(x-4)^2 + 4(x-3)^2 = 36$$

Now divide each term by 36.

$$\frac{9(x-4)^2}{36} + \frac{4(y-3)^2}{36} = \frac{36}{36} \rightarrow \frac{(x-4)^2}{4} + \frac{(y-3)^2}{9} = 1$$

Q. Determine which conic you have and write its equation in the standard form: $2x^2+2y^2+12x-20y+24=0$.

A. This is a circle, because $A=B$. You find the standard form for this equation by rearranging the terms and completing the squares.

$$2x^2+2y^2+12x-20y+24=0 \rightarrow 2x^2+12x+2y^2-20y=-24 \rightarrow$$
$$2\left(x^2+6x\right)+2\left(y^2-10y\right)=-24 \rightarrow 2\left(x^2+6x+9\right)+2\left(y^2-10y+25\right)=-24+18+50 \rightarrow$$
$$2(x+3)^2+2(y-5)^2=44$$

Dividing each term by 2, you have its standard form: $(x+3)^2+(y-5)^2=22$.

Q. Determine which conic you have and write its equation in the standard form: $9y^2-8x^2-18y-16x-71=0$.

A. This is a hyperbola, because A and B have opposite signs. You find the standard form for this equation by rearranging the terms and completing the squares.

$$9y^2-8x^2-18y-16x-71=0 \rightarrow 9y^2-18y-8x^2-16x=71 \rightarrow$$
$$9\left(y^2-2y\right)-8\left(x^2+2x\right)=71 \rightarrow 9\left(y^2-2y+1\right)-8\left(x^2+2x+1\right)=71+9-8 \rightarrow$$
$$9(y-1)^2-8(x+1)^2=72$$

Dividing each term by 72, $\frac{9(y-1)^2}{72}-\frac{8(x+1)^2}{72}=\frac{72}{72} \rightarrow \frac{(y-1)^2}{8}-\frac{(x+1)^2}{9}=1$

Q. Determine which conic you have and write its equation in the standard form: $x^2+8x-6y+10=0$.

A. This is a parabola. You see only an x^2 term — the only variable raised to the second power. You find the standard form for this equation by rearranging the terms and completing the square.

$$x^2+8x-6y+10=0 \rightarrow x^2+8x=6y-10 \rightarrow x^2+8x+16=6y-10+16 \rightarrow$$
$$(x+4)^2=6y+6 \rightarrow (x+4)^2=6(y+1)$$

YOUR TURN

22 Identify which type of conic this is and write the standard form: $x^2+y^2+6x+4y-3=0$.

23 Identify which type of conic this is and write the standard form: $2y^2-x+16y+30=0$.

Practice Questions Answers and Explanations

1. **The points and information are as follows.** The value of a is -2. The focus falls on the point $(0,-2)$. The directrix forms a line at $y=2$. The vertex is at $(0,0)$. The axis of symmetry is $x=0$. The graph opens downward.

   The parabola is wide because $|4(a)|=|4(-2)|=|8|>1$.

2. **The points and information are as follows.** The value of a is $\frac{1}{6}$. The focus falls on the point $\left(0,\frac{1}{6}\right)$. The directrix forms a line at $x=-\frac{1}{6}$. The vertex is at $(0,0)$. The axis of symmetry is $y=0$. The graph opens to the right.

   The parabola is relatively steep because $|4(a)|=\left|4\left(\frac{1}{6}\right)\right|=\left|\frac{2}{3}\right|<1$.

3. **The points and information are as follows.** The vertex is $(1,-7)$, the focus is $(1,-9)$, the axis of symmetry is $x=1$, the directrix is $y=-5$, and the parabola opens downward and is fairly wide. The value of a is -2.

4. **The points and information are as follows.** The vertex is $(5,-3)$, the focus is $(7,-3)$, the axis of symmetry is $y=-3$, the directrix is $x=3$, and the parabola opens to the right and is fairly wide, because $a=2$.

5. $\mathbf{(x-1)^2=\frac{8}{5}(y-2)}$**.** Rewrite the equation with the x-terms on the left: $5x^2-10x=8y-21$.

   Factor 5 from each term on the left: $5\left(x^2-2x\right)=8y-21$.

   Add 1 inside the parentheses to complete the square. Add 5 to the other side of the equation: $5\left(x^2-2x+1\right)=8y-21+5$.

   Write the trinomial as a binomial squared and factor the terms on the right: $5(x-1)^2=8(y-2)$.

   Divide each side by 5: $(x-1)^2=\frac{8}{5}(y-2)$.

6. $\mathbf{(y+1)^2=-\frac{1}{2}(x+3)}$**.** Rewrite the equation with the y-terms on the left: $2y^2+4y=-x-5$.

   Factor 2 from each term on the left: $2\left(y^2+2y\right)=-x-5$.

   Add 1 inside the parentheses to complete the square. Add 2 to the other side of the equation: $2\left(y^2+2y+1\right)=-x-5+2$.

   Write the trinomial as a binomial squared and factor -1 from the terms on the right: $2(y+1)^2=-1(x+3)$.

   Divide each side by 2: $(y+1)^2=-\frac{1}{2}(x+3)$.

(7) **See the figure.** The vertex is (–4,–2). The axis of symmetry is $y = -2$. The parabola opens to the left and is relatively wide. Two other points on the graph are (–6,2) and $\left(-\frac{9}{2},0\right)$. Their partner points on the other side of the axis of symmetry are (–6,–6) and $\left(-\frac{9}{2},-4\right)$.

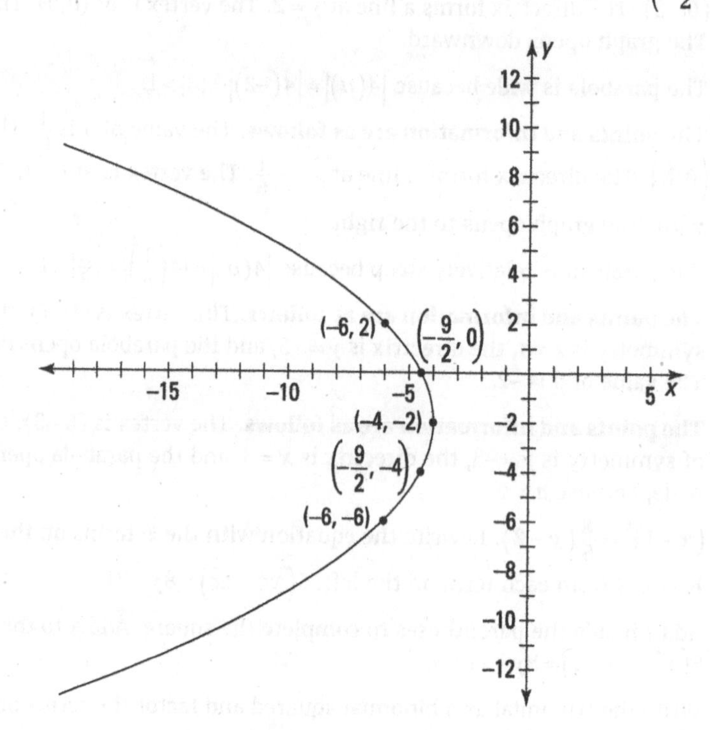

(8) **See the figure.** The vertex is (1,–3). The axis of symmetry is $x = 1$. The parabola opens upward and is relatively steep. Two other points on the graph are (0,1) and (3,13). Their partner points on the other side of the axis of symmetry are (2,1) and (–1,13).

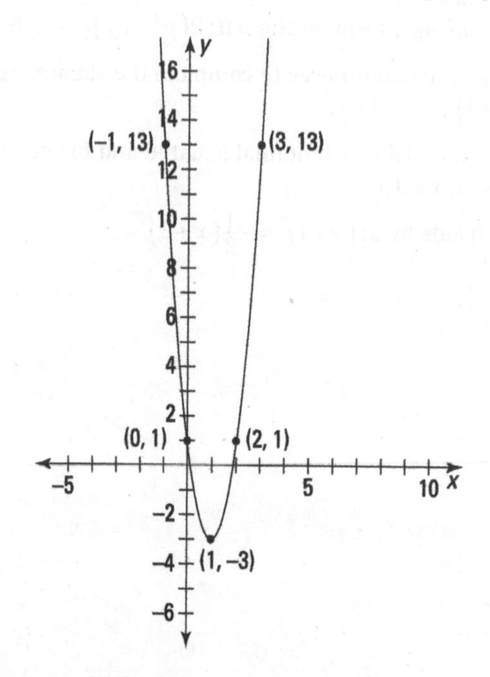

9. **See the figure.** First, write the equation in standard form by isolating the x-terms on the left, completing the square, and factoring on the right: $x^2 - 4x = 12y - 40 \rightarrow x^2 - 4x + 4 = 12y - 36 \rightarrow (x-2)^2 = 12(y-3)$. The vertex is (2,3). The axis of symmetry is $x = 2$. The parabola opens upward and is wide. Two other points on the graph are $\left(4, \frac{10}{3}\right)$ and (8,6). Their partner points on the other side of the axis of symmetry are $\left(0, \frac{10}{3}\right)$ and (−4,6).

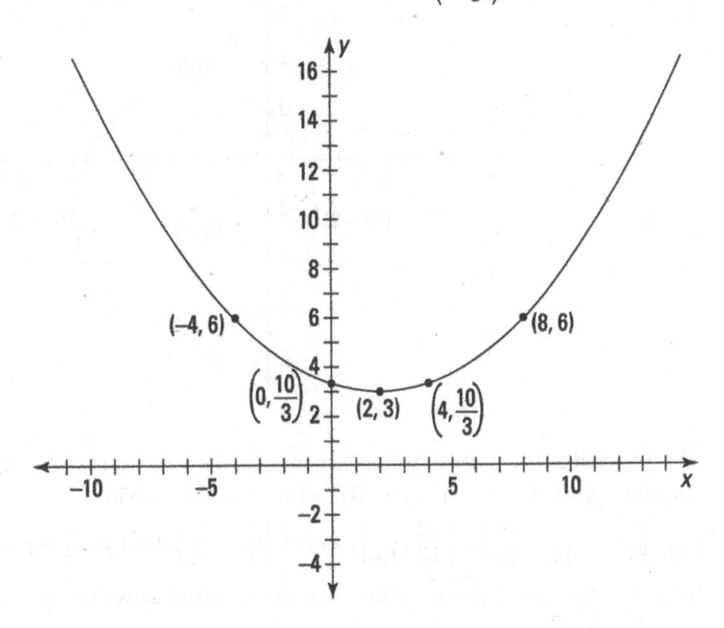

10. **See the figure.** The center is (−2,3), and the radius is 3. Going left, right, up, and down from the center, some points on the circle are: (−5,3), (1,3), (−2,6), and (−2,0).

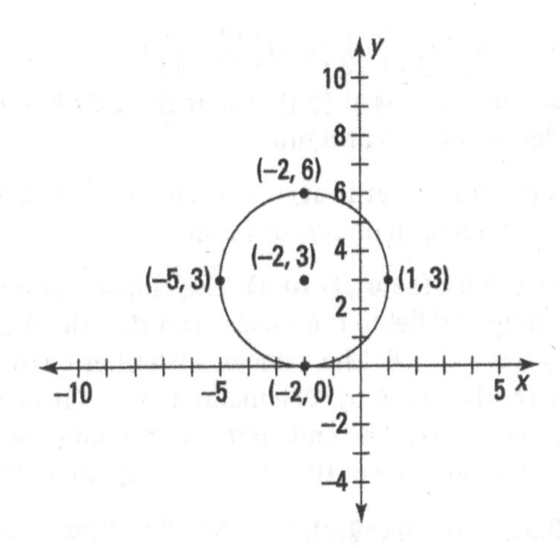

(11) **See the figure.** First, write the equation in the standard form by completing two squares: $x^2+y^2-6x+4y-12=0 \rightarrow x^2-6x+9+y^2+4y+4=12+9+4 \rightarrow (x-3)^2+(y+2)^2=25$. The center of the circle is (3,–2) and the radius is 5. Going left, right, up, and down from the center, some points on the circle are: (–2,–2), (8,–2), (3,3), and (3,–7).

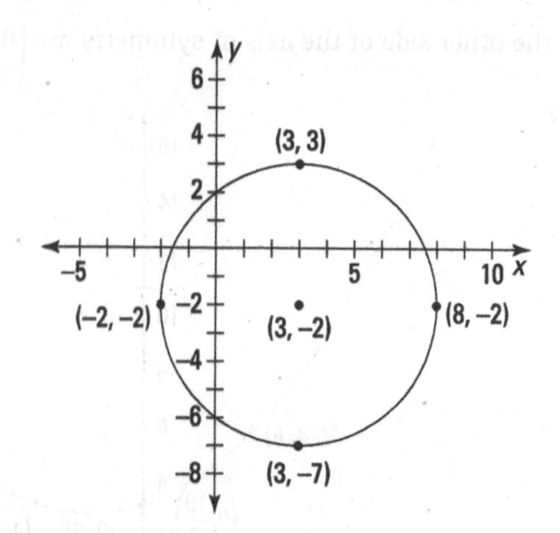

(12) $\mathbf{x^2+y^2=256}$. Using the standard form, $(x-h)^2+(y-k)^2=r^2$, replace the h and k with zeros and the r with 16: $(x-0)^2+(y-0)^2=16^2 \rightarrow x^2+y^2=256$.

(13) $\left(\frac{5}{13},\frac{12}{13}\right)$, $\left(\frac{5}{13},-\frac{12}{13}\right)$, $\left(\frac{12}{13},\frac{5}{13}\right)$, **and** $\left(\frac{12}{13},-\frac{5}{13}\right)$. Using the equation $x^2+y^2=1$ and solving for y, you have $y=\pm\sqrt{1-x^2}$. Next, use the 5 and 13 to write the value for x, $\frac{5}{13}$. Solving for y, $y=\pm\sqrt{1-\left(\frac{5}{13}\right)^2}=\pm\sqrt{1-\frac{25}{169}}=\pm\sqrt{\frac{144}{169}}=\pm\frac{12}{13}$. The two points created are $\left(\frac{5}{13},\frac{12}{13}\right)$ and $\left(\frac{5}{13},-\frac{12}{13}\right)$. If the x value chosen had been $\frac{12}{13}$, then the two y-values would turn out to be $\pm\frac{5}{13}$ and the two points would be $\left(\frac{12}{13},\frac{5}{13}\right)$ and $\left(\frac{12}{13},-\frac{5}{13}\right)$.

(14) **(7,1), 10, 8.** The center is at (7,1). The major axis has a length of 2(5) or 10 units. The minor axis has a length of 2(4) or 8 units.

(15) **(0,–3), 20, 14.** The center is at (0,–3). The major axis has a length of 2(10) or 20 units. The minor axis has a length of 2(7) or 14 units.

(16) **(0,0); (6,0), (-6,0); (10,0), (-10,0).** The ellipse has its center at (0,0). The major axis is horizontal, running parallel to the x-axis. You find the foci by solving $c^2=a^2-b^2$, which, in this case, is $c^2=100-64=36$. The value of c is either 6 (the root of 36) or –6 from the center, so the foci are (6,0) and (–6, 0). The major axis is $2(10)=20$ units, and the minor axis is $2(8)=16$ units. And the vertices, the endpoints of the major axis, are found by adding and subtracting 10 from the x-coordinate of the center: (10,0) and (–10,0).

(17) **(-9,2); (-9,5), (-9,-1); (-9,7), (-9,-3).** The ellipse has its center at (–9,2). The major axis is vertical, running parallel to the y-axis. You find the foci by solving $c^2=b^2-a^2$, which, in this case, is $c^2=25-16=9$. The value of c is either 3 (the root of 9) or –3 from the center, so the foci are (–9,5) and (–9,–1). The major axis is $2(5)=10$ units, and the minor axis is $2(4)=8$ units. And the vertices, the endpoints of the major axis, are found by adding and subtracting 5 from the y-coordinate of the center: (–9,7) and (–9,–3).

18. **See the figure.** The center is at (1,–2). The horizontal major axis is 10 units long; 5 units on either side of the center are the vertices (6,–2) and (–4,–2). The minor axis is 6 units long — ending 3 units above and below the center.

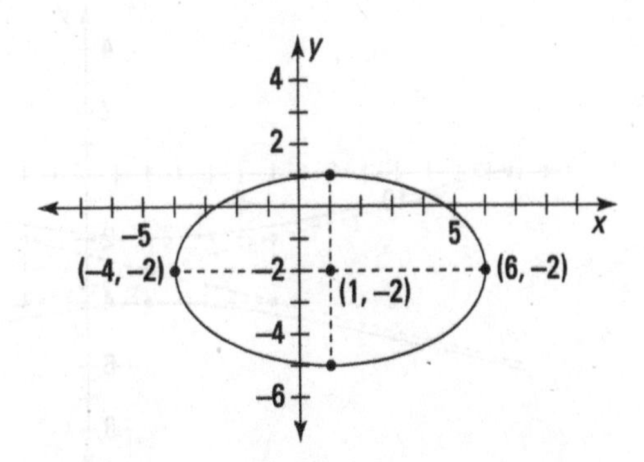

19. **See the figure.** The center is at (–3,0). The vertical major axis is 8 units long; 4 units above and below the center are the vertices (–3,4) and (–3,–4). The minor axis is 4 units long — ending 2 units to the left and right of the center.

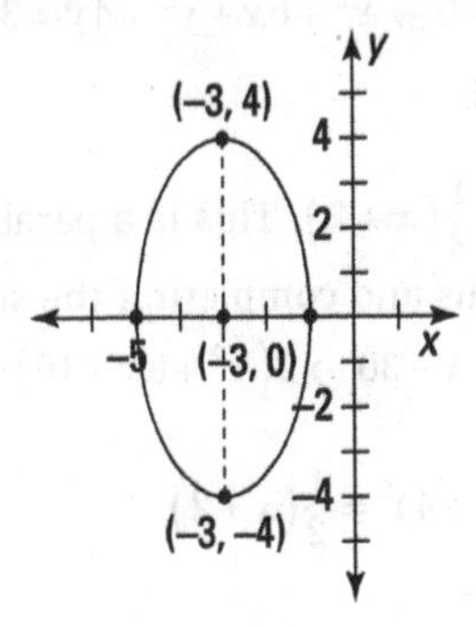

20. **See the figure.** The center is at (–1,2). The asymptotes are $y=2\pm\frac{3}{2}(x+1)$. The foci are at $\left(-1+\sqrt{13},2\right)$ and $\left(-1-\sqrt{13},2\right)$. The vertices are at (1,2) and (–3,2).

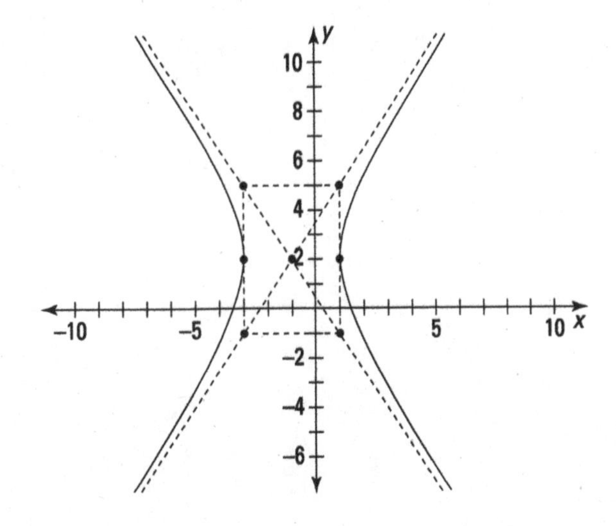

(21) **See the figure.** The center is at (–2,–3). The asymptotes are $y=-3\pm\frac{1}{4}(x+2)$. The foci are at $\left(-2,-3+\sqrt{17}\right)$ and $\left(-2,-3-\sqrt{17}\right)$. The vertices are at (–2,–2) and (–2,–4).

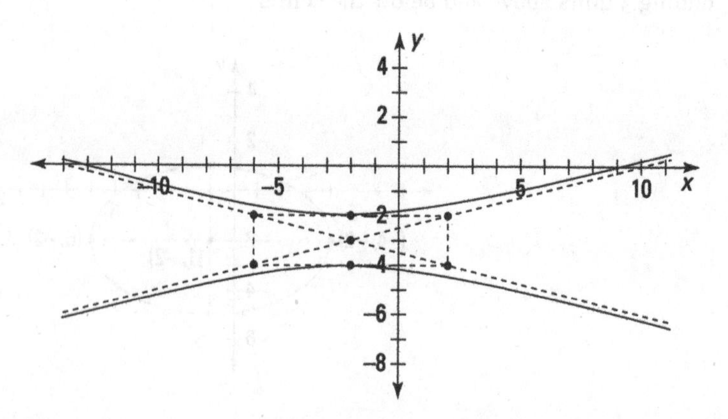

(22) **Circle, $(x+3)^2+(y+2)^2=16$.** This is a circle because $A=B$. Rearranging the terms and completing the squares:

$$x^2+y^2+6x+4y-3=0\rightarrow x^2+6x+y^2+4y=3\rightarrow x^2+6x+9+y^2+4y+4=3+9+4\rightarrow$$
$$(x+3)^2+(y+2)^2=16$$

(23) **Parabola, $(y+4)^2=\frac{1}{2}(x+2)$.** This is a parabola because only the y variable is squared. Rearranging the terms and completing the square: $2y^2-x+16y+30=0\rightarrow 2y^2+16y=$ $x-30\rightarrow 2\left(y^2+8y\right)=x-30\rightarrow 2\left(y^2+8y+16\right)=x-30+32\rightarrow$

$$2(y+4)^2=x+2\rightarrow(y+4)^2=\frac{1}{2}(x+2)$$

Whaddya Know? Chapter 13 Quiz

Quiz time! Complete each problem to test your knowledge on the various topics covered in this chapter. You can then find the solutions and explanations in the next section.

1. **Given $x^2 + y^2 - 6x + 10y + 30 = 0$, identify which conic this is and put the equation in standard form.**

2. **Given the equation of the parabola $(x-3)^2 = -8(y+6)$, identify the vertex, focus, directrix, and the direction in which it opens. Then sketch the graph, indicating these features.**

3. **Write the equation of the parabola in standard form: $y = 2x^2 - 8x + 3$.**

4. **Write the equation of the circle in standard form and graph it: $x^2 + y^2 - 8x + 6y = 0$.**

5. **Given the equation of the ellipse $x^2 + 4y^2 + 2x - 24y + 33 = 0$, write it in standard form, and identify the center, vertices, foci, length of the major axis, and length of the minor axis. Then sketch the ellipse.**

6. **Given the equation of the hyperbola $\frac{(y+1)^2}{16} - \frac{(x-2)^2}{4} = 0$, identify the center, vertices, foci, and asymptotes, and sketch a graph of the hyperbola, indicating these properties.**

7. **Given the equation of the parabola $(y-2)^2 = 4(x+3)$, identify the vertex, focus, directrix, and the direction in which it opens. Then sketch the graph, indicating these features.**

8. **Given the equation of the ellipse $\frac{(x-1)^2}{16} + \frac{(y-4)^2}{25} = 1$, identify the center, vertices, foci, length of the major axis, and length of the minor axis. Then sketch the ellipse.**

9. **Given the equation of the hyperbola $\frac{(x-4)^2}{16} - \frac{(y-2)^2}{9} = 1$, identify the center, vertices, foci, and asymptotes, and sketch a graph of the hyperbola, indicating these properties.**

10. **Given $y^2 - 4x^2 - 2y + 16x - 19 = 0$, identify which type of conic this is and write its equation in standard form.**

Answers to Chapter 13 Quiz

1. **Circle, $(x-3)^2+(y+5)^2=4$.** Because $A=B$, this is a circle. Rearrange the terms and complete the squares:

$$x^2+y^2-6x+10y+30=0 \rightarrow x^2-6x+y^2+10y=-30 \rightarrow x^2-6x+9+y^2+10y+25=-30+9+25 \rightarrow (x-3)^2+(y+5)^2=4.$$

The center is at (3,–5), and the radius is 2.

2. **The vertex is (3,–6), the focus is (3,–8), the directrix is $y=-4$, and the direction is downward.**

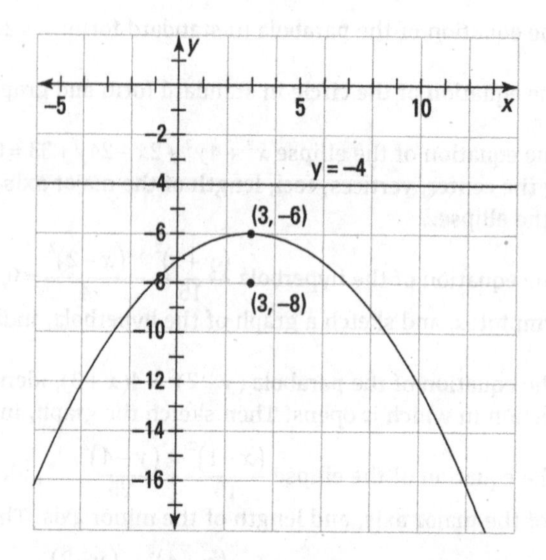

The focus is found with $(h,k+a)=(3,-6-2)=(3,-8)$. The directrix is $y=-6-(-2)=-4$.

3. **$(x-2)^2=\frac{1}{2}(y+5)$.** First, isolate the x-terms on the right, factor out the 2, and complete the square.

$$y=2x^2-8x+3 \rightarrow y-3=2x^2-8x \rightarrow y-3=2\left(x^2-4x\right) \rightarrow y-3+8=2\left(x^2-4x+4\right)$$

Factor the trinomial, and then divide both sides by 2.

$$y+5=2(x-2)^2 \rightarrow (x-2)^2=\frac{1}{2}(y+5)$$

4. **$(x-4)^2+(y+3)^2=25$.** Rearrange the terms and then complete the squares.

$$x^2+y^2-8x+6y=0 \rightarrow x^2-8x+y^2+6y=0 \rightarrow \left(x^2-8x+16\right)+\left(y^2+6y+9\right)=16+9 \rightarrow (x-4)^2+(y+3)^2=25$$

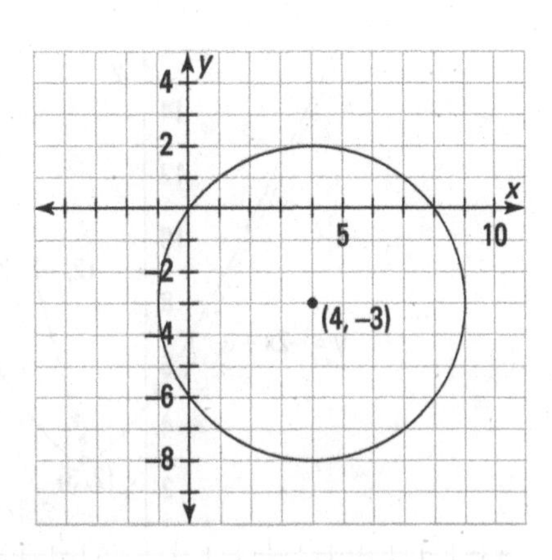

(5) **The standard form is $\frac{(x+1)^2}{4}+\frac{(y-3)^2}{1}=1$; the center is $(-1,3)$; the vertices are $(-3,3)$ and $(1,3)$; the foci are $\left(-1-\sqrt{3},3\right)$ and $\left(-1+\sqrt{3},3\right)$; the major axis is 4; and the minor axis is 1.** Rearrange the terms and then complete the squares. Finally, divide each term by 4.

$$x^2+4y^2+2x-24y+33=0 \rightarrow x^2+2x+4y^2-24y=-33 \rightarrow \left(x^2+2x+1\right)+4\left(y^2-6y+9\right)=-33+1+36 \rightarrow$$

$$(x+1)^2+4(y-3)^2=4 \rightarrow \frac{(x+1)^2}{4}+\frac{(y-3)^2}{1}=1$$

The vertices are $(-1\pm 2,3)$ or $(1,3)$ and $(-3,3)$. The foci are $\left(-1\pm\sqrt{4-1},3\right)$ or $\left(-1+\sqrt{3},3\right)$ and $\left(-1-\sqrt{3},3\right)$. The length of the major axis is 2(2) = 4, and the length of the minor axis is 2(1) = 2.

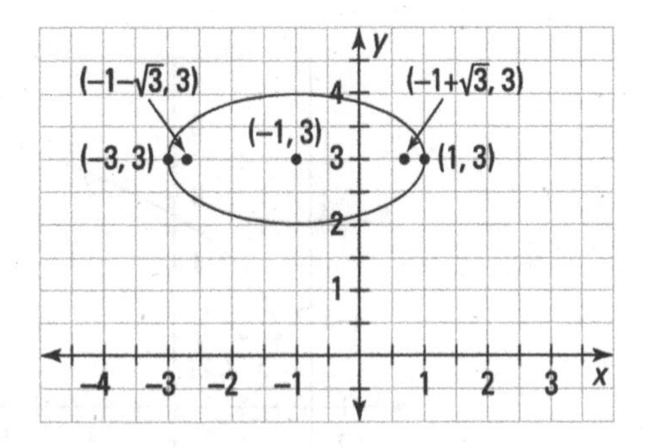

(6) **The center is $(2,-1)$; the vertices are $(2,3)$ and $(2,-5)$; the foci are $\left(2,-1+2\sqrt{5}\right)$ and $\left(2,-1-2\sqrt{5}\right)$; and the asymptotes are $y=2x-5$ and $y=-2x+3$.** The vertices are $(2,-1+4)$ and $(2,-1-4)$ or $(2,3)$ and $(2,-5)$. The foci are $\left(2,-1+2\sqrt{5}\right)$ and $\left(2,-1-2\sqrt{5}\right)$. The asymptotes are found by solving $\frac{(y+1)^2}{16}-\frac{(x-2)^2}{4}=0$ for y.

$$\frac{(y+1)^2}{16}=\frac{(x-2)^2}{4} \rightarrow \sqrt{\frac{(y+1)^2}{16}}=\pm\sqrt{\frac{(x-2)^2}{4}}\ \frac{y+1}{4}=\pm\frac{x-2}{2} \rightarrow$$

$$y+1=\pm 2(x-2)$$

$$y=-1\pm(2x-4)$$

$$y=2x-5 \text{ or } y=-2x+3$$

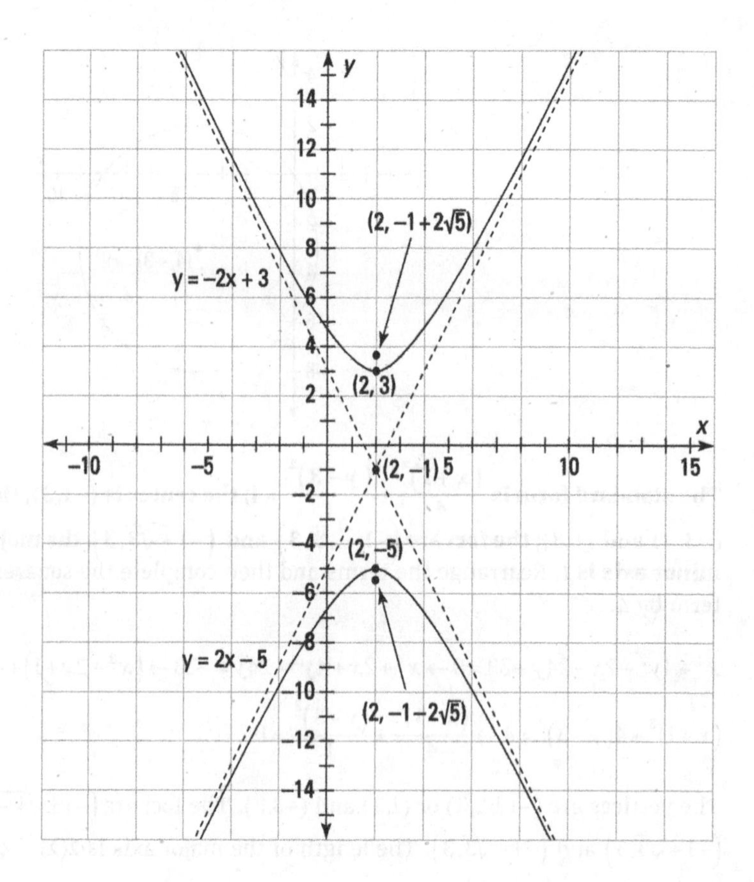

7. **The vertex is (−3,2); the focus is (−2,2); the directrix is $x = -4$; and the direction is right.**

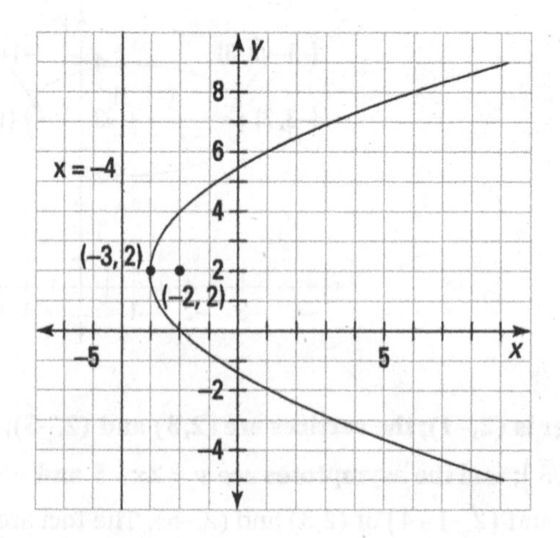

8. **The center is (1,4); the vertices are (1,−1) and (1,9); the foci are (1,1) and (1,7); the major axis is 10; and the minor axis is 8.**

   The vertices are $(1, 4 \pm 5)$ or (1,9) and (1,−1). The foci are $(1, 4 \pm 3)$ or (1,7) and (1,1). The length of the major axis is 2(5) = 10, and the minor axis is 2(4) = 8.

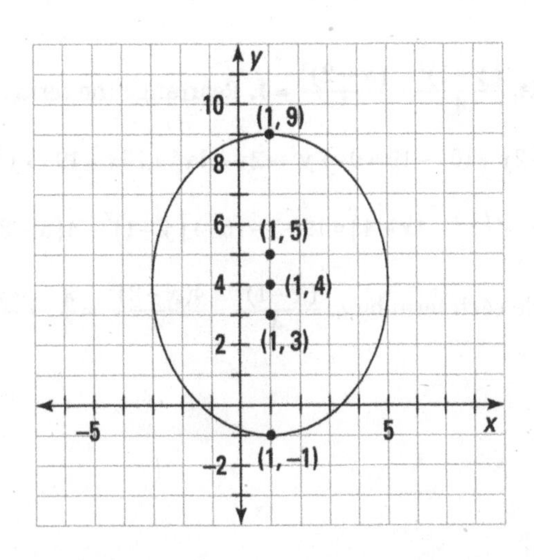

(9) **The center is (4,2); the vertices are (8,2) and (0,2); the foci are (9,2) and (−1,2); and the asymptotes are $y=\frac{3}{4}x-1$ and $y=-\frac{3}{4}x+5$.**

The vertices are $(4+4,2)$ and $(4-4,2)$ or $(8,2)$ and $(0,2)$. The foci are $(4+5,2)$ and $(4-5,2)$ or $(9,2)$ and $(0,-1)$. The asymptotes are found by solving $\frac{(x-4)^2}{16}-\frac{(y-2)^2}{9}=0$ for y.

$$\frac{(y-2)^2}{9}=\frac{(x-4)^2}{16}\rightarrow\sqrt{\frac{(y-2)^2}{9}}=\pm\sqrt{\frac{(x-4)^2}{16}}\rightarrow\frac{y-2}{3}=\pm\frac{x-4}{4}\rightarrow y-2=\pm\frac{3}{4}(x-4)\rightarrow$$

$$y=2\pm\left(\frac{3}{4}x-3\right)$$

$$y=\frac{3}{4}x-1 \text{ or } y=\frac{3}{4}x+5$$

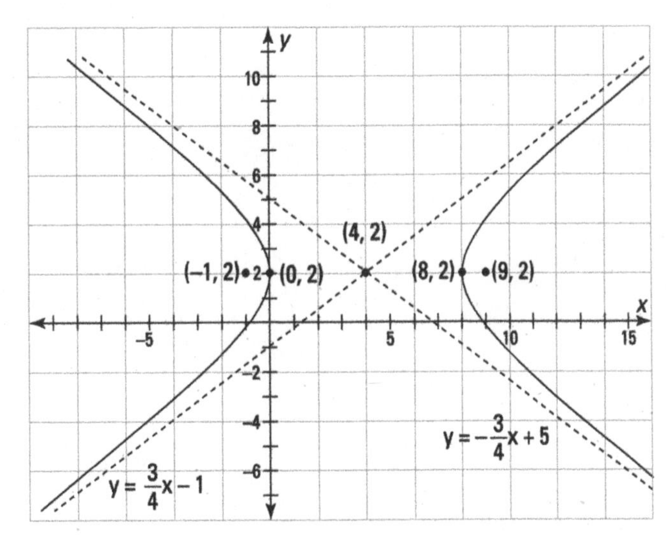

(10) **Hyperbola, $\frac{(y-1)^2}{4}-\frac{(x-2)^2}{1}=1$.** Rearrange the terms and complete the squares.

$$y^2-4x^2-2y+16x-19=0\rightarrow y^2-2y-4x^2+16x=19\rightarrow y^2-2y-4\left(x^2-4x\right)=19\rightarrow$$

$$y^2-2y+1-4\left(x^2-4x+4\right)=19+1-16\rightarrow(y-1)^2-4(x-2)^2=4$$

Now divide each term by 4: $\frac{(y-1)^2}{4}-\frac{4(x-2)^2}{4}=\frac{4}{4}\rightarrow\frac{(y-1)^2}{4}-\frac{(x-2)^2}{1}=1$.

IN THIS CHAPTER

» Processing the possible solutions of a linear system of equations

» Transferring linear equations to graph paper

» Breaking down and substituting systems of two equations

» Handling systems of three or more linear equations

Chapter 14
Solving Systems of Linear Equations

A *system of equations* consists of a number of equations with shared variables — variables that are linked in a specific way. The solution of a system of equations uncovers these links in one of two ways: with a list of numbers that makes each equation in the system a true statement or a list of relationships between numbers that makes each equation in the system a true statement.

In this chapter, I cover systems of *linear* equations. As I explain in Chapter 2, linear equations feature variables that reach only the first degree, meaning that the highest power of any variable you solve for is 1. You have a number of techniques at your disposal to solve systems of linear equations, including graphing lines, adding multiples of one equation to another, substituting one equation into another, and taking advantage of a neat rule that Gabriel Cramer (an 18th-century Swiss mathematician) developed. I cover each of the many different ways to solve a linear system of equations in this chapter.

Looking at the Standard Linear-Systems Form

The standard form for a system of linear equations is as follows:

$$\begin{cases} a_1x_1 + a_2x_2 + a_3x_3 + \cdots = k_1 \\ b_1x_1 + b_2x_2 + b_3x_3 + \cdots = k_2 \\ c_1x_1 + c_2x_2 + c_3x_3 + \cdots = k_3 \\ \vdots \end{cases}$$

The x_n factors are variables; the c_n factors are coefficients; the k_n terms are constants.

REMEMBER

If a system has only two equations, the equations usually appear in the $Ax + By = C$ form I introduce in Chapter 2, and a brace groups them together. But don't let this fool you — a system of equations can contain any number of equations (I show you how to work through larger systems toward the end of the chapter). When your task is to find the solution of a system of equations, you want to find the variables that make each of the equations true at the same time. The techniques used to find the values of the two variables when there are two equations can be adjusted to use on three variables in three equations, and so on.

Solving Systems of Two Linear Equations by Using Elimination

Even though graphing lines to solve systems of equations can be heaps of fun (see the later section, "Graphing Solutions of Linear Systems," if you don't believe me), the graphing method has one big drawback: You'll find it almost impossible to find non-integer answers. Graphing requires careful plotting of points and drawing lines, but it is often preferred for the visual effect. The methods mathematicians usually prefer, though, for solving systems of linear equations involve using algebraic operations. The two most preferred (and common) methods for solving systems of two linear equations are *elimination*, which I cover in this section, and *substitution*, which I cover in the section, "Making Substitution the Choice," later in the chapter. Determining which method you should use depends on what form the equations start out in and, often, just personal preference.

TIP

The elimination method may also be called *linear combinations* or just add/subtract. The word *elimination* describes exactly what you accomplish with this method, but *add/subtract* tells you how you accomplish the elimination.

Getting to the point with elimination

To carry out the elimination method, you want to add the two equations together, or subtract one from the other, resulting in eliminating (getting rid of) one of the variables. Sometimes you have to multiply one or both of the equations by a carefully selected number before you add them together (or subtract them).

EXAMPLE

Q. Solve the system of equations using elimination.

$$\begin{cases} 3x-5y=2 \\ 2x+5y=18 \end{cases}$$

A. By adding the two equations together, you eliminate the y variable. The two y terms are opposites in the two different equations. The resulting expression is $5x=20$. Dividing each side by 5, you get $x=4$. When you put $x=4$ into the first equation, you get $3(4)-5y=2$. Solving that equation for y, you get $y=2$. The solution, therefore, is $x=4$, $y=2$. If you graph the two lines corresponding to the equations, you see them intersecting at the point (4,2).

Q. You should always check to be sure that you have the right answer. One way to check the previous example is to replace the x and y in the second equation (the one you didn't use to solve for the second variable) with the 4 and the 2 to see if the statement you get is true. Substituting, you get $2(4)+5(2)=18 \rightarrow 8+10=18 \rightarrow 18=18$. It works!

Q. Solve the system of equations using elimination.

$$\begin{cases} 3x-2y=17 \\ 2x-5y=26 \end{cases}$$

A. This system requires some adjustments before you add or subtract the two equations. If you add the two equations in their current forms, you get just another equation: $5x-7y=43$. You create a perfectly nice equation, and it has the same solution as the other two, but it doesn't help you find the solution of the system. Before you add or subtract, you need to make sure that one of the variables in the two equations has the same or opposite coefficient as its counterpart; this way, when you add or subtract the variable, you eliminate it. You have several different options to choose from to make the equations in this example system ready for elimination:

- You can multiply the first equation by 2 and the second by 3 and then subtract to eliminate the x's.
- You can multiply the first equation by 2 and the second by -3 and then add to eliminate the x's.
- You can multiply the first equation by 5 and the second by 2 and then subtract to eliminate the y's.
- You can multiply the first equation by 5 and the second by -2 and then add to eliminate the y's.

When faced with multiplying an equation through by a negative number, be sure that you multiply every term in the equation by the negative sign; the sign of every term will change.

Let's say you choose to solve the system by multiplying the first equation by 2 and the second by -3; you get a new version of the system:

$$\begin{cases} 3x-2y=17 \\ 2x-5y=26 \end{cases} \Rightarrow \begin{matrix} 2R_1 \\ -3R_2 \end{matrix} \Rightarrow \begin{cases} 6x-4y=34 \\ -6x+15y=-78 \end{cases}$$

(Note that I use R_1 and R_2 to denote the equations. Think of these as saying "Row 1" and "Row 2." This will make more sense when I show you systems of three or more equations — where you see a lot more rows.)

Adding the two equations together, you get $11y = -44$, eliminating the x's. Dividing each side of the new equation by 11, you get $y = -4$. Substitute this value into the first original equation. (Always go back to the original equations to solve for the other variable or to check your work. You have a better chance of catching errors that way.)

Substituting -4 for the y value, you get $3x - 2(-4) = 17$. Solving for x, you get $x = 3$. Now check your work by putting the 3 and -4 into the second original equation. You get $2(3) - 5(-4) = 26 \rightarrow 6 + 20 = 26 \rightarrow 6 + 20 = 26 \rightarrow 26 = 26$. The solution is the point $(3, -4)$.

YOUR TURN

Solve the system of equations using elimination.

$$\begin{cases} 2x - 3y = 7 \\ -2x - y = 5 \end{cases}$$

Solve the system of equations using elimination.

$$\begin{cases} 4x + 3y = 5 \\ 7x + 2y = 12 \end{cases}$$

Recognizing solutions indicating parallel or coexisting lines

When you graph systems of linear equations, it becomes pretty apparent when the systems produce parallel lines or have equations that represent the same lines. You don't have to graph the lines to recognize these situations algebraically, however; you just have to know what to look for. And that warning signal is a statement that is impossible!

EXAMPLE

Q. Solve the system of equations.

$$\begin{cases} 3x+5y=12 \\ 3x+5y=7 \end{cases}$$

A. Multiply the second equation through by –1 and add the equations together.

$$\begin{cases} 3x+5y=12 \\ 3x+5y=7 \end{cases} \underset{-1R_2}{\Rightarrow} \Rightarrow \begin{cases} 3x+5y=12 \\ -3x-5y=-7 \end{cases}$$

The result that you get is $0=5$. That just isn't so. The false statement is your signal that the system doesn't have a solution and that the lines are parallel.

Q. Solve the system of equations.

$$\begin{cases} 8x-2y=6 \\ y=4x-3 \end{cases}$$

A. First, rewrite the second equation so that the common terms line up. Then multiply the second equation by –2 and add them together.

$$\begin{cases} 8x-2y=6 \\ y=4x-3 \end{cases} \Rightarrow \begin{cases} 8x-2y=6 \\ 4x-y=3 \end{cases} \underset{-2R_2}{\Rightarrow} \Rightarrow \begin{cases} 8x-2y=6 \\ -8x+2y=-6 \end{cases}$$

When you add the terms, you get $0=0$. If you get an equation that's always true, such as $0=0$ or $5=5$, it means that these are just two different versions of the same equation. Everything works!

YOUR TURN

3 Determine if the system of equations has no answer or too many answers.

$$\begin{cases} 3x=21-12y \\ 4y+x+7=0 \end{cases}$$

Making Substitution the Choice

Another method used to solve systems of linear equations is called *substitution*. Some people prefer to use this method most of the time because it's used when solving systems with higher exponents; that way, you only have to master one method. Substitution in algebra works something like substitution in a basketball game — you replace a player with another player who can play that position and hope for better results. Isn't that what you hope for in algebra, too? One drawback of substitution is that you may have to work with fractions (oh, the horror), which you can avoid by using elimination (see the previous section). The method used is often a matter of personal choice.

Variable substitution made easy

Executing substitution in systems of two linear equations is a two-step process:

1. **Solve one of the equations for one of the variables, *x* or *y*.**
2. **Substitute the value of the variable into the other equation and solve for the remaining variable.**

TIP

Before substituting, look for a variable with a coefficient of 1 or −1 to use when solving one of the equations. By sticking with terms that have coefficients of 1 or −1, you avoid having to substitute fractions into the other equation. Sometimes fractions are unavoidable; in those cases, you should choose a term with a small coefficient so the fractions don't get too unwieldy.

EXAMPLE

Q. Solve the system by substitution.

$$\begin{cases} 2x - y = 1 \\ 3x - 2y = 8 \end{cases}$$

A. You first look for a variable you can identify as a likely candidate for the first step. In other words, you want to solve for it. The y term in the first equation has a coefficient of −1, so you should solve this equation for y (rewrite it so y is alone on one side of the equation). You get $y = 2x - 1$. Now you can substitute the $2x - 1$ for the y in the other equation: Replace the y in $3x - 2y = 8$ with $2x - 1$, giving you the equation $3x - 2(2x - 1) = 8$. Simplify the terms on the left side: $3x - 4x + 2 = 8 \rightarrow -x + 2 = 8$. Subtract 2 from each side, and you have $-x = 6$. Multiply each side by −1, and the solution is $x = -6$.

You've already created the equation $y = 2x - 1$, so you can put the value $x = -6$ into the equation to get $y: y = 2(-6) - 1 = -12 - 1 = -13$. To check your work, put both values, $x = -6$ and $y = -13$, into the equation that you didn't change (the second equation, in this case): $3(-6) - 2(-13) = 8 \rightarrow -18 + 26 = 8 \rightarrow 8 = 8$. Your work checks out.

Q. Solve the system by substitution.

$$\begin{cases} 4x+7y=15 \\ 3x+4y=10 \end{cases}$$

A. None of the variables has a coefficient of 1 or −1, so you just make a choice. Going with the smallest coefficient, I look at the 3 in the second equation. Solving for x in that equation, you get $x=\frac{10-4y}{3}$. Substituting into the first equation, $4\left(\frac{10-4y}{3}\right)+7y=15 \rightarrow \frac{40-16y}{3}+7y=15$. Get rid of the fraction by multiplying each term by 3: $\cancel{3}\cdot\frac{40-16y}{\cancel{3}}+3\cdot 7y=3\cdot 15 \rightarrow 40-16y+21y=45$. Solving for y, $40+5y=45 \rightarrow 5y=5 \rightarrow y=1$. Now use that x equivalence to find its value: $x=\frac{10-4(1)}{3}=\frac{6}{3}=2$. Yes, the solution is (2,1).

Identifying parallel and coexisting lines

As I mention in the section, "Recognizing solutions indicating parallel or coexisting lines," earlier in the chapter, your job is well and good when you come up with a simple point of intersection for your solution. But you also have to identify the impossible (parallel lines) and always possible (coexisting lines) when using the substitution method to find solutions.

TIP

Here are some hints for recognizing these two special cases:

- When lines are parallel, the algebraic result is an impossible statement. You get an equation that can't be true, such as $2=6$.
- When lines are coexistent (the same), the algebraic result is a statement that's always true (the equation is always correct). An example is the equation $7=7$.

EXAMPLE

Q. Determine whether the system of equations represents parallel lines or coexisting lines.

$$\begin{cases} 3x-2y=4 \\ y=\frac{3}{2}x+2 \end{cases}$$

A. Using substitution to solve the system, you insert the equivalence of y from the second equation into the first:

$$3x-2y=4 \rightarrow 3x-2\left(\frac{3}{2}x+2\right)=4 \rightarrow 3x-3x-4=4 \rightarrow -4=4$$

Substitution produces an incorrect statement. This equation is always wrong, so you can never find a solution. Thus, the system of equations doesn't have a solution. If you graph the lines, you see that the graphs the equations represent are parallel.

Q. Determine whether the system of equations represents parallel lines or coexisting lines.

$$\begin{cases} 3x - 2y = 4 \\ y = \frac{3}{2}x - 2 \end{cases}$$

A. Using substitution, you substitute the equivalent of y into the first equation:

$$3x - 2y = 4 \rightarrow 3x - 2\left(\frac{3}{2}x - 2\right) = 4 \rightarrow 3x - 3x + 4 = 4 \rightarrow 4 = 4$$

Substitution creates an equation that's always true. So, any pair of values that works for one equation will work for the other. The system of equations represents two ways of showing the same equation — two equations that represent the same line. When you graph the equations, you produce one identical line. You can write the solution of this system in the (x,y) form for the coordinates of a point by using a parameter for one variable — in this case, the x — and writing the other variable in terms of the parameter. When you let x be represented by the parameter k, the solutions are $\left(k, \frac{3}{2}k - 2\right)$. Just pick a number for the k value, and replace the k's with that number to form a point in the solution. For instance, if you choose $k = 6$, plug it in the solution to get $\left(6, \frac{3}{2}(6) - 2\right) = (6, 9 - 2) = (6, 7)$. The point (6,7) works for both of the original equations, as do infinitely more pairs of values.

A *parameter* in this case is a value that replaces one or more variables. It acts somewhat like a variable but is different in that you choose the values that the parameter takes on in order to evaluate the equations.

YOUR TURN

Solve the systems of equations using substitution.

4 $\begin{cases} x + 3y = 4 \\ 2x + 5y = 5 \end{cases}$	5 $\begin{cases} 8x - 3y = 41 \\ 3x + 2y = 6 \end{cases}$
6 $\begin{cases} x - 2y = 5 \\ 8x = 16y + 40 \end{cases}$	7 $\begin{cases} 3x - y = 8 \\ 6x - 2y + 2 = 0 \end{cases}$

Graphing Solutions of Linear Systems

Linear equations with two variables, like $Ax+By=C$, have lines as graphs. In order to solve a system of linear equations like these with a graph, you need to determine what values for x and y are true at the same time. Your job is to account for three possible solutions (if you count "no solution" as a solution) that can make this happen:

- *One solution:* The solution appears at the point where the lines intersect — the same x and the same y work at the same time in all the equations.
- *An infinite number of solutions:* The equations are describing the same line; the graphs overlap.
- *No solution:* This occurs when the lines are parallel — no value for (x,y) that works in one equation also works in the other.

To solve a system of two linear equations (with integers as solutions), you can graph both equations on the same axes (x and y). (Check out Chapter 6 for instructions on graphing lines.) With the graphs on paper, you see one of three things: intersecting lines (one solution), identical lines (infinite solutions), or parallel lines (no solution).

REMEMBER

Solving linear systems by graphing the lines created by the equations is visually satisfying but beware: Using this method to find a solution requires careful plotting of the lines. Therefore, it works best for lines that have integers for solutions. If the lines don't intersect at a place where the grid on the graph crosses, you have a problem. The task of determining rational (fractions) or irrational (square roots) solutions from graphs on graph paper is too difficult, if not impossible. (In non-integer cases, you have to use substitution, elimination, or Cramer's Rule, which I show you in this chapter.) You can estimate or approximate fractional or irrational values but you don't get an exact answer.

Pinpointing the intersection

Lines are made up of many, many points. When two lines cross one another, they share just one of those points. Sketching the graphs of two intersecting lines allows you to determine that one special point by observing where, on the graph, the two lines cross. You need to graph very carefully, using a sharpened pencil and a ruler with no bumps or holes. The resulting graph is very gratifying.

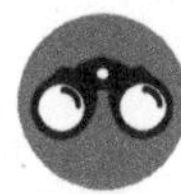

EXAMPLE

Q. Find the point of intersection of the lines:

$$\begin{cases} 2x+3y=12 \\ x-y=11 \end{cases}$$

A. A quick way to sketch these lines is to find their intercepts — where they cross the axes. For the first equation, if you let $x=0$ and solve for y, you get $y=4$, so the y-intercept is (0,4); in the same equation, when you let $y=0$, you get $x=6$, so the x-intercept is (6,0). Plot those two points on a graph and draw a line through them. You do the same for the other equation, $x-y=11$, finding the intercepts $(0,-11)$ and $(11,0)$. Figure 14-1 shows the two lines graphed using their intercepts.

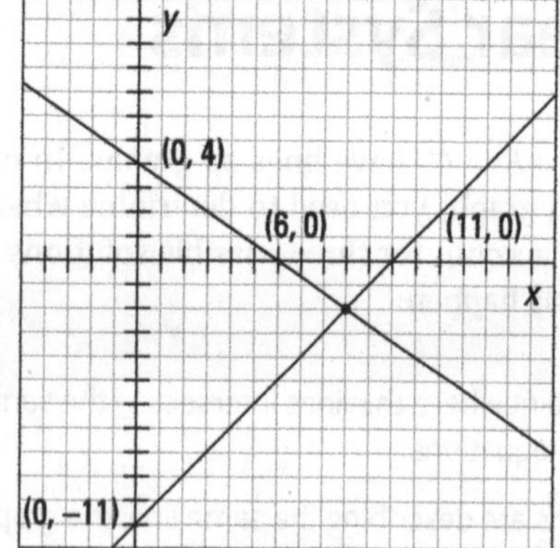

FIGURE 14-1: Two lines from a linear system crossing at a single point.

The two lines intersect at the point (9,–2). You mark the point by counting the grid marks in the figure. This method shows you how important it is to graph the lines very carefully!

What exactly does the point (9,–2) mean in terms of a solution of the system? It means that if you let $x = 9$ and $y = -2$, then both equations in the system are true statements. Try putting these values to work. Insert them into the first equation: $2(9) + 3(-2) = 12 \rightarrow 18 - 6 = 12 \rightarrow 12 = 12$. In the second equation, $9 - (-2) = 11 \rightarrow 9 + 2 = 11 \rightarrow 11 = 11$. The solution $x = 9$ and $y = 2$ is the only one that works for both equations.

Toeing the same line twice and dealing with parallel lines

A unique situation that occurs with systems of linear equations happens when everything seems to work. Every point you find that works for one equation works for the other, too. This match-made-in-heaven scenario plays out when the equations are just two different ways of describing the same line. Kind of like finding out you're dating a set of twins, no? When two equations in a system of linear equations represent the same line, the equations are multiples of one another. When they are multiples of each other, their graphs are the same line.

And then there is the other situation. Parallel lines never intersect and never have anything in common except the direction they move in (their slope; see Chapter 6 for more on a line's slope). So, when you solve systems of equations that have no solutions at all, you should know right away that the lines represented by the equations are parallel.

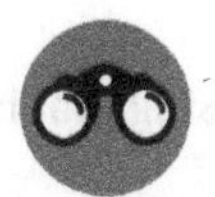

EXAMPLE

Q. Determine the point of intersection of the system of equations:

$$\begin{cases} x+3y=7 \\ 2x+6y=14 \end{cases}$$

A. You can tell that the second equation is twice the first. Sometimes, however, this sameness is disguised when the equations appear in different forms. Here's the same system as before only with the second equation written in slope-intercept form:

$$\begin{cases} x+3y=7 \\ y=-\frac{1}{3}x+\frac{7}{3} \end{cases}$$

The sameness isn't as obvious here, but when you graph the two equations, you can't tell one graph from the other because they're the same line (for more on graphing lines, see Chapter 6).

Q. Determine the point of intersection of the system of equations:

$$\begin{cases} x+2y=8 \\ 3x+6y=7 \end{cases}$$

A. When you graph the two lines, the first with intercepts of (0,4) and (8,0) and the second with intercepts $\left(0,\frac{7}{6}\right)$ and $\left(\frac{7}{3},0\right)$, you see that they never touch — even if you extend the graph forever and ever. The lines are parallel. Figure 14-2 shows what the lines look like.

One way you can determine (with more confidence) that two lines are parallel — and that no solution exists for the system of equations — is by checking the slopes of the lines. You can write each equation from this system in slope-intercept form (see Chapter 5 for a refresher on this form). The slope-intercept form for the line $x+2y=8$, for example, is $y=-\frac{1}{2}x+4$, and the slope-intercept form for $3x+6y=7$ is $y=-\frac{1}{2}x+\frac{7}{6}$. The lines both have the slope $-\frac{1}{2}$, and their y-intercepts are different, so you know the lines are parallel.

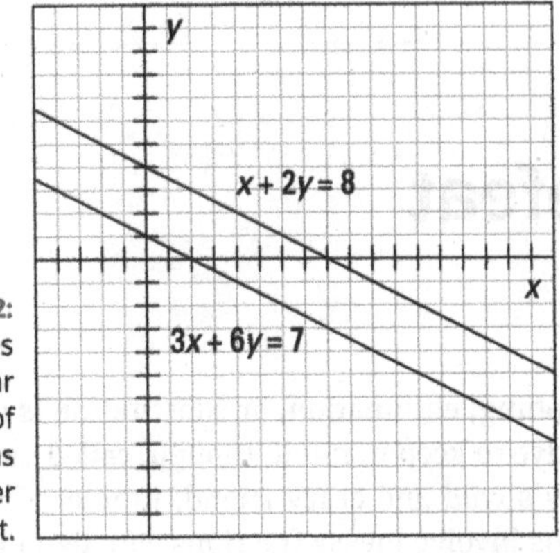

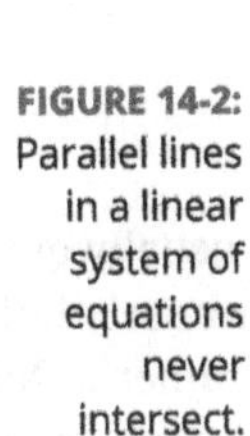

FIGURE 14-2: Parallel lines in a linear system of equations never intersect.

YOUR TURN

Determine the point of intersection (if there is one) of the systems of equations using graphing.

8 $\begin{cases} y = 3x - 8 \\ 2x - 5y = 1 \end{cases}$

9 $\begin{cases} 4x - 3y = 2 \\ x + 2y = 6 \end{cases}$

10 $\begin{cases} 6x - 3y = 9 \\ y = 2x - 3 \end{cases}$

11 $\begin{cases} x - 5y = 11 \\ 2x + 22 = 10y \end{cases}$

Using Cramer's Rule to Defeat Unwieldy Fractions

Solving systems of linear equations with graphing, elimination, or substitution is usually quite doable and simple. You have another alternative to keep in mind, though, called Cramer's Rule. For Seinfeld fans, this Cramer is unlike Jerry's wild and crazy friend, Kramer. Cramer's Rule remains very nice and calm when the solutions involve messy fractions with denominators like 47 or 319, or some equally nasty value. Cramer's Rule gives you the exact fractional value of the solutions you find — not some rounded-off decimal value you're likely to get with a calculator

or computer solution — and you can use the rule for any system of linear equations. It isn't the method of choice for most linear systems, however, because it's more complicated and takes longer than the other methods. The extra effort is worth it, though, when the answers are huge fractions.

Setting up the linear system for Cramer

To use Cramer's Rule, you first have to write the two linear equations in the following form:

$$a_1x + b_1y = c_1$$
$$a_2x + b_2y = c_2$$

The two equations have the variables on one side — in x, y order — and the constant, c, on the other side. The coefficients bear the subscripts 1 or 2 to identify which equation they come from. Your next step is to assign d to represent the difference of the two products of the coefficients of x and y: $d = a_1b_2 - b_1a_2$. The order of the subtraction is very important here.

TIP

An easy way to remember the order of the products and difference is to imagine the coefficients in a square where you multiply and subtract diagonally. This rectangular array is actually a *determinant*. In Chapter 18, you find lots of information on matrices, which are the basis of determinants.

Putting the coefficients of the variables in a square array, you multiply the top left times the bottom right and subtract the product of the top right times the bottom left:

$$\begin{vmatrix} a_1 & b_1 \\ a_2 & b_2 \end{vmatrix} \rightarrow \begin{vmatrix} a_1 & b_1 \\ a_2 & b_2 \end{vmatrix} \rightarrow \begin{vmatrix} a_1 & b_1 \\ a_2 & b_2 \end{vmatrix} \rightarrow d = a_1b_2 - b_1a_2$$

You find the solution of the system, the values of x and y, by dividing two other cleverly created differences by the value of d. You can memorize these differences, or you can picture the squares you form by replacing the a's and b's with the c's and doing the criss-cross multiplication and subtraction:

To solve for x, use this array and d: $\begin{vmatrix} c_1 & b_1 \\ c_2 & b_2 \end{vmatrix} = c_1b_2 - b_1c_2$

To solve for y, use this array and d: $\begin{vmatrix} a_1 & c_1 \\ a_2 & c_2 \end{vmatrix} = a_1c_2 - c_2a_2$

To solve for x and y by using Cramer's Rule when you have two linear equations written in the correct form, use the following equations:

$$x = \frac{c_1b_2 - b_1c_2}{d} = \frac{c_1b_2 - b_1c_2}{a_1b_2 - b_1a_2} \text{ and } y = \frac{a_1c_2 - c_1a_2}{d} = \frac{a_1c_2 - c_1a_2}{a_1b_2 - b_1a_2}$$

Applying Cramer's Rule to a linear system

You can solve most systems of equations with elimination or substitution, but some systems can get pretty gruesome because of the fractions that pop up. If you use Cramer's Rule, the work is much easier (see the previous section for the necessary setup).

Q. Solve the system of equations with Cramer's Rule.

EXAMPLE

$$\begin{cases} 13x + 7y = 25 \\ 10x - 9y = 13 \end{cases}$$

A. You first find the value of the denominator you'll create, d. Using the formula $d = a_1b_2 - b_1a_2$ (the subscripts point to which equations the a and b values come from), $d = 13(-9) - 7(10) = -117 - 70 = -187$.

With your d value in tow, try solving for x first:

$$x = \frac{c_1b_2 - b_1c_2}{d} = \frac{25(-9) - 7(13)}{-187} = \frac{-225 - 91}{-187} = \frac{-316}{-187} = \frac{316}{187}$$

You don't want to substitute such a scary number into one of the equations to solve for y, so move on to the formula for y:

$$y = \frac{a_1c_2 - c_1a_2}{d} = \frac{13(13) - 25(10)}{-187} = \frac{169 - 250}{-187} = \frac{-81}{-187} = \frac{81}{187}$$

The solution is $\left(\frac{316}{187}, \frac{81}{187}\right)$.

Checking your solution isn't much fun because of the fractions, but substituting the values of x and y into both equations shows you that you do, indeed, have the solution.

Q. Solve the system of equations with Cramer's Rule.

$$\begin{cases} 11x - 9y = 2 \\ 13x + 6y = 8 \end{cases}$$

A. Using the formula $d = a_1b_2 - b_1a_2$, $d = 11(6) - (-9)(13) = 66 + 117 = 183$.

Solving for x, $x = \frac{c_1b_2 - b_1c_2}{d} = \frac{2(6) - (-9)(8)}{183} = \frac{12 + 72}{183} = \frac{84}{183} = \frac{28}{61}$.

And solving for y, $y = \frac{a_1c_2 - c_1a_2}{d} = \frac{11(8) - 2(13)}{183} = \frac{88 - 26}{183} = \frac{62}{183}$. The solution is $\left(\frac{28}{61}, \frac{62}{183}\right)$.

If you get $d = 0$, you have to stop. You can't divide by zero. The zero value for d indicates that you have either no solution or an infinite number of solutions — the lines are parallel or you have two equations for the same line. In any case, Cramer's Rule has failed you, and you have to go back and use a different method to determine exactly what the situation is.

Use Cramer's Rule to find the solutions of the systems of equations.

YOUR TURN

12 $\begin{cases} 8x - 19y = 20 \\ 23x + 14y = 11 \end{cases}$	**13** $\begin{cases} 22x + 15y = 89 \\ 19x + 33y = 137 \end{cases}$

Tackling Linear Systems with Three Linear Equations

Systems of three linear equations also have solutions: sets of numbers (all the same for each equation) that make each of the equations true. When a system has three variables rather than two, you no longer graph the equations as lines. To graph these equations, you have to do a three-dimensional graph of the planes represented by the equations containing the three variables. In other words, you really can't find the solution by graphing. The best method for solving systems with three linear equations involves using your algebra skills.

Solving three-equation systems with algebra

When you have a system of three linear equations and three unknown variables, you solve the system by reducing the three equations with three variables to a system of two equations with two variables. At that point, you're back to familiar territory and have all sorts of methods at your disposal to solve the system (see the previous sections in this chapter). After you determine the values of the two variables in the new system, you back-substitute into one of the original equations to solve for the value of the third variable.

EXAMPLE

Q. Solve the system of equations:

$$\begin{cases} 3x - 2y + z = 17 \\ 2x + y + 2z = 12 \\ 4x - 3y - 3z = 6 \end{cases}$$

A. First, choose a variable to eliminate. The prime two candidates for elimination are the y and z because of the coefficients of 1 or −1 that occur in their equations. It makes your job easier if you can avoid larger coefficients on the variables when you have to multiply an equation through by a number to create sums of zero. Assume that you choose to eliminate the z variable.

To eliminate the z's from the equations, you add two of the equations together — after multiplying by an appropriate number — to get a new equation. You then repeat the process with a different combination of two equations. Your result is two equations that contain only the x and y variables.

For this problem, you start by multiplying the terms in the top equation by –2 and adding them to the terms in the middle equation:

$$\begin{array}{r} -2(3x-2y+z=17)\rightarrow -6x+4y-2z=-34 \\ \underline{2x+\ \ y+2z=\ 12} \\ -4x+5y\qquad =-22 \end{array}$$

Next, you multiply the terms in the top equation (the original top equation, not the one that you multiplied earlier) by 3 and add them to the terms in the bottom equation (again, the original one):

$$\begin{array}{r} 3(3x-2y+z=17)\rightarrow 9x-6y+3z=51 \\ \underline{4x-3y-3z=6} \\ 13x-9y\qquad =57 \end{array}$$

The two equations you create by adding comprise a new system of equations with just two variables:

$$\begin{cases} -4x+5y=-22 \\ 13x-9y=57 \end{cases}$$

To solve this new system, you can multiply the terms in the first equation by 9 and the terms in the second equation by 5 to create coefficients of 45 and –45 on the y terms. You add the two equations together, getting rid of the y terms, and solve for x:

$$\begin{aligned} -36x+45y&=-198 \\ \underline{65x-45y}&\underline{=\ 285} \\ 29x\qquad&=\ 87 \\ \frac{29x}{29}&=\frac{87}{29} \\ x&=3 \end{aligned}$$

Now you substitute $x=3$ into one of the two equations. I choose the equation $-4(3)+5y=-22$. Choosing this equation is just an arbitrary choice — either equation will do. When you substitute $x=3$, you get $-4(3)+5y=-22$. Adding 12 to each side, you get $5y=-10$, or $y=-2$.

You can check your work by taking the $x=3$ and $y=-2$ and substituting them into one of the original equations. A good habit is to substitute the values into the first equation and then check by substituting all three answers into the other two.

Putting $x=3$ and $y=-2$ into the first equation, you get $3(3)-2(-2)+z=17$, giving you $9+4+z=17$. You subtract 13 from each side for a result of $z=4$. Now check these three values in the other two equations:

$$2(3)+(-2)+2(4)=6-2+8=12$$
$$4(3)-3(-2)-3(4)=12+6-12=6$$

They both check — of course!

You can write the solution of the system as $x=3$, $y=-2$, $z=4$, or you can write it as an ordered triple. An ordered triple consists of three numbers in parentheses, separated by commas. The order of the numbers matters. The first value represents x, the second y, and the third z. For example, you would write the previous solution as (3, −2, 4). The ordered triple is a simpler and neater method — as long as everyone agrees what the numbers stand for.

YOUR TURN

14 Solve the system of equations.

$$\begin{cases} x+2y+z=1 \\ 3x-y-z=2 \\ x+y+2z=4 \end{cases}$$

15 Solve the system of equations.

$$\begin{cases} x-y-4z=4 \\ 2x+y-z=3 \\ 3x+2y-2z=7 \end{cases}$$

Generalizing multiple solutions for linear equations

When dealing with three linear equations and three variables, you may come across a situation where one of the equations is a linear combination of the other two. This means you won't find one single solution for the system, such as (3, −2, 4). A more generalized solution may be $(-k, 2k, k)$, where you write all the possible solutions in terms of a *parameter*, k. In this case, where the solution is $(-k, 2k, k)$, if you let $z=7$, the ordered triple becomes $(-7, 14, 7)$. You can find an infinite number of solutions for this particular system of equations, but the solutions are very specific in form — the variables all have a relationship. You first get an inkling that a system has a generalized answer when you find out that one of the reduced equations you create is a multiple of another.

Q. Solve the system of equations:

EXAMPLE

$$\begin{cases} 2x+3y-z=12 \\ x-3y+4z=-12 \\ 5x-6y+11z=-24 \end{cases}$$

A. **To solve this system, you can eliminate the z's by multiplying the terms in the first equation by 4 and adding them to the second equation. You then multiply the terms in the first equation by 11 and add them to the third equation:**

$$\begin{aligned} 4(2x+3y-z=12) \rightarrow 8x+12y-4z &= 48 \\ x-3y+4z &= -12 \\ \hline 9x+9y \quad\quad &= 36 \end{aligned}$$

$$\begin{aligned} 11(2x+3y-z=12) \rightarrow 22x+33y-11z &= 132 \\ 5x-6y+11z &= -24 \\ \hline 27x+27y \quad\quad &= 108 \end{aligned}$$

The second equation, $27x+27y=108$, is three times the first equation. Because these equations are multiples of one another, you know that the system has no single solution; it has an infinite number of solutions.

To find all those solutions, you take one of the equations and solve for a variable. You may choose to solve for y in $9x+9y=36$. Dividing through by 9, you get $x+y=4$. Solving for y, you get $y=4-x$. Now replace the x with a parameter; I choose the letter k. Because you already have y in terms of x, you can write $y=4-k$. You can then write all the variables in terms of that parameter. You substitute k for x and $4-k$ for y into one of the original equations in the system to solve for z in terms of k. After you solve for z this way, you have the three variables all written as some version of k. Substituting k for x and $4-k$ for y into $2x+3y-z=12$, for example, you get

$$2k+3(4-k)-z=12 \rightarrow 2k+12-3k-z=12 \rightarrow -k-z=0 \rightarrow -k=z$$

The ordered triple giving the solutions of the system is $(k, 4-k, -k)$. You can find an infinite number of solutions, all determined by this pattern. Just pick a value for k, such as $k=3$. The solution is $(3, 1, -3)$. These values of x, y, and z all work in the equations of the original system.

Upping the Ante with Larger Systems

Systems of linear equations can be any size. You can have 2, 3, 4, or even 100 linear equations. (After you get past 3 or 4, you resort to technology.) Some of these systems have solutions and others don't. You have to dive in to determine whether you can find a solution or not. You can try to solve a system of just about any number of linear equations, but you find a single, unique solution (one set of numbers for the answer) only when the number of variables in the system has at least that many equations. If a system has three different variables, you need at least three different equations. Having enough equations for the variables doesn't guarantee a unique solution but you have to at least start out that way.

The general process for solving n equations with n variables is to keep eliminating variables. A systematic way is to start with the first variable, eliminate it, move to the second variable, eliminate it, and so on until you create a reduced system with two equations and two variables. You solve for the solutions of that system and then start substituting values into the original equations. This process can be long and tedious, and errors are easy to come by, but if you have to do it by hand, this is a nice, systematic method. Technology, however, is most helpful when systems get unmanageable.

EXAMPLE

Q. Solve the system of five equations and five variables:

$$\begin{cases} x+y+z+w+t=3 \\ 2x-y+z-w+3t=28 \\ 3x+y-2z+w+t=-8 \\ x-4y+z-w+2t=28 \\ 2x+3y+z-w+t=6 \end{cases}$$

A. Begin the process by eliminating the x's. Multiply the terms in the first equation by –2 and add them to the second equation. Then multiply the first equation through by –3 and add the terms to the third equation. Next, multiply the first equation through by –1 and add the terms to the fourth equation. And, finally, multiply the first equation through by –2 and add the terms to the last equation.

$$\begin{cases} x+y+z+w+t=3 \\ 2x-y+z-w+3t=28 \\ 3x+y-2z+w+t=-8 \\ x-4y+z-w+2t=28 \\ 2x+3y+z-w+t=6 \end{cases} \Rightarrow \begin{matrix} -2R_1+R_2 \\ -3R_1+R_3 \\ -1R_1+R_4 \\ -2R_1+R_5 \end{matrix} \Rightarrow \begin{cases} -3y-z-3w+t=22 \\ -2y-5z-2w-2t=-17 \\ -5y-2w+t=25 \\ y-z-3w-t=0 \end{cases}$$

After you finish (whew!), you get a system with the x's eliminated. Now you eliminate the y's in the new system by multiplying the last equation by 3, 2, and 5 and adding the results to the first, second, and third equations, respectively.

$$\begin{cases} -3y-z-3w+t=22 \\ -2y-5z-2w-2t=-17 \\ -5y-2w+t=25 \\ y-z-3w-t=0 \end{cases} \Rightarrow \begin{matrix} 3R_4+R_1 \\ 2R_4+R_2 \\ 5R_4+R_3 \end{matrix} \Rightarrow \begin{cases} -4z-12w-2t=22 \\ -7z-8w-4t=-17 \\ -5z-17w-4t=25 \end{cases}$$

You eliminate the z's in the latest system by multiplying the terms in the first equation by 7 and the second by –4 and adding them together. You then multiply the terms in the second equation by 5 and the third by –7 and add them together. The new system you create has only two variables and two equations.

$$\begin{cases} -4z-12w-2t=22 \\ -7z-8w-4t=-17 \\ -5z-17w-4t=25 \end{cases} \Rightarrow \begin{matrix} 7R_1-4R_2 \\ 5R_2-7R_3 \end{matrix} \Rightarrow \begin{cases} -52w+2t=222 \\ 79w+8t=-260 \end{cases}$$

To solve the two-variable system in the most convenient way, you multiply the terms in the first equation by –4 and add the terms to the second.

$$\begin{cases} -52w + 2t = 222 \\ 79w + 8t = -260 \end{cases} \Rightarrow -4R_1 + R_2 \Rightarrow 287w = -1148$$

Solving for w, $w = -4$. Now do several back-substitutions to find the values of the other variables.

Back-substituting w into $-52w + 2t = 222$: $-52(-4) + 2t = 222 \rightarrow 208 + 2t = 222 \rightarrow 2t = 14 \rightarrow t = 7$.

Plug these two values into $-4z - 12w - 2t = 22$: $-4z - 12(-4) - 2(7) = 22 \rightarrow -4z + 34 = 22 \rightarrow -4z = -12 \rightarrow z = 3$.

Put the three values into $y - z - 3w - t = 0$: $y - (3) - 3(-4) - 7 = 0 \rightarrow y + 2 = 0 \rightarrow y = -2$.

Move back to the equation $x + y + z + w + t = 3$, and plug in values: $x + (-2) + 3 + (-4) + 7 = 3 \rightarrow x + 4 = 3 \rightarrow x = -1$. The solution reads $x = -1$, $y = -2$, $z = 3$, $w = -4$, and $t = 7$.

You can put this into an ordered quintuple (five numbers in parentheses), as long as everyone knows the proper order: (x, y, z, w, t). The order isn't alphabetical, in this case, but this is pretty typical — to have the x, y, and z first and then list other variables. You know what order to use from the way the problem is stated. The ordered quintuple is $(-1, -2, 3, -4, 7)$.

YOUR TURN

16 Solve the system of equations:

$$\begin{cases} x + 2y + z = 2 \\ x - y - z = 2 \\ 4x - y - 2z = 8 \end{cases}$$

17 Solve the system of equations:

$$\begin{cases} x + y - z + w = 10 \\ 2x - y + z + 2w = 11 \\ x + 2y + z - w = 4 \\ 3x + 3y - z + 2w = 25 \end{cases}$$

Practice Questions Answers and Explanations

1. **$x=-1$, $y=-3$ or $(-1,-3)$.** By adding the two equations together, you eliminate the x-variable. The sum of the two equations is $-4y=12$. Then, dividing each side by -4, you have $y=-3$. Substitute this back into the first equation: $2x-3(-3)=7 \rightarrow 2x+9=7 \rightarrow 2x=-2 \rightarrow x=-1$. Your solution is $x=-1$ and $y=-3$. Substitute these values into the second equation, and it checks!

2. **$x=2$, $y=-1$ or $(2,-1)$.** Multiply the top equation by -2 and the bottom by 3. Then add the two equations together.

$$\begin{cases} 4x+3y=5 \\ 7x+2y=12 \end{cases} \Rightarrow \begin{matrix} -2R_1 \\ 3R_2 \end{matrix} \Rightarrow \begin{cases} -8x-6y=-10 \\ 21x+6y=36 \end{cases}$$

You have $13x=26$. Dividing both sides by 13, the result is $x=2$. Substitute this back into the first original equation and $4(2)+3y=5 \rightarrow 8+3y=5 \rightarrow 3y=-3 \rightarrow y=-1$. Substituting $x=2$ and $y=-1$ into the second original equation, the solution checks!

3. **No answer.** First, rearrange the equations so the like-terms are lined up under one another. Then multiply the bottom equation by -3.

$$\begin{cases} 3x=21-12y \\ 4y+x+7=0 \end{cases} \Rightarrow \begin{cases} 3x+12y=21 \\ x+\ 4y=-7 \Rightarrow -3R_2 \end{cases} \Rightarrow \begin{cases} 3x+12y=21 \\ -3x-12y=21 \end{cases}$$

Adding the two equations together, you get 0 = 42. This is impossible. There is no solution.

4. **(-5, 3).** Solving for x in the first equation: $x=4-3y$. Now substitute the equivalence of x into the second equation and solve for y: $2x+5y=5 \rightarrow 2(4-3y)+5y=5 \rightarrow 8-6y+5y=5 \rightarrow -y=-3 \rightarrow y=3$. Solving for x using that x-substitution: $x=4-3y \rightarrow x=4-3(3) \rightarrow x=4-9 \rightarrow x=-5$.

5. **(4, -3).** Neither variable has a coefficient of 1 or –1, so, choosing to solve for y in the second equation, $3x+2y=6 \rightarrow 2y=6-3x \rightarrow y=3-\frac{3}{2}x$. Substituting into the first equation, $8x-3\left(3-\frac{3}{2}x\right)=41 \rightarrow 8x-9+\frac{9}{2}x=41 \rightarrow \frac{25}{2}x=50$. Multiply each side by the reciprocal: $\frac{\cancel{2}}{\cancel{25}}\cdot\frac{\cancel{25}}{\cancel{2}}x=\frac{2}{\cancel{25}}\cdot\cancel{50}^{2} \rightarrow x=4$. Now solve for y using the y-substitution: $y=3-\frac{3}{2}x \rightarrow y=3-\frac{3}{2}(4)=3-6=-3$.

6. **$(2k+5,\ k)$.** Solving for x in the first equation, $x=2y+5$. Substituting this into the second equation: $8x=16y+40 \rightarrow 8(2y+5)=16y+40 \rightarrow 16y+40=16y+40$. These are the same expression. There are an infinite number of solutions. Subtracting $16y$ and 40 from each side, you have 0 = 0. Let y be represented by k, and you can write the solution as $(2k+5,\ k)$.

7. **No solution.** Solving for y in the first equation: $3x-8=y$. Substituting this into the second equation: $6x-2y+2=0 \rightarrow 6x-2(3x-8)+2=0 \rightarrow 6x-6x+16+2=0 \rightarrow 18=0$. This is a false statement. The two lines are parallel.

8. **(3,1).** Graphing $y = 3x - 8$, the y-intercept is (0,–8) and another point on the line is (4,4). The line $2x - 5y = 1$ has the points (8,3) and (–2,–1). Graphing the lines, you find the point of intersection at (3,1). Refer to the graph on the left.

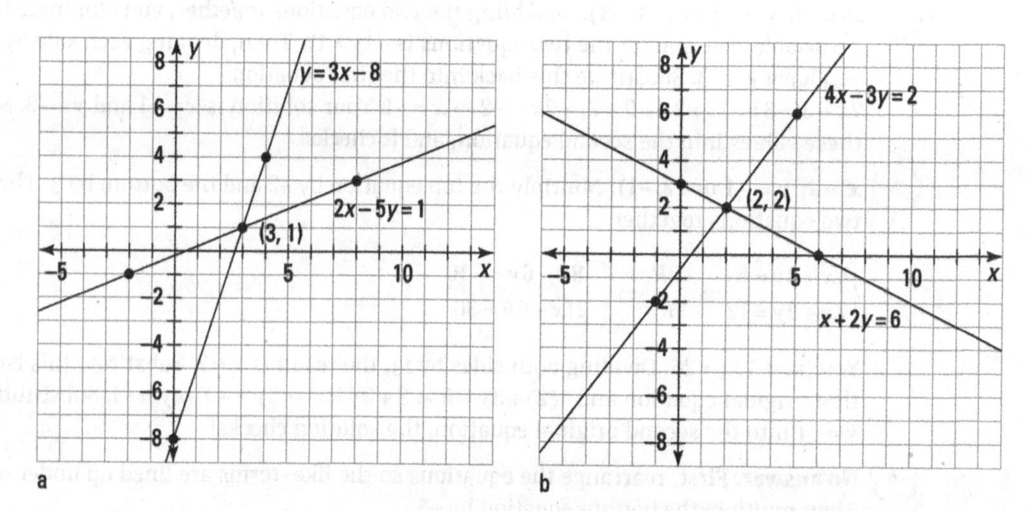

9. **(2,2).** Graphing $4x - 3y = 2$, the line has the points (–1,–2) and (5,6). The line $x + 2y = 6$ has the intercepts (0,3) and (6,0). Graphing the lines, you find the point of intersection at (2,2). Refer to the graph on the above right.

10. **Same line; coexist.** Graphing $6x - 3y = 9$, the line has the points (2,1) and (1,–1). The line $y = 2x - 3$ has the y-intercept (0,–3) and the point (5,7). Graphing the lines, you find that the lines are the same. These are equations of the same graph. Refer to the graph on the left.

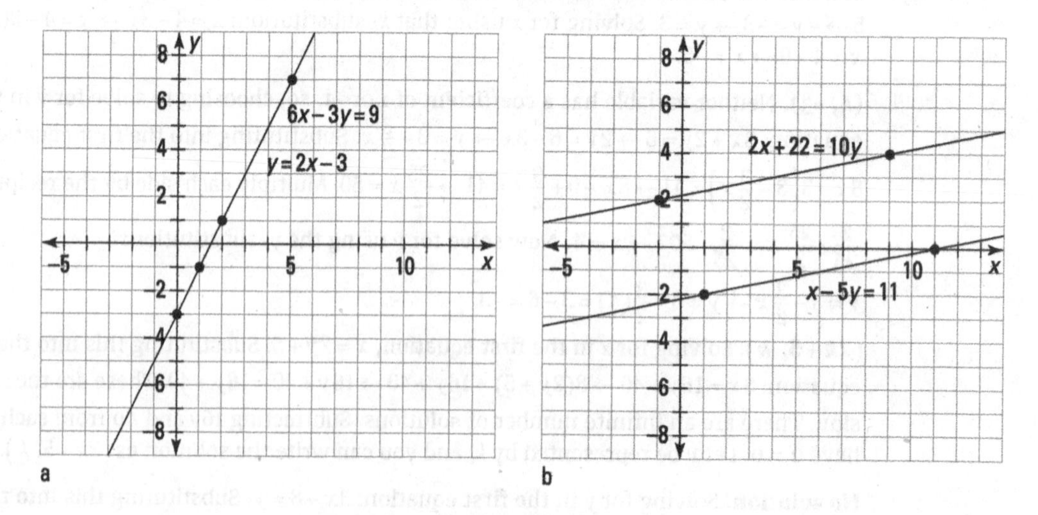

11. **Parallel lines; no intersection.** Graphing $x - 5y = 11$, the line has the points (11,0) and (1,–2). The line $2x + 22 = 10y$ has the points (–1,2) and (9,4). Graphing the lines, you find that the lines are parallel. They have no point of intersection. Refer to the graph above right.

12. $\left(\frac{163}{183}, -\frac{124}{183}\right)$. Using the formula $d = a_1b_2 - b_1a_2$, $d = 8(14) - (-19)(23) = 112 + 437 = 549$.

Solving for x, $x = \frac{c_1b_2 - b_1c_2}{d} = \frac{20(14) - (-19)(11)}{549} = \frac{280 + 209}{549} = \frac{489}{549} = \frac{163}{183}$.

And solving for y, $y = \frac{a_1c_2 - c_1a_2}{d} = \frac{8(11) - 20(23)}{549} = \frac{88 - 460}{549} = \frac{-372}{549} = -\frac{124}{183}$.

13. **(2,3).** Using the formula $d = a_1b_2 - b_1a_2$, $d = 22(33) - (15)(19) = 726 - 285 = 441$.

Solving for x, $x = \frac{c_1b_2 - b_1c_2}{d} = \frac{89(33) - (15)(137)}{441} = \frac{2937 - 2055}{441} = \frac{882}{441} = 2$.

And solving for y, $y = \frac{a_1c_2 - c_1a_2}{d} = \frac{22(137) - 89(19)}{441} = \frac{3014 - 1691}{441} = \frac{1323}{441} = 3$. Yes, you sometimes do get very nice answers.

14. **(1, −1, 2).** Either y or z is a good choice for elimination. Choosing z, add the first two equations together and then add 2 times the second equation to the third equation.

$$\begin{array}{rcrcrcr} x & + & 2y & + & z & = & 1 \\ 3x & - & y & - & z & = & 2 \\ \hline 4x & + & y & & & = & 3 \end{array} \qquad \begin{array}{rcrcrcr} 6x & - & 2y & - & 2z & = & 4 \\ x & + & y & + & 2z & = & 4 \\ \hline 7x & - & y & & & = & 8 \end{array}$$

The resulting system of equations is in x and y. Add the two equations together to eliminate the y term.

$$\begin{array}{rcrcr} 4x & + & y & = & 3 \\ 7x & - & y & = & 8 \\ \hline 11x & & & = & 11 \end{array}$$

From the result, you have that $x = 1$. Substitute that back into the top equation you just worked with and you have $4(1) + y = 3 \rightarrow y = -1$. Now put both the x and y values into the first of the original equations to get $1 + (-1) + z = 1 \rightarrow z = 2$. Substituting the three values back into the other two original equations, they both check.

15. **(−1, 3, −2).** Choosing y to be the variable to eliminate, add the first two equations together and then add 2 times the first equation to the third equation.

$$\begin{array}{rcrcrcr} x & - & y & - & 4z & = & 4 \\ 2x & + & y & - & z & = & 3 \\ \hline 3x & & & - & 5z & = & 7 \end{array} \qquad \begin{array}{rcrcrcr} 2x & - & 2y & - & 8z & = & 8 \\ 3x & + & 2y & - & 2z & = & 7 \\ \hline 5x & & & - & 10z & = & 15 \end{array}$$

An inclination might be to divide each term in the second resulting equation by 5, but, looking ahead, you see that multiplying the first resulting equation through by −2 will get rid of the z terms when the equations are added together.

$$\begin{array}{rcrcr} -6x & + & 10z & = & -14 \\ 5x & - & 10z & = & 15 \\ \hline -x & & & = & 1 \end{array}$$

From the result, you see that $x = -1$. Substitute that back into the bottom equation you just worked with and you have $5(-1) - 10z = 15 \to -10z = 20 \to z = -2$. Put both the x and z values into the first of the original equations to solve for y: $-1 - y - 4(-2) = 4 \to -y + 7 = 4 \to -y = -3 \to y = 3$. And, checking these solutions in the second and third original equations, the answer checks.

16 **(k, 4-2k, 3k-6).** Start by eliminating z. Add the first two equations together, and then add twice the first equation to the third equation.

$$\begin{array}{rcl} x + 2y + z &=& 2 \\ x - y - z &=& 2 \\ \hline 2x + y &=& 4 \end{array} \qquad \begin{array}{rcl} 2x + 4y + 2z &=& 4 \\ 4x - y - 2z &=& 8 \\ \hline 6x + 3y &=& 12 \end{array}$$

The second equation is exactly 3 times the first equation. There are multiple solutions to this system, because one of the equations is equal to a combination of the others. Solve for y in the first result: $y = 4 - 2x$. Substitute this into the first equation and solve for z: $x + 2y + z = 2 \to x + 2(4 - 2x) + z = 2 \to x + 8 - 4x + z = 2 \to z = 3x - 6$. Replace x with the parameter k, and the solution is: $(k, 4 - 2k, 3k - 6)$.

17 **(4, 2, -1, 3).** Start by eliminating the z variable. Add the first and second equations, the first and third equations, and the third and fourth equations.

$$\begin{cases} x + y - z + w = 10 \\ 2x - y + z + 2w = 11 \\ x + 2y + z - w = 4 \\ 3x + 3y - z + 2w = 25 \end{cases} \begin{array}{c} R_1 + R_2 \\ \Rightarrow R_1 + R_3 \Rightarrow \\ R_3 + R_4 \end{array} \begin{cases} 3x \qquad + 3w = 21 \\ 2x + 3y \qquad = 14 \\ 4x + 5y + w = 29 \end{cases}$$

Only two of the equations have the w variable, so eliminate it by adding the first equation to -3 times the third equation.

The resulting equation can be reduced by dividing each term by -3.

$$\begin{cases} 3x \qquad + 3w = 21 \\ 2x + 3y \qquad = 14 \\ 4x + 5y + w = 29 \end{cases} \Rightarrow R_1 - 3R_3 \Rightarrow \begin{cases} -9x - 15y = -66 \\ 2x + 3y = 14 \end{cases} \Rightarrow \begin{cases} 3x + 5y = 22 \\ 2x + 3y = 14 \end{cases}$$

Now solve for x by adding 3 times the first equation to -5 times the second equation.

$$\begin{cases} 3x + 5y = 22 \\ 2x + 3y = 14 \end{cases} \Rightarrow 3R_1 - 5R_2 \Rightarrow -x = -4 \to x = 4$$

Now, back-solving, when $x = 4$, $3x + 5y = 22 \to 3(4) + 5y = 22 \to 5y = 10 \to y = 2$.

When $x = 4$ and $y = 2$, $4x + 5y + w = 29 \to 4(4) + 5(2) + w = 29 \to 26 + w = 29 \to w = 3$.

And when $x = 4$, $y = 2$, and $w = 3$, $x + y - z + w = 10 \to 4 + 2 - z + 3 = 10 \to 9 - z = 10 \to z = -1$.

Whaddya Know? Chapter 14 Quiz

Quiz time! Complete each problem to test your knowledge on the various topics covered in this chapter. You can then find the solutions and explanations in the next section.

1. Use Cramer's Rule to find the solution of the system of equations.

$$\begin{cases} 5x - 6y = 8 \\ 8x + 7y = 46 \end{cases}$$

2. Find the solution of the system of equations.

$$\begin{cases} x - 3y + z = 5 \\ 2x + y - z = 9 \\ 5x - y - z = -13 \end{cases}$$

3. Find the solution of the system of equations.

$$\begin{cases} 3x + 4y = 8 \\ x + 2y = 6 \end{cases}$$

4. Find the solution of the system of equations by graphing.

$$\begin{cases} y = -2x + 7 \\ x + 3y = 6 \end{cases}$$

5. Find the solution of the system of equations.

$$\begin{cases} y = 3x + 5 \\ 3x - y = 4 \end{cases}$$

6. Solve the system of equations.

$$\begin{cases} x + 3y - 2z = 13 \\ 2x - y + z = 4 \\ 4x - 2y - z = 17 \end{cases}$$

7. Find the solution of the system of equations.

$$\begin{cases} y = \frac{1}{4}x - \frac{3}{2} \\ x - 4y = 6 \end{cases}$$

8. **Write a generalized solution for the system of equations.**

$$\begin{cases} y = 3x - 2 \\ 6x - 2y = 4 \end{cases}$$

9. **Find the solution of the system of equations using substitution.**

$$\begin{cases} y = 2x + 7 \\ y = -3x + 2 \end{cases}$$

10. **Find the solution of the system of equations.**

$$\begin{cases} x + y - z + w + v = 7 \\ 2x - y + 2z + w - v = -10 \\ x + y + z - w + 2v = 4 \\ x + 2y - z + 2w + v = 10 \\ 2x - 2y + z - w + 3v = 0 \end{cases}$$

Answers to Chapter 14 Quiz

1. **(4, 2).** First, find $d = a_1b_2 - b_1a_2 = 5(7) - (-6)8 = 35 + 48 = 83$.

   Then $x = \frac{56-(-276)}{35-(-48)} = \frac{332}{83} = 4$ and $y = \frac{230-64}{35-(-48)} = \frac{166}{83} = 2$.

2. **No solution.** Eliminate the z variable by adding together the first two equations and then the first and third equation.

   $$\begin{cases} x-3y+z=5 \\ 2x+y-z=9 \\ 5x-y-z=-13 \end{cases} \Rightarrow \begin{matrix} R_1+R_2 \\ R_1+R_3 \end{matrix} \Rightarrow \begin{cases} 3x-2y=14 \\ 6x-4y=-8 \end{cases}$$

   Multiply the first equation by –2 and add it to the second equation.

   $$\begin{cases} 3x-2y=14 \\ 6x-4y=-8 \end{cases} \Rightarrow -2R_1+R_2 \Rightarrow 0+0=-36$$

   This is an impossible statement. There is no solution.

3. **(–4, 5).** Using elimination, multiply the second equation by –3 and add the two equations together. Solve for y.

   $$\begin{array}{rcrcr} 3x & + & 4y & = & 8 \\ -3x & - & 6y & = & -18 \\ \hline & - & 2y & = & -10 \\ & & y & = & 5 \end{array}$$

   Substitute $y = 5$ into the first equation and solve for x.

   $3x + 4(5) = 8 \rightarrow 3x = -12 \rightarrow x = -4$

4. **(3,1).** You can graph the first line using the y-intercept of (0,7) and slope of –2. One possibility for the second line is to find the intercepts: (0,2) and (6,0).

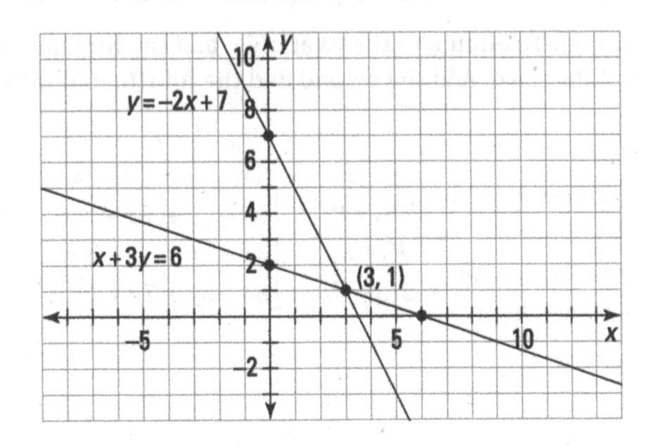

5. **Parallel lines; no solution.** Using substitution, replace the y in the second equation with its equivalent in the first equation: $3x-(3x+5)=4 \to 3x-3x-5=4 \to -5=4$. This is an impossible statement. The lines are parallel.

6. **(4, 1, −3).** Eliminate the z variable by adding the first equation to twice the second and then adding the second and third equations.

$$\begin{cases} x+3y-2z=13 \\ 2x-y+z=4 \\ 4x-2y-z=17 \end{cases} \Rightarrow \begin{matrix} R_1+2R_2 \\ R_2+R_3 \end{matrix} \Rightarrow \begin{cases} 5x+y=21 \\ 6x-3y=21 \end{cases}$$

Add three times the first equation to the second equation to eliminate the y variable.

$$\begin{cases} 5x+y=21 \\ 6x-3y=21 \end{cases} \Rightarrow 3R_1+R_2 \Rightarrow 21x=84 \to x=4$$

Substitute the value of x back into the top equation that you just used: $5(4)+y=21 \to y=1$.

Now substitute the values of x and y into the original first equation:

$$4+3(1)-2z=13 \to 7-2z=13 \to -2z=6 \to z=-3$$

7. **Infinite solutions; same line.** Substitute the equivalent of y from the first equation into the second equation.

$$x-4\left(\frac{1}{4}x-\frac{3}{2}\right)=6 \to x-x+6=6 \to 0=0$$

8. **$(k, 3k-2)$ or $\left(\frac{k+2}{3}, k\right)$.** Letting $x = k$, a parameter, the solution is $(k, 3k-2)$. Solving for x in the first equation, $y=3x-2 \to y+2=3x \to \frac{y+2}{3}=x$. Then, letting $y = k$, the solution is $\left(\frac{k+2}{3}, k\right)$.

9. **(−1, 5).** Using substitution, replace the y in the first equation with its equivalent in the second equation:

$$-3x+2=2x+7 \to -5x=5 \to x=-1$$

Now substitute $x=-1$ into the first equation and solve for y: $y=2(-1)+7=5$.

10. **(−1, 2, −2, 1, 3).** To eliminate the v variable, add the first and second equations. Add twice the second and the third. Add the second and the fourth. Add three times the second to the fifth.

$$\begin{cases} x+y-z+w+v=7 \\ 2x-y+2z+w-v=-10 \\ x+y+z-w+2v=4 \\ x+2y-z+2w+v=10 \\ 2x-2y+z-w+3v=0 \end{cases} \Rightarrow \begin{matrix} R_1+R_2 \\ 2R_2+R_3 \\ R_2+R_4 \\ 3R_2+R_5 \end{matrix} \Rightarrow \begin{cases} 3x+z+2w=-3 \\ 5x-y+5z+w=-16 \\ 3x+y+z+3w=0 \\ 8x-5y+7z+2w=-30 \end{cases}$$

To eliminate the *y* variable, add the second and third equations. Add five times the third equation to the fourth.

$$\begin{cases} 3x + z + 2w = -3 \\ 5x - y + 5z + w = -16 \\ 3x + y + z + 3w = 0 \\ 8x - 5y + 7z + 2w = -30 \end{cases} \Rightarrow \begin{matrix} R_2 + R_3 \\ 5R_3 + R_4 \end{matrix} \Rightarrow \begin{cases} 3x + z + 2w = -3 \\ 8x + 6z + 4w = -16 \\ 23x + 12z + 17w = -30 \end{cases}$$

To eliminate the *z* variable, multiply the first equation by –6 and add it to the second equation. Then add –12 times the first equation to the third equation.

$$\begin{cases} 3x + z + 2w = -3 \\ 8x + 6z + 4w = -16 \\ 23x + 12z + 17w = -30 \end{cases} \Rightarrow \begin{matrix} -6R_1 + R_2 \\ -12R_1 + R_3 \end{matrix} \Rightarrow \begin{cases} -10x - 8w = 2 \\ -13x - 7w = 6 \end{cases}$$

Now, to avoid dealing with nasty fractions, I could go to Cramer's Rule. Instead, I choose to multiply the top equation by 7 and the bottom equation by –8 to make the coefficients on the *w* terms the opposite. Then I can add the two equations together.

$$\begin{cases} -10x - 8w = 2 \\ -13x - 7w = 6 \end{cases} \Rightarrow \begin{matrix} 7R_1 \\ -8R_2 \end{matrix} \Rightarrow \begin{cases} -70x - 56w = 14 \\ 104x + 56w = -48 \end{cases} \Rightarrow R_1 + R_2 \Rightarrow 34x = -34$$

Divide by 34, and you have $x = -1$.

Do back-solving, and you have $w = 1,\ z = -2,\ y = 2,\ v = 3$.

IN THIS CHAPTER

- » Pinpointing solutions of parabola/line systems
- » Combining parabolas and circles to find intersections
- » Tackling polynomial, exponential, and rational systems

Chapter 15
Solving Systems of Nonlinear Equations

In systems of linear equations, the variables have exponents of 1 and you typically find only one solution (see Chapter 14). The possibilities for multiple solutions in systems seem to grow as the exponents of the equations get larger, creating systems of nonlinear equations. For example, a line and parabola may intersect in two points, at one point, or at no point at all. A circle and ellipse can intersect in four different points.

One of the most important parts of solving nonlinear systems is planning. If you have an inkling as to what's coming, you'll have an easy time planning for the solution, and you'll be more convinced when your predictions come true. In this chapter, you find out at how many points a line and a parabola can cross and how many ways a parabola and a circle can cross. I also help you visualize a circle and an ellipse; when you put one on top of the other, you can plan on how many points of intersection you expect to find.

Crossing Parabolas with Lines

A parabola is a predictable, smooth, U-shaped curve (which I first introduce in detail in Chapter 6). A line is also very predictable; it goes up or down and left or right at the same rate forever and ever. If you put these two characteristics together, you can predict with a fair amount of accuracy what will happen when a line and a parabola share the same place in space.

When you combine the equations of a line and a parabola, you get one of three results (which you can check out in Figure 15-1):

- Two common solutions (Figure 15-1a)
- One common solution (Figure 15-1b)
- No solution at all (Figure 15-1c)

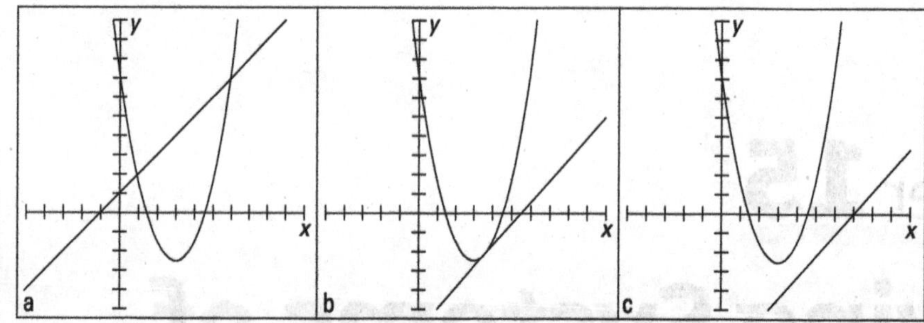

FIGURE 15-1: A line and a parabola sharing space on a graph.

The easiest way to find the common solutions, or sets of values, for a line and a parabola is to solve their system of equations algebraically. A graph is helpful for confirming your work and putting the problem into perspective. When solving a system of equations involving a line and a parabola, most mathematicians use the substitution method. For a complete look at how to use substitution, see Chapter 14. You can also pick up on the method by following the work in the coming pages.

REMEMBER

You almost always substitute *x*'s for the *y*'s in an equation, because you often see functions written with the *y*'s equal to so many *x*'s. You may have to replace *x*'s with *y*'s, but that's the exception. Just be flexible. (If you want to see an exception, check out the section, "Sorting out the solutions," later in this chapter.)

Determining the points where a line and parabola cross paths

The graphs of a line and a parabola can cross in two places, one place, or no place at all (refer to Figure 15-1). In terms of equations, these assertions translate to two common solutions, one solution, or no solution at all. Doesn't that fit together nicely?

Finding two solutions

You could always graph the curve and line in question to find their common point, but using substitution is usually the best plan, especially when fractions can be involved.

Q. Find the intersection of the parabola $y = 3x^2 - 4x - 1$ and the line $x + y = 5$.

EXAMPLE **A.** To solve for the two solutions by using the substitution method, you first solve for y in the equation of the line: $y = -x + 5$. Substitute this equivalent of y into the first equation: $-x + 5 = 3x^2 - 4x - 1$. Next, set the new equation equal to zero, and factor: $0 = 3x^2 - 3x - 6 = 3(x^2 - x - 2) = 3(x - 2)(x + 1)$. Setting each of the binomial factors equal to zero, you get $x = 2$ and $x = -1$. When you substitute those values into the equation $y = -x + 5$, you find that when $x = 2$, $y = 3$, and when $x = -1$, $y = 6$. The two points of intersection, therefore, are (2,3) and (–1,6). Figure 15-2 shows the graphs of the parabola ($y = 3x^2 - 4x - 1$), the line ($y = -x + 5$), and the two points of intersection.

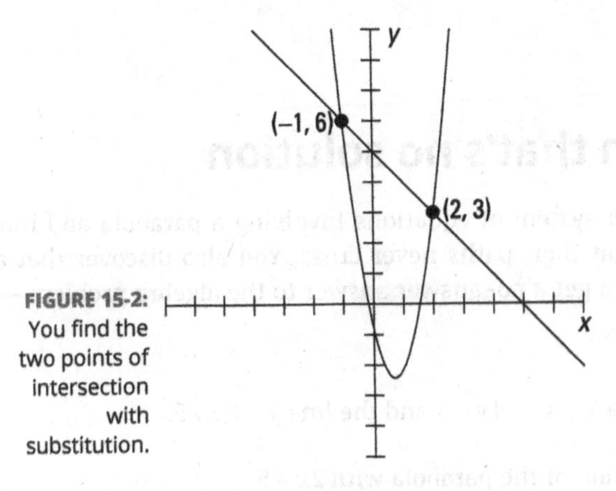

FIGURE 15-2: You find the two points of intersection with substitution.

Settling for one solution

When a line and a parabola have one point of intersection, and their equations share one common solution, the line is tangent to the parabola. A line and a curve can be tangent to one another if they touch or share exactly one point and if the line appears to follow the curvature at that point. (Two curves can also be tangent to one another — they touch at a point and then go their own merry ways.)

Q. Find the intersection of the parabola $y = -x^2 + 5x + 6$ and the line $y = 3x + 7$.

EXAMPLE **A.** Substitute the equivalent of y in the line equation into the parabolic equation and solve for x (see Chapter 14): $3x + 7 = -x^2 + 5x + 6$ is set equal to 0, giving you $0 = -x^2 + 2x - 1$. Factor out –1, and then factor the trinomial: $0 = -1(x^2 - 2x + 1) = -1(x - 1)^2$. Setting the binomial $x - 1 = 0$, you have $x = 1$, which is the only solution, at (1,10). They have only one point in common — at their point of tangency. Figure 15-3 shows how a line and a parabola can be tangent.

The dead giveaway that the parabola and line are tangent is the quadratic equation that results from the substitution. It has a double root — the same solution appears twice — when the binomial factor is squared.

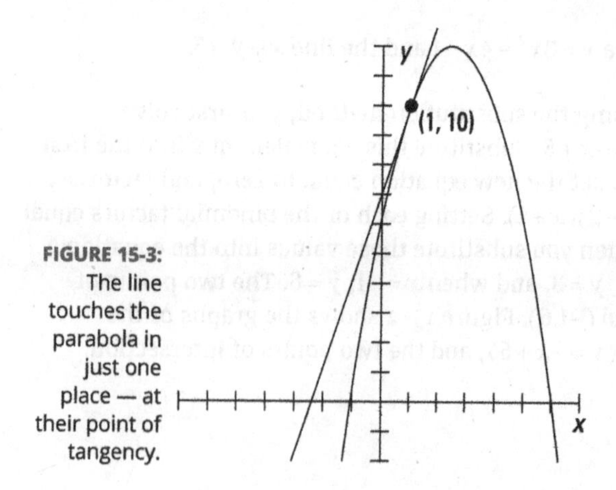

FIGURE 15-3: The line touches the parabola in just one place — at their point of tangency.

Dealing with a solution that's no solution

You can see when no solution exists in a system of equations involving a parabola and line if you graph the two figures and find that their paths never cross. You also discover that a parabola and line don't intersect when you get a no-answer answer to the algebra problem — there's no need to even graph the figures.

EXAMPLE

Q. Find the intersection of the parabola $x = y^2 - 4y + 3$ and the line $y = 2x + 5$.

A. First, replace all the y's in the equation of the parabola with $2x + 5$: $x = (2x+5)^2 - 4(2x+5) + 3$. Square the binomial and distribute the -4 to get $x = 4x^2 + 20x + 25 - 8x - 20 + 3$. When you simplify on the right and set the equation equal to 0, the equation becomes $0 = 4x^2 + 11x + 8$. The equation looks perfectly good so far, even though the quadratic doesn't factor. You have to resort to the quadratic formula. (You can find details on using the quadratic formula in Chapter 3 if you need a refresher.) Substituting the numbers from the quadratic equation into the formula, you get the following:

$$x = \frac{-11 \pm \sqrt{121 - 4(4)(8)}}{2(4)} = \frac{-11 \pm \sqrt{121 - 128}}{8} = \frac{-11 \pm \sqrt{-7}}{8}$$

Whoa! You can stop right there. You see that a negative value sits under the radical. The square root of -7 isn't real, so no real answer exists for x (for more on nonreal numbers, see Chapter 17). The nonexistent answer is your big clue that the system of equations doesn't have a common solution, meaning that the parabola and line never intersect (hey, even Sherlock Holmes had to dig around a bit before finding his clues). Figure 15-4 shows the graphs of the parabola and line. You can see why you found no solution. I wish I could give you an easy way to tell that a system has no solution before you go to all that work. Think of it this way: An answer of no solution is a perfectly good answer.

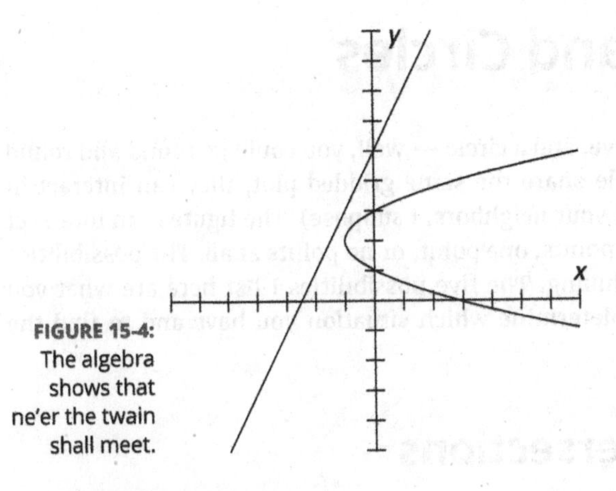

FIGURE 15-4: The algebra shows that ne'er the twain shall meet.

YOUR TURN

1 Find the intersection of the parabola $y = x^2 + 4x + 7$ and the line $y = 3x + 9$.

2 Find the intersection of the parabola $x = 2y^2 - y$ and the line $x + 3y = 12$.

3 Find the intersection of the parabola $y = x^2 - 4x$ and the line $2x + y + 1 = 0$.

4 Find the intersection of the parabola $y = -x^2 - x + 6$ and the line $y = -4x + 9$.

Intertwining Parabolas and Circles

The graph of a parabola is a U-shaped curve, and a circle — well, you could go round and round about a circle. When a parabola and circle share the same gridded plot, they can interact in several different ways (much like you and your neighbors, I suppose). The figures can intersect at four different points, three points, two points, one point, or no points at all. The possibilities may seem endless, but that's wishful thinking. The five possibilities I list here are what you have to work with. Your challenge is to determine which situation you have and to find the solutions of the system of equations.

Managing multiple intersections

A parabola and a circle can intersect at up to four different points, meaning that their system of equations can have up to four common solutions. To solve for the common solution or solutions, you apply pretty much the same technique as you would use for the intersections of lines and parabolas. The steps for creating a solution are given in the next example. These steps can be generalized for other intersection situations.

EXAMPLE

Q. Solve for the common solutions of the parabola $y=-x^2+6x+8$ and the circle $x^2+y^2-6x-8y=0$.

A. Use the following steps:

1. **Put the equations in a usable form.**

   The parabola is already in standard form. Leave the circle in its current form. You can put it in its standard form if you want to graph it.

2. **Do a substitution of one curve's equation into the other.**

   To solve for common points, replace each y in the equation of the circle with the equivalent of y in the parabola. Starting with the general form $x^2+y^2-6x-8y=0$, replace each y with $-x^2+6x+8$. This process requires squaring a trinomial, unfortunately.

$$\begin{aligned} x^2+(-x^2+6x+8)^2-6x-8(-x^2+6x+8)&=0\\ x^2+x^4-12x^3+20x^2+96x+64-6x+8x^2-48x-64&=0\\ x^4-12x^3+29x^2+42x&=0 \end{aligned}$$

   When squaring a trinomial, you may find it easier to distribute the terms instead of stacking the terms like in a multiplication problem. To find $(-x^2+6x+8)^2$, for example, think of the product $(-x^2+6x+8)(-x^2+6x+8)$. You multiply each term by $-x^2$, and then by $6x$, and lastly by 8. Finish by combining the like terms (see Chapter 9 for more on polynomials):

$$\begin{aligned} &-x^2(-x^2+6x+8)+6x(-x^2+6x+8)+8(-x^2+6x+8)\\ &=x^4-6x^3-8x^2-6x^3+36x^2+48x-8x^2+48x+64\\ &=x^4-12x^3+20x^2+96x+64 \end{aligned}$$

3. **Set the resulting terms from Step 2 equal to zero and solve for x.**

   The terms in the equation have a common factor of x. Factoring out the x, you get $x(x^3 - 12x^2 + 29x + 42) = 0$. The expression in the parentheses factors into the product of three binomials. You can do this factorization and find these binomials by using the Rational Root Theorem, which leads you to try the factors of 42 — 1, 6, and 7 — and synthetic division. (Chapter 9 has a full explanation of the Rational Root Theorem and factoring.) The final factorization of the equation is $x(x+1)(x-6)(x-7) = 0$. The solutions are $x = 0, -1, 6$, and 7.

4. **Substitute the solutions you find into the equation of the curve with the smaller exponents to find the coordinates of the points of intersection.**

   In this case, you substitute into the equation of the parabola. You find that when $x = 0, y = 8$; when $x = -1, y = 1$; when $x = 6, y = 8$; and when $x = 7, y = 1$. The points of intersection are, therefore, (0,8), (−1,1), (6,8), and (7,1). See Figure 15-5 for a graph of the parabola, circle, and points of intersection.

   If you want to graph one or more of the curves, you need to change them to their standard form. In this case, you need to change the circle to its standard form, $(x-h)^2 + (y-k)^2 = r^2$, by completing the square. The circle's equation, written in standard form, is $(x-3)^2 + (y-4)^2 = 25$.

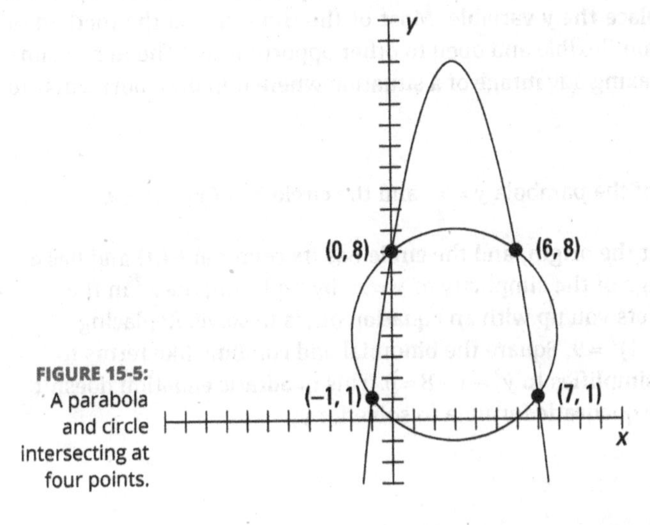

FIGURE 15-5: A parabola and circle intersecting at four points.

A parabola and a circle can also intersect at three points, two points, one point, or no points. Figure 15-6 shows what the three-point and two-point situations look like. In Figure 15-6a, the parabola's vertex is tangent to a point on the circle, and the parabola cuts the circle at two other points. In Figure 15-6b, the parabola cuts the circle at only two points.

You use the same substitution method to solve systems of equations with fewer than four intersections. The algebra leads you to the solutions, but beware the false promises. You have to watch out for extraneous solutions by checking your answers.

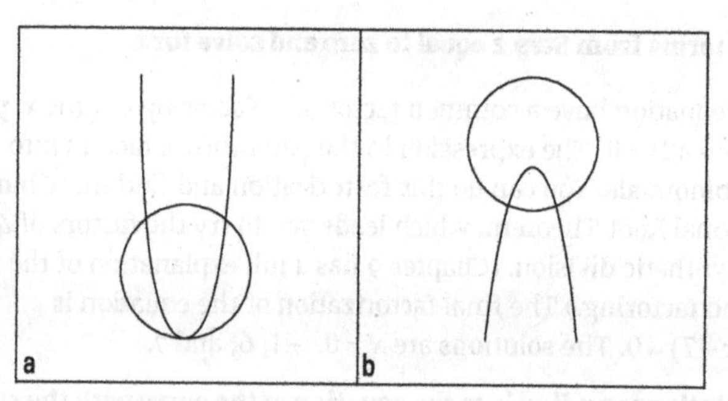

FIGURE 15-6: Parabolas and circles tangling, offering up different solutions.

After substituting one equation into another, take a look at the resulting equation. The highest power of the equation tells you what to expect for the number of common solutions. When the power is three or four, you can have as many as three or four solutions, respectively. When the power is two, you can have up to two common solutions. A power of one indicates only one possible solution. If you end up with an equation that has no solutions, you know the system has no points of intersection — the graphs just pass by like ships in the night.

Sorting out the solutions

In the section, "Crossing Parabolas with Lines," earlier in the chapter, the examples I provide use substitution where the *x*'s replace the *y* variable. Most of the time, this is the method of choice, but I suggest that you remain flexible and open to other opportunities. The next example is just such an opportunity — taking advantage of a situation where it makes more sense to replace the *x* term with the *y* term.

EXAMPLE

Q. Find the common solutions of the parabola $y = x^2$ and the circle $x^2 + (y-1)^2 = 9$.

A. The parabola has its vertex at the origin, and the circle has its center at (0,1) and has a radius of 3. You take advantage of the simplicity of $y = x^2$ by replacing the x^2 in the circle equation with *y*. That sets you up with an equation of *y*'s to solve. Replacing the x^2 term, you have $y + (y-1)^2 = 9$. Square the binomial and combine like terms to get $y + y^2 - 2y + 1 = 9$, which simplifies to $y^2 - y - 8 = 0$. This quadratic equation doesn't factor, so you have to use the quadratic formula to solve for *y*:

$$y = \frac{1 \pm \sqrt{1-4(1)(-8)}}{2(1)} = \frac{1 \pm \sqrt{33}}{2}$$

You find two different values for *y*, according to this solution. When you use the positive part of the ±, you find that *y* is close to 3.37. When you use the negative part, you find that *y* is about −2.37. Something doesn't seem right. What is it that's bothering you? It has to be the negative value for *y*. The common solutions of a system should work in both equations, and $y = -2.37$ doesn't work in $y = x^2$, because when you square *x*, you don't get a negative number. So, only the positive part of the solution, where $y \approx 3.37$, works. Substitute $\frac{1+\sqrt{33}}{2}$ into the equation $y = x^2$ to get *x*. If $y = x^2 = \frac{1+\sqrt{33}}{2}$, then, taking the square root of both sides, $x = \pm\sqrt{\frac{1+\sqrt{33}}{2}}$. The value of *x* comes out to about ±1.84. The parabola and circle have points of intersection at about (1.84,3.37) and about (−1.84,3.37). When $y = -2.37$, you get points that lie on the circle, but these points don't fall on the parabola. The algebra shows that and the picture agrees. See Figure 15-7.

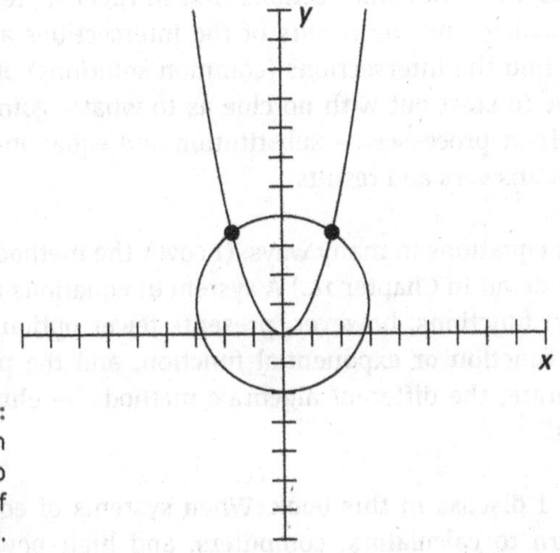

FIGURE 15-7: This system has only two points of intersection.

TIP

When substituting into one of the original equations to solve for the other variable, always substitute into the simpler equation — the one with smaller exponents. This helps you catch any extraneous solutions.

YOUR TURN

Find the points of intersection of the circles and parabolas.

5 $x^2 + y^2 = 25$ and $y = x^2 - 13$

6 $x^2 + y^2 = 100$ and $y^2 - 2x = 20$

7 $x^2 + y^2 - 2x + 4y - 4 = 0$ and $y = x^2 - 2x + 2$

Planning Your Attack on Other Systems of Equations

I deal with intersections of lines and different conic sections first in this chapter because the curves of the conics are easy to visualize and the results of the intersections are somewhat predictable. However, you can also find the intersections (common solutions) of other functions and curves; you may just have to start out with no clue as to what's going to happen. Not to worry, though; using the correct processes — substitution and equation-solving (see Chapter 14) — assures you of honest answers and results.

You can deal with a system of linear equations in many ways. (I cover the methods for solving systems of linear equations in great detail in Chapter 14.) A system of equations that contains one or more polynomial (nonlinear) functions, however, presents fewer options for finding the solutions. Throw in a rational function or exponential function, and the plot thickens. But as long as the equations cooperate, the different algebraic methods — elimination and substitution — will work. Lucky you!

Cooperative equations are the ones I discuss in this book. When systems of equations defy algebraic methods, you have to turn to calculators, computers, and high-powered college mathematics courses. In the meantime, you can concentrate on the nicely defined, manageable systems I present in this section for nonlinear entities.

Mixing polynomials and lines

A polynomial is a continuous, smooth curve. (Chapter 9 gives you plenty of information on the behavior of polynomial curves and how to graph them.) The more the curve of a polynomial changes direction and moves up and down across a graph, the more opportunities a line has to cross it. Substitution that replaces the *y* in the polynomial with the equivalent of *y* in the line is usually your most effective route for solving for the common solutions of the intersections. (You apply this method when the solutions are cooperative integers. If the common points involve fractions and radicals, technology is the only way to go.)

EXAMPLE

Q. Find the points of intersection of the line $y = 3x + 21$ and the polynomial $y = -x^3 + 5x^2 + 20x$.

A. Start by replacing the *y* in the polynomial with the equivalent of *y* in the line: $y = -x^3 + 5x^2 + 20x$ becomes $3x + 21 = -x^3 + 5x^2 + 20x$. Move the terms to the left to set the equation equal to 0: $x^3 - 5x^2 - 17x + 21 = 0$. This result factors into $(x+3)(x-1)(x-7) = 0$. (If you need help with this factorization, refer to the Rational Root Theorem and synthetic division information in Chapter 9.) The zeros, or solutions, of the equation are $x = -3, 1$, and 7. You solve for the *y*-values of the intersection points by plugging these *x*-values into the equation of the line. The points of intersection that you find are $(-3,12)$, $(1,24)$, and $(7,42)$. The line crosses the polynomial three times. Figure 15-8 shows you the graphs of the line, the polynomial, and the points of intersection.

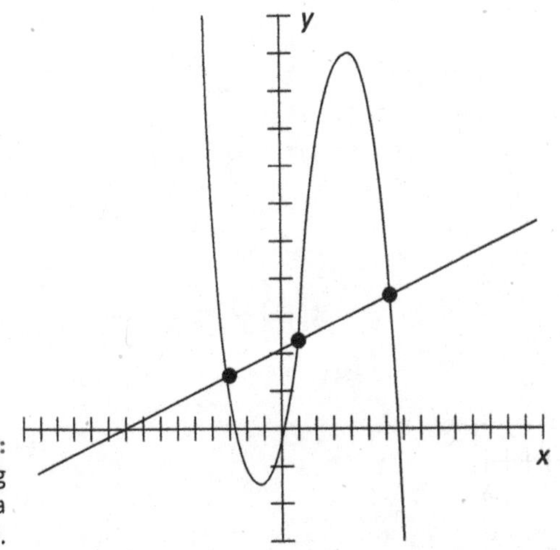

FIGURE 15-8: A line crossing the curves of a polynomial.

Always use the equation with the smaller exponents when you solve for the full solution. The substitution is easier with smaller exponents, and, more importantly, you won't end up with extraneous solutions — nonexistent answers for the problem (for more on imaginary numbers, see Chapter 17).

Crossing polynomials

"Crossing polynomials" almost sounds like you're doing a genetics experiment and creating a new, hybrid curve — a nonlinear monster of sorts. But before I start my sinister laugh, I must admit that crossing polynomials is bloomin' wonderful stuff. (Hey, beauty is in the eye of the beholder.) And, just to show you how intersecting polynomials can provide several solutions, I've chosen a quartic and a cubic to intersect in the following example. A fourth-degree (quartic) polynomial (the power 4 is the highest power) and a third-degree (cubic) polynomial can share as many as four common solutions!

EXAMPLE

Q. Find the points of intersection of $y = x^4 + 2x^3 - 13x^2 - 14x + 24$ and $y = x^3 + 8x^2 - 13x + 4$.

A. To solve systems of equations containing two polynomials, you use the substitution method (see Chapter 14). Set y equal to y, move all the terms to the left, and simplify: $x^4 + 2x^3 - 13x^2 - 14x + 24 = x^3 + 8x^2 - 13x + 4$ becomes $x^4 + x^3 - 21x^2 - x + 20 = 0$.

This equation factors into $(x+5)(x+1)(x-1)(x-4) = 0$ (see Chapter 9), which gives you the solutions $x = -5, -1, 1$, and 4. Substituting these values into the cubic (third-degree) equation (you should always substitute into the equation with the smallest exponential values), you get $y = 144$ when $x = -5$; $y = 24$ when $x = -1$; $y = 0$ when $x = 1$; and $y = 144$ when $x = 4$. You now have all the points of intersection. Figure 15-9 shows you the W-shaped curve (the fourth-degree polynomial) and the sideways S (the third-degree polynomial) and their points of intersection.

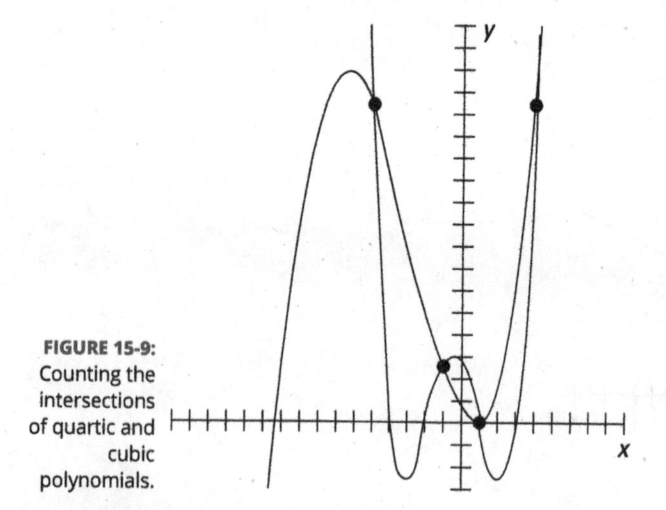

FIGURE 15-9: Counting the intersections of quartic and cubic polynomials.

YOUR TURN

Find the points of intersection of the functions.

8 $y = x^3 - x^2 - 6x$ and $y = 6x$

9 $y = 9x^2 - x^4$ and $y = 4x^2 + 4$

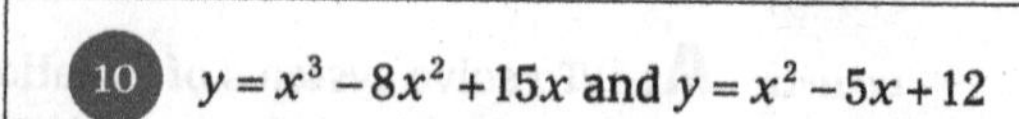

10 $y = x^3 - 8x^2 + 15x$ and $y = x^2 - 5x + 12$

Navigating Exponential Intersections

Exponential functions are flattened, C-shaped curves when graphed (I cover exponentials in Chapter 11). When exponentials intersect with one another, they usually do so in only one place, creating one common solution. Mixing exponential curves with other types of curves produces results similar to those you see when mixing lines and parabolas — you may get more than one solution.

You can solve some systems that contain exponential functions by using algebraic techniques. What should you look for? When the bases of the exponential functions are the same number or are powers of the same number, an algebraically found solution is possible. To solve a system of exponential functions when the bases of the functions are the same number or are powers of the same number, you set the two y-values equal to one another, set the exponents equal to one another, and solve for x. You change it from the exponential form by setting those exponents equal to one another and discarding the bases.

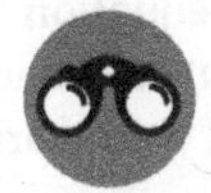

EXAMPLE

Q. Solve the system $y = 4^x$ and $y = 2^{x+1}$.

A. The base 4 in the first equation is a power of 2, the base in the second equation. You can write the number 4 as 2^2. Setting the two y-values equal to one another, you then change 4 to a power of 2 and simplify the terms: $4^x = 2^{x+1} \rightarrow (2^2)^x = 2^{x+1} \rightarrow 2^{2x} = 2^{x+1}$. Now you can set the exponents equal to one another. The solution of $2x = x + 1$ is $x = 1$. When $x = 1$, $y = 4$ in both equations.

Q. Solve the system $y = 9 \cdot 2^x - 3$ and $y = 2^{2x} + 5$.

A. Set the two y-values equal to one another to get $2^{2x} + 5 = 9 \cdot 2^x - 3$. Move the two terms to the right, and you have a quadratic-like equation. Just substitute z for 2^x: $2^{2x} + 5 = 9 \cdot 2^x - 3 \rightarrow 2^{2x} - 9 \cdot 2^x + 8 = 0 \rightarrow z^2 - 9z + 8 = 0$. Factor the quadratic and put the 2^x back in for z: $(z-8)(z-1) = 0 \rightarrow (2^x - 8)(2^x - 1) = 0$. Solving for the solutions, when $2^x - 8 = 0$, you can write $2^x = 8 \rightarrow 2^x = 2^3$, which gives you the solution $x = 3$. And when $2^x - 1 = 0$, you can write $2^x = 1 \rightarrow 2^x = 2^0$, which gives you the solution $x = 0$. To find the points of intersection, substitute the x-values back into the second equation. When $x = 3$, $y = 2^{2x} + 5 = 2^{2 \cdot 3} + 5 = 2^6 + 5 = 64 + 5 = 69$. This gives you the point (3,69). And then, when $x = 0$, $y = 2^{2x} + 5 = 2^{2 \cdot 0} + 5 = 2^0 + 5 = 1 + 5 = 6$. This gives you the point (0,6).

YOUR TURN

11 Solve the system $y = 3^{x+2}$ and $y = 27^x$.

12 Solve the system $y = 8^{x^2}$ and $y = 2^{4x-1}$.

Rounding Up Rational Functions

A rational function is a fraction that contains a polynomial expression in both its numerator and denominator. A polynomial has one or more terms that have whole-number exponents, so a rational function has all whole-number exponents — just in fractional form. The graph of a rational function typically has vertical and/or horizontal asymptotes that reveal its shape. Also, rational functions usually have pieces of curves that resemble hyperbolas in their graphs. (You can find plenty of information on rational functions in Chapter 10.)

Solving and graphing systems of equations that include rational functions means dealing with fractions — every student and teacher's favorite task. No worries, though. I prepare you in the following sections.

REMEMBER

A rational function and a line can intersect in, at most, two points. The same is true of two rational functions. You use the same technique that I used in the previous sections: set one equation equal to another. But be careful! When you change the form of an equation that contains fractions, radicals, or exponentials, you have to be cautious about extraneous solutions — answers that satisfy the new form but not the original. Always check your work by substituting your answers into the original equation.

EXAMPLE

Q. Find the points of intersection of $y=\frac{x-1}{x+2}$ and the line $3x+4y=7$.

A. Don't confuse the intersections of the line with the asymptotes of the rational function as parts of the solution. You consider only the intersections with the curves of the rational function. The algebraic solution you find also confirms that you use only the points on the curve.

To solve this system of equations, you solve the equation of the line for y, and then you substitute this equivalent for y into the equation of the rational function. First solving for y in the line, you get $4y=7-3x$ and then $y=\frac{7}{4}-\frac{3}{4}x$. Setting y equal to y, you get $\frac{7}{4}-\frac{3}{4}x=\frac{x-1}{x+2}$. The equation that remains looks like a bit of a mess, doesn't it? You can make the equation look much nicer by multiplying each side by $4(x+2)$, the common denominator of the fractions in the equation:

$$4(x+2)\left(\frac{7}{4}-\frac{3}{4}x\right)=\left(\frac{x-1}{x+2}\right)4(x+2)$$

$$\cancel{4}(x+2)\frac{7}{\cancel{4}}-\cancel{4}(x+2)\frac{3}{\cancel{4}}x=\left(\frac{x-1}{\cancel{x+2}}\right)4\cancel{(x+2)}$$

$$7(x+2)-3x(x+2)=4(x-1)$$

Now you can simplify the resulting equation by distributing and combining like terms: $7x+14-3x^2-6x=4x-4$ becomes $x+14-3x^2=4x-4$. Now set the whole equation equal to zero: $3x^2+3x-18=0$. The result is a quadratic equation that you can factor: $3\left(x^2+x-6\right)=3(x+3)(x-2)=0$.

The solutions of the quadratic equation are $x = -3$ and $x = 2$. You now substitute these values into the rational function to check your work. When $x = -3$, you get $y = 4$. When $x = 2$, you get $y = \frac{1}{4}$. These values represent the common solutions (coordinates of intersection) of the rational function and the line (you can see them in Figure 15-10).

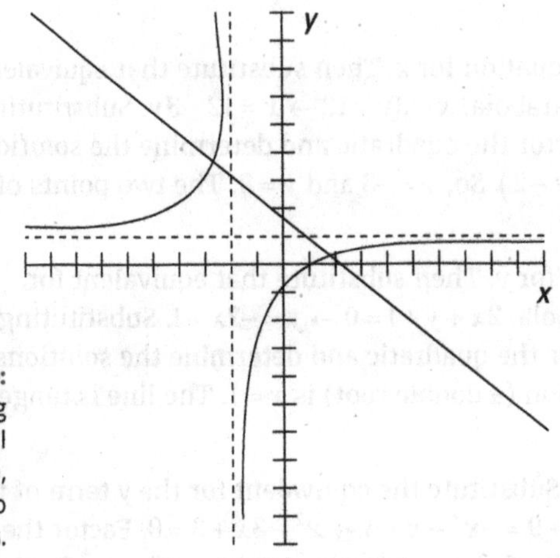

FIGURE 15-10: A line crossing a rational function, forming two solutions.

YOUR TURN

13 Find the points of intersection of $y = \frac{2}{x}$ and $y = x - 1$.

14 Find the points of intersection of $y = \frac{1}{x+6}$ and $y = \frac{1}{x^2}$.

Practice Questions Answers and Explanations

1. **(−2,3), (1,12).** Substitute the equivalent of y from the line's equation into the parabola and simplify the equation: $3x+9=x^2+4x+7 \rightarrow x^2+x-2=0$. Factor and solve for x: $x^2+x-2=0 \rightarrow (x+2)(x-1)=0$. The solutions are $x=-2$ and $x=1$. The two points of intersection are: (−2,3) and (1,12).

2. **(21,−3), (6,2).** First, solve the linear equation for x. Then substitute that equivalent for the x term in the equation of the parabola: $x+3y=12 \rightarrow x=12-3y$. Substituting: $12-3y=2y^2-y \rightarrow 0=2y^2+2y-12$. Factor the quadratic and determine the solutions: $2y^2+2y-12=2\left(y^2+y-6\right)=2(y+3)(y-2)$. So, $y=-3$ and $y=2$. The two points of intersection are: (21,−3) and (6,2).

3. **(1,−3).** First, solve the linear equation for y. Then substitute that equivalent for the y term in the equation of the parabola: $2x+y+1=0 \rightarrow y=-2x-1$. Substituting: $-2x-1=x^2-4x \rightarrow x^2-2x+1=0$. Factor the quadratic and determine the solutions: $x^2-2x+1=(x-1)^2=0$. The only solution (a double root) is $x=1$. The line is tangent to the parabola at the point (1,−3).

4. **No solution; no point of intersection.** Substitute the equivalent for the y term of the line into the equation of the parabola: $-4x+9=-x^2-x+6 \rightarrow x^2-3x+3=0$. Factor the quadratic and determine the solutions: $x^2-3x+3=0$ doesn't factor. Applying the quadratic formula, $x=\dfrac{-(-3)\pm\sqrt{(-3)^2-4(1)(3)}}{2(1)}=\dfrac{3\pm\sqrt{9-12}}{2}$. There is a negative number under the radical. There is no real solution, so the parabola and line do not intersect.

5. **(3,−4), (−3,−4), (4,3), and (−4,3).** Replace the y-term in the equation of the circle with the equivalent from the equation of the parabola. Simplify the equation: $x^2+y^2=25 \rightarrow x^2+\left(x^2-13\right)^2=25 \rightarrow x^2+x^4-26x^2+169=25 \rightarrow x^4-25x^2+144=0$. Factor the polynomial and solve for the roots: $x^4-25x^2+144=\left(x^2-9\right)\left(x^2-16\right)=(x-3)(x+3)(x-4)(x+4)=0$. The solutions are 3, −3, 4, −4. The points of intersection are (3,−4), (−3,−4), (4,3), and (−4,3).

6. **(−10,0), (8,6).** Solve for the y^2 term in the parabola, and then replace the y^2 term in the circle with that equivalent: $y^2-2x=20 \rightarrow y^2=2x+20$. Substituting: $x^2+y^2=100 \rightarrow x^2+2x+20=100 \rightarrow x^2+2x-80=0 \rightarrow (x+10)(x-8)=0$. Solving for x, you have $x=-10$ or $x=8$. The points of intersection are (−10,0) and (8,6).

7. **(1,1).** Replace the y-terms in the equation of the circle with the y-equivalent from the equation of the parabola.

$$x^2+\left(x^2-2x+2\right)^2-2x+4\left(x^2-2x+2\right)-4=0$$
$$x^2+x^4-4x^3+8x^2-8x+4-2x+4x^2-8x+8-4=0$$
$$x^4-4x^3+13x^2-18x+8=0$$

The possible solutions, using the Rational Root Theorem and synthetic division, are $\pm1, \pm2, \pm4, \pm8$. The number 1 works twice, and then the resulting quotient is an unfactorable quadratic. There is a double root at (1,1). The circle can be written $(x-1)^2+(y+2)^2=9$, showing you a circle with its center at (1,−2) and a radius of 3. The parabola can be written $y-1=(x-1)^2$, a parabola with its vertex at (1,1) and opening upward.

8 **(-3,-18), (0,0), (4,24).** Substitute the y equivalent of the line into the equation of the cubic and solve for x: $6x = x^3 - x^2 - 6x \rightarrow 0 = x^3 - x^2 - 12x = x\left(x^2 - x - 12\right) = x\left(x-4\right)\left(x+3\right) = 0$. You have $x = 0$, $x = 4$, $x = -3$. Substituting these values into the equation of the line, the points of intersection are (0,0), (4,24), and (−3,−18).

9 **(-2,20), (-1,8), (1,8), (2,20).** Substitute the y equivalent of the parabola into the equation of the quartic and solve for x: $4x^2 + 4 = 9x^2 - x^4 \rightarrow x^4 - 5x^2 + 4 = 0 \rightarrow \left(x^2 - 4\right)\left(x^2 - 1\right) = (x-2)(x+2)(x-1)(x+1) = 0$. The solutions are $x = 2$, $x = -2$, $x = 1$, $x = -1$. Substitute these solutions into the equation of the parabola, and you get the points of intersection (2,20), (−2,20), (1,8), and (−1,8).

10 **(1,8), (2,6), (6,18).** Substitute the y-equivalent of the parabola into the equation of the cubic and solve for x: $x^2 - 5x + 12 = x^3 - 8x^2 + 15x \rightarrow x^3 - 9x^2 + 20x - 12 = 0$. Using the Rational Root Theorem and synthetic division, you find the roots are $x = 1,\ 2,\ 6$. The points of intersection are: (1,8), (2,6), and (6,18).

11 **(1,27).** Write the 27 as a power of 3 and set the two expressions equal to one another: $3^{x+2} = \left(3^3\right)^x \rightarrow 3^{x+1} = 3^{3x}$. Now set the exponents equal to one another and solve for x: $x + 2 = 3x \rightarrow 2 = 2x \rightarrow 1 = x$. Substituting this value into $y = 3^{x+2}$, you have $y = 3^3 = 27$. The solution is the point (1,27).

12 $\left(\frac{1}{3}, \sqrt[3]{2}\right)$**, (1,8).** Write the 8 as a power of 2 and set the two expressions equal to one another: $\left(2^3\right)^{x^2} = 2^{4x-1} \rightarrow 2^{3x^2} = 2^{4x-1}$. Now set the exponents equal to one another and solve for x: $3x^2 = 4x - 1 \rightarrow 3x^2 - 4x + 1 = 0 \rightarrow (3x-1)(x-1) = 0$. The two solutions are $x = \frac{1}{3}$ and $x = 1$. Solving for the y-values in the points of intersection, use $y = 2^{1-4x}$. When $x = \frac{1}{3}$, $y = 2^{4(1/3)-1} = 2^{1/3}$, and when $x = 1$, $y = 2^{4(1)-1} = 2^3 = 8$.

13 **(2,1), (-1,-2).** Set the y-equivalents equal to one another. Then multiply through by x to get rid of the denominator: $\frac{2}{x} = x - 1 \rightarrow 2 = x^2 - x$. Rewrite and solve the quadratic equation: $x^2 - x - 2 = 0 \rightarrow (x-2)(x+1) = 0$. The two solutions are $x = 2$ and $x = -1$. Substituting into the linear function to solve for y, when $x = 2$, $y = 2 - 1 = 1$, giving you the point $(2,1)$. And when $x = -1$, $y = -1 - 1 = -2$, giving you the point $(-1,-2)$.

14 $\left(3, \frac{1}{9}\right)$, $\left(-2, \frac{1}{4}\right)$. Set the y-equivalents equal to one another. Then flip the fractions in the resulting proportion: $\frac{1}{x+6} = \frac{1}{x^2} \rightarrow x + 6 = x^2$. Rewrite and solve the quadratic equation: $x^2 - x - 6 = 0 \rightarrow (x-3)(x+2) = 0$. The two solutions are $x = 3$ and $x = -2$. Substituting into the first function to solve for y, when $x = 3$, $y = \frac{1}{3+6} = \frac{1}{9}$, giving you the point $\left(3, \frac{1}{9}\right)$. And when $x = -2$, $y = \frac{1}{-2+6} = \frac{1}{4}$, giving you the point $\left(-2, \frac{1}{4}\right)$.

Whaddya Know? Chapter 15 Quiz

Quiz time! Complete each problem to test your knowledge on the various topics covered in this chapter. You can then find the solutions and explanations in the next section.

1. **Find the points of intersection of the functions $y = x^3 - 21x + 20$ and $y = 2x^2 - 2x$.**
2. **Find the intersection(s) of the parabola and line $y = x^2 - 2x - 1$ and $y = -3x + 5$.**
3. **Find the intersection(s) of the cubic and parabola $y = x^3 - 3x^2 + 6x + 4$ and $y = -x^2 + 11x - 2$.**
4. **Find the intersections of the circle and the line $x^2 + y^2 = 25$ and $y = x + 1$.**
5. **Find the intersection(s) of the two exponential curves $y = 2^{x^2 - 15}$ and $y = 4^x$.**
6. **Find the intersection of the circle and parabola $x^2 + y^2 = 16$ and $y = \frac{1}{4}(x - 4)^2$.**
7. **Find the intersections of the exponential functions $y = e^{x^2 + 1}$ and $y = e^5$.**
8. **Intersection of the two rational functions $y = \frac{1}{x + 4}$ and $y = -\frac{1}{x}$.**
9. **Find the intersection(s) of the parabola and line $y = 2x^2 + 3x - 4$ and $3x + y + 4 = 0$.**
10. **Find the intersection(s) of the cubic and parabola $y = -x^3 + x^2 - 2x + 3$ and $y = 2x^2 - 4x + 3$.**
11. **Find the intersection of the rational function and line $y = -\frac{3}{x^2} + 2$ and $y = -x$.**

Answers to Chapter 15 Quiz

1. **(1,0), (5,40), (–4,40).** Set the two y-equivalents equal to one another and rewrite the equation to solve for x: $x^3-21x+20=2x^2-2x \rightarrow x^3-2x^2-19x+20=0$. Using the Rational Root Theorem and synthetic division, the cubic factors: $(x-1)(x-5)(x+4)=0$. This gives you $x=1$, $x=5$, and $x=-4$. Substituting those values back into the quadratic equation, you find the coordinates of the points of intersection: (1,0), (5,0), and (–4,0).

2. **(2,–1), (–3,14).** Set the two expressions that are equivalent to y equal to one another and solve for x: $x^2-2x-1=-3x+5 \rightarrow x^2+x-6=0 \rightarrow (x+3)(x-2)=0$. This gives you $x=-3$ or $x=2$. Substitute the values into the line equation to solve for y.

3. **(–2,–28), (1,8), (3,22).** Set the two expressions that are equivalent to y equal to one another and solve for x: $x^3-3x^2+6x+4=-x^2+11x-2 \rightarrow x^3-2x^2-5x+6=0$. The cubic factors into $(x+2)(x-1)(x-3)=0$, so $x=-2$ or $x=1$ or $x=3$. Substitute the x-values into the equation of the parabola to solve for x.

4. **(3,4), (–4,–3).** Substitute the equivalent for y from the first equation into the second equation. Then solve for x.

$$x^2+(x+1)^2=25 \rightarrow x^2+x^2+2x+1=25 \rightarrow 2x^2+2x-24=0 \rightarrow$$
$$2\left(x^2+x-12\right)=0 \rightarrow 2(x-3)(x+4)=0 \rightarrow x=3 \text{ or } x=-4$$

Substitute the x-values into the equation of the circle to solve for x.

5. **(5,1024), $\left(-3,\frac{1}{64}\right)$.** Rewrite the second equation as a power of 2: $y=4^x=\left(2^2\right)^x=2^{2x}$. Set the two y-equivalents equal to one another. Then, setting the exponents equal to one another, solve for x:

$$2^{x^2-15}=2^{2x} \rightarrow x^2-15=2x \rightarrow x^2-2x-15=0 \rightarrow (x-5)(x+3)=0 \rightarrow x=5 \text{ or } x=-3$$

Substitute the x values into the original second equation to solve for the corresponding y values: $x=5,\ y=4^5=1024 \rightarrow x=-3,\ y=4^{-3}=\frac{1}{64}$

6. **(0,4), (4,0).** Replace the y in the circle equation with the equivalent of y in the parabola. Simplify the new equation: $x^2+\left(\frac{1}{4}(x-4)^2\right)^2=16 \rightarrow x^2+\frac{1}{16}(x-4)^4=16 \rightarrow 16x^2+(x-4)^4=256$. Use Pascal's Triangle to expand the fourth power of the binomial. Then apply the Rational Root Theorem and synthetic division to factor the equation: $16x^2+x^4-16x^3+96x^2-256x+256=256 \rightarrow x^4-16x^3+112x^2-256x=0 \rightarrow$ $x\left(x^3-16x^2+112x-256\right)=0 \rightarrow x(x-4)\left(x^2-12x+64\right)=0$. The only two points of intersection occur when $x=0$ or $x=4$. Substituting into the equation of the parabola, the points of intersection are (0,4) and (4,0).

7. **$(2,e^5)$, $(-2,e^5)$.** Set the two y-equivalents equal to one another. Then, setting the exponents equal to one another, solve for x: $e^{x^2+1}=e^5 \rightarrow x^2+1=5 \rightarrow x^2=4 \rightarrow x=\pm 2$. Substitute the x values into the original first equation to solve for the corresponding y values.

(8) $\left(-2,\frac{1}{2}\right)$. Set the two y-equivalents equal to one another. Then flip the fractions using the rule for proportions. Finally, solve for x.

$\frac{1}{x+4}=-\frac{1}{x}\rightarrow x+4=-x\rightarrow 2x=-4\rightarrow x=-2$

(9) $(0,-4)$, $(-3,5)$. Solve for y in the second equation, and then set the two expressions that are equivalent to y equal to one another and solve for x: $2x^2+3x-4=-3x-4\rightarrow 2x^2+6x=0\rightarrow 2x(x+3)=0\rightarrow x=0$ or $x=-3$. Substitute the values into the line equation to solve for y.

(10) $(0,3)$, $(-2,19)$, $(1,1)$. Set the two expressions that are equivalent to y equal to one another and solve for x: $-x^3+x^2-2x+3=2x^2-4x+3\rightarrow -x^3-x^2+2x=0\rightarrow -x(x^2+x-2)=0\rightarrow -x(x+2)(x-1)=0$. Setting the factors equal to 0, $x=0$ or $x=-2$ or $x=1$. Substitute the x-values into the equation of the parabola to solve for x.

(11) $(1,-1)$. Set the y-equivalents equal to one another and solve for x: $-x=-\frac{3}{x^2}+2\rightarrow -x^3=-3+2x^2\rightarrow x^3+2x^2-3=0\rightarrow (x-1)(x^2+3x+3)=0$. The only real root occurs when $x=1$. Substituting into the first equation, $y=-1$.

IN THIS CHAPTER

» Determining where the solutions lie

» Starting with linear systems

» Adding in polynomials and conics

» Looking at some real-life applications

Chapter 16
Solving Systems of Inequalities

Systems of equations and systems of inequalities have several similarities. The biggest one is that you're looking for common solutions — what the statements have in common. But there's a huge difference in the number of solutions. With systems of equations, if there is a solution or solutions, you can have one or two or a few more points that the equations have in common. With systems of inequalities, the solutions can go into infinity!

One advantage of knowing how to work with systems of equations (you find these in Chapters 14 and 15) is that this is the starting point for systems of inequalities. You start with a graph of the lines and curves that help designate where the solutions are lying, and then you describe where the solutions fall with respect to those helpers.

Playing Fair with Inequalities

Systems of inequalities appear in applications used for business ventures and calculus problems. A system of inequalities, for instance, can represent a set of constraints in a problem that involves production of some item — the constraints put limits on the resources being used or the time available. In calculus, systems of inequalities represent areas between curves that you need to compute. Graphically, the solutions of systems of inequalities appear as shaded areas between curves. This gives you a visual solution and helps you determine values of x and y that work.

You find so many answers for systems of inequalities — infinitely many solutions — that you can't list them all; you just give rules in terms of the inequality statements. Algebraically, the solutions are statements that involve inequalities, telling what x or y is bigger than or smaller than. Often, the graph of a system gives you more information than the listing of the inequalities shown in an algebraic solution. You can see that all the points in the solution lie above a certain line, so you pick numbers that work in the system based on what you see.

Keeping It Linear with Inequalities

The simplest inequality to graph and solve is one that falls above or below a horizontal line or to the right or left of a vertical line. A system of inequalities that involves two such lines (one vertical and one horizontal) has a graph that appears as one-quarter of the plane.

EXAMPLE

Q. Graph the system of inequalities $\begin{cases} x \geq 2 \\ y \leq 3 \end{cases}$

A. Your answer is the graph of the two lines and the shading on the sides that contain solutions for each line. You're looking for where the shading overlaps. First, draw the vertical line $x = 2$. Everything to the right of the line $x = 2$ represents the graph of $x \geq 2$, so lightly shade in that entire side of the line. Then draw the horizontal line $y = 3$. Everything below this line represents the graph of $y \leq 3$. Lightly shade in that area. The intersection of the shading should be a heavier or darker area. This is in the lower-right quadrant formed by the intersecting lines. The solution consists of all the points in that shaded area and all the points on the lines that outline the area. You can find an infinite number of solutions. Some examples are the points (3,1), (4,2), and (2,−1). You could never list all the answers. See Figure 16-1 for a solution.

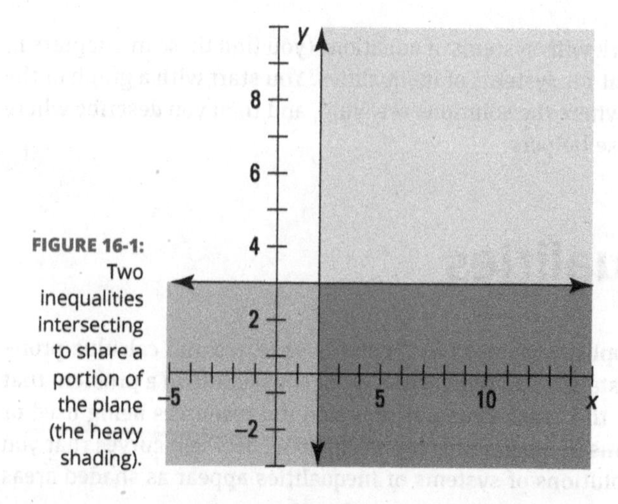

FIGURE 16-1: Two inequalities intersecting to share a portion of the plane (the heavy shading).

Getting a new slant on linear inequalities

Lines play an important role in systems of inequalities. Lines are often very descriptive of situations in business, education, agriculture, and so on. But they don't have to be horizontal or vertical. Take any line and use it for your purposes.

EXAMPLE

Q. Graph the solution of the system $\begin{cases} y \geq x+1 \\ y \geq 4 \end{cases}$

A. First, graph the line $y = x+1$. Then shade everything above it. If needed, you can try a test point to be sure you're on the correct side. The point (2,5) satisfies the inequality. Next, graph the horizontal line $y = 4$, and then shade everything above it. The solution consists of all the points in the darkened area and all the points on the line bordering that area. You see the graph in Figure 16-2a.

Q. Graph the solution of the system $\begin{cases} y \leq 3x+3 \\ y > 3-x \end{cases}$

A. First, graph the line $y = 3x+3$. Then shade everything below it. A test point would be (1,1); it satisfies the inequality. Next, graph the horizontal line $y = 3-x$, but make it a dotted or dashed line to indicate that the points on the line are not part of the solution. Then shade everything above it; use the test point (1,3). The solution consists of all the points in the darkened area and all the points on the upper line bordering that area, but not the points on the lower line. You see the graph in Figure 16-2b.

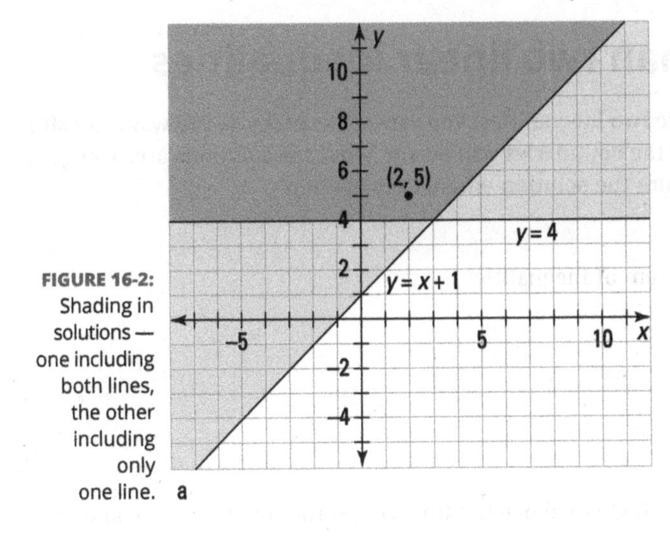

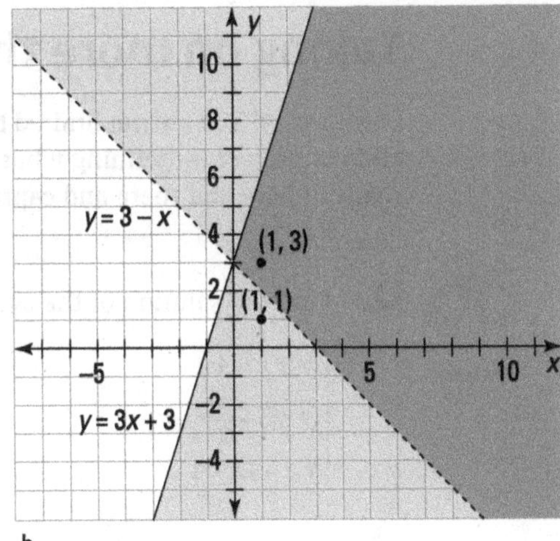

FIGURE 16-2: Shading in solutions — one including both lines, the other including only one line.

Graph the solutions of the systems of inequalities.

YOUR TURN

1 $\begin{cases} x > 0 \\ y \leq \frac{1}{2}x + 3 \end{cases}$

2 $\begin{cases} y \geq 1 - x \\ y \geq 2x - 6 \end{cases}$

3 $\begin{cases} y < 2x + 1 \\ x \leq 5 \end{cases}$

Taking on more than two linear inequalities

Don't worry! You are not limited to two inequalities. You can add as many as you want to a situation or project — defining what the boundaries will be and what the solutions are. You get to shade in here and there and capture the solution where they overlap.

EXAMPLE

Q. Find the solution of the system of inequalities.

$$\begin{cases} y \geq x - 4 \\ y \leq 5 - \frac{1}{2}x \\ y \geq -4x - 4 \end{cases}$$

A. Graph the line $y = x - 4$ and shade in above the line. Graph the line $y = 5 - \frac{1}{2}x$ and shade in below the line. Graph the line $y = -4x - 4$ and shade in above the line. You see the solution in Figure 16-3. It's all the points in the middle of the triangle formed by the lines and the points on the lines forming the border of that triangle. Notice that I've used arrows to show which side of the line is to be shaded. With more than two lines, the shading can get pretty "busy."

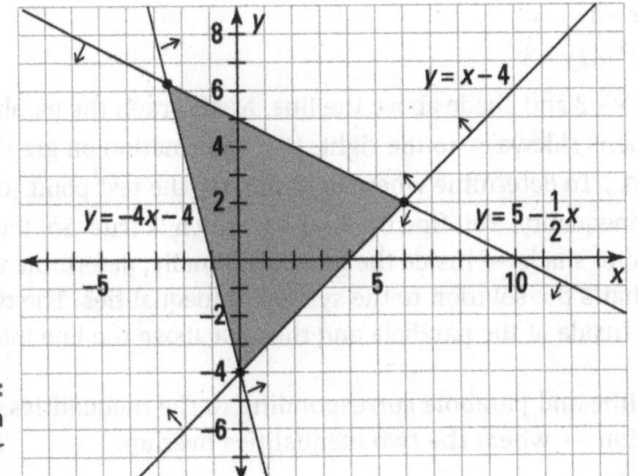

FIGURE 16-3: Creating a triangular solution.

YOUR TURN

4 **Find the solution of the system of inequalities.**

$$\begin{cases} y \le 6 \\ y \ge 6 - x \\ y \ge 3x - 18 \end{cases}$$

Find the solution of the system of inequalities.

$$\begin{cases} y \le -\frac{2}{5}x + 8 \\ y \ge 2x - 4 \\ y \ge -2x + 8 \end{cases}$$

Graphing areas with curves and lines

You find the solution of a system of inequalities involving a line and a curve (such as a parabola), two curves, or any other such combination by graphing the individual equations, determining which side to shade for each curve, and identifying where the equations share the shading.

EXAMPLE

Q. Solve the system $\begin{cases} y \geq x-3 \\ x \geq y^2+2y-3 \end{cases}$

A. First, graph the line $y = x-3$ and shade above the line. Next, graph the parabola $x \geq y^2+2y-3$, which opens sideways to the right. (For information on graphing parabolas, see Chapter 13.) To determine where to shade, try the test point (0,0) and see whether it satisfies the inequality. You find $0 \geq 0+0-3$, which is true. So, the point (0,0) falls in the area you need to shade — inside the parabola. Finally, determine where the two shaded areas overlap. That's the solution to the system of inequalities. The two shaded areas overlap where the inside of the parabola and the area above the line intersect.

Figure 16-4 shows the line and parabola corresponding to the inequalities. The shaded area indicates the solution — where the two inequalities overlap.

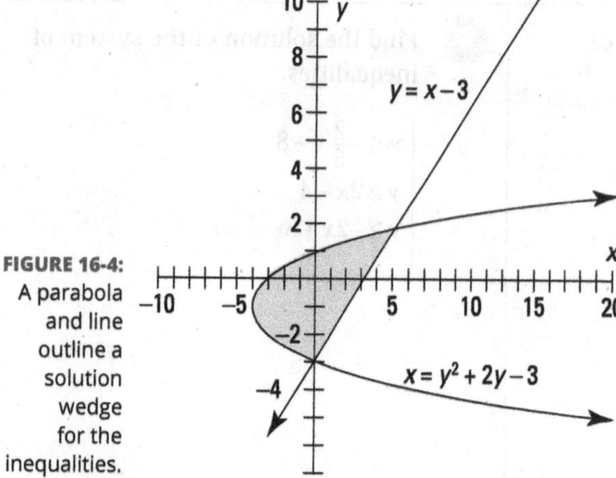

FIGURE 16-4: A parabola and line outline a solution wedge for the inequalities.

YOUR TURN

6 Solve the system of inequalities.

$$\begin{cases} y \geq 2x \\ y \leq -2(x-2)^2+8 \end{cases}$$

7 Solve the system of inequalities.

$$\begin{cases} (x-6)^2+(y-6)^2 \leq 25 \\ y \leq \frac{4}{3}x-2 \end{cases}$$

Applying the Systems to Real Life

Yes, there is a really good reason for producing these systems of equations and finding their solutions. In business and economics and even in medicine, this process is very valuable.

EXAMPLE

Q. You're in charge of your club's latest taco feast. The tickets are selling at $10 if bought in advance and $15 if bought at the door. You have room for a total of 50 people at the event, and you want to collect at least $600. How many of each ticket do you need to sell in order to reach your goal?

A. Let x represent the number of tickets sold in advance and y the number sold at the door. Then you can write the inequalities $10x + 15y \geq 600$ in terms of the money and $x + y \leq 50$. Both x and y have to be numbers greater than or equal to 0, so you also have $x \geq 0$ and $y \geq 0$, the two axes. Figure 16-5a shows you a graph with the inequality $x + y \leq 50$ shaded in. This represents the limits to the number of tickets that can be sold. In Figure 16-5b, you see the intersection of $x + y \leq 50$ and $10x + 15y \geq 600$. The coordinates of the points inside the solution tell you what the possibilities are for meeting the two constraints. Some possibilities are: (0,50), where there would be the greatest income because everyone is a walk-in, (5,45), (10,35), and (30,20). This gives you the goals for your ticket sales.

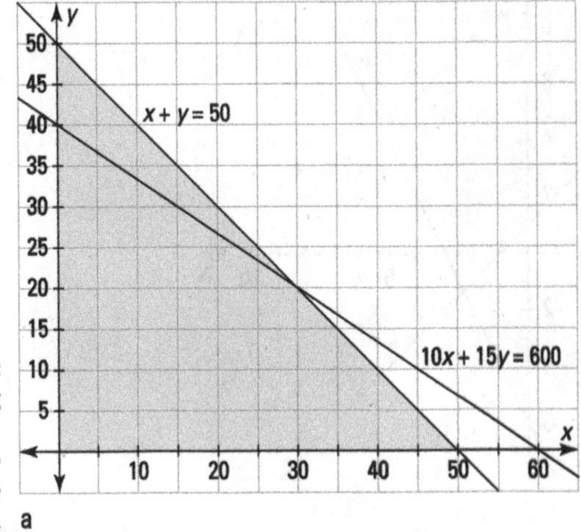

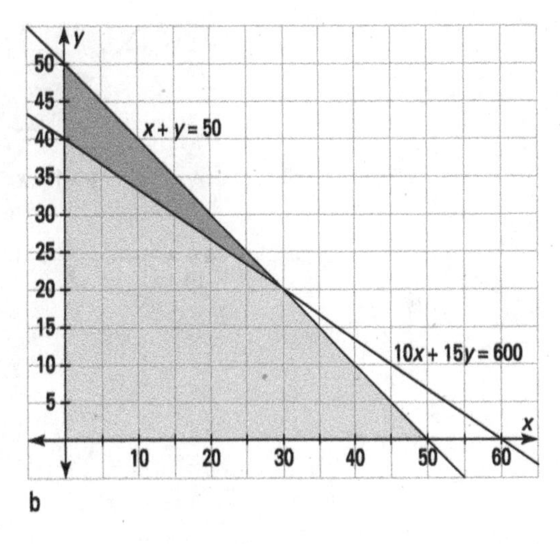

FIGURE 16-5: Selling enough tickets to make a profit.

YOUR TURN

8 You are making batches of chocolate cookies and macaroons. Each batch of chocolate cookies requires 4 cups of flour and 2 cups of sugar. The macaroons require 3 cups of flour and 3 cups of sugar. You have on hand 120 cups of flour and 90 cups of sugar. How many batches of each can you make if you don't need to use all of the available ingredients?

Practice Questions Answers and Explanations

1. **See the figure.** Graph the vertical line $x = 0$ using dots or dashes and shade to the right of the line. Graph the line $y = \frac{1}{2}x + 3$ and shade underneath it. The solution is to the lower right.

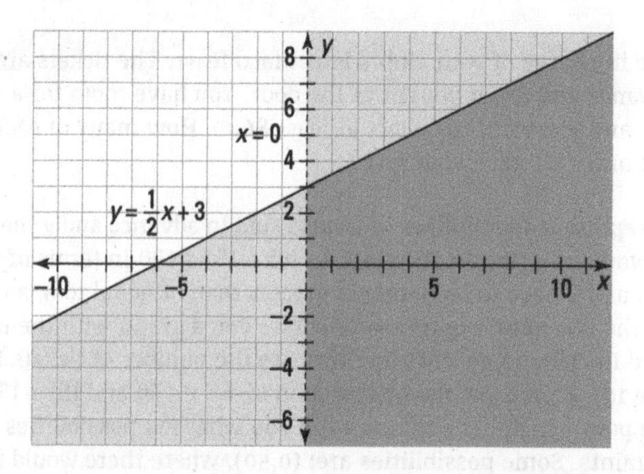

2. **See the figure.** Graph the line $y = 1 - x$ and shade above the line. Graph the line $y = 2x - 6$ and shade above the line. The solution is above and between the lines.

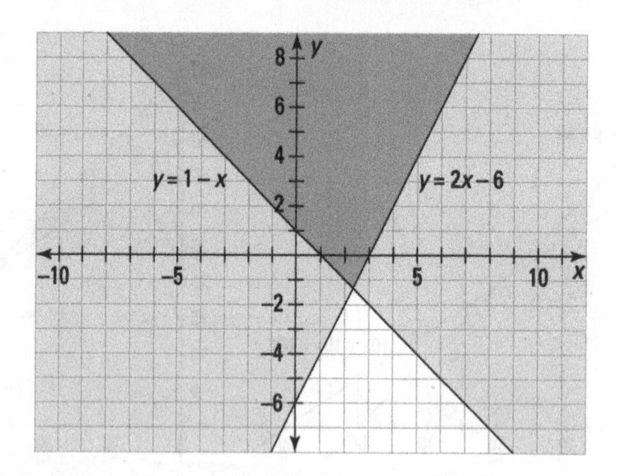

3) **See the figure.** Graph the line $y = 2x + 1$ using dots or dashes and shade below the line. Graph the vertical line $x = 5$ and shade to the left of the line. The solution comes down through the center of the graph.

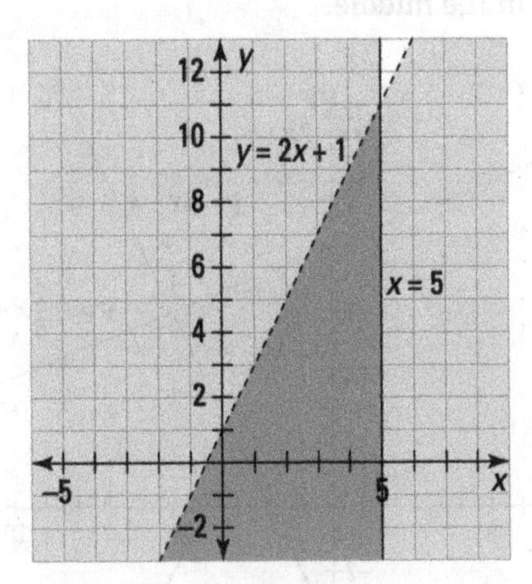

4) **See the figure.** Graph the horizontal line $y = 6$ and shade below the line. Graph the line $y = 6 - x$ and shade above the line. Graph the line $y = 3x - 18$ and shade above the line. The solution is the triangular shape in the middle.

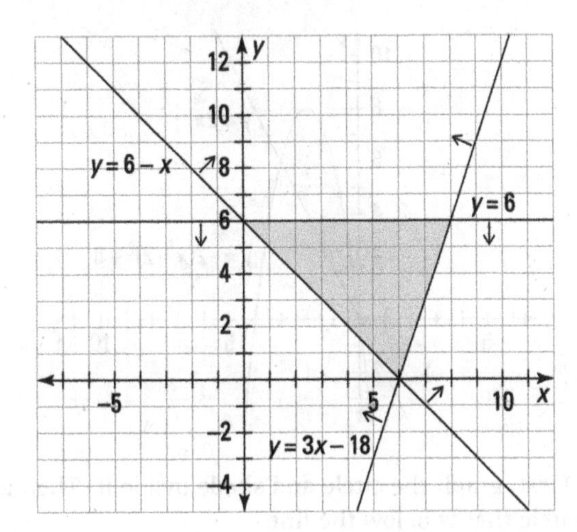

5. **See the figure.** Graph the line $y = -\frac{2}{5}x + 8$ and shade below the line. Graph the line $y = 2x - 4$ and shade above the line. Graph the line $y = -2x + 8$ and shade above the line. The solution is the triangular shape in the middle.

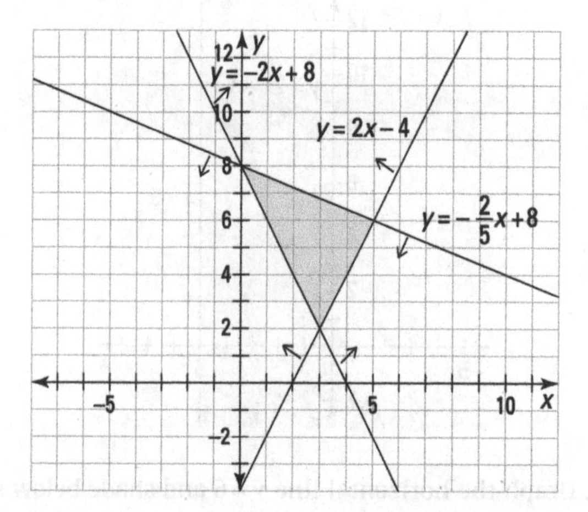

6. **See the figure.** Graph the parabola and shade inside down to the line.

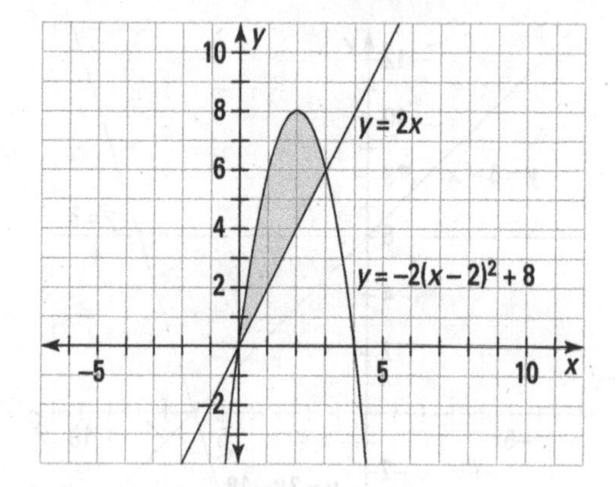

7. **See the figure.** First, graph the circle and shade inside it. Then graph the line and find the portion of the circle that is below the line.

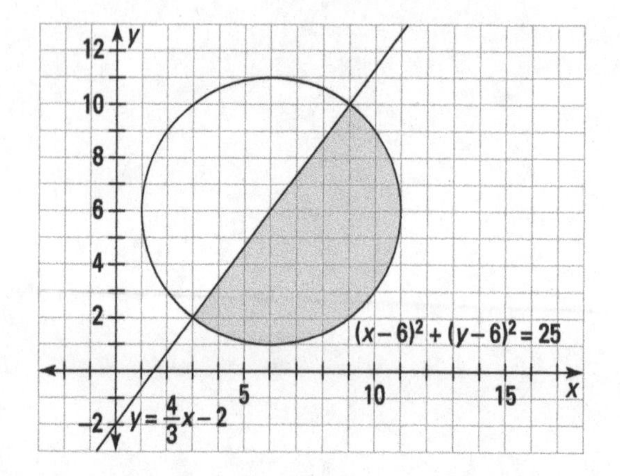

8. **See the figure.** Let x represent the batches of chocolate cookies and y the batches of macaroons. Writing the inequalities involved: the flour is $4x+3y\le 120$ and the sugar is $2x+3y\le 90$. Graphing the inequalities (and only using points in the first quadrant, because x and y both have to be 0 or greater), you have the intersection of the two inequalities. Looking at points in the intersection, you see many combinations where you don't use all the supplies, such as at (10,20) and (25,5). But, if you're going to want to use all of the ingredients on hand, go with (20,15): 20 batches of chocolate cookies and 15 batches of macaroons.

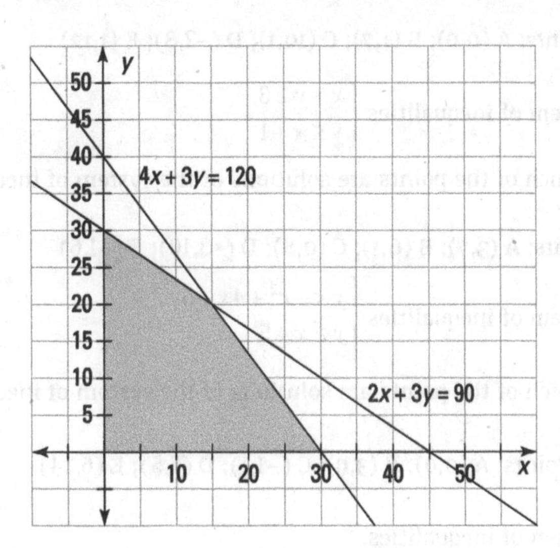

Whaddya Know? Chapter 16 Quiz

Quiz time! Complete each problem to test your knowledge on the various topics covered in this chapter. You can then find the solutions and explanations in the next section.

1. **Determine which of the points are solutions of the system of inequalities.**

   $\begin{cases} x+y \le 11 \\ y > x+1 \end{cases}$ **Points: A (0,0); B (1,7); C (10,1); D (−2,8); E (1,12)**

2. **Solve the system of inequalities** $\begin{cases} x+y \ge 3 \\ y \le x+1 \end{cases}$

3. **Determine which of the points are solutions of the system of inequalities.**

   $\begin{cases} y < 8 \\ x+y \ge 5 \end{cases}$ **Points: A (3,7); B (6,1); C (0,8); D (−3,10); E (−1,6)**

4. **Solve the system of inequalities** $\begin{cases} y < -x^2+4x+5 \\ y \ge x+5 \end{cases}$

5. **Determine which of the points are solutions of the system of inequalities.**

   $\begin{cases} y \le 4x \\ y \ge (x-3)^2 \end{cases}$ **Points: A (0,0); B (3,0); C (−1,1); D (5,5); E (6,24)**

6. **Solve the system of inequalities.**

   $$\begin{cases} y \le \frac{1}{9}x+8 \\ y \ge -2x+6 \\ x \le 9 \\ y \ge 0 \end{cases}$$

7. **Determine which of the points are solutions of the system of inequalities.**

   $\begin{cases} 2x+y \le 6 \\ 2x+y \ge 1 \\ x \ge 1 \end{cases}$ **Points: A (2,3); B (10,1); C (0,0); D (5,5); E (3,0)**

8. **Solve the inequality.**

   $$\begin{cases} y \le 11 \\ y > x-3 \end{cases}$$

9. **Solve the inequality.**

   $$\begin{cases} y \le x^2 \\ x^2+y^2 \le 9 \end{cases}$$

10. **Solve the inequality.**

    $$\begin{cases} x \ge 0 \\ y \ge 1 \\ x+y < 8 \end{cases}$$

Answers to Chapter 16 Quiz

1. **B, D.** Substitute the coordinates of the points into the inequalities. The points B (1,7) and D (–2,8) are part of the solution.

2. **See the figure.** Graph the line corresponding to the first inequality: $x + y = 3$. Using the test point, (0,0), $0 + 0 \not\geq 3$, so all the points above and on the line satisfy the inequality. See Figure a. Graph the line corresponding to the second inequality: $y = x + 1$. Using the test point, (0,0), $0 \leq 0 + 1$, so all the points below and on the line satisfy the inequality. See Figure b. The intersection of the two sets of points is to the right of the intersecting lines. See Figure c. Shade in the area corresponding to the set of all points satisfying both inequalities. All points in that area and on the lines bordering the area are solutions. See Figure d.

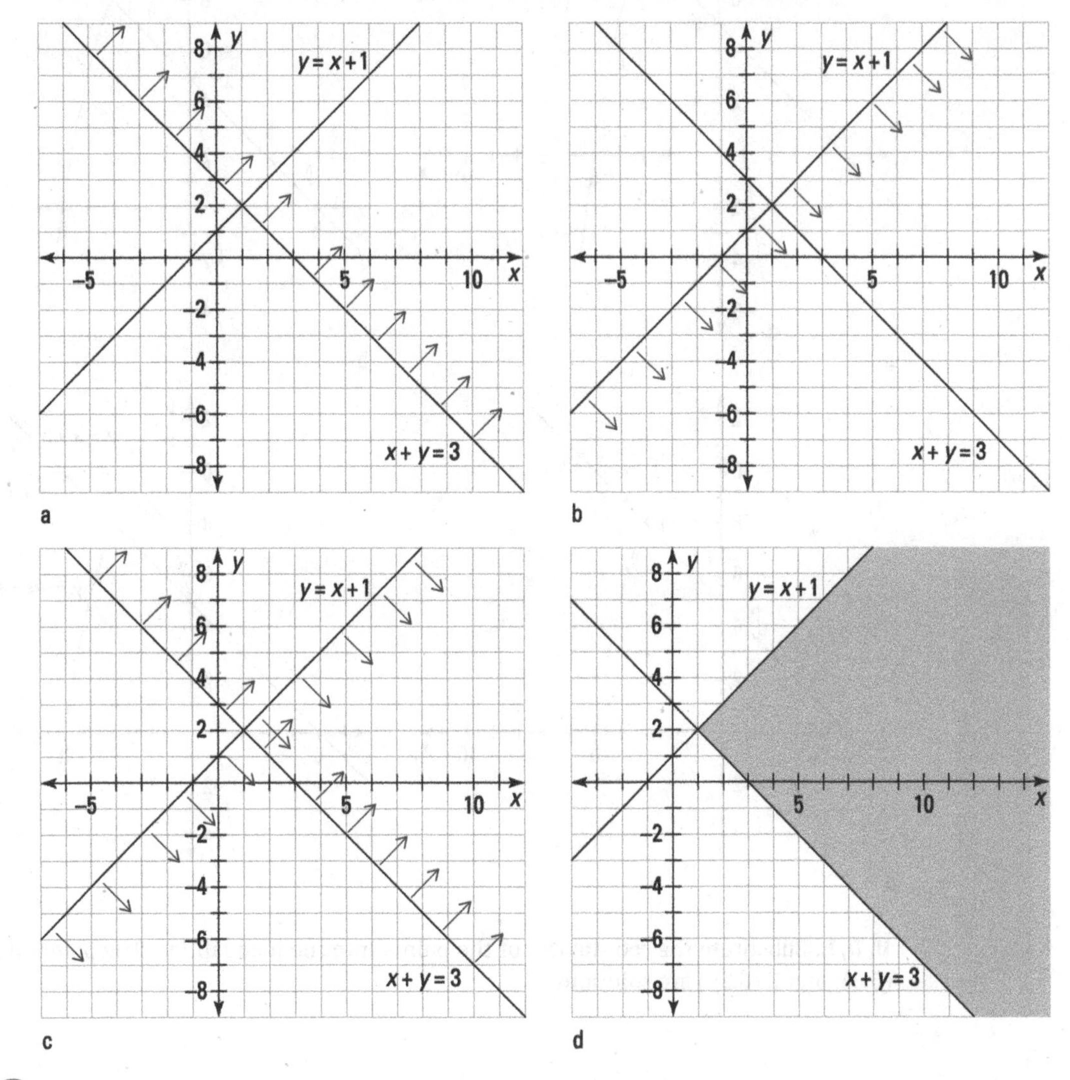

3. **A, B, E.** Substitute the coordinates of the points into the inequalities. The points A (3,7), B (6,1), and E (–1,6) are solutions.

4. **See the figure.** Graph the parabola corresponding to the first inequality: $y = -x^2 + 4x + 5$. Use a dashed curve, because the points on the parabola are not a part of the solution. Using the test point, (0,0), $0 < -0^2 + 4 \cdot 0 + 5$, so all the points below the parabola satisfy the inequality. See Figure a. Graph the line corresponding to the second inequality: $y = x + 5$. Using the test point, (0,0), $0 \geq 0 + 5$, so all the points below and on the line satisfy the inequality. See Figure b. The intersection of the two sets of points is below the parabola and above the line. See Figure c. Shade in the area corresponding to the set of all points satisfying both inequalities. All points in that area and on the line are solutions. See Figure d.

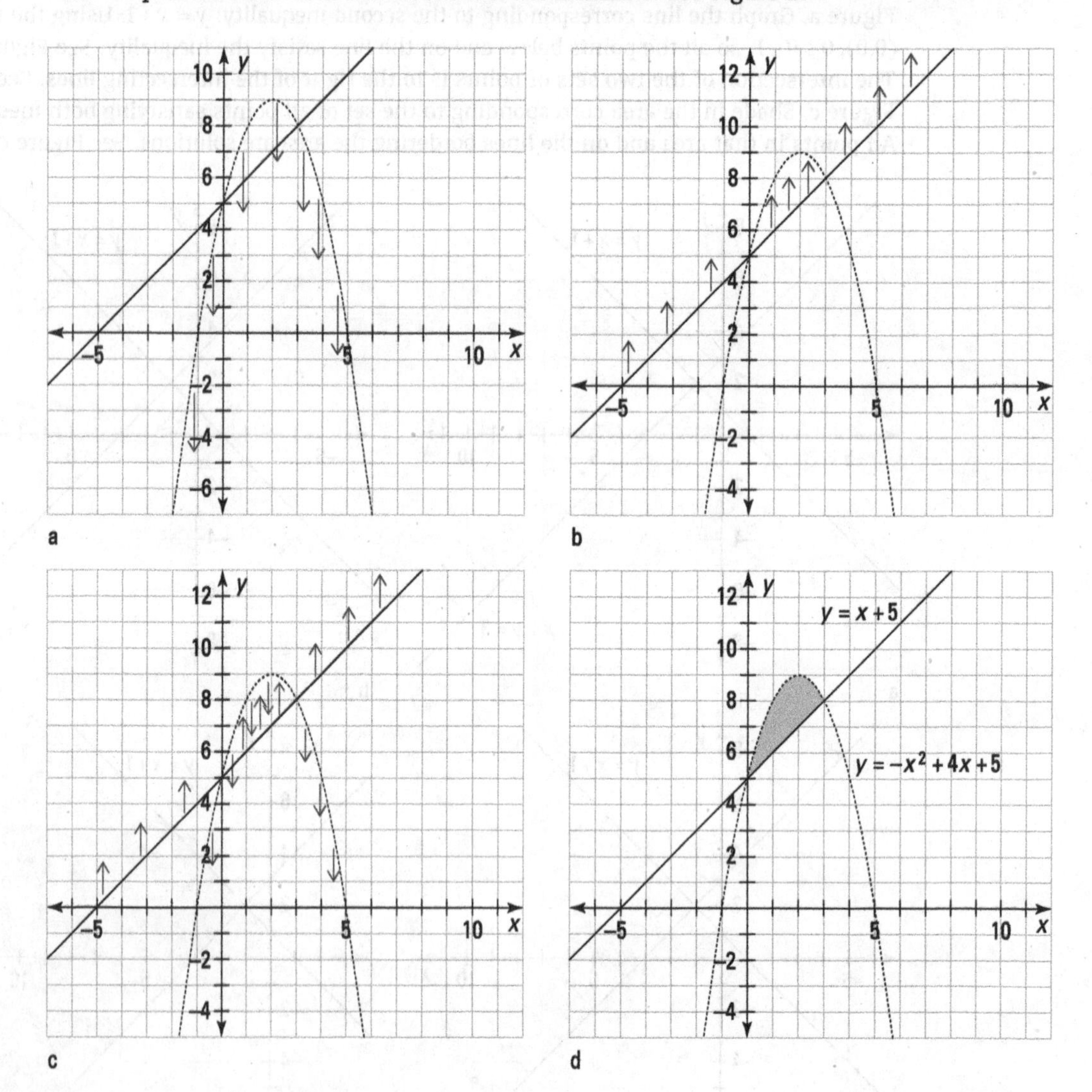

5. **B, D, E.** Substitute the coordinates of the points into the inequalities. The points B (3,0), D (5,5), and E (6,24) are solutions.

6. **See the figure.** Graph the line corresponding to the first inequality: $y = \frac{1}{9}x + 8$. Using the test point, (0,0), $0 \le \frac{1}{9} \cdot 0 + 8$, so all the points below and on the line satisfy the inequality. See Figure a. Graph the line corresponding to the second inequality: $y = -2x + 6$. Using the test point, (0,0), $0 \not\ge -2 \cdot 0 + 6$, so all the points to the right and on the line satisfy the inequality. See Figure b. Graph the lines corresponding to the third and fourth inequalities: $x = 9$ and $y = 0$. Using the test point, (0,0), $0 \le 9$ and $0 \ge 0$, so all the points to the left of and on the line $x = 9$ and all the points above and on the line $y = 0$ satisfy the inequalities. See Figure c. The intersection of the four sets of points is contained within the quadrilateral formed by the intersecting lines. Shade in the area corresponding to the set of all points satisfying the inequalities. All points in that area and on the lines are solutions. See Figure d.

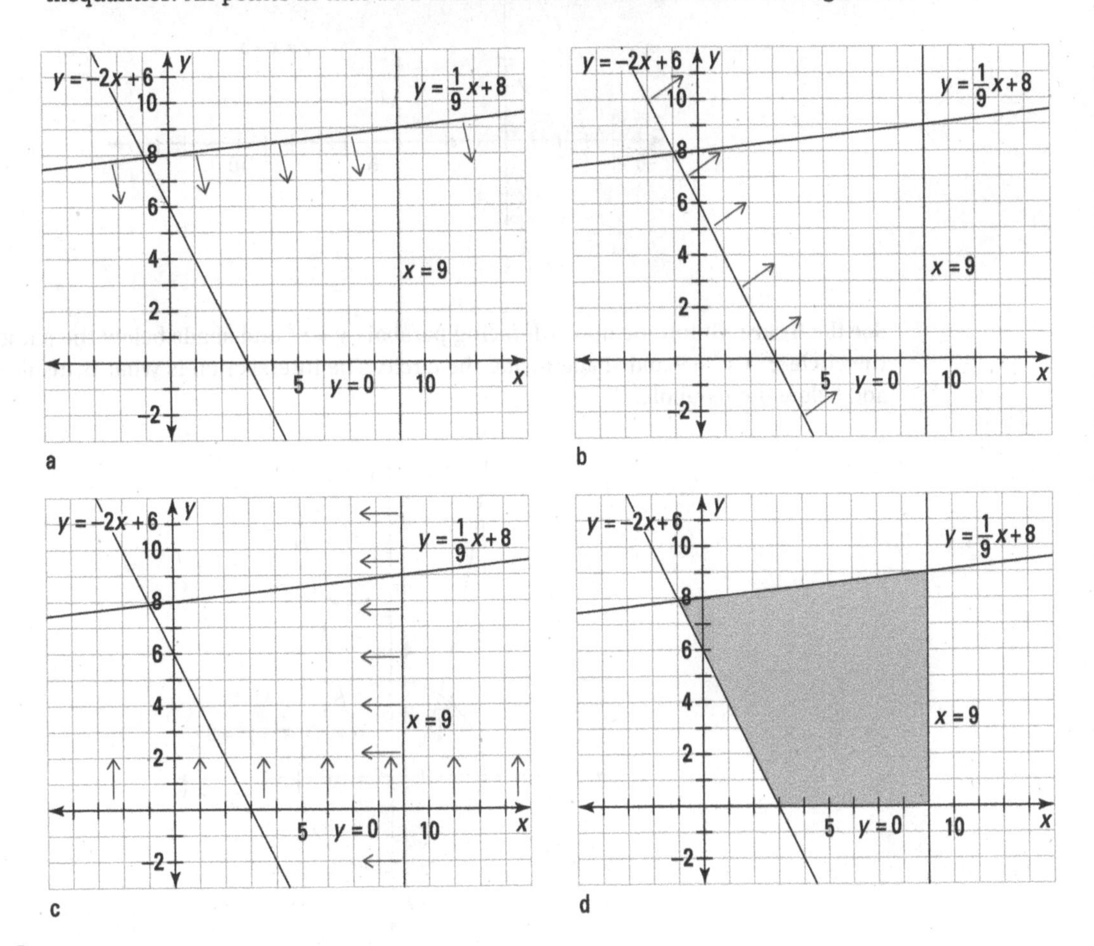

7. **E.** Substitute the coordinates of the points into the inequalities. Only the point E (3,0) satisfies all the inequalities.

8. **See the figure.** Graph the horizontal line $y = 11$ and shade in below it. Graph the line $y = x - 3$ using a dotted or dashed line and shade in above it. The intersection is to the left of the dashed line and below the horizontal line.

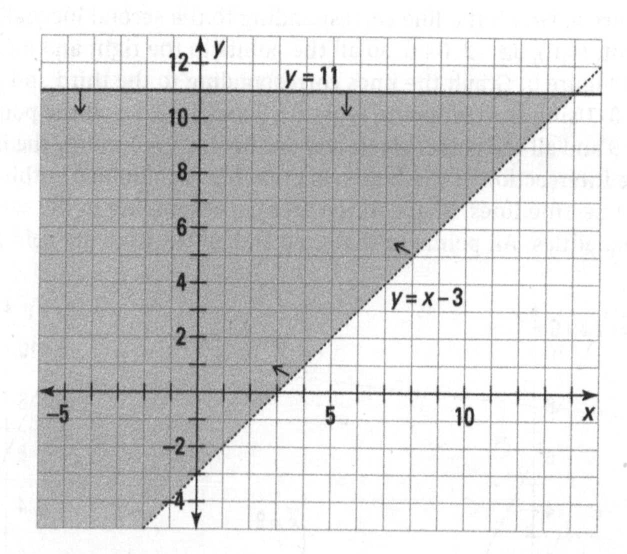

9. **See the figure.** Graph the upward-facing parabola $y = x^2$ and shade below the parabola. Graph the circle $x^2 + y^2 = 9$ and shade inside the circle. The intersection is what is inside the circle and below the parabola.

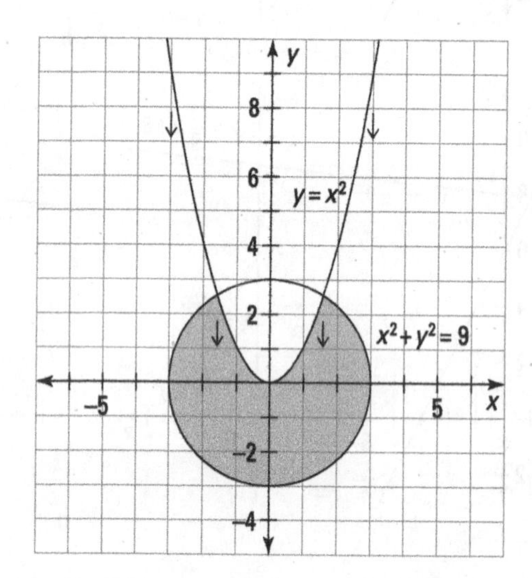

10 **See the figure.** Graph the vertical line $x = 0$, which is the y-axis, and shade to the right. Graph the horizontal line $y = 1$ and shade in above it. Graph the line $x + y = 8$ using a dotted or dashed line. Then, changing the inequality to read $y < 8 - x$, shade in below the line.

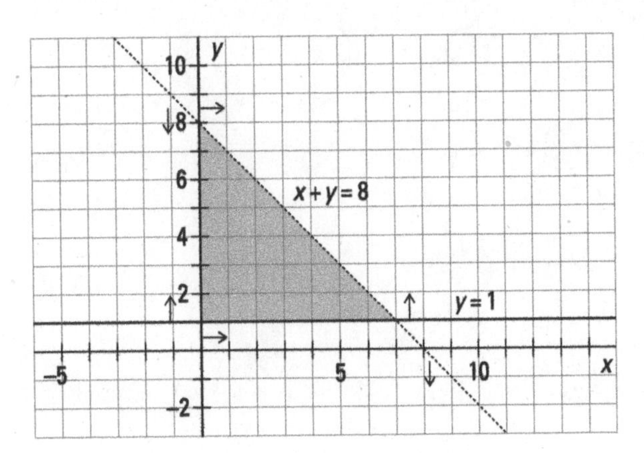

4

Making Lists and Checking for Imaginary Numbers

In This Unit . . .

IN THIS CHAPTER

» Sparking your math imagination with imaginary numbers

» Performing operations with complex numbers

» Dividing and conquering complex numbers and radicals

» Facing complex solutions in quadratic equations

» Identifying complex roots of polynomials

Chapter 17 Getting More Complex with Imaginary Numbers

Imaginary numbers are the results of mathematicians' imaginations. No, imaginary numbers aren't real — although is any number real? Can you touch a number (if you don't count playing with a toddler's educational toy set)? Can you feel it? Who decided a 9 should be shaped the way it is, and what makes that person right? Does your brain hurt yet?

Mathematicians define real numbers as all the whole numbers, negative and positive numbers, fractions and decimals, radicals — anything you can think of to use in counting, graphing, and comparing amounts. Mathematicians introduced imaginary numbers when they couldn't finish problems without them. For example, when solving for roots of quadratic equations or polynomials, they found that they needed to take the square root of a negative number. That can't be done! So the answer was always, "No solution."

The answers seem sort of final, like you have nowhere to go. You can't find the square root of a negative number, because no real number multiplies itself and ends up with a negative result. So, rather than staying stuck there, with no final solution, mathematicians came up with a new rule: They said, "Let $i^2 = -1$."

They made up a number to replace $\sqrt{-1}$, and they called it i (didn't take much imagination to come up with that). When you take the square root of each side of $i^2 = -1$, you get $\sqrt{i^2} = \sqrt{-1}$ or $i = \sqrt{-1}$. Using this equivalence, you can complete the problems and write the answers with i's in them.

In this chapter, you find out how to create, work with, and analyze imaginary numbers and the complex expressions they appear in. Just remember to use your imagination!

Simplifying Powers of *i*

The powers of x (representing real numbers) — x^2, x^3, x^4, and so on — follow the rules of exponents, such as adding exponents when multiplying the powers of x together or subtracting exponents when dividing them (see Chapter 1). The powers of i (representing imaginary numbers) are rule-followers, too. However, the powers of i also have some neat features that set them apart from other numbers.

You can write all the powers of i as one of four different numbers: $i, -i, 1$, and -1; all it takes is some simplifying of products, using the properties of exponents, to rewrite the powers of i:

- $i = i$: Just plain old i
- $i^2 = -1$: From the definition of an imaginary number (see the introduction to this chapter)
- $i^3 = -i$: Use the rule for exponents ($i^3 = i^2 \cdot i$) and then replace i^2 with -1; so, $i^3 = (-1) \cdot i = -i$
- $i^4 = 1$: Because $i^4 = i^2 \cdot i^2 = (-1)(-1) = 1$
- $i^5 = i$: Because $i^5 = i^4 \cdot i = (1) \cdot i = i$
- $i^6 = -1$: Because $i^6 = i^4 \cdot i^2 = (1)(-1) = -1$
- $i^7 = -i$: Because $i^7 = i^4 \cdot i^2 \cdot i = (1)(-1) \cdot i = -i$
- $i^8 = 1$: Because $i^8 = i^4 \cdot i^4 = (1)(1) = 1$

Consider the two powers of i presented in the following list, and how you determine the rewritten values; if you want to find a power of i, you make use of the rules for exponents and the first four powers of i.

- $i^{41} = i$: Because $i^{41} = i^{40} \cdot i = \left(i^4\right)^{10} \cdot i = (1)^{10} \cdot i = 1 \cdot i = i$.
- $i^{935} = -i$: Because $i^{935} = i^{932} \cdot i^3 = \left(i^4\right)^{233}\left(i^3\right) = (1)^{233}(-i) = 1(-i) = -i$.

The process of changing the powers of i seems like a lot of work — plus, you need to figure out what to multiply 4 by to get a high power (you want to find a multiple of 4, the biggest possible value that's smaller than the exponent). But you really don't need to go through all the raising of powers if you recognize a particular pattern in the powers of i.

Every power of i that's a multiple of 4 is equal to 1. If the power is one value greater than a multiple of 4, the power of i is equal to i. And so the process goes. Here's the list in full:

$$i^{4n} = 1$$
$$i^{4n+1} = i$$
$$i^{4n+2} = -1$$
$$i^{4n+3} = -i$$

EXAMPLE

Q. Simplify the power of $i^{5,001}$.

A. The number 5,001 is one more than 5,000, which is a multiple of 4. So $i^{5,001} = i^1 = i$.

Q. Simplify the power of $i^{85,000,000,000}$.

A. The number 85,000,000,000 is a multiple of 4. So $i^{85,000,000,000} = 1$.

YOUR TURN

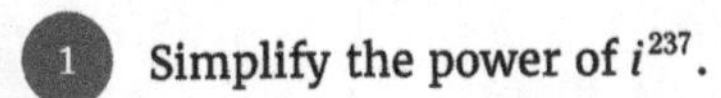

1 Simplify the power of i^{237}.

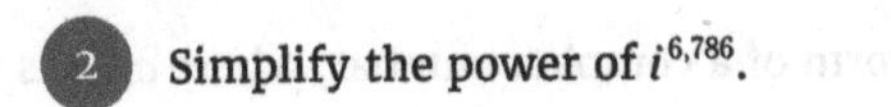

2 Simplify the power of $i^{6,786}$.

Understanding the Complexity of Complex Numbers

An imaginary number, i, is a part of the numbers called complex numbers, which arose after mathematicians established imaginary numbers. The standard form of complex numbers is $a + bi$, where a and b are real numbers, and i^2 is -1. The fact that i^2 is equal to -1 and i is equal to $\sqrt{-1}$ is the foundation of the complex numbers.

If you didn't have the imaginary numbers with the i's, you'd have no need for complex numbers with imaginaries in them. For instance, some complex numbers include $3 + 2i$, $-6 + 4.45i$, and $7i$. In the last number, $7i$, the value of a is zero, so the number is *pure imaginary*. If the value of b is zero as well, you have a complex number that is a *real* number.

So a is the real part of a complex number and the bi is the imaginary part (even though b is a real number). Is that complex enough for you?

Complex numbers have many applications, and mathematicians study them extensively. In fact, entire math courses and fields of study are devoted to complex numbers. And imagine this, you get a glimpse of this ethereal world right here and now in this section.

Operating on complex numbers

REMEMBER

You can add, subtract, multiply, and divide complex numbers — in a very careful manner. The rules used to perform operations on complex numbers look very much like the rules used for any algebraic expression, with two big exceptions:

- You simplify the powers of *i*, change them to their equivalent in the first four powers of *i* (see the section, "Simplifying Powers of *i*," earlier in the chapter), and then combine like terms.
- You don't really divide complex numbers; you multiply by the conjugate (I tell you all about this in the section, "Multiplying by the conjugate to perform division," later in the chapter).

Adding complex numbers

When you add two complex numbers $a+bi$ and $c+di$, you find the sum of the real parts and the sum of the imaginary parts:

$$(a+bi)+(c+di)=(a+c)+(b+d)i$$

The result of the addition is now in the form of a complex number, where $a+c$ is the real part and $(b+d)i$ is the imaginary part.

Subtracting complex numbers

When you subtract the complex numbers $a+bi$ and $c+di$, you find the difference of the real parts and the difference of the imaginary parts:

$$(a+bi)-(c+di)=(a-c)+(b-d)i$$

The result of the subtraction is now in the form of a complex number, where $a-c$ is the real part and $(b-d)i$ is the imaginary part.

Multiplying complex numbers

To multiply complex numbers, you can't just multiply the real parts together and the complex parts together; you have to treat the numbers like binomials and distribute both of the terms of one complex number over the other. Another way to look at it is that you have to FOIL the terms (for the details on FOIL, see Chapter 1):

$$(a+bi)(c+di)=(ac-bd)+(ad+bc)i$$

The result of the multiplication shown here is in the form of a complex number, with $ac-bd$ as the real part and $(ad+bc)i$ as the imaginary part. You can see from the following distribution where the values in this rule come from. First, FOILing the binomials, you have $(a+bi)(c+di)=ac+adi+bci+bdi^2$. You combine the two middle terms by factoring out the $i: ac+(ad+bc)i+bdi^2$. Replace the i^2 with $-1: ac+(ad+bc)i+bd(-1)$. And now the first and last terms are both real numbers, and you can write them together: $ac-bd+(ad+bc)i$.

EXAMPLE

Q. Add $(-4+5i)+(3+2i)$.

A. $-1+7i=(-4+5i)+(3+2i)=(-4+3)+(5+2)i$

Q. Subtract $(-4+5i)-(3+2i)$.

A. $-7+3i=(-4+5i)-(3+2i)=(-4-3)+(5-2)i$

Q. Find the product of $(-4+5i)(3+2i)$.

A. FOIL to get $(-4+5i)(3+2i)=-12-8i+15i+10i^2$. You simplify the last term to -10 and combine it with the first term; the two middle terms also combine. Your result is $-22+7i$, a complex number.

Q. Simplify: $i(4-3i)$.

A. Distribute the i over the terms in the parentheses: $4i-3i^2=4i-3(-1)=3+4i$.

YOUR TURN

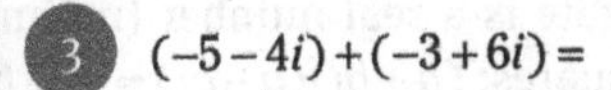

3 $(-5-4i)+(-3+6i)=$

4 $(3-2i)-(-3-9i)=$

5 $(6-i)(-3+4i)=$

Multiplying by the conjugate to perform division

The complex thing about dividing complex numbers is that you don't really divide. Do you remember when you first found out how to multiply and divide fractions? You never really divided fractions; you changed the second fraction to its reciprocal, and then you changed the problem to multiplication. You found that the answer to the multiplication problem you created was the same as the answer to the original division problem. You avoid division in much the same way with complex numbers. You do a multiplication problem — one that has the same answer as the division problem. But before you tackle the "division," you should know more about the conjugate of a complex number.

Defining the conjugate

A complex number and its conjugate have opposite signs between the two terms. The conjugate of the complex number $a+bi$ is $a-bi$, for instance. The conjugate of $-3+2i$ is $-3-2i$, and the conjugate of $5-3i$ is $5+3i$. Seems simple enough, because you don't see the special trait attributed to the conjugate of a complex number until you multiply the complex number and its conjugate together.

The product of an imaginary number and its conjugate is a real number (no imaginary part) and it takes the following form of the sum of two squares: $(a+bi)(a-bi)=a^2+b^2$.

And here's the product of the complex number and its conjugate, all worked out by using FOIL (see Chapter 1): $(a+bi)(a-bi)=a^2-abi+abi-b^2i^2=a^2-b^2(-1)=a^2+b^2$. The middle terms are opposites of one another, and $i^2=-1$ (the definition of an imaginary number) gets rid of the i^2 factor.

Using conjugates to divide complex numbers

When a problem calls for you to divide one complex number by another, you write the problem as a fraction and then multiply by the number 1. You don't actually multiply by 1; you multiply by a fraction that has the conjugate of the denominator in both numerator and denominator (because the same value appears in the numerator and denominator, the fraction is equal to 1):

$$(a+bi)\div(c+di)=\frac{a+bi}{c+di}\cdot\frac{c-di}{c-di}=\frac{(ac+bd)+(bc-ad)i}{c^2+d^2}$$

To write the result of the division of complex numbers in a strictly real-and-imaginary-parts format, you break up the fraction:

$$\frac{(ac+bd)}{c^2+d^2}+\frac{(bc-ad)}{c^2+d^2}i$$

The form of the fraction that results looks awfully complicated, and you don't really want to have to memorize it. When doing a complex division problem, you can use the same process that I use to get the previous form for dividing complex numbers.

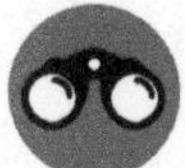

Q. Divide $(-4+5i)$ by $(3+2i)$.

EXAMPLE

A. **Write the problem as a fraction. Then multiply the numerator and denominator by the conjugate of the denominator. Simplify the result.**

$$\frac{-4+5i}{3+2i}\cdot\frac{3-2i}{3-2i}=\frac{-12+8i+15i-10i^2}{3^2+2^2}=\frac{-12+(8+15)i-10(-1)}{9+4}$$

$$=\frac{-12+10+(8+15)i}{13}=\frac{-2+23i}{13}=-\frac{2}{13}+\frac{23}{13}i$$

Q. $\frac{4}{3-i}=$

A. **Multiply the numerator and denominator by the conjugate of the denominator and simplify.**

$$\frac{4}{3-i}=\frac{4}{3-i}\cdot\frac{3+i}{3+i}=\frac{12+4i}{9-i^2}=\frac{12+4i}{9-(-1)}=\frac{12+4i}{10}=\frac{6}{5}+\frac{2}{5}i$$

YOUR TURN

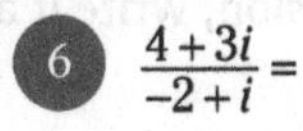

6 $\frac{4+3i}{-2+i}=$

7 $\frac{8-3i}{10+4i}=$

Simplifying radicals

Until mathematicians defined imaginary numbers, many problems had no answers because the answers involved square roots of negative numbers, or radicals. After the definition of an imaginary number, $i^2=-1$, came into being, doors opened; windows were flung wide; parades were held; children danced in the streets; and problems were solved. Eureka!

To simplify the square root of a negative number, you write the square root as the product of square roots and simplify: $\sqrt{-a}=\sqrt{-1}\sqrt{a}=i\sqrt{a}$.

EXAMPLE

Q. Simplify $\sqrt{-24}$.

A. You first split up the radical into the square root of –1 and the square root of the rest of the number, and then you do any simplifying by factoring out perfect squares:

$$\sqrt{-24} = \sqrt{-1}\sqrt{24} = \sqrt{-1}\sqrt{4}\sqrt{6} = i \cdot 2\sqrt{6}$$

By convention, you write the previous solution as $2i\sqrt{6}$.

Q. Simplify $\sqrt{-144}$.

A. Write the term as the product of two radicals — one with the –1 inside. Then simplify: $\sqrt{-144} = \sqrt{144}\sqrt{-1} = 12i$.

When writing complex numbers, most mathematicians just want to put the numerical part of the coefficient first, the variables or other letters next (in alphabetical order), and the radicals last. Technically, in complex form, you write a number with the i at the end, after all the other numbers — even after the radical. That's the strict $a + bi$ form. If you write the number this way, be sure not to tuck the i under the radical — keep it clearly to the right. For instance, $\sqrt{6} \cdot i \neq \sqrt{6i}$. To avoid confusion, write it as $i\sqrt{6}$.

YOUR TURN

8 Simplify $\sqrt{-75}$.

9 Simplify $\sqrt{-500}$.

Solving Quadratic Equations with Complex Solutions

You can always solve quadratic equations with the quadratic formula. It may be easier to solve quadratic equations by factoring, but when you can't factor, the formula comes in handy. (See Chapter 3 if you need a refresher on the quadratic formula.)

Until mathematicians began recognizing imaginary numbers, however, they couldn't complete many results of the quadratic formula. Whenever a negative value appeared under a radical, the equation stumped the mathematicians. The modern world of imaginary numbers to the rescue!

EXAMPLE

Q. Solve the quadratic equation $2x^2 + x + 8 = 0$.

A. Apply the quadratic formula to get the following:

$$x = \frac{-1 \pm \sqrt{1^2 - 4(2)(8)}}{2(2)} = \frac{-1 \pm \sqrt{1-64}}{4} = \frac{-1 \pm \sqrt{-63}}{4} = \frac{-1 \pm \sqrt{-1}\sqrt{9}\sqrt{7}}{4} = \frac{1 \pm 3i\sqrt{7}}{4}$$

Q. Solve the quadratic equation $x^2 + 16 = 0$.

A. Applying the quadratic equation: $x = \frac{0 \pm \sqrt{0 - 4(1)(16)}}{2(1)} = \frac{\pm\sqrt{-64}}{2} = \frac{\pm 8i}{2} = \pm 4i$.
But an even easier route is to apply the square root rule and write:
$x^2 + 16 = 0 \rightarrow x^2 = -16 \rightarrow x = \pm\sqrt{-16} = \pm 4i$. Your choice.

YOUR TURN

10 Solve $x^2 - 3x + 12 = 0$.

11 Solve $3x^2 + 2x + 9 = 0$.

Along with the quadratic formula that produces a complex solution, it helps to look at the curve that corresponds to the equation on a graph (see Chapter 8).

REMEMBER

A parabola that opens upward or downward always has a y-intercept; these parabolas are functions and have domains that contain all real numbers. However, a parabola that opens upward or downward doesn't necessarily have any x-intercepts.

To solve a quadratic function for its x-intercepts, you set the equation equal to zero. When no real solution for the equation exists, you find no x-intercepts. You still have a y-intercept, because all quadratic functions cross the y-axis somewhere, but its graph may stay above or below the x-axis without crossing it.

Q. Sketch the graph of $2x^2 + x + 8 = 0$.

EXAMPLE

A. This equation corresponds to the parabola $2x^2 + x + 8 = y$. Substituting zero for the y variable allows you to solve for the x-intercepts of the parabola: $2x^2 + x + 8 = 0 \rightarrow x = \frac{-1 \pm \sqrt{1 - 4(2)(8)}}{2(2)} = \frac{-1 \pm \sqrt{-63}}{4} = \frac{-1 \pm 3i\sqrt{7}}{4}$. The fact that you find no real solutions for the equation tells you that the parabola has no x-intercepts. (Figure 17-1 shows you a graph of this situation.)

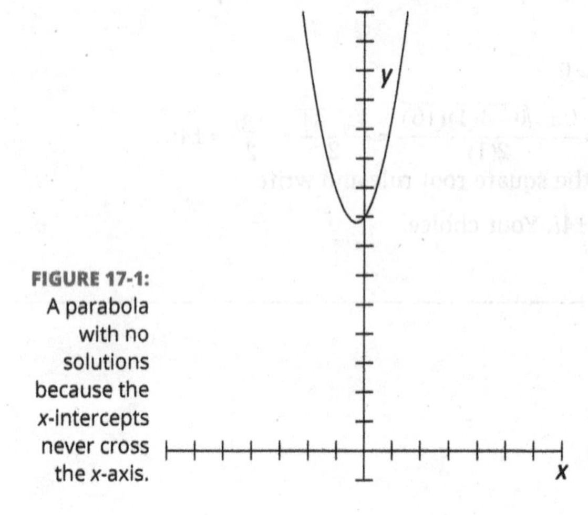

FIGURE 17-1: A parabola with no solutions because the x-intercepts never cross the x-axis.

Parabolas that open to the left or right aren't functions. A function has only one y-value for every x-value, so these curves violate that requirement. A parabola opening to the left or right may never cross the y-axis if its vertex is to the left or right of that axis, so you find no real solution for the y-intercept. A parabola that opens to the left or right always has an x-intercept, however.

To solve for any y-intercept, you set x equal to zero and then solve the quadratic equation. If the equation has no solution, it has no y-intercept(s).

Q. Sketch $x = -y^2 + 6y - 12$.

EXAMPLE

A. Solving the equation $x = -y^2 + 6y - 12$ for its y-intercepts, you let $x = 0$ and solve for y:

$$y = \frac{-6 \pm \sqrt{36 - 4(-1)(-12)}}{2(-1)} = \frac{-6 \pm \sqrt{36 - 48}}{-2} = \frac{-6 \pm \sqrt{-12}}{-2} = \frac{-6 \pm 2i\sqrt{3}}{-2} = 3 \pm i\sqrt{3}$$

The parabola $x = -y^2 + 6y - 12$ opens to the left and never crosses the y-axis. The graph of this parabola is shown in Figure 17-2.

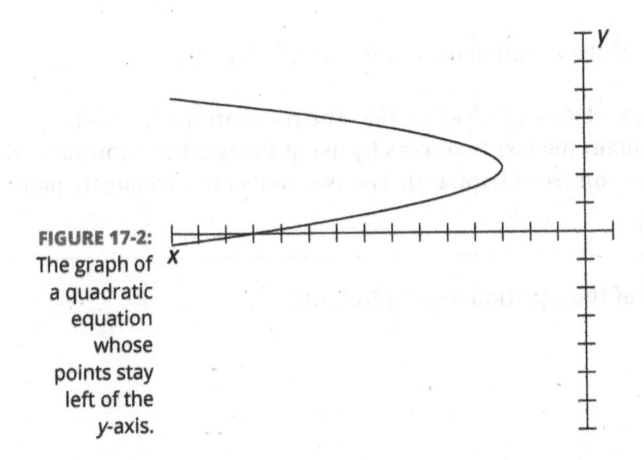

FIGURE 17-2: The graph of a quadratic equation whose points stay left of the y-axis.

Working Polynomials with Complex Solutions

Polynomials are functions whose graphs are nice, smooth curves that may or may not cross the x-axis. If the degree (or highest power) of a polynomial is an odd number, its graph must cross the x-axis, and it must have a real root or solution. (In Chapter 9, you discover how to find the real solutions or roots of polynomial equations.) When solving equations formed by setting polynomials equal to zero, you plan ahead as to how many solutions you can expect to find. The highest power tells you the maximum number of solutions you can find. In the case of polynomial equations where the highest powers are even numbers, you may find that the equations have absolutely no real solutions, whereas if the highest powers are odd, you're guaranteed at least one solution in each equation.

Identifying conjugate pairs

A polynomial of degree (or power) n can have as many as n real zeros (also known as solutions or x-intercepts). If the polynomial doesn't have n real zeros, it has $n-2$ zeros, $n-4$ zeros, or some number of zeros decreased two at a time. (See Chapter 9 for how to count the number of zeros by using Descartes' Rule of Signs.) The reason that the number of zeros decreases by two is that complex zeros always come in conjugate pairs — a complex number and its conjugate.

REMEMBER

Complex zeros, or solutions of polynomials, come in conjugate pairs: $a+bi$ and $a-bi$. The product of a complex number and its conjugate, $(a+bi)(a-bi)$, is a^2+b^2, so you don't see any i's in the equation of the polynomial.

EXAMPLE

Q. Determine any complex zeros of the polynomial $0 = x^5 - x^4 + 14x^3 - 16x^2 - 32x$.

A. Applying the Rational Root Theorem and Descartes' Rule of Signs, you factor or use synthetic division and have $0 = x(x-2)(x+1)(x^2+16)$. This equation has three real roots and two complex roots. The three real zeros are 0, 2, and –1 and the two complex zeros are $4i$ and $-4i$. You say that the two complex zeros are a conjugate pair, and you get the roots by solving the equation $x^2+16=0$.

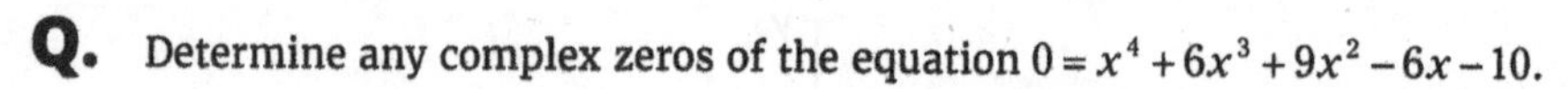

Q. Determine any complex zeros of the equation $0 = x^4 + 6x^3 + 9x^2 - 6x - 10$.

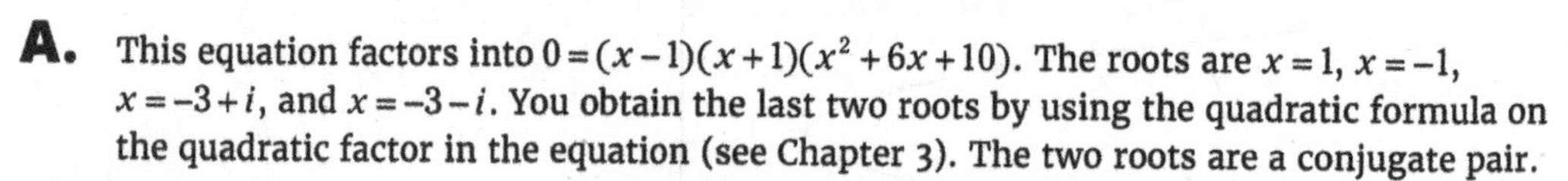

A. This equation factors into $0 = (x-1)(x+1)(x^2+6x+10)$. The roots are $x = 1$, $x = -1$, $x = -3 + i$, and $x = -3 - i$. You obtain the last two roots by using the quadratic formula on the quadratic factor in the equation (see Chapter 3). The two roots are a conjugate pair.

YOUR TURN

12 Determine any complex zeros of the equation $0 = x^4 + 5x^2 - 36$.

13 Determine any complex zeros of the equation $0 = x^3 + 64$.

Interpreting complex zeros

The polynomial function can have all real roots or all complex roots. It can also have some of each type. You count how many roots are possible using the Rational Root Theorem, and then you try to factor the polynomial. Finding the real roots helps you with the graphing of the polynomial. And you can also get some graphing help from the complex roots. At a complex root, you find a "flattening" of the curve. This is what often prevents the curve from crossing or touching an axis.

EXAMPLE

Q. Determine any flattening of $y = x^4 + 7x^3 + 9x^2 - 28x - 52$ by finding its complex roots.

A. According to Descartes' Rule of Signs (see Chapter 9), this function could have as many as four real roots. You can determine the number of complex roots in two different ways: by factoring the polynomial or by looking at the graph of the function. The function factors into $y = (x-2)(x+2)(x^2+7x+13)$. The first two factors give you real roots, or x-intercepts. When you set $x - 2$ equal to zero, you get the intercept (2, 0). When you set $x + 2$ equal to zero, you get the intercept $(-2, 0)$. Setting the last factor, $x^2 + 7x + 13$, equal to zero doesn't give you a real root, as you see here when you use the quadratic formula:

$$x = \frac{-7 \pm \sqrt{49 - 4(1)(13)}}{2(1)} = \frac{-7 \pm \sqrt{49 - 52}}{2} = \frac{-7 \pm i\sqrt{3}}{2}$$

You can also tell that a polynomial function has complex roots by looking at its graph. Figure 17-3 shows the graph of the function $y = x^4 + 7x^3 + 9x^2 - 28x - 52$. You can see the two x-intercepts, which represent the two real zeros. You also see the graph flattening on the left.

TIP

A flattening-out behavior can represent a change of direction, or a point of inflection, where the curvature of the graph changes. Areas like this on a graph indicate that complex zeros exist for the polynomial.

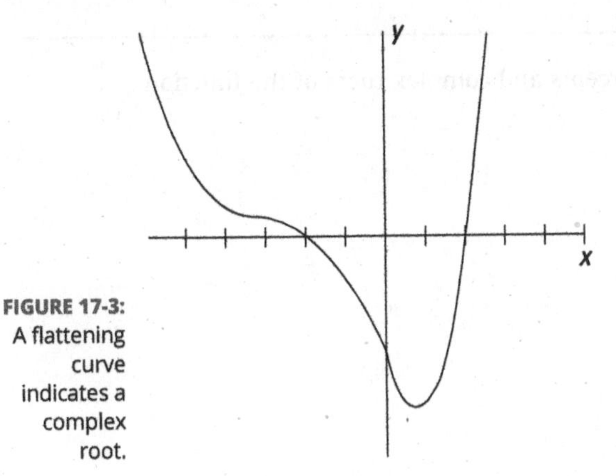

FIGURE 17-3: A flattening curve indicates a complex root.

Figure 17-4 can tell you plenty about the number of real zeros and complex zeros the graph of the polynomial has before you even see the equation it represents.

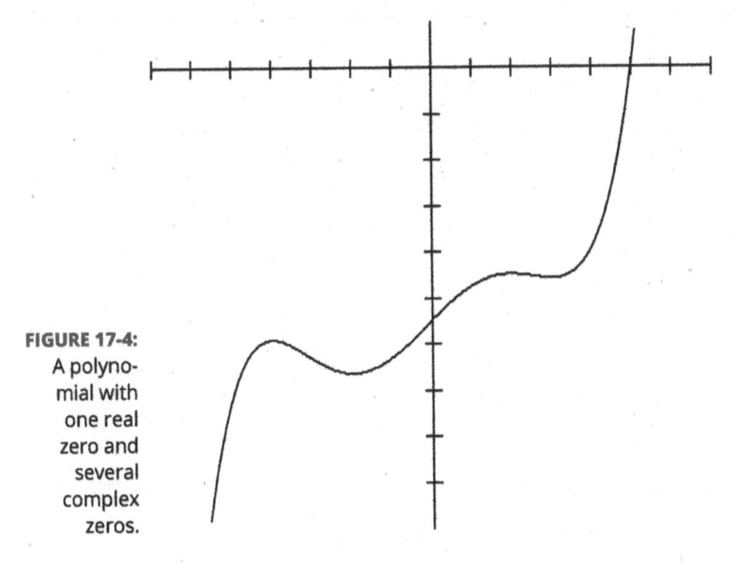

FIGURE 17-4: A polynomial with one real zero and several complex zeros.

The polynomial in Figure 17-4 appears to have one real zero and several complex zeros. Do you see how it changes direction all over the place under the x-axis? It goes up, down, up, down, and then up. These changes indicate the presence of complex zeros. The graph represents the polynomial function $y = 12x^5 + 15x^4 - 320x^3 - 120x^2 + 2{,}880x - 18{,}275$. The function has four complex zeros, two complex (conjugate) pairs and one real zero (when $x = 5$). To solve these high-order equations, you need the efforts and capabilities of good algebra; the Rational Root Theorem and Descartes' Rule of Signs combined with a graphing calculator or computer; and some luck and common sense. When the zeros are nice integers, life is good. When the zeros are irrational or complex, life is still good, but it's a bit more complicated. Assume the best, and work through the challenges if they arise.

YOUR TURN

14. Determine the number of x-intercepts and complex roots of the function $y = (x^2 + 6)(x^2 - x - 2)(x^3 + 1)$.

Practice Questions Answers and Explanations

1. ***i*.** The number 237 is equal to $4 \cdot 59 + 1$. So $i^{237} = i^{4 \cdot 59 + 1} = i^1 = i$.

2. **-1.** The number 6,786 is equal to $4 \cdot 1{,}696 + 2$. So $i^{6{,}786} = i^{4 \cdot 1{,}696 + 2} = i^2 = -1$.

3. **-8 + 2*i*.** Rewrite the sum, grouping the like terms: $(-5-3)+(-4i+6i)$. Combine the like terms: $= -8 + 2i$.

4. **6 + 7*i*.** Change the signs of the two terms being subtracted and change the operation to addition. Then combine the like terms and add: $(3-2i)-(-3-9i) = (3-2i)+(3+9i) = (3+3)+(-2i+9i) = 6+7i$.

5. **-14 + 27*i*.** Use FOIL to multiply: $(6-i)(-3+4i) = -18+24i+3i-4i^2$. Change the last term to +4 by multiplying the -4 times the value of $i^2 = -1$. Then combine the like terms: $-18+24i+3i-4i^2 = -18+24i+3i+4 = -14+27i$.

6. **-1 – 2*i*.** Multiply the numerator and denominator by the conjugate of the denominator. Then simplify: $\frac{4+3i}{-2+i} \cdot \frac{-2-i}{-2-i} = \frac{-8-4i-6i-3i^2}{4-i^2} = \frac{-8-4i-6i-3(-1)}{4-(-1)} = \frac{-8-4i-6i+3}{4+1} = \frac{-5-10i}{5} = -1-2i$.

7. $\mathbf{\frac{17}{29} - \frac{31}{58}i}$. Multiply the numerator and denominator by the conjugate of the denominator. Then simplify: $\frac{8-3i}{10+4i} \cdot \frac{10-4i}{10-4i} = \frac{80-32i-30i+12i^2}{100-16i^2} = \frac{80-32i-30i-12}{100+16} = \frac{68-62i}{116} = \frac{17}{29} - \frac{31}{58}i$.

8. $\mathbf{5i\sqrt{3}}$. Rewrite as the product of a perfect square under a radical, a non-perfect square under a radical, and -1 under a radical. Then simplify each: $\sqrt{-75} = \sqrt{25}\sqrt{3}\sqrt{-1} = 5i\sqrt{3}$.

9. $\mathbf{10i\sqrt{5}}$. Rewrite as the product of a perfect square under a radical, a non-perfect square under a radical, and -1 under a radical. Then simplify each: $\sqrt{-500} = \sqrt{100}\sqrt{5}\sqrt{-1} = 10i\sqrt{5}$.

10. $\mathbf{\frac{3 \pm i\sqrt{39}}{2}}$. Using the quadratic formula, $x = \frac{3 \pm \sqrt{9-4(1)(12)}}{2(1)} = \frac{3 \pm \sqrt{9-48}}{2} = \frac{3 \pm \sqrt{-39}}{2} = \frac{3 \pm i\sqrt{39}}{2}$.

11. $\mathbf{\frac{-1 \pm i\sqrt{26}}{3}}$. Using the quadratic formula, $x = \frac{-2 \pm \sqrt{4-4(3)(9)}}{2(3)} = \frac{-2 \pm \sqrt{4-108}}{6} = \frac{-2 \pm \sqrt{-104}}{6} = \frac{-2 \pm \sqrt{4}\sqrt{26}\sqrt{-1}}{6} = \frac{-2 \pm 2i\sqrt{26}}{6} = \frac{-1 \pm i\sqrt{26}}{3}$.

12. **3*i*, -3*i*.** Factoring, you have $0 = (x^2-4)(x^2+9) = (x-2)(x+2)(x^2+9)$. There are two complex zeros: $x = \pm 3i$.

13. $\mathbf{2 \pm 2i\sqrt{3}}$. Factoring, $0 = x^3 + 64 = (x+4)\left(x^2-4x+16\right)$. Using the quadratic formula on the quadratic factor, $x = \frac{4 \pm \sqrt{16-4(1)(16)}}{2(1)} = \frac{4 \pm \sqrt{-48}}{2} = \frac{4 \pm 4i\sqrt{3}}{2} = 2 \pm 2i\sqrt{3}$.

14. **Two intercepts, four complex roots.** There are seven, five, three, or one possible real roots. Factoring the polynomial (the quadratic factors and the sum of cubes factors), you have $y = (x-2)(x+1)^2(x^2+6)(x^2-x+1)$. There are intercepts when $x = 2$ and $x = -1$. At the intercept (-1,0), the curve just touches the axis and goes back down. But that still counts as two of the three real roots. The other four are complex roots because the quadratic formula yields negatives under the radicals.

Whaddya Know? Chapter 17 Quiz

Quiz time! Complete each problem to test your knowledge on the various topics covered in this chapter. You can then find the solutions and explanations in the next section.

1. **Simplify:** $i^7 =$

2. **Simplify:** $i^{77} =$

3. **Simplify:** $(4+2i)+(-3+i)-(-i)=$

4. **Simplify:** $(5-3i)(-2+4i)=$

5. **Simplify:** $(3-2i)^3 =$

6. **Divide:** $\dfrac{10-15i}{2+i}=$

7. **Divide:** $\dfrac{6}{i}=$

8. **Solve for** x**:** $x^2+4x+5=0$

9. **Solve for** x**:** $3x^2-6x+4=0$

10. **Solve for** x**:** $x^4+17x^2-18=0$

11. **Determine the real and complex zeros:** $x^3-64=0$

12. **Determine the real and complex zeros:** $x^5+4x^3+8x^2+32=0$

Answers to Chapter 17 Quiz

1. **$-i$.** Rewrite as a product of a power of 4 times a basic power of i: $i^7 = i^4 \cdot i^3 = 1(-i) = -i$.

2. **i.** Rewrite as a product of a power of 4 times a basic power of i: $i^{77} = \left(i^4\right)^{19} \cdot i = 1 \cdot i = i$.

3. **$1+4i$.** Distribute the subtraction and combine like terms.

$$(4+2i)+(-3+i)-(-i) = 4+2i-3+i+i = (4-3)+(2i+i+i) = 1+4i$$

4. **$2+26i$.** Multiply using FOIL and simplify the complex term.

$$(5-3i)(-2+4i) = -10+20i+6i-12i^2 = -10+26i-12(-1) = -10+12+26i = 2+26i$$

5. **$18-46i$.** Use the coefficients from Pascal's Triangle and the decreasing and increasing powers of the terms. Perform the operations and combine like terms.

$$\begin{aligned}(3-2i)^3 &= 1(3)^3+3(3)^2(-2i)^1+3(3)^1(-2i)^2+1(-2i)^3\\ &= 2(27)+3(9)(-2i)+3(3)\left(4i^2\right)+1\left(-8i^3\right)\\ &= 54-54i+36i^2-8i^3\\ &= 54-54i+36(-1)-8(-i)\\ &= 54-54i-36+8i\\ &= 18-46i\end{aligned}$$

6. **$1-8i$.** Multiply the numerator and denominator by the conjugate of the denominator. Simplify.

$$\frac{10-15i}{2+i}\cdot\frac{2-i}{2-i} = \frac{(10-15i)(2-i)}{(2+i)(2-i)} = \frac{20-10i-30i+15i^2}{4-i^2} = \frac{20-40i+15(-1)}{4-(-1)} = \frac{5-40i}{5} = 1-8i$$

7. **$-6i$.** Multiply the numerator and denominator by $-i$. Simplify.

$$\frac{6}{i}\cdot\frac{-i}{-i} = \frac{-6i}{-i^2} = \frac{-6i}{-(-1)} = \frac{-6i}{1} = -6i$$

8. **$-2 \pm i$.** Apply the quadratic formula:

$$x = \frac{-4\pm\sqrt{16-4(1)(5)}}{2(1)} = \frac{-4\pm\sqrt{16-20}}{2} = \frac{-4\pm\sqrt{-4}}{2} = \frac{-4\pm2\sqrt{-1}}{2} = \frac{-4\pm2i}{2} = -2\pm i$$

9. **$1\pm\frac{\sqrt{3}}{3}i$.** Apply the quadratic formula:

$$x = \frac{6\pm\sqrt{36-4(3)(4)}}{2(3)} = \frac{6\pm\sqrt{36-48}}{6} = \frac{6\pm\sqrt{-12}}{6} = \frac{6\pm\sqrt{4}\sqrt{3}\sqrt{-1}}{6} = \frac{6\pm2i\sqrt{3}}{6} = 1\pm\frac{\sqrt{3}}{3}i$$

(10) **$\pm 1, \pm 3i\sqrt{2}$.** Factor the trinomial. Then set the two factors equal to zero. Use the square root rule of the third factor.

$$x^4 + 17x^2 - 18 = \left(x^2 - 1\right)\left(x^2 + 18\right) = 0$$
$$x^2 - 1 = 0,\ x = \pm 1$$
$$x^2 + 18 = 0,\ x = \pm\sqrt{-18} = \pm\sqrt{9}\sqrt{-2} = \pm 3 \cdot i\sqrt{2}$$

(11) **Real: 4, complex: $-2 \pm 2i\sqrt{3}$.** First, factor the difference of cubes: $x^3 - 64 = (x-4)\left(x^2 + 4x + 16\right)$.

Setting the factors equal to zero, the second factor requires the quadratic formula $x - 4 = 0,\ x = 4$.

$$x = \frac{-4 \pm \sqrt{16 - 4(1)(16)}}{2(1)} = \frac{-4 \pm \sqrt{16 - 64}}{2} = \frac{-4 \pm \sqrt{-48}}{2} = \frac{-4 \pm 4i\sqrt{3}}{2} = -2 \pm 2i\sqrt{3}$$

(12) **Real: -2, complex: $\pm 2i$, $1 \pm i\sqrt{3}$.** First, factor by grouping. Then factor the terms created.

$$x^5 + 4x^3 + 8x^2 + 32 = x^3\left(x^2 + 4\right) + 8\left(x^2 + 4\right) = \left(x^2 + 4\right)\left(x^3 + 8\right) = \left(x^2 + 4\right)\left(x + 2\right)\left(x^2 - 2x + 4\right)$$

Set each factor equal to zero and solve for x.

$$x^2 + 4 = 0,\ x^2 = -4,\ x = \pm\sqrt{-4} = \pm 2i$$
$$x + 2 = 0,\ x = -2$$
$$x^2 - 2x + 4 = 0,\ x = \frac{2 \pm \sqrt{-2^2 - 4(1)(4)}}{2(1)} = \frac{2 \pm \sqrt{-12}}{2} = \frac{2 \pm 2i\sqrt{3}}{2} = 1 \pm i\sqrt{3}$$

IN THIS CHAPTER

» Getting lined up with matrices

» Operating on matrices

» Singling out single rows for operations

» Identifying the inverses of matrices

» Employing matrices to solve systems of equations

Chapter 18 Making Moves with Matrices

A matrix is a rectangular array of numbers. It has the same number of elements in each of its rows, and each column shares the same number of elements. If you've seen the *Matrix* films, you may remember the lines and columns of green code scrolling down the characters' computer screens. This matrix of codes represented the abstract "matrix" of the films. Putting numbers or elements in an orderly array allows you to organize information, access information quickly, do computations involving some of the entries in the matrix, and communicate your results efficiently.

The word *matrix* is singular — you have just one of them. When you have more than one matrix, you have *matrices*, the plural form. Bet you weren't expecting an English (or Latin) lesson in this book.

In this chapter, you discover how to add and subtract matrices, how to multiply matrices, and how to solve systems of equations by using matrices. The processes in this chapter are easily adaptable for use in technology (as evidenced by the machines in *The Matrix*!); you can transfer the information to a spreadsheet, phone app, or graphing calculator for help with computing large amounts of data.

Describing the Different Types of Matrices

A matrix has a size, or dimension, that you have to recognize before you can proceed with any matrix operations.

You give the dimensions of a matrix in a particular order. You identify the number of rows in the matrix first, and then you mention the number of columns. Usually, you put the numbers for the rows and columns on either side of an × sign: rows × columns.

In the following list, you see six matrices. I've indicated their dimensions and some other classifications for several of them. You see more on the classifications in the next few sections. The brackets around the array of numbers serve as clear indicators that you're dealing with matrices.

- **Dimension 2 × 4:** $A = \begin{bmatrix} 1 & 3 & -2 & 4 \\ 0 & -3 & 5 & 9 \end{bmatrix}$
- **Dimension 3 × 3 (square matrix):** $B = \begin{bmatrix} 4 & -2 & -3 \\ -1 & 0 & 0 \\ 1 & 3 & 1 \end{bmatrix}$
- **Dimension 1 × 4 (row matrix):** $C = \begin{bmatrix} 1 & 2 & 4 & -2 \end{bmatrix}$
- **Dimension 3 × 1 (column matrix):** $D = \begin{bmatrix} 2 \\ 0 \\ -4 \end{bmatrix}$
- **Dimension 3 × 2 (zero matrix):** $E = \begin{bmatrix} 0 & 0 \\ 0 & 0 \\ 0 & 0 \end{bmatrix}$
- **Dimension 2 × 2 (square matrix, identity matrix):** $F = \begin{bmatrix} 1 & 0 \\ 0 & 1 \end{bmatrix}$

When you have to deal with more than one matrix, you can keep track of them by labeling the matrices with different names. You don't call the matrices Bill or Ted; that wouldn't be very mathematical of you! Matrices are traditionally labeled with capital letters, such as matrix A or matrix B, to avoid confusion.

The numbers that appear in the rectangular array of a matrix are called its *elements*. You refer to each element in a matrix by listing the same letter as the matrix's name, in lowercase form, and following it with subscript numbers for the row and then the column. For example, the item in the first row and third column of matrix B is b_{13}, which, in the matrix B shown here, is −3. If the number of rows or columns grows to a number bigger than 9, you put a comma between the two numbers in the subscript.

Row and column matrices

Matrices come in many sizes (or dimensions), just like rectangles, but instead of measuring width and length, you count a matrix's rows and columns. You call matrices that have only one row or one column *row matrices* or *column matrices*, respectively. A row matrix has the dimension $1 \times n$, where n is the number of columns. The matrix C in the preceding section is a row matrix with dimension 1×4.

A column matrix has the dimension $m \times 1$, where m is the number of rows. Matrix D from the previous section is a column matrix with dimension 3×1.

Square matrices

A square matrix has the same number of rows and columns. Square matrices have dimensions such as 2×2, 3×3, 8×8, and so on. The elements in square matrices can take on any number — although some special square matrices are called *identity matrices* (coming up later in this section). All matrices are rectangular arrays, and a square is a special type of rectangle. In the earlier list, both B and F are square matrices.

Zero matrices

Zero matrices can have any dimension; they can have any number of rows or columns. The matrix E in the earlier listing is a zero matrix because zeros make up all the elements.

Zero matrices may not look very impressive — after all, they don't have much in them — but they're necessary for matrix arithmetic. Just like you need a zero to add and subtract numbers, you need zero matrices to do matrix addition and matrix subtraction.

Identity matrices

Identity matrices add a couple of characteristics to the zero-matrix format (see the previous section) in terms of their dimensions and elements. An identity matrix has to

- Be a square matrix
- Have a diagonal strip of 1s that goes from the top left to the bottom right of the matrix
- Consist of zeros outside of the diagonal strip of 1s

The matrix F in the earlier listing is an identity matrix, although you will come across many, many more sizes, all square.

Identity matrices are instrumental to matrix multiplication and matrix inverses. Identity matrices act pretty much like the number 1 in the multiplication of numbers. What happens when you multiply a number times 1? The number keeps its identity. You see the same behavior when you multiply a matrix by an identity matrix — the matrix stays the same.

EXAMPLE

Q. **Determine all the ways the matrix can be described.**

$$\begin{bmatrix} 1 & 0 & 0 \\ 0 & 1 & 0 \\ 0 & 0 & 1 \end{bmatrix}$$

A. **This is a square matrix that is also an identity matrix.**

Q. **Determine all the ways the matrix can be described.**

$$\begin{bmatrix} 0 & 0 & 0 & 0 \end{bmatrix}$$

Q. **This is a row matrix that is also a zero matrix.**

YOUR TURN

Given matrices *A, B, C, and D*, determine which match the descriptions.

$$A = \begin{bmatrix} 0 & 0 & 0 \\ 0 & 0 & 0 \\ 0 & 0 & 0 \end{bmatrix}, B = \begin{bmatrix} 0 \\ 1 \\ 0 \end{bmatrix}, C = \begin{bmatrix} 1 & 0 & 0 \\ 0 & 1 & 0 \end{bmatrix}, D = \begin{bmatrix} 1 & 1 \\ 1 \\ 1 \end{bmatrix}$$

1 **Dimension 2 × 3**

2 **Rectangular matrix**

3 **Zero matrix**

4 **Identity matrix**

Performing Operations on Matrices

You can add matrices, subtract one from another, multiply them by numbers, multiply them by each other, and divide them. Well, actually, you don't divide matrices; you change the division problem to a multiplication problem. You can't add, subtract, or multiply just any matrices, though. Each operation has its own set of rules. I cover the rules for addition, subtraction, and multiplication in this section. (Division comes later in the chapter, after I discuss matrix inverses.)

Adding and subtracting matrices

To add or subtract matrices, you have to make sure that the matrices are the same size. In other words, they need to have identical dimensions. You find the resulting matrix by performing the operation on the corresponding elements in the matrices. If two matrices don't have the same dimension, you can't add or subtract them, and you can do nothing to fix the situation.

Here I show you two matrices and how their elements are added and subtracted:

$$A=\begin{bmatrix} a_{11} & a_{12} & a_{13} \\ a_{21} & a_{22} & a_{23} \end{bmatrix},\quad B=\begin{bmatrix} b_{11} & b_{12} & b_{13} \\ b_{21} & b_{22} & b_{23} \end{bmatrix}$$

$$A+B=\begin{bmatrix} a_{11}+b_{11} & a_{12}+b_{12} & a_{13}+b_{13} \\ a_{21}+b_{21} & a_{22}+b_{22} & a_{23}+b_{23} \end{bmatrix}$$

$$A-B=\begin{bmatrix} a_{11}-b_{11} & a_{12}-b_{12} & a_{13}-b_{13} \\ a_{21}-b_{21} & a_{22}-b_{22} & a_{23}-b_{23} \end{bmatrix}$$

You can see why matrices have to have the same dimensions before you can add or subtract. Differing matrices would have some elements without partners.

EXAMPLE

Using the matrices C and D: $C=\begin{bmatrix} 2 & 5 & -3 & 8 \\ -1 & 0 & 7 & -4 \end{bmatrix},\quad D=\begin{bmatrix} -2 & 4 & 3 & -6 \\ 0 & 2 & 7 & 3 \end{bmatrix}$

Q. Compute $C + D$.

A. $C+D=\begin{bmatrix} 2+(-2) & 5+4 & -3+3 & 8+(-6) \\ -1+0 & 0+2 & 7+7 & -4+3 \end{bmatrix}=\begin{bmatrix} 0 & 9 & 0 & 2 \\ -1 & 2 & 14 & -1 \end{bmatrix}$

Q. Compute $C - D$.

A. $C-D=\begin{bmatrix} 2-(-2) & 5-4 & -3-3 & 8-(-6) \\ -1-0 & 0-2 & 7-7 & -4-3 \end{bmatrix}=\begin{bmatrix} 4 & 1 & -6 & 14 \\ -1 & -2 & 0 & -7 \end{bmatrix}$

Multiplying matrices by scalars

TIP

Scalar is just a fancy word for number. Algebra uses the word scalar with regard to matrix multiplication to contrast a number with a matrix, which has a dimension. A scalar has no dimension, so you can use it uniformly throughout the matrix.

Scalar multiplication of a matrix signifies that you multiply every element in the matrix by a number. To multiply matrix A by the number k, for example, you multiply each element in A by k. The matrix A and scalar k illustrate how this scalar multiplication works:

$$A=\begin{bmatrix} a_{11} & a_{12} \\ a_{21} & a_{22} \\ a_{31} & a_{32} \\ a_{41} & a_{42} \end{bmatrix}, \quad k\cdot A=k\cdot\begin{bmatrix} a_{11} & a_{12} \\ a_{21} & a_{22} \\ a_{31} & a_{32} \\ a_{41} & a_{42} \end{bmatrix}=\begin{bmatrix} ka_{11} & ka_{12} \\ ka_{21} & ka_{22} \\ ka_{31} & ka_{32} \\ ka_{41} & ka_{42} \end{bmatrix}$$

EXAMPLE

Q. Given matrix $F=\begin{bmatrix} 4 & -3 & 1 \\ 2 & 0 & -1 \\ 5 & 10 & -4 \end{bmatrix}$, find $3F$.

A. $3\cdot F=3\cdot\begin{bmatrix} 4 & -3 & 1 \\ 2 & 0 & -1 \\ 5 & 10 & -4 \end{bmatrix}=\begin{bmatrix} 12 & -9 & 3 \\ 6 & 0 & -3 \\ 15 & 30 & -12 \end{bmatrix}$

Q. Given matrix $G=\begin{bmatrix} 1 & 0 & 1 \\ 4 & 2 & 6 \\ 3 & 1 & -4 \end{bmatrix}$, find $-1G$.

A. $-1\cdot G=-1\cdot\begin{bmatrix} 1 & 0 & 1 \\ 4 & 2 & 6 \\ 3 & 1 & -4 \end{bmatrix}=\begin{bmatrix} -1 & 0 & -1 \\ -4 & -2 & -6 \\ -3 & -1 & 4 \end{bmatrix}$

YOUR TURN

5 Add $\begin{bmatrix} -1 & 3 \\ 0 & 4 \\ -2 & 5 \end{bmatrix}+\begin{bmatrix} 1 & -4 \\ 6 & -4 \\ 3 & 3 \end{bmatrix}$.

6 Subtract $\begin{bmatrix} 3 & 5 & -9 & 9 \end{bmatrix}-\begin{bmatrix} 0 & 2 & -1 & 6 \end{bmatrix}$.

7 Multiply $-\frac{1}{2}\cdot\begin{bmatrix}-1 & 2\\ 0 & 4\\ -2 & 8\end{bmatrix}$.

Multiplying two matrices

Matrix multiplication requires that the number of columns in the first matrix be equal to the number of rows in the second matrix. This means, for instance, that a matrix with 3 rows and 11 columns can multiply a matrix with 11 rows and 4 columns — but it has to be in that order. The 11 columns in the first matrix must match up with the 11 rows in the second matrix.

Matrix multiplication requires some strict rules about the dimensions and the order in which the matrices are multiplied. Even when the matrices are square and can be multiplied in either order, you don't get the same answer when multiplying them in the different orders. In general, the product *AB* does not equal the product *BA*.

Determining dimensions

If you want to multiply matrices, the number of columns in the first matrix has to equal the number of rows in the second matrix. After you multiply matrices, you get a whole new matrix that features the number of rows the first matrix had and the number of columns from the original second matrix. The process is sort of like cross-pollinating white and red petunias and getting pink.

In algebraic terms, if matrix A has the dimension $m\times n$ and matrix B has the dimension $p\times q$, to multiply $A\cdot B$, n must equal p. The dimension of the resulting matrix will be $m\times q$, the number of rows in the first matrix and the number of columns in the second.

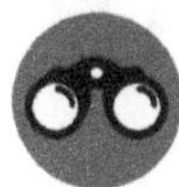
EXAMPLE

Q. What is the dimension of the matrix if you multiply a 2×3 matrix times a 3×7 matrix?

A. You get a 2×7 matrix.

Q. What is the dimension of the matrix if you multiply a 7×2 matrix by a 2×2 matrix?

A. You get a 7×2 matrix.

Defining the process

Multiplying matrices is not necessarily an easy matter, but it isn't all that difficult, if you can multiply and add correctly. When you multiply two matrices, you compute the elements with the following rule:

> If you find the element c_{ij} after multiplying matrix A times matrix B, c_{ij} is the sum of the products of the elements in the ith row of matrix A and the jth column of matrix B.

Okay, so that rule may sound like a lot of hocus pocus. How about I give you something a bit more concrete? I'm a visual learner, myself, so I hope this helps.

EXAMPLE

Q. Multiply matrices A and B: $A=\begin{bmatrix} a_{11} & a_{12} \\ a_{21} & a_{22} \\ a_{31} & a_{32} \end{bmatrix}$, $B=\begin{bmatrix} b_{11} & b_{12} & b_{13} & b_{14} \\ b_{21} & b_{22} & b_{23} & b_{24} \end{bmatrix}$

A. Name the resulting matrix C. Matrix C will have the dimension 3×4 because matrix A is 3×2 and B is 2×4.

$$A\cdot B=\begin{bmatrix} a_{11} & a_{12} \\ a_{21} & a_{22} \\ a_{31} & a_{32} \end{bmatrix}\times\begin{bmatrix} b_{11} & b_{12} & b_{13} & b_{14} \\ b_{21} & b_{22} & b_{23} & b_{24} \end{bmatrix}=\begin{bmatrix} c_{11} & c_{12} & c_{13} & c_{14} \\ c_{21} & c_{22} & c_{23} & c_{24} \\ c_{31} & c_{32} & c_{33} & c_{34} \end{bmatrix}$$

You grab the first row of A and multiply the two elements times the two elements in the first column of B. Then add the two products together. So, $c_{11}=a_{11}\cdot b_{11}+a_{12}\cdot b_{21}$. That product and sum is what it takes to create just one element in the final product. Here's one more element: Look at the element c_{32} in the product matrix. That element is equal to the sum of the two products when you multiply the third row of A times the second column of B: $c_{32}=a_{31}\cdot b_{12}+a_{32}\cdot b_{22}$. Here's the complete product:

$$\begin{bmatrix} c_{11} & c_{12} & c_{13} & c_{14} \\ c_{21} & c_{22} & c_{23} & c_{24} \\ c_{31} & c_{32} & c_{33} & c_{34} \end{bmatrix}=\begin{bmatrix} a_{11}\cdot b_{11}+a_{12}\cdot b_{21} & a_{11}\cdot b_{12}+a_{12}\cdot b_{22} & a_{11}\cdot b_{13}+a_{12}\cdot b_{23} & a_{11}\cdot b_{14}+a_{12}\cdot b_{24} \\ a_{21}\cdot b_{11}+a_{22}\cdot b_{21} & a_{21}\cdot b_{12}+a_{22}\cdot b_{22} & a_{21}\cdot b_{13}+a_{22}\cdot b_{23} & a_{21}\cdot b_{14}+a_{22}\cdot b_{24} \\ a_{31}\cdot b_{11}+a_{32}\cdot b_{21} & a_{31}\cdot b_{12}+a_{32}\cdot b_{22} & a_{31}\cdot b_{13}+a_{32}\cdot b_{23} & a_{31}\cdot b_{14}+a_{32}\cdot b_{24} \end{bmatrix}$$

Q. Multiply matrices $J=\begin{bmatrix} 1 & 2 & -3 \\ 0 & 4 & 2 \end{bmatrix}$ and $K=\begin{bmatrix} 4 & 5 \\ 1 & -1 \\ 2 & 3 \end{bmatrix}$.

A. The resulting matrix will be 2×2:

$$J\cdot K=\begin{bmatrix} 1\cdot 4+2\cdot 1+(-3)\cdot 2 & 1\cdot 5+2\cdot(-1)+(-3)\cdot 3 \\ 0\cdot 4+4\cdot 1+2\cdot 2 & 0\cdot 5+4\cdot(-1)+2\cdot 3 \end{bmatrix}=\begin{bmatrix} 4+2-6 & 5-2-9 \\ 0+4+4 & 0-4+6 \end{bmatrix}=\begin{bmatrix} 0 & -6 \\ 8 & 2 \end{bmatrix}$$

8 Multiply matrices $M = \begin{bmatrix} 4 & -1 \\ 0 & 3 \end{bmatrix}$ and $N = \begin{bmatrix} -1 & 3 & 0 \\ 2 & 5 & -2 \end{bmatrix}$.

9 Multiply matrices $P = \begin{bmatrix} 1 & 2 & 3 \\ 4 & 5 & 6 \\ 7 & 8 & 9 \end{bmatrix}$ and $Q = \begin{bmatrix} -1 & 0 & -1 \\ 1 & -1 & 1 \\ -1 & 0 & 1 \end{bmatrix}$.

Defining Row Operations

Along with the matrix operations I list in the previous sections of this chapter, you can perform row operations on the individual rows of a matrix. You perform a row operation one matrix at a time; you don't combine one matrix with another matrix when performing these operations. A row operation changes the look of a matrix by altering some of the elements, but proper operations allow the matrix to retain the properties that enable you to use it in other applications, such as solving systems of equations. (See the section, "Using Matrices to Find Solutions for Systems of Equations," later in this chapter.)

The business of changing matrices to equivalent matrices is sort of like changing fractions to equivalent fractions so that they have a common denominator — the change makes the fractions more useful. The same goes for matrices.

You have several different row operations at your disposal:

- You can exchange two rows.
- You can multiply the elements in a row by a constant (not zero).
- You can add the elements in one row to the elements in another row.
- You can add a row that you multiply by some number to another row.

Next I show you a matrix that has experienced the following row operations, in order, each on the previous result. I use a special notation or shorthand to indicate which operations are being performed (it's the one I introduce in Chapter 14 when performing operations on linear equations):

- a: I exchange the first and third rows, indicated by $R_1 \leftrightarrow R_3$.
- b: I multiply the second row through by –1, indicated with $(-1)R_2$.
- c: I add the first and third rows and put the result in the third row, shown with $R_1 + R_3 \rightarrow R_3$.
- d: I add twice the first row to the second row and put the result in the second row, shown by $(2)R_1 + R_2 \rightarrow R_2$.

$$\begin{bmatrix} -5 & 2 & 4 & 1 \\ -4 & 1 & -2 & 0 \\ 1 & 0 & 3 & 2 \end{bmatrix} \xrightarrow{R_1 \leftrightarrow R_3} \begin{bmatrix} 1 & 0 & 3 & 2 \\ -4 & 1 & -2 & 0 \\ -5 & 2 & 4 & 1 \end{bmatrix} \xrightarrow{(-1)R_2} \begin{bmatrix} 1 & 0 & 3 & 2 \\ 4 & -1 & 2 & 0 \\ -5 & 2 & 4 & 1 \end{bmatrix}$$

$$\xrightarrow{R_1 + R_3 \rightarrow R_3} \begin{bmatrix} 1 & 0 & 3 & 2 \\ 4 & -1 & 2 & 0 \\ -4 & 2 & 7 & 3 \end{bmatrix} \xrightarrow{(2)R_1 + R_2 \rightarrow R_2} \begin{bmatrix} 1 & 0 & 3 & 2 \\ 6 & -1 & 8 & 4 \\ -4 & 2 & 7 & 3 \end{bmatrix}$$

Performing row operations correctly results in a matrix that's equivalent to the original matrix. The rows themselves aren't equivalent to one another; the whole matrix and the relationships between its rows are preserved with the operations.

Row operations may seem pointless and aimless. For these examples of row operations, I have no particular goal other than to illustrate the possibilities. But you will choose more wisely when you do row operations to perform a task, such as solving for an inverse matrix (see the following section) or solving systems of equations, later in this chapter.

YOUR TURN

10 Perform the row operation $R_2 \leftrightarrow R_4$ on the matrix $\begin{bmatrix} 6 & -9 & 0 \\ 1 & 1 & 0 \\ 1 & 1 & 2 \\ 6 & 7 & -3 \end{bmatrix}$.

11 Perform the row operation $(-1)R_2 + R_3 \rightarrow R_3$ on the matrix $\begin{bmatrix} 6 & -9 & 0 \\ 1 & 1 & 0 \\ 1 & 1 & 2 \\ 6 & 7 & -3 \end{bmatrix}$.

Finding Inverse Matrices

Inverses of matrices act somewhat like inverses of numbers. The additive inverse of a number is what you have to add to the number to get zero; it's often called its opposite. For example, the additive inverse of the number 2 is –2, and the additive inverse of –3.14159 is 3.14159. Simple enough.

Algebra also provides you with the multiplicative inverse. Multiplicative inverses give you the number 1 when you multiply the inverse times the original number. A multiplicative inverse is also know as the number's reciprocal. For example, the multiplicative inverse of 2 is $\frac{1}{2}$.

Adding or multiplying inverses always gives you the identity element for a particular operation. Before you can find inverses of matrices, you need to understand identities, so I cover those in detail here.

Determining additive inverses

You can label the number zero as the additive identity, because adding zero to a number allows that number to keep its identity. The additive identity for matrices is a zero matrix. When you add a zero matrix to any matrix with the same dimension, the original matrix doesn't change.

Inverse matrices associated with addition are easy to spot and easy to create. The additive inverse of a matrix is another matrix with the same dimension but every element has the opposite sign. When you add the matrix and its additive inverse, the sums of the corresponding elements are all zero, and you have a zero matrix — the additive identity for matrices (see the section, "Zero matrices," earlier in the chapter).

EXAMPLE

Q. Find the additive inverse of the matrix $\begin{bmatrix} 3 & 7 & -2 & 0 \\ -1 & 4 & -4 & 5 \end{bmatrix}$.

A. Replace each element with its additive inverse: $\begin{bmatrix} -3 & -7 & 2 & 0 \\ 1 & -4 & 4 & -5 \end{bmatrix}$.

Q. Find the additive inverse of the matrix $[x \quad y \quad z]$.

A. Replace each element with its additive inverse: $[-x \quad -y \quad -z]$.

All matrices have additive inverses no matter what the dimension of the matrix. This isn't the case with multiplicative inverses of matrices. Some matrices have multiplicative inverses and some do not. Read on if you're intrigued by this situation (or if the topic will be on your next test).

Determining multiplicative inverses

You can label the number 1 as the multiplicative identity, because multiplying a number by 1 doesn't change the number — it keeps its identity.

Matrices have additive identities that consist of nothing but zeros but multiplicative identities for matrices are a bit more picky. A multiplicative identity for a matrix has to be a square matrix, and this square matrix has to have a diagonal of 1s; the rest of the elements are zeros. This arrangement ensures that when you multiply any matrix times a multiplicative identity, you don't change the original matrix. Here is the process of multiplying two different matrices by identities — one with the identity first and the other with it last. The multiplication rule for matrices still holds here (see the earlier section, "Multiplying two matrices"): The number of columns in the first matrix must match the number of rows in the second matrix.

EXAMPLE

Q. Multiply the matrix $\begin{bmatrix} -3 \\ 6 \end{bmatrix}$ by a multiplicative identity.

A. Use the 2 × 2 identity, placing it first.

$$\begin{bmatrix} 1 & 0 \\ 0 & 1 \end{bmatrix} \times \begin{bmatrix} -3 \\ 6 \end{bmatrix} = \begin{bmatrix} 1\cdot(-3)+0\cdot 6 \\ 0\cdot(-3)+1\cdot 6 \end{bmatrix} = \begin{bmatrix} -3+0 \\ 0+6 \end{bmatrix} = \begin{bmatrix} -3 \\ 6 \end{bmatrix}$$

Q. Multiply the matrix $\begin{bmatrix} 3 & 0 & -2 \\ 1 & 5 & 9 \end{bmatrix}$ by a multiplicative identity.

A. Use the 3 × 3 identity, placing it last.

$$\begin{bmatrix} 3 & 0 & -2 \\ 1 & 5 & 9 \end{bmatrix} \times \begin{bmatrix} 1 & 0 & 0 \\ 0 & 1 & 0 \\ 0 & 0 & 1 \end{bmatrix} = \begin{bmatrix} 3\cdot1+0\cdot0+(-2)\cdot0 & 3\cdot0+0\cdot1+(-2)\cdot0 & 3\cdot0+0\cdot0+(-2)\cdot1 \\ 1\cdot1+5\cdot0+9\cdot0 & 1\cdot0+5\cdot1+9\cdot0 & 1\cdot0+5\cdot0+9\cdot1 \end{bmatrix}$$

$$= \begin{bmatrix} 3+0+0 & 0+0+0 & 0+0+(-2) \\ 1+0+0 & 0+5+0 & 0+0+9 \end{bmatrix} = \begin{bmatrix} 3 & 0 & -2 \\ 1 & 5 & 9 \end{bmatrix}$$

Like the multiplicative identity of a matrix, the multiplicative inverse of a matrix isn't quite as accommodating as its additive cousin. When you multiply two matrices, you perform plenty of multiplication and addition, and you see plenty of change in dimension. Because of this, matrices and their inverses are always square matrices; non-square matrices don't have multiplicative inverses.

If matrix A and matrix A^{-1} are multiplicative inverses, then $A \cdot A^{-1} = I$, and $A^{-1} \cdot A = I$. The superscript −1 on matrix A identifies it as the inverse matrix of A; it isn't a reciprocal, which the exponent −1 usually represents. Also, the capital letter I identifies the identity matrix associated with these matrices — with dimensions 2×2, 3×3, 4×4, and so on.

Multiplying a matrix B and its inverse B^{-1} in one order and then in the other order always results in the identity matrix.

Not all square matrices have inverses. But for those that do, you have a way to find the inverse matrices. You don't always know ahead of time which matrices will fail you but it becomes apparent as you go through the steps. The first process, or *algorithm* (a process or routine that produces a result), you can use works for any size square matrix. You can also use a neat, quick method that's special for 2×2 matrices and works only for matrices with that dimension.

Identifying an inverse for any size square matrix

The general method you use to find a matrix's inverse involves writing down the matrix, inserting an identity matrix, and then changing the original matrix into an identity matrix.

To solve for the inverse of a matrix, follow these steps:

1. **Create a double matrix — consisting of the target matrix and the identity matrix of the same size — with the identity matrix to the right of the original.**

2. **Perform row operations until the elements on the left become an identity matrix (see the section, "Defining Row Operations," earlier in the chapter).**

   The elements created in the new left-hand matrix need to feature a diagonal of 1s with zeros above and below the 1s. Upon completion of this step, the elements on the right become the elements of the inverse matrix.

EXAMPLE

Q. Solve for the inverse of matrix $M = \begin{bmatrix} 1 & 2 & 4 \\ -3 & -5 & -6 \\ 2 & -3 & -36 \end{bmatrix}$.

A. First, put the 3×3 identity matrix to the right of the elements in matrix M. The goal is to make the elements on the left look like an identity matrix by using matrix row operations.

$$\text{Double-wide:} \left[\begin{array}{ccc} 1 & 2 & 4 \\ -3 & -5 & -6 \\ 2 & -3 & -36 \end{array} \vdots \begin{array}{ccc} 1 & 0 & 0 \\ 0 & 1 & 0 \\ 0 & 0 & 1 \end{array}\right]$$

The identity matrix has a diagonal slash of 1s, and zeros lie above and below the 1s. The first thing you do to create your identity matrix is to get the zeros below the 1 in the upper-left corner of the original identity. Here are the row operations you use:

1. Multiply Row 1 times 3 and add the result to Row 2, making the result the new Row 2.
2. Multiply Row 1 by –2 and add the result to Row 3, putting the resulting answer in Row 3.

$$\left[\begin{array}{ccc} 1 & 2 & 4 \\ -3 & -5 & -6 \\ 2 & -3 & -36 \end{array} \vdots \begin{array}{ccc} 1 & 0 & 0 \\ 0 & 1 & 0 \\ 0 & 0 & 1 \end{array}\right] \underset{(-2)R_1+R_3 \to R_3}{\overset{(3)R_1+R_2 \to R_2}{\rightarrow}} \left[\begin{array}{ccc} 1 & 2 & 4 \\ 0 & 1 & 6 \\ 0 & -7 & -44 \end{array} \vdots \begin{array}{ccc} 1 & 0 & 0 \\ 3 & 1 & 0 \\ -2 & 0 & 1 \end{array}\right]$$

You see a 1 in the second column and second row of the resulting matrix, along the diagonal of 1s that you want. Getting the 1 in this position is a nice coincidence; it doesn't always work out as well. If the 1 hadn't fallen in that position, you'd have to divide the whole row by whatever number makes that element a 1.

Now, you want zeros above and below the 1 in the middle, so you follow these steps:

1. Multiply Row 2 by –2 and add the result to Row 1.

2. Multiply Row 2 by 7 and add the result to Row 3.

$$\left[\begin{array}{ccc:ccc} 1 & 2 & 4 & 1 & 0 & 0 \\ 0 & 1 & 6 & 3 & 1 & 0 \\ 0 & -7 & -44 & -2 & 0 & 1 \end{array}\right] \underset{(7)R_2+R_3\to R_3}{\overset{(-2)R_2+R_1\to R_1}{\to}} \left[\begin{array}{ccc:ccc} 1 & 0 & -8 & -5 & -2 & 0 \\ 0 & 1 & 6 & 3 & 1 & 0 \\ 0 & 0 & -2 & 19 & 7 & 1 \end{array}\right]$$

You now need to turn the element in the third row, third column of the matrix into a 1, so you multiply the row through by –0.5; this multiplication is the same as dividing through by –2.

$$\left[\begin{array}{ccc:ccc} 1 & 0 & -8 & -5 & -2 & 0 \\ 0 & 1 & 6 & 3 & 1 & 0 \\ 0 & 0 & -2 & 19 & 7 & 1 \end{array}\right] \overset{(-.5)R_3\to R_3}{\to} \left[\begin{array}{ccc:ccc} 1 & 0 & -8 & -5 & -2 & 0 \\ 0 & 1 & 6 & 3 & 1 & 0 \\ 0 & 0 & 1 & -9.5 & -3.5 & -0.5 \end{array}\right]$$

You have one last set of row operations to perform to get zeros above the last 1 on the diagonal:

1. Multiply Row 3 by 8 and add the elements in the row to Row 1.

2. Multiply Row 3 by –6 and add the result to Row 2.

The following shows the final two steps and the result:

$$\left[\begin{array}{ccc:ccc} 1 & 0 & -8 & -5 & -2 & 0 \\ 0 & 1 & 6 & 3 & 1 & 0 \\ 0 & 0 & 1 & -9.5 & -3.5 & -0.5 \end{array}\right] \underset{(-6)R_3+R_2\to R_2}{\overset{(8)R_3+R_1\to R_1}{\to}} \left[\begin{array}{ccc:ccc} 1 & 0 & 0 & -81 & -30 & -4 \\ 0 & 1 & 0 & 60 & 22 & 3 \\ 0 & 0 & 1 & -9.5 & -3.5 & -0.5 \end{array}\right]$$

$$M^{-1} = \begin{bmatrix} -81 & -30 & -4 \\ 60 & 22 & 3 \\ -9.5 & -3.5 & -0.5 \end{bmatrix}$$

The operations gave you a 3×3 identity matrix on the left and a new 3×3 matrix on the right. The elements in the matrix on the right form the inverse matrix, M^{-1}. The product of M and its inverse, M^{-1}, is the identity matrix.

The following checks your work — showing that $M\cdot M^{-1}$ is the identity matrix:

$$M\times M^{-1} = \begin{bmatrix} 1 & 2 & 4 \\ -3 & -5 & -6 \\ 2 & -3 & -36 \end{bmatrix} \times \begin{bmatrix} -81 & -30 & -4 \\ 60 & 22 & 3 \\ -9.5 & -3.5 & -.5 \end{bmatrix} = \begin{bmatrix} 1 & 0 & 0 \\ 0 & 1 & 0 \\ 0 & 0 & 1 \end{bmatrix}$$

How can you tell when a matrix doesn't have an inverse? You can't get 1s along the diagonal by using the row operations. You usually get a whole row of zeros as a result of your row operations. A row of zeros is the warning that no inverse is possible.

Using a quick-and-slick rule for 2×2 matrices

You have a special rule at your disposal to find inverses of 2×2 matrices. To implement the rule for a 2×2 matrix, you have to switch two elements, negate two elements, and divide all the elements by the difference of the cross products of the elements. This may sound complicated, but the math is really neat and sweet, and the process is much quicker than the general method (see the previous section).

Here is the general formula for the 2×2 matrix rule.

$$K = \begin{bmatrix} a & b \\ c & d \end{bmatrix},\ K^{-1} = \begin{bmatrix} \frac{d}{ad-bc} & \frac{-b}{ad-bc} \\ \frac{-c}{ad-bc} & \frac{a}{ad-bc} \end{bmatrix}$$

As you can see from this special rule, the upper-left and lower-right corner elements are switched; the upper-right and lower-left corner elements are negated (changed to the opposite sign); and all the elements are divided by the result of doing two cross products and subtracting.

EXAMPLE

Q. Find the inverse of matrix Z using the 2×2 method: $Z = \begin{bmatrix} 5 & 6 \\ 9 & 11 \end{bmatrix}$.

A. Switch the 5 and the 11, and change the 6 to –6 and the 9 to –9. Then divide by the difference of the cross products: $(5 \cdot 11) - (6 \cdot 9) = 55 - 54 = 1$. Watch the order in which you do the subtraction — the order does matter. Dividing each element by 1 doesn't change the elements, as you can see.

$$Z = \begin{bmatrix} 5 & 6 \\ 9 & 11 \end{bmatrix} \to \begin{bmatrix} 11 & 6 \\ 9 & 5 \end{bmatrix} \to \begin{bmatrix} 11 & -6 \\ -9 & 5 \end{bmatrix} \to \begin{bmatrix} \frac{11}{1} & \frac{-6}{1} \\ \frac{-9}{1} & \frac{5}{1} \end{bmatrix} \to \begin{bmatrix} 11 & -6 \\ -9 & 5 \end{bmatrix},\ Z^{-1} = \begin{bmatrix} 11 & -6 \\ -9 & 5 \end{bmatrix}$$

YOUR TURN

12 Find the inverse of the matrix: $W = \begin{bmatrix} 3 & -2 \\ -7 & 5 \end{bmatrix}$.

13 Find the inverse of the matrix: $V = \begin{bmatrix} 1 & 2 & -3 \\ 3 & 5 & -7 \\ -1 & -3 & 6 \end{bmatrix}$.

Dividing Matrices by Using Inverses

Until this point in the chapter, I've avoided the topic of division of matrices. I haven't spent much time on this topic because you don't really divide matrices — you multiply one matrix by the inverse of the other (see the previous section for information on inverses). The division process resembles what you can do with real numbers. Instead of dividing 27 by 2, for example, you can multiply 27 by 2's inverse, $\frac{1}{2}$.

To perform the division $\frac{\text{Matrix } A}{\text{Matrix } B}$, you change the problem to $(\text{Matrix } A)\cdot(\text{Matrix } B^{-1})$.

EXAMPLE

Q. Perform the matrix division:

$$\frac{\begin{bmatrix} 3 & -2 \\ 4 & -3 \end{bmatrix}}{\begin{bmatrix} 6 & -10 \\ 1 & -2 \end{bmatrix}}$$

A. First, find the inverse of the matrix in the denominator.

$$\begin{bmatrix} 6 & -10 \\ 1 & -2 \end{bmatrix}^{-1} = \begin{bmatrix} \frac{-2}{-12-(-10)} & \frac{10}{-12-(-10)} \\ \frac{-1}{-12-(-10)} & \frac{6}{-12-(-10)} \end{bmatrix} = \begin{bmatrix} 1 & -5 \\ .5 & -3 \end{bmatrix}$$

You then multiply the matrix in the numerator by the inverse of the matrix in the denominator (see the section, "Multiplying two matrices," earlier in this chapter).

$$\frac{\begin{bmatrix} 3 & -2 \\ 4 & -3 \end{bmatrix}}{\begin{bmatrix} 6 & -10 \\ 1 & -2 \end{bmatrix}} = \begin{bmatrix} 3 & -2 \\ 4 & -3 \end{bmatrix} \times \begin{bmatrix} 1 & -5 \\ .5 & -3 \end{bmatrix} = \begin{bmatrix} 3-1 & -15+6 \\ 4-1.5 & -20+9 \end{bmatrix} = \begin{bmatrix} 2 & -9 \\ 2.5 & -11 \end{bmatrix}$$

YOUR TURN

14 Perform the matrix division:

$$\frac{\begin{bmatrix} 6 & -12 \\ -18 & 9 \end{bmatrix}}{\begin{bmatrix} 2 & 6 \\ -1 & 3 \end{bmatrix}}$$

Using Matrices to Find Solutions for Systems of Equations

One of the nicest applications of matrices is that you can use them to solve systems of linear equations. In Chapter 14, you find out how to solve systems of two, three, four, and more linear equations. The methods you use in that chapter involve elimination of variables and substitution. When you use matrices, you deal only with the coefficients of the variables in the problem. This way is less messy, and you can enter the matrices into graphing calculators or computer programs for an even more relaxing process.

This method using matrices is the most desirable when you have some technology available to you. Finding inverse matrices can be nasty work if fractions and decimals crop up. Simple graphing calculators can make the process quite nice.

REMEMBER

You can use matrices to solve systems of linear equations as long as the number of equations and variables is the same. To solve a system, adhere to the following steps:

1. **Make sure all the variables in the equations appear in the same order.**

   Replace any missing variables with zeros, and write all the variable terms on one side set equal to the constant term.

2. **Create a square coefficient matrix, *A*, by using the coefficients of the variables.**
3. **Create a column constant matrix, *B*, by using the constants in the equations.**
4. **Find the inverse of the coefficient matrix, A^{-1}.**

   You can perform this step by using the procedures shown in the earlier section, "Finding Inverse Matrices," or by using a graphing calculator.

5. **Multiply the inverse of the coefficient matrix times the constant matrix — $A^{-1} \cdot B$.**

   The resulting column matrix has the solutions, or the values of the variables, in order from top to bottom.

EXAMPLE

Q. Solve the following system of equations using matrices:

$$\begin{cases} x - 2y + 8z = 5 \\ 2x + 15z = 3y + 6 \\ 8y + 22 = 4x + 30z \end{cases}$$

A. You first rewrite the equations so that the variables appear in order and the constants appear on the right side of the equations:

$$\begin{cases} x - 2y + 8z = 5 \\ 2x - 3y + 15z = 6 \\ -4x + 8y - 30z = -22 \end{cases}$$

Now you follow Steps 2 and 3 by writing in the coefficient matrix, *A*, and the constant matrix, *B* (shown here):

$$A=\begin{bmatrix}1 & -2 & 8\\ 2 & -3 & 15\\ -4 & 8 & -30\end{bmatrix},\ B=\begin{bmatrix}5\\ 6\\ -22\end{bmatrix}$$

Step 4 is to find the inverse of matrix *A*. Use the steps from the previous section on finding an inverse.

$$\left[\begin{array}{ccc:ccc}1 & -2 & 8 & 1 & 0 & 0\\ 2 & -3 & 15 & 0 & 1 & 0\\ -4 & 8 & -30 & 0 & 0 & 1\end{array}\right]\begin{matrix}(-2)R_1+R_2\to R_2\\ (4)R_1+R_3\to R_3\end{matrix}\left[\begin{array}{ccc:ccc}1 & -2 & 8 & 1 & 0 & 0\\ 0 & 1 & -1 & -2 & 1 & 0\\ 0 & 0 & 2 & 4 & 0 & 1\end{array}\right]$$

$$\left[\begin{array}{ccc:ccc}1 & -2 & 8 & 1 & 0 & 0\\ 0 & 1 & -1 & -2 & 1 & 0\\ 0 & 0 & 2 & 4 & 0 & 1\end{array}\right]\begin{matrix}(2)R_2+R_1\to R_1\\ (0.5)R_3\to R_3\end{matrix}\left[\begin{array}{ccc:ccc}1 & 0 & 6 & -3 & 2 & 0\\ 0 & 1 & -1 & -2 & 1 & 0\\ 0 & 0 & 1 & 2 & 0 & 0.5\end{array}\right]$$

$$\left[\begin{array}{ccc:ccc}1 & 0 & 6 & -3 & 2 & 0\\ 0 & 1 & -1 & -2 & 1 & 0\\ 0 & 0 & 1 & 2 & 0 & 0.5\end{array}\right]\begin{matrix}(-6)R_3+R_1\to R_1\\ R_3+R_2\to R_2\end{matrix}\left[\begin{array}{ccc:ccc}1 & 0 & 0 & -15 & 2 & -3\\ 0 & 1 & 0 & 0 & 1 & 0.5\\ 0 & 0 & 1 & 2 & 0 & 0.5\end{array}\right]$$

Voila! The inverse matrix.

$$A^{-1}=\begin{bmatrix}-15 & 2 & -3\\ 0 & 1 & 0.5\\ 2 & 0 & 0.5\end{bmatrix}$$

Now you multiply the inverse of matrix *A* times the constant matrix, *B* (see the earlier section, "Multiplying two matrices"); you get a column matrix with all the solutions of *x*, *y*, and *z* listed in order from top to bottom:

$$A^{-1}\times B=\begin{bmatrix}-15 & 2 & -3\\ 0 & 1 & 0.5\\ 2 & 0 & 0.5\end{bmatrix}\times\begin{bmatrix}5\\ 6\\ -22\end{bmatrix}=\begin{bmatrix}3\\ -5\\ -1\end{bmatrix}$$

The column matrix tells you that $x=3$, $y=-5$, and $z=-1$.

YOUR TURN

15 Solve the system of equations using matrices:

$$\begin{cases}x+\ y+\ z=-3\\ \quad\ \ y-3z=5\\ 2x+5y\quad\ \ =2\end{cases}$$

Practice Questions Answers and Explanations

1. **C.** Matrix *C* has 2 rows and 3 columns.

2. **B, C, D.** Although, technically, a square is a special matrix, only *B*, *C*, and *D* would be designated as rectangular matrices.

3. **A.** All the elements in matrix *A* are zeros.

4. **None.** None of the matrices are square with a diagonal of 1s and 0s for all the other elements.

5. $\begin{bmatrix} \mathbf{0} & \mathbf{-1} \\ \mathbf{6} & \mathbf{0} \\ \mathbf{1} & \mathbf{8} \end{bmatrix}$. Add the like-elements: $\begin{bmatrix} -1 & 3 \\ 0 & 4 \\ -2 & 5 \end{bmatrix} + \begin{bmatrix} 1 & -4 \\ 6 & -4 \\ 3 & 3 \end{bmatrix} = \begin{bmatrix} -1+1 & 3+(-4) \\ 0+6 & 4+(-4) \\ -2+3 & 5+3 \end{bmatrix} = \begin{bmatrix} 0 & -1 \\ 6 & 0 \\ 1 & 8 \end{bmatrix}$.

6. $\begin{bmatrix} \mathbf{3} & \mathbf{3} & \mathbf{-8} & \mathbf{3} \end{bmatrix}$. Subtract the like-elements:
$\begin{bmatrix} 3 & 5 & -9 & 9 \end{bmatrix} - \begin{bmatrix} 0 & 2 & -1 & 6 \end{bmatrix} = \begin{bmatrix} 3-0 & 5-2 & -9-(-1) & 9-6 \end{bmatrix} = \begin{bmatrix} 3 & 3 & -8 & 3 \end{bmatrix}$.

7. $\begin{bmatrix} \mathbf{\frac{1}{2}} & \mathbf{-1} \\ \mathbf{0} & \mathbf{-2} \\ \mathbf{1} & \mathbf{-4} \end{bmatrix}$. Multiply each element by the scalar: $-\frac{1}{2} \cdot \begin{bmatrix} -1 & 2 \\ 0 & 4 \\ -2 & 8 \end{bmatrix} = \begin{bmatrix} -\frac{1}{2}\cdot(-1) & -\frac{1}{2}\cdot(2) \\ -\frac{1}{2}\cdot(0) & -\frac{1}{2}\cdot(4) \\ -\frac{1}{2}\cdot(-2) & -\frac{1}{2}\cdot(8) \end{bmatrix} = \begin{bmatrix} \frac{1}{2} & -1 \\ 0 & -2 \\ 1 & -4 \end{bmatrix}$.

8. $\begin{bmatrix} \mathbf{-6} & \mathbf{7} & \mathbf{2} \\ \mathbf{6} & \mathbf{15} & \mathbf{-6} \end{bmatrix}$. The resulting matrix will be 2×3.

$$\begin{bmatrix} 4 & -1 \\ 0 & 3 \end{bmatrix} \times \begin{bmatrix} -1 & 3 & 0 \\ 2 & 5 & -2 \end{bmatrix} = \begin{bmatrix} 4(-1)+(-1)2 & 4(3)+(-1)5 & 4(0)+(-1)(-2) \\ 0(-1)+(3)2 & 0(3)+(3)5 & 0(0)+(3)(-2) \end{bmatrix} = \begin{bmatrix} -6 & 7 & 2 \\ 6 & 15 & -6 \end{bmatrix}$$

9. $\begin{bmatrix} \mathbf{-2} & \mathbf{-2} & \mathbf{4} \\ \mathbf{-5} & \mathbf{-5} & \mathbf{7} \\ \mathbf{-8} & \mathbf{-8} & \mathbf{10} \end{bmatrix}$. The resulting matrix will be 3×3.

$$\begin{bmatrix} 1 & 2 & 3 \\ 4 & 5 & 6 \\ 7 & 8 & 9 \end{bmatrix} \times \begin{bmatrix} -1 & 0 & -1 \\ 1 & -1 & 1 \\ -1 & 0 & 1 \end{bmatrix} = \begin{bmatrix} 1(-1)+2(1)+3(-1) & 1(0)+2(-1)+3(0) & 1(-1)+2(1)+3(1) \\ 4(-1)+5(1)+6(-1) & 4(0)+5(-1)+6(0) & 4(-1)+5(1)+6(1) \\ 7(-1)+8(1)+9(-1) & 7(0)+8(-1)+9(0) & 7(-1)+8(1)+9(1) \end{bmatrix}$$
$$= \begin{bmatrix} -2 & -2 & 4 \\ -5 & -5 & 7 \\ -8 & -8 & 10 \end{bmatrix}$$

10. $\begin{bmatrix} \mathbf{6} & \mathbf{-9} & \mathbf{0} \\ \mathbf{6} & \mathbf{7} & \mathbf{-3} \\ \mathbf{1} & \mathbf{1} & \mathbf{2} \\ \mathbf{1} & \mathbf{1} & \mathbf{0} \end{bmatrix}$. Switch the second and fourth rows: $\begin{bmatrix} 6 & -9 & 0 \\ 1 & 1 & 0 \\ 1 & 1 & 2 \\ 6 & 7 & -3 \end{bmatrix} R_2 \leftrightarrow R_4 \begin{bmatrix} 6 & -9 & 0 \\ 6 & 7 & -3 \\ 1 & 1 & 2 \\ 1 & 1 & 0 \end{bmatrix}$.

11 $\begin{bmatrix} 6 & -9 & 0 \\ 1 & 1 & 0 \\ 0 & 0 & 2 \\ 6 & 7 & -3 \end{bmatrix}$. **Perform the operations and replace the elements in the third row:**

$$\begin{bmatrix} 6 & -9 & 0 \\ 1 & 1 & 0 \\ 1 & 1 & 2 \\ 6 & 7 & -3 \end{bmatrix} (-1)R_2 + R_3 \rightarrow R_3 \begin{bmatrix} 6 & -9 & 0 \\ 1 & 1 & 0 \\ 0 & 0 & 2 \\ 6 & 7 & -3 \end{bmatrix}.$$

12 $\begin{bmatrix} 5 & 2 \\ 7 & 3 \end{bmatrix}$. **Using the 2 × 2 rule, first switch the 3 and 5, and then change the signs of the other two elements. The difference between the cross products is 1, and dividing the new matrix by 1 doesn't change the elements.**

$$\begin{bmatrix} 3 & -2 \\ -7 & 5 \end{bmatrix} \rightarrow \begin{bmatrix} 5 & -2 \\ -7 & 3 \end{bmatrix} \rightarrow \begin{bmatrix} 5 & 2 \\ 7 & 3 \end{bmatrix}$$

13 $\begin{bmatrix} -9 & 3 & -1 \\ 11 & -3 & 2 \\ 4 & -1 & 1 \end{bmatrix}$. **Write the double-wide matrix with the original matrix on the left and the identity matrix on the right.**

$$\left[\begin{array}{ccc|ccc} 1 & 2 & -3 & 1 & 0 & 0 \\ 3 & 5 & -7 & 0 & 1 & 0 \\ -1 & -3 & 6 & 0 & 0 & 1 \end{array}\right]$$

Now create zeros in the first column under the 1.

$$\left[\begin{array}{ccc|ccc} 1 & 2 & -3 & 1 & 0 & 0 \\ 3 & 5 & -7 & 0 & 1 & 0 \\ -1 & -3 & 6 & 0 & 0 & 1 \end{array}\right] \begin{array}{c} (-3)R_1 + R_2 \rightarrow R_2 \\ R_1 + R_3 \rightarrow R_3 \end{array} \left[\begin{array}{ccc|ccc} 1 & 2 & -3 & 1 & 0 & 0 \\ 0 & -1 & 2 & -3 & 1 & 0 \\ 0 & -1 & 3 & 1 & 0 & 1 \end{array}\right]$$

Next create a +1 in the second row, second column.

$$\left[\begin{array}{ccc|ccc} 1 & 2 & -3 & 1 & 0 & 0 \\ 0 & -1 & 2 & -3 & 1 & 0 \\ 0 & -1 & 3 & 1 & 0 & 1 \end{array}\right] (-1)R_2 \rightarrow R_2 \left[\begin{array}{ccc|ccc} 1 & 2 & -3 & 1 & 0 & 0 \\ 0 & 1 & -2 & 3 & -1 & 0 \\ 0 & -1 & 3 & 1 & 0 & 1 \end{array}\right]$$

Create zeros above and below the 1 in the second row.

$$\left[\begin{array}{ccc|ccc} 1 & 2 & -3 & 1 & 0 & 0 \\ 0 & 1 & -2 & 3 & -1 & 0 \\ 0 & -1 & 3 & 1 & 0 & 1 \end{array}\right] \begin{array}{c} (-2)R_2 + R_1 \rightarrow R_1 \\ R_2 + R_3 \rightarrow R_3 \end{array} \left[\begin{array}{ccc|ccc} 1 & 0 & 1 & -5 & 2 & 0 \\ 0 & 1 & -2 & 3 & -1 & 0 \\ 0 & 0 & 1 & 4 & -1 & 1 \end{array}\right]$$

Create zeros above the 1 in the last row.

$$\left[\begin{array}{ccc|ccc} 1 & 0 & 1 & -5 & 2 & 0 \\ 0 & 1 & -2 & 3 & -1 & 0 \\ 0 & 0 & 1 & 4 & -1 & 1 \end{array}\right] \begin{array}{l} (-1)R_3 + R_1 \to R_1 \\ (2)R_3 + R_2 \to R_2 \end{array} \left[\begin{array}{ccc|ccc} 1 & 0 & 0 & -9 & 3 & -1 \\ 0 & 1 & 0 & 11 & -3 & 2 \\ 0 & 0 & 1 & 4 & -1 & 1 \end{array}\right]$$

$$V^{-1} = \begin{bmatrix} -9 & 3 & -1 \\ 11 & -3 & 2 \\ 4 & -1 & 1 \end{bmatrix}$$

(14) $\begin{bmatrix} \frac{1}{2} & -5 \\ -3\frac{3}{4} & 10\frac{1}{2} \end{bmatrix}$**. First, find the inverse of the matrix in the denominator.**

$$\begin{bmatrix} 2 & 6 \\ -1 & 3 \end{bmatrix}^{-1} = \frac{1}{12}\begin{bmatrix} 3 & -6 \\ 1 & 2 \end{bmatrix} = \begin{bmatrix} \frac{1}{4} & -\frac{1}{2} \\ \frac{1}{12} & \frac{1}{6} \end{bmatrix}$$

Now multiply this inverse times the numerator.

$$\begin{bmatrix} 6 & -12 \\ -18 & 9 \end{bmatrix} \times \begin{bmatrix} \frac{1}{4} & -\frac{1}{2} \\ \frac{1}{12} & \frac{1}{6} \end{bmatrix} = \begin{bmatrix} \frac{1}{2} & -5 \\ -3\frac{3}{4} & 10\frac{1}{2} \end{bmatrix}$$

(15) **(-4, 2, -1).** Write the coefficient matrix and constant matrix. Then find the inverse of the coefficient matrix.

$$A = \begin{bmatrix} 1 & 1 & 1 \\ 0 & 1 & -3 \\ 2 & 5 & 0 \end{bmatrix}, \quad B = \begin{bmatrix} -3 \\ 5 \\ 2 \end{bmatrix}$$

$$\left[\begin{array}{ccc|ccc} 1 & 1 & 1 & 1 & 0 & 0 \\ 0 & 1 & -3 & 0 & 1 & 0 \\ 2 & 5 & 0 & 0 & 0 & 1 \end{array}\right] (-2)R_1 + R_3 \to R_3 \left[\begin{array}{ccc|ccc} 1 & 1 & 1 & 1 & 0 & 0 \\ 0 & 1 & -3 & 0 & 1 & 0 \\ 0 & 3 & -2 & -2 & 0 & 1 \end{array}\right]$$

$$\left[\begin{array}{ccc|ccc} 1 & 1 & 1 & 1 & 0 & 0 \\ 0 & 1 & -3 & 0 & 1 & 0 \\ 0 & 3 & -2 & -2 & 0 & 1 \end{array}\right] \begin{array}{l} (-1)R_2 + R_1 \to R_1 \\ (-3)R_2 + R_3 \to R_3 \end{array} \left[\begin{array}{ccc|ccc} 1 & 0 & 4 & 1 & -1 & 0 \\ 0 & 1 & -3 & 0 & 1 & 0 \\ 0 & 0 & 7 & -2 & -3 & 1 \end{array}\right]$$

$$\left[\begin{array}{ccc|ccc} 1 & 0 & 4 & 1 & -1 & 0 \\ 0 & 1 & -3 & 0 & 1 & 0 \\ 0 & 0 & 7 & -2 & -3 & 1 \end{array}\right] \left(\frac{1}{7}\right)R_3 \to R_3 \left[\begin{array}{ccc|ccc} 1 & 0 & 4 & 1 & -1 & 0 \\ 0 & 1 & -3 & 0 & 1 & 0 \\ 0 & 0 & 1 & -\frac{2}{7} & -\frac{3}{7} & \frac{1}{7} \end{array}\right]$$

$$\left[\begin{array}{ccc:ccc} 1 & 0 & 4 & 1 & -1 & 0 \\ 0 & 1 & -3 & 0 & 1 & 0 \\ 0 & 0 & 1 & -\frac{2}{7} & -\frac{3}{7} & \frac{1}{7} \end{array}\right] \begin{array}{l} (-4)R_3 + R_1 \rightarrow R_1 \\ (3)R_3 + R_2 \rightarrow R_2 \end{array} \left[\begin{array}{ccc:ccc} 1 & 0 & 0 & \frac{15}{7} & \frac{5}{7} & -\frac{4}{7} \\ 0 & 1 & 0 & -\frac{6}{7} & -\frac{2}{7} & \frac{3}{7} \\ 0 & 0 & 1 & -\frac{2}{7} & -\frac{3}{7} & \frac{1}{7} \end{array}\right]$$

The inverse of matrix A is $A^{-1} = \begin{bmatrix} \frac{15}{7} & \frac{5}{7} & -\frac{4}{7} \\ -\frac{6}{7} & -\frac{2}{7} & \frac{3}{7} \\ -\frac{2}{7} & -\frac{3}{7} & \frac{1}{7} \end{bmatrix}$. Multiply this times the constant matrix:

$$\begin{bmatrix} \frac{15}{7} & \frac{5}{7} & -\frac{4}{7} \\ -\frac{6}{7} & -\frac{2}{7} & \frac{3}{7} \\ -\frac{2}{7} & -\frac{3}{7} & \frac{1}{7} \end{bmatrix} \times \begin{bmatrix} -3 \\ 5 \\ 2 \end{bmatrix} = \begin{bmatrix} -4 \\ 2 \\ -1 \end{bmatrix}.$$

Whaddya Know? Chapter 18 Quiz

Quiz time! Complete each problem to test your knowledge on the various topics covered in this chapter. You can then find the solutions and explanations in the next section.

1. Find the sum of the matrices *A* and *B*.

$$A=\begin{bmatrix}2 & -3 & 1 & 0\\ 0 & -5 & 2 & -2\\ 3 & 1 & 4 & 0\end{bmatrix}\quad B=\begin{bmatrix}4 & -2 & 0 & 3\\ -3 & -5 & 1 & 1\\ 0 & -1 & -2 & 6\end{bmatrix}$$

2. Perform the row operations indicated.

$$\begin{bmatrix}15 & 5 & -4\\ -6 & -2 & 3\\ -2 & -3 & 1\end{bmatrix}(-2)R_3+R_2\rightarrow R_2$$

3. Solve the system of equations using matrices: $\begin{cases}x+2y+3z=9\\ 2x+5y+5z=12.\\ 3x+8y+4z=3\end{cases}$

4. Multiply the matrices: $A\cdot B$.

$$A=\begin{bmatrix}3 & -2\\ 4 & 1\end{bmatrix},\ B=\begin{bmatrix}2 & 1\\ -3 & 0\end{bmatrix}$$

5. Find the inverse of the matrix $A=\begin{bmatrix}4 & -1\\ -7 & 2\end{bmatrix}$.

6. Perform the operations on the matrices: $A+3B-C$.

$$A=\begin{bmatrix}2 & -1 & 3\\ 3 & 2 & -5\\ -2 & 0 & 2\end{bmatrix},\ B=\begin{bmatrix}-1 & 0 & -2\\ 2 & 3 & -2\\ -2 & 4 & -5\end{bmatrix},\ C=\begin{bmatrix}3 & -1 & 2\\ 0 & -5 & 3\\ -1 & -1 & 0\end{bmatrix}$$

7. Divide matrix *A* by matrix *B*.

$$A=\begin{bmatrix}3 & -2\\ 4 & 1\end{bmatrix},\ B=\begin{bmatrix}2 & 1\\ -3 & -1\end{bmatrix}$$

8. Solve the system of equations using matrices: $\begin{cases}x+y-z=7\\ 2x+y+z=9.\\ x-y+z=1\end{cases}$

9. Multiply the matrices: $A\cdot B$.

$$A=\begin{bmatrix}1 & 3\\ 4 & -2\\ 0 & 7\end{bmatrix},\ B=\begin{bmatrix}2 & -3 & 0\\ 1 & 1 & 4\end{bmatrix}$$

10 Find the additive inverse of the matrix C.

$$C = \begin{bmatrix} 2 & -1 & 3 \\ 3 & 2 & -5 \\ -2 & 0 & 2 \end{bmatrix}$$

11 Multiply the matrices: $E \cdot F$.

$$E = \begin{bmatrix} 2 & 7 & -3 & -1 & 5 \\ 1 & -1 & 3 & 3 & 6 \end{bmatrix}, F = \begin{bmatrix} 4 \\ 0 \\ 1 \\ 3 \\ -2 \end{bmatrix}$$

12 Find the inverse of the matrix $D = \begin{bmatrix} 1 & 2 & 3 \\ 1 & 1 & 4 \\ 2 & 5 & 4 \end{bmatrix}$.

Answers to Chapter 18 Quiz

1. $\begin{bmatrix} 6 & -5 & 1 & 3 \\ -3 & -10 & 3 & -1 \\ 3 & 0 & 2 & 6 \end{bmatrix}$. **Find the sums of the corresponding elements.**

$$A+B=\begin{bmatrix} 2+4 & -3+(-2) & 1+0 & 0+3 \\ 0+(-3) & -5+(-5) & 2+1 & -2+1 \\ 3+0 & 1+(-1) & 4+(-2) & 0+6 \end{bmatrix}=\begin{bmatrix} 6 & -5 & 1 & 3 \\ -3 & -10 & 3 & -1 \\ 3 & 0 & 2 & 6 \end{bmatrix}$$

2. $\begin{bmatrix} 15 & 5 & -4 \\ -2 & 4 & 1 \\ -2 & -3 & 1 \end{bmatrix}$. **Multiply –2 times Row 3 and add this to Row 2.**

$$\begin{bmatrix} 15 & 5 & -4 \\ -6 & -2 & 3 \\ -2 & -3 & 1 \end{bmatrix} (-2)R_3+R_2 \rightarrow R_2 \begin{bmatrix} 15 & 5 & -4 \\ -6+4 & -2+6 & 3+(-2) \\ -2 & -3 & 1 \end{bmatrix}=\begin{bmatrix} 15 & 5 & -4 \\ -2 & 4 & 1 \\ -2 & -3 & 1 \end{bmatrix}$$

3. $x=1, y=-2, z=4$. **First, write the matrix of the coefficients and then put that together with the identity matrix to solve for the inverse.**

$$\left[\begin{array}{ccc:ccc} 1 & 2 & 3 & 1 & 0 & 0 \\ 2 & 5 & 5 & 0 & 1 & 0 \\ 3 & 8 & 4 & 0 & 0 & 1 \end{array}\right] \begin{matrix} (-2)R_1+R_2 \rightarrow R_2 \\ (-3)R_1+R_3 \rightarrow R_3 \end{matrix} \left[\begin{array}{ccc:ccc} 1 & 2 & 3 & 1 & 0 & 0 \\ 0 & 1 & -1 & -2 & 1 & 0 \\ 0 & 2 & -5 & -3 & 0 & 1 \end{array}\right]$$

$$\left[\begin{array}{ccc:ccc} 1 & 2 & 3 & 1 & 0 & 0 \\ 0 & 1 & -1 & -2 & 1 & 0 \\ 0 & 2 & -5 & -3 & 0 & 1 \end{array}\right] \begin{matrix} (-2)R_2+R_1 \rightarrow R_1 \\ (-2)R_2+R_3 \rightarrow R_3 \end{matrix} \left[\begin{array}{ccc:ccc} 1 & 0 & 5 & 5 & -2 & 0 \\ 0 & 1 & -1 & -2 & 1 & 0 \\ 0 & 0 & -3 & 1 & -2 & 1 \end{array}\right]$$

$$\left[\begin{array}{ccc:ccc} 1 & 0 & 5 & 5 & -2 & 0 \\ 0 & 1 & -1 & -2 & 1 & 0 \\ 0 & 0 & -3 & 1 & -2 & 1 \end{array}\right] \left(-\frac{1}{3}\right)R_3 \rightarrow R_3 \left[\begin{array}{ccc:ccc} 1 & 0 & 5 & 5 & -2 & 0 \\ 0 & 1 & -1 & -2 & 1 & 0 \\ 0 & 0 & 1 & -\frac{1}{3} & \frac{2}{3} & -\frac{1}{3} \end{array}\right]$$

$$\left[\begin{array}{ccc:ccc} 1 & 0 & 5 & 5 & -2 & 0 \\ 0 & 1 & -1 & -2 & 1 & 0 \\ 0 & 0 & 1 & -\frac{1}{3} & \frac{2}{3} & -\frac{1}{3} \end{array}\right] \begin{matrix} (-5)R_3+R_1 \rightarrow R_1 \\ R_3+R_2 \rightarrow R_2 \end{matrix} \left[\begin{array}{ccc:ccc} 1 & 0 & 0 & \frac{20}{3} & -\frac{16}{3} & \frac{5}{3} \\ 0 & 1 & 0 & -\frac{7}{3} & \frac{5}{3} & -\frac{1}{3} \\ 0 & 0 & 1 & -\frac{1}{3} & \frac{2}{3} & -\frac{1}{3} \end{array}\right]$$

Now multiply the inverse matrix times the column matrix of constants from the system of equations.

$$\begin{bmatrix} \frac{20}{3} & -\frac{16}{3} & \frac{5}{3} \\ -\frac{7}{3} & \frac{5}{3} & -\frac{1}{3} \\ -\frac{1}{3} & \frac{2}{3} & -\frac{1}{3} \end{bmatrix} \times \begin{bmatrix} 9 \\ 12 \\ 3 \end{bmatrix} = \begin{bmatrix} 1 \\ -2 \\ 4 \end{bmatrix}$$

④ $\begin{bmatrix} 12 & 3 \\ 5 & 4 \end{bmatrix}$. $A \cdot B = \begin{bmatrix} 3\cdot 2+(-2)(-3) & 3\cdot 1+1\cdot 0 \\ 4\cdot 2+1(-3) & 4\cdot 1+1\cdot 0 \end{bmatrix} = \begin{bmatrix} 12 & 3 \\ 5 & 4 \end{bmatrix}$

⑤ $A^{-1} = \begin{bmatrix} 2 & 1 \\ 7 & 4 \end{bmatrix}$. Using the formula, $A^{-1} = \begin{bmatrix} \frac{2}{1} & \frac{-(-1)}{1} \\ \frac{-(-7)}{1} & \frac{4}{1} \end{bmatrix} = \begin{bmatrix} 2 & 1 \\ 7 & 4 \end{bmatrix}$.

⑥ $\begin{bmatrix} -4 & 0 & -5 \\ 9 & 16 & -14 \\ -7 & 13 & -13 \end{bmatrix}$. First, perform the scalar multiplication on matrix B. Then find the sum and difference of the corresponding elements.

$$A+3B-C = \begin{bmatrix} 2 & -1 & 3 \\ 3 & 2 & -5 \\ -2 & 0 & 2 \end{bmatrix} + \begin{bmatrix} -3 & 0 & -6 \\ 6 & 9 & -6 \\ -6 & 12 & -15 \end{bmatrix} - \begin{bmatrix} 3 & -1 & 2 \\ 0 & -5 & 3 \\ -1 & -1 & 0 \end{bmatrix} = \begin{bmatrix} -4 & 0 & -5 \\ 9 & 16 & -14 \\ -7 & 13 & -13 \end{bmatrix}$$

⑦ $\begin{bmatrix} -9 & -7 \\ -1 & -2 \end{bmatrix}$. Find the inverse of matrix B using the formula. Then multiply A times that inverse.

$$B^{-1} = \begin{bmatrix} \frac{-1}{1} & \frac{-1}{1} \\ \frac{3}{1} & \frac{2}{1} \end{bmatrix} = \begin{bmatrix} -1 & -1 \\ 3 & 2 \end{bmatrix} \text{ then } A \cdot B^{-1} = \begin{bmatrix} 3 & -2 \\ 4 & 1 \end{bmatrix} \times \begin{bmatrix} -1 & -1 \\ 3 & 2 \end{bmatrix} = \begin{bmatrix} -3+(-6) & -3+(-4) \\ -4+3 & -4+2 \end{bmatrix} = \begin{bmatrix} -9 & -7 \\ -1 & -2 \end{bmatrix}$$

⑧ $x = 4, y = 2, z = -1$. First, write the matrix of the coefficients and then put that together with the identity matrix to solve for the inverse.

$$\left[\begin{array}{ccc|ccc} 1 & 1 & -1 & 1 & 0 & 0 \\ 2 & 1 & 1 & 0 & 1 & 0 \\ 1 & -1 & 1 & 0 & 0 & 1 \end{array}\right] \begin{array}{l} (-2)R_1 + R_2 \to R_2 \\ (-1)R_1 + R_3 \to R_3 \end{array} \left[\begin{array}{ccc|ccc} 1 & 1 & -1 & 1 & 0 & 0 \\ 0 & -1 & 3 & -2 & 1 & 0 \\ 0 & -2 & 2 & -1 & 0 & 1 \end{array}\right]$$

$$\left[\begin{array}{ccc|ccc} 1 & 1 & -1 & 1 & 0 & 0 \\ 0 & -1 & 3 & -2 & 1 & 0 \\ 0 & -2 & 2 & -1 & 0 & 1 \end{array}\right] (-1)R_2 \to R_2 \left[\begin{array}{ccc|ccc} 1 & 1 & -1 & 1 & 0 & 0 \\ 0 & 1 & -3 & 2 & -1 & 0 \\ 0 & -2 & 2 & -1 & 0 & 1 \end{array}\right]$$

$$\left[\begin{array}{ccc|ccc} 1 & 1 & -1 & 1 & 0 & 0 \\ 0 & 1 & -3 & 2 & -1 & 0 \\ 0 & -2 & 2 & -1 & 0 & 1 \end{array}\right] \begin{array}{l} (-1)R_2 + R_1 \to R_1 \\ (2)R_2 + R_3 \to R_3 \end{array} \left[\begin{array}{ccc|ccc} 1 & 0 & 2 & -1 & 1 & 0 \\ 0 & 1 & -3 & 2 & -1 & 0 \\ 0 & 0 & -4 & 3 & -2 & 1 \end{array}\right]$$

$$\left[\begin{array}{ccc|ccc} 1 & 0 & 2 & -1 & 1 & 0 \\ 0 & 1 & -3 & 2 & -1 & 0 \\ 0 & 0 & -4 & 3 & -2 & 1 \end{array}\right] \left(-\frac{1}{4}\right)R_3 \to R_3 \left[\begin{array}{ccc|ccc} 1 & 0 & 2 & -1 & 1 & 0 \\ 0 & 1 & -3 & 2 & -1 & 0 \\ 0 & 0 & 1 & -\frac{3}{4} & \frac{1}{2} & -\frac{1}{4} \end{array}\right]$$

$$\left[\begin{array}{ccc|ccc} 1 & 0 & 2 & -1 & 1 & 0 \\ 0 & 1 & -3 & 2 & -1 & 0 \\ 0 & 0 & 1 & -\frac{3}{4} & \frac{1}{2} & -\frac{1}{4} \end{array}\right] \begin{array}{l} (-2)R_3 + R_1 \to R_1 \\ (3)R_3 + R_2 \to R_2 \end{array} \left[\begin{array}{ccc|ccc} 1 & 0 & 0 & \frac{1}{2} & 0 & \frac{1}{2} \\ 0 & 1 & 0 & -\frac{1}{4} & \frac{1}{2} & -\frac{3}{4} \\ 0 & 0 & 1 & -\frac{3}{4} & \frac{1}{2} & -\frac{1}{4} \end{array}\right]$$

Now multiply the inverse matrix times the column matrix of constants from the system of equations.

$$\begin{bmatrix} \frac{1}{2} & 0 & \frac{1}{2} \\ -\frac{1}{4} & \frac{1}{2} & -\frac{3}{4} \\ -\frac{3}{4} & \frac{1}{2} & -\frac{1}{4} \end{bmatrix} \times \begin{bmatrix} 7 \\ 9 \\ 1 \end{bmatrix} = \begin{bmatrix} 4 \\ 2 \\ -1 \end{bmatrix}$$

(9) $\begin{bmatrix} \mathbf{5} & \mathbf{0} & \mathbf{12} \\ \mathbf{6} & \mathbf{-14} & \mathbf{-8} \\ \mathbf{7} & \mathbf{7} & \mathbf{28} \end{bmatrix}$

$$\begin{bmatrix} 1 & 3 \\ 4 & -2 \\ 0 & 7 \end{bmatrix} \times \begin{bmatrix} 2 & -3 & 0 \\ 1 & 1 & 4 \end{bmatrix} = \begin{bmatrix} 1\cdot2+3\cdot1 & 1(-3)+3\cdot1 & 1\cdot0+3\cdot4 \\ 4\cdot2+(-2)\cdot1 & 4(-3)+(-2)\cdot1 & 4\cdot0+(-2)\cdot4 \\ 0\cdot2+7\cdot1 & 0(-3)+7\cdot1 & 0\cdot0+7\cdot4 \end{bmatrix} = \begin{bmatrix} 5 & 0 & 12 \\ 6 & -14 & -8 \\ 7 & 7 & 28 \end{bmatrix}$$

(10) $\begin{bmatrix} \mathbf{-2} & \mathbf{1} & \mathbf{-3} \\ \mathbf{-3} & \mathbf{-2} & \mathbf{5} \\ \mathbf{2} & \mathbf{0} & \mathbf{-2} \end{bmatrix}$. **Change each element to the opposite number.**

(11) $\begin{bmatrix} \mathbf{-8} \\ \mathbf{4} \end{bmatrix}$. $E\cdot F = \begin{bmatrix} 2 & 7 & -3 & -1 & 5 \\ 1 & -1 & 3 & 3 & 6 \end{bmatrix} \times \begin{bmatrix} 4 \\ 0 \\ 1 \\ 3 \\ -2 \end{bmatrix} = \begin{bmatrix} 2\cdot4+7\cdot0+(-3)\cdot1+(-1)\cdot3+5(-2) \\ 1\cdot4+(-1)\cdot0+3\cdot1+3\cdot3+6(-2) \end{bmatrix} = \begin{bmatrix} -8 \\ 4 \end{bmatrix}$

(12) $\boldsymbol{D^{-1}} = \begin{bmatrix} \mathbf{-1} & \mathbf{0} & \mathbf{-5} \\ \mathbf{1} & \mathbf{0} & \mathbf{-1} \\ \mathbf{0} & \mathbf{0} & \mathbf{-1} \end{bmatrix}$. **Set up the original plus identity matrix:** $\begin{bmatrix} 1 & 2 & 3 & \vdots & 1 & 0 & 0 \\ 1 & 1 & 4 & \vdots & 0 & 1 & 0 \\ 2 & 5 & 4 & \vdots & 0 & 0 & 1 \end{bmatrix}$.

$$\begin{bmatrix} 1 & 2 & 3 & \vdots & 1 & 0 & 0 \\ 1 & 1 & 4 & \vdots & 0 & 1 & 0 \\ 2 & 5 & 4 & \vdots & 0 & 0 & 1 \end{bmatrix} \begin{matrix} (-1)R_1 + R_2 \to R_2 \\ (-2)R_1 + R_3 \to R_3 \end{matrix} \begin{bmatrix} 1 & 2 & 3 & \vdots & 1 & 0 & 0 \\ 0 & -1 & 1 & \vdots & -1 & 1 & 0 \\ 0 & 1 & -2 & \vdots & -2 & 0 & 1 \end{bmatrix}$$

$$\begin{bmatrix} 1 & 2 & 3 & \vdots & 1 & 0 & 0 \\ 0 & -1 & 1 & \vdots & -1 & 1 & 0 \\ 0 & 1 & -2 & \vdots & -2 & 0 & 1 \end{bmatrix} (-1)R_2 \to R_2 \begin{bmatrix} 1 & 2 & 3 & \vdots & 1 & 0 & 0 \\ 0 & 1 & -1 & \vdots & 1 & -1 & 0 \\ 0 & 1 & -2 & \vdots & -2 & 0 & 1 \end{bmatrix}$$

$$\begin{bmatrix} 1 & 2 & 3 & \vdots & 1 & 0 & 0 \\ 0 & 1 & -1 & \vdots & 1 & -1 & 0 \\ 0 & 1 & -2 & \vdots & -2 & 0 & 1 \end{bmatrix} \begin{matrix} (-2)R_2 + R_1 \to R_1 \\ (-1)R_2 + R_3 \to R_3 \end{matrix} \begin{bmatrix} 1 & 0 & 5 & \vdots & -1 & 2 & 0 \\ 0 & 1 & -1 & \vdots & 1 & -1 & 0 \\ 0 & 0 & -1 & \vdots & -3 & 1 & 1 \end{bmatrix}$$

$$\begin{bmatrix} 1 & 0 & 5 & \vdots & -1 & 2 & 0 \\ 0 & 1 & -1 & \vdots & 1 & -1 & 0 \\ 0 & 0 & -1 & \vdots & -3 & 1 & 1 \end{bmatrix} (-1)R_3 \to R_3 \begin{bmatrix} 1 & 0 & 5 & \vdots & -1 & 2 & 0 \\ 0 & 1 & -1 & \vdots & 1 & -1 & 0 \\ 0 & 0 & 1 & \vdots & 3 & -1 & -1 \end{bmatrix}$$

$$\begin{bmatrix} 1 & 0 & 5 & \vdots & -1 & 2 & 0 \\ 0 & 1 & -1 & \vdots & 1 & -1 & 0 \\ 0 & 0 & 1 & \vdots & 3 & -1 & -1 \end{bmatrix} \begin{matrix} (-5)R_3 + R_1 \to R_1 \\ R_3 + R_2 \to R_2 \end{matrix} \begin{bmatrix} 1 & 0 & 0 & \vdots & -16 & 7 & 5 \\ 0 & 1 & 0 & \vdots & 4 & -2 & -1 \\ 0 & 0 & 1 & \vdots & 3 & -1 & -1 \end{bmatrix}$$

$$D^{-1} = \begin{bmatrix} -16 & 7 & 5 \\ 4 & -2 & -1 \\ 3 & -1 & -1 \end{bmatrix}$$

IN THIS CHAPTER

» Familiarizing yourself with sequence terminology

» Working with arithmetic and geometric sequences

» Coming to terms with recursively defined functions

» Stepping up from sequences to series

» Recognizing sequences at work in the real world

» Discussing and using special formulas for sequences and series

Chapter 19 Seeking Out Sequences and Series

A sequence is a list of items or individuals — and because this is an algebra book, the sequences you see feature lists of numbers. A series is the sum of the numbers in a list. These concepts pop up in many areas of life. For example, you can make a list of the number of seats in each row at a movie theater. With this list, you can add the numbers to find the total number of seats. You may prefer to see situations where the number of items in a list isn't random; you may like it when the number follows a pattern or rule. You can describe the patterns formed by elements in a sequence with mathematical expressions containing mathematical symbols and operations. In this chapter, you discover how to describe the terms in sequences and, when you get lucky, how to add as many of the terms as you want without too much fuss or bother.

Understanding Sequence Terminology

A sequence of events consists of two or more happenings in which one item or event follows another, which follows another, and so on. In mathematics, a sequence is a list of terms, or numbers, created with some sort of mathematical rule. For instance, consider the following rule given in the braces (where the n represents the counting numbers [1, 2, 3,...]). The rule $\{3+4n\}$ says that the numbers in a particular sequence start with the number 7 and increase by four with each additional term. The numbers in the sequence are $7, 11, 15, \ldots$

You call the three dots following a short list of terms an *ellipsis* (not to be confused with the ellipse from Chapter 13). You use them in place of et cetera (abbreviated etc.) or and so on. To get even more specific, here's the formal definition of a sequence: a function whose domain consists of positive integers $(1, 2, 3, \ldots)$. This is really a nice feature — having to deal with only the positive integers in the domain. The rest of this section has many more useful tidbits about sequences, guaranteed to leave your brain satisfied.

Using sequence notation

One big clue that you're dealing with a sequence is when you see something like $\{7, 10, 13, 16, 19, \ldots\}$ or $\{a_n\}$. The braces, $\{\,\}$, indicate that you have a list of items, called *terms;* commas usually separate the terms in a list from one another. The term a_n is the notation for the rule that represents a particular sequence. When identifying a sequence, you can list the terms in the sequence, showing enough terms to establish a pattern, or you can give the rule that creates the terms.

Because the terms in a sequence are linked to positive integers, you can refer to them by their positions in the listing of the integers. If the rule for a sequence is $\{a_n\}$, for example, the terms in the sequence are named a_1, a_2, a_3, a_4, and so on. This nice, orderly arrangement allows you to ask for the tenth term in the sequence.

EXAMPLE

Q. Given $\{a_n\} = \{2n+1\}$, write the first seven terms of the sequence.

A. The sequence consists of the terms $\{3, 5, 7, 9, 11, 13, \ldots\}$. The expression $2n+1$ is the rule that creates the sequence when you insert all the positive integers in place of the n. When $n=1$, $2(1)+1=3$; when $n=2$, $2(2)+1=5$; and so on. The domain of a sequence is all positive integers (counting numbers), so the process is easy as 1, 2, 3.

Q. Given $\{a_n\} = \{n^2-1\}$, what is the tenth term?

A. $a_{10} = 10^2 - 1 = 99$. You don't have to write the first nine terms to get to the tenth one. Sequence notation is a timesaver!

No-fear factorials in sequences

A mathematical operation you see in many sequences is the factorial. The symbol for the factorial is an exclamation mark. I first introduce the factorial operation in Chapter 1.

Here's the formula for the factorial in a sequence: $n! = n(n-1)(n-2)(n-3)\ldots 3\cdot 2\cdot 1$. When you perform a factorial, you multiply the number you're operating on times every positive integer smaller than that number.

EXAMPLE

Q. Determine the values of 6! and 9!

A. $6! = 6\cdot 5\cdot 4\cdot 3\cdot 2\cdot 1 = 720$, and $9! = 9\cdot 8\cdot 7\cdot 6\cdot 5\cdot 4\cdot 3\cdot 2\cdot 1 = 362{,}880$.

Q. If $\{c_n\} = \{n! - n\}$, then what are the first four terms of the sequence?

A. You write that $c_1 = 1! - 1 = 1 - 1 = 0$, $c_2 = 2! - 2 = 2\cdot 1 - 2 = 0$, $c_3 = 3! - 3 = 3\cdot 2\cdot 1 - 3 = 6 - 3 = 3$, $c_4 = 4! - 4 = 4\cdot 3\cdot 2\cdot 1 - 4 = 24 - 4 = 20$. You write the terms in the sequence as $\{0, 0, 3, 20, \ldots\}$.

REMEMBER

You apply a special rule for 0! (zero factorial). The rule is that $0! = 1$. "What?" I hear you asking, "How can that possibly be?" Well, it is. Mathematicians discovered that assigning the number 1 to 0! makes every operation involving factorials work out better. (Why you need this rule for 0! becomes more apparent in Chapter 20. For now, just be content to use the rule when writing the terms in a sequence.)

YOUR TURN

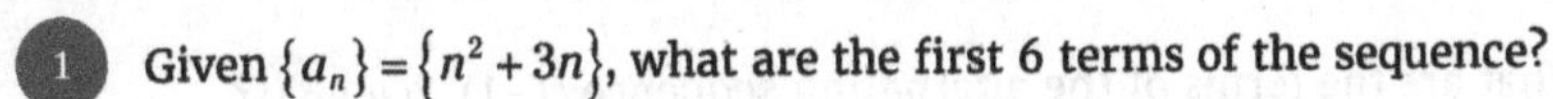

1 Given $\{a_n\} = \{n^2 + 3n\}$, what are the first 6 terms of the sequence?

2 Given $\{b_n\} = \{2^n\text{I} - n\}$, what are the first 6 terms of the sequence?

3 Given $\{c_n\} = \left\{\frac{(n+1)!}{2}\right\}$, what are the first 6 terms of the sequence?

4 Given $\{d_n\} = \{2 + 3(n+1)\}$, what are the first 6 terms of the sequence?

Alternating sequential patterns

One special type of sequence is an *alternating* sequence. An alternating sequence has terms that forever alternate back and forth from positive to negative to positive. The terms in an alternating sequence have a multiplier of −1, which is raised to some power such as n, $n-1$, or $n+1$. Adding the power, which is related to the number of the term, to the −1 causes the terms to alternate because the positive integers alternate between even and odd. Even powers of −1 are equal to +1, and odd powers of −1 are equal to −1.

EXAMPLE

Q. What are the terms of the alternating sequence $\{(-1)^n 2(n+3)\}$?

A. $a_1 = (-1)^1 \cdot 2(1+3) = -1 \cdot 2(4) = -8$ $\quad a_2 = (-1)^2 \cdot 2(2+3) = 1 \cdot 2(5) = 10$

$a_3 = (-1)^3 \cdot 2(3+3) = -1 \cdot 2(6) = -12$ $\quad a_3 = (-1)^4 \cdot 2(4+3) = 1 \cdot 2(7) = 14$, and so on.

So, the terms of the alternating sequence $\{(-1)^n 2(n+3)\} = \{-8, 10, -12, 14, \ldots\}$.

Q. Write the first four terms of the sequence $\left\{(-1)^n \frac{(n+1)!}{n}\right\}$.

$a_1 = (-1)^1 \frac{(1+1)!}{1} = -1\left[\frac{2!}{1}\right] = -1\left[\frac{2}{1}\right] = -2$ $\quad a_2 = (-1)^2 \frac{(2+1)!}{2} = +1\left[\frac{3!}{2}\right] = 1\left[\frac{6}{2}\right] = 3$

$a_3 = (-1)^3 \frac{(3+1)!}{3} = -1\left[\frac{4!}{3}\right] = -1\left[\frac{24}{3}\right] = -8$ $\quad a_4 = (-1)^4 \frac{(4+1)!}{4} = +1\left[\frac{5!}{4}\right] = 1\left[\frac{120}{4}\right] = 30$

So, the sequence is $\{-2, 3, -8, 30, \ldots\}$. You can see how the absolute value of the terms keeps getting larger while the terms alternate between positive and negative.

YOUR TURN

5 Write the first four terms of the sequence $\left\{(-1)^n \frac{n!}{n+1}\right\}$.

6 Write the first four terms of the sequence $\{(-1)^{n+1}(n-1)!\}$.

Looking for sequential patterns

The list of terms in a sequence may or may not display an apparent pattern. Of course, if you see the function rule — the rule that tells you how to create all the terms in the sequence — you have a huge hint about the pattern of the terms. You can always list the terms of a sequence if you have the rule, and you can often create the rule when you have enough terms in the sequence to figure out the pattern.

The patterns you can look for range from simple to a bit tricky:

- A single-number difference between each term, such as: 4, 9, 14, 19, ..., where the difference between each term is 5
- A multiplier separating the terms, such as multiplying by 5 to get: 2, 10, 50, 250, ...
- A pattern within a pattern, such as with the numbers: 2, 5, 9, 14, 20, ..., where the differences between the numbers get bigger by 1 each time

When you have to figure out a pattern and write a rule for a sequence of numbers, you can refer to your list of possibilities — the ones I mention and others — and see which type of rule applies.

Difference between terms

The quickest, easiest pattern to find features a common difference between the terms. A difference between two numbers is the result of subtracting the previous term from the term in question. You can usually tell when you have a sequence of this type by inspecting it — looking at how far apart the numbers sit on the number line.

The following three sequences have something in common: The terms in the sequences have a common first difference, a common second difference, or a common third difference.

EXAMPLE

Q. Given the sequence $\{2, 7, 12, 17, 22, 27, \ldots\}$, find the common difference between the terms and write the rule that applies to the sequence.

A. Writing the terms and the differences between them:

$$\begin{array}{ccccccccccc} 2 & & 7 & & 12 & & 17 & & 22 & & 27 \\ & \vee & & \vee & & \vee & & \vee & & \vee & \\ & 5 & & 5 & & 5 & & 5 & & 5 & \end{array}$$

The rule for this example sequence is $\{5n-3\}$. You use the multiplier 5 to make the terms in the sequence each 5 more than the previous term. You subtract the 3 because when you replace n with 1, you get a number that's too big; you want to start with the number 2, so you subtract 3 from the first multiple. Sequences with a common first difference are called *arithmetic sequences*. (I cover these sequences thoroughly in the later section, "Taking Note of Arithmetic and Geometric Sequences".)

Q. Given the sequence $\{-2, 1, 6, 13, 22, 33, 46, \ldots\}$, find the common *second* difference and write the rule that applies.

A. When the second difference of the terms in a sequence is a constant, like 2, the rule for that sequence is usually quadratic (see Chapter 3); it contains the term n^2. The terms and the second differences are

−2 1 6 13 22 33 46

V V V V V V

3 5 7 9 11 13

V V V V V

2 2 2 2 2

The rule used to create this example sequence is $\{n^2-3\}$.

I can't give you a quick, easy way to find the specific rules, but if you know that a rule should be quadratic, you have a place to start. You can try squaring the numbers 1, 2, 3, and so on, and then see how you have to adjust the squares by subtracting or adding so that the numbers in the sequence appear according to the rule.

Q. Given the sequence $\{0, 6, 24, 60, 120, 210, 336, \ldots\}$, find the common *third* difference and write the rule that applies.

A. The sequence features a common third difference of 6. You see that the row below the sequences shows the first differences; under the first differences are the second differences; and, finally, the third row contains the third differences:

0 6 24 60 120 210 336

V V V V V V

6 18 36 60 90 126

V V V V V

12 18 24 30 36

V V V V

6 6 6 6

The rule for this example sequence is $\{n^3-n\}$, which contains a cubed term.

The rule for this sequence doesn't necessarily leap off the page when you look at the terms. You have to play around with the terms a bit to determine the rule. Start with a cubed term and then try subtracting or adding constant numbers. If that doesn't help, try adding or subtracting squares of numbers or just multiples of numbers. Sounds sort of haphazard but you have limited options without the use of some calculus. Graphing calculators have curve-fitting features that take data and figure out the rules for you, but you still have to choose which kinds of rules (what powers) make the data work.

Multiples and powers

Some sequences have fairly apparent rules that generate their terms, because each term is a multiple or a power of some constant number.

But sometimes sequences aren't quite so obvious. They may start with a multiple that isn't 1 times the constant. And then you'll have sequences that have more than one thing going on to create the terms!

Another type of sequence is one that can have terms that are all powers of the same number. These sequences are called *geometric sequences*. (I discuss them at great length in the later section, "Taking Note of Arithmetic and Geometric Sequences.") If you're familiar with the powers of the first few integers, this will be most helpful.

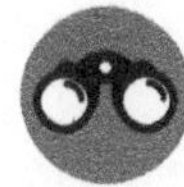

EXAMPLE

Q. Find a rule for the sequence $\{3, 6, 9, 12, 15, 18, \ldots\}$.

A. This sequence consists of multiples of 3, and its rule is $\{3n\}$.

Q. Find a rule for the sequence $\{21, 24, 27, 30, 33, 36, \ldots\}$.

A. The terms are all multiples of 3, but $\{3n\}$ doesn't work because you have to start with $n=1$. Remember, the domain or input of a sequence is made up of positive integers $(1, 2, 3, \ldots)$, so you can't use anything smaller than 1 for n. The way to get around starting sequences with smaller numbers is to add a constant to n (which is like a counter). The number 21 is $3 \cdot 7$, so add 6 to n in the parentheses to form the rule $\{3(n+6)\}$.

Q. Find a rule for the sequence $\left\{1, -\frac{1}{2}, \frac{1}{3}, -\frac{1}{4}, \frac{1}{5}, -\frac{1}{6}, \ldots\right\}$.

A. This sequence has two interesting features: The consecutive terms alternate their signs and have consecutive integers in their denominators. To write the rule for this sequence, consider the two features. The alternating terms suggest a multiplier of -1. The first, third, fifth, and all other odd terms are positive, so you can create alternating multiples of -1 that are positive by raising a -1 factor to $(n+1)$. This makes those exponents even when n is odd and all the odd terms positive. For the fractions, you can put n, the term's number, in the denominator. The rule for this sequence, therefore, is as follows: $\left\{(-1)^{n+1}\frac{1}{n}\right\} = \left\{\frac{(-1)^{n+1}}{n}\right\}$.

Q. Find a rule for the sequence $\{2, 4, 8, 16, 32, 64, 128, \ldots\}$.

A. This is an example of a geometric sequence. You can see that these terms are powers of the number 2, and the rule for the terms is 2^n.

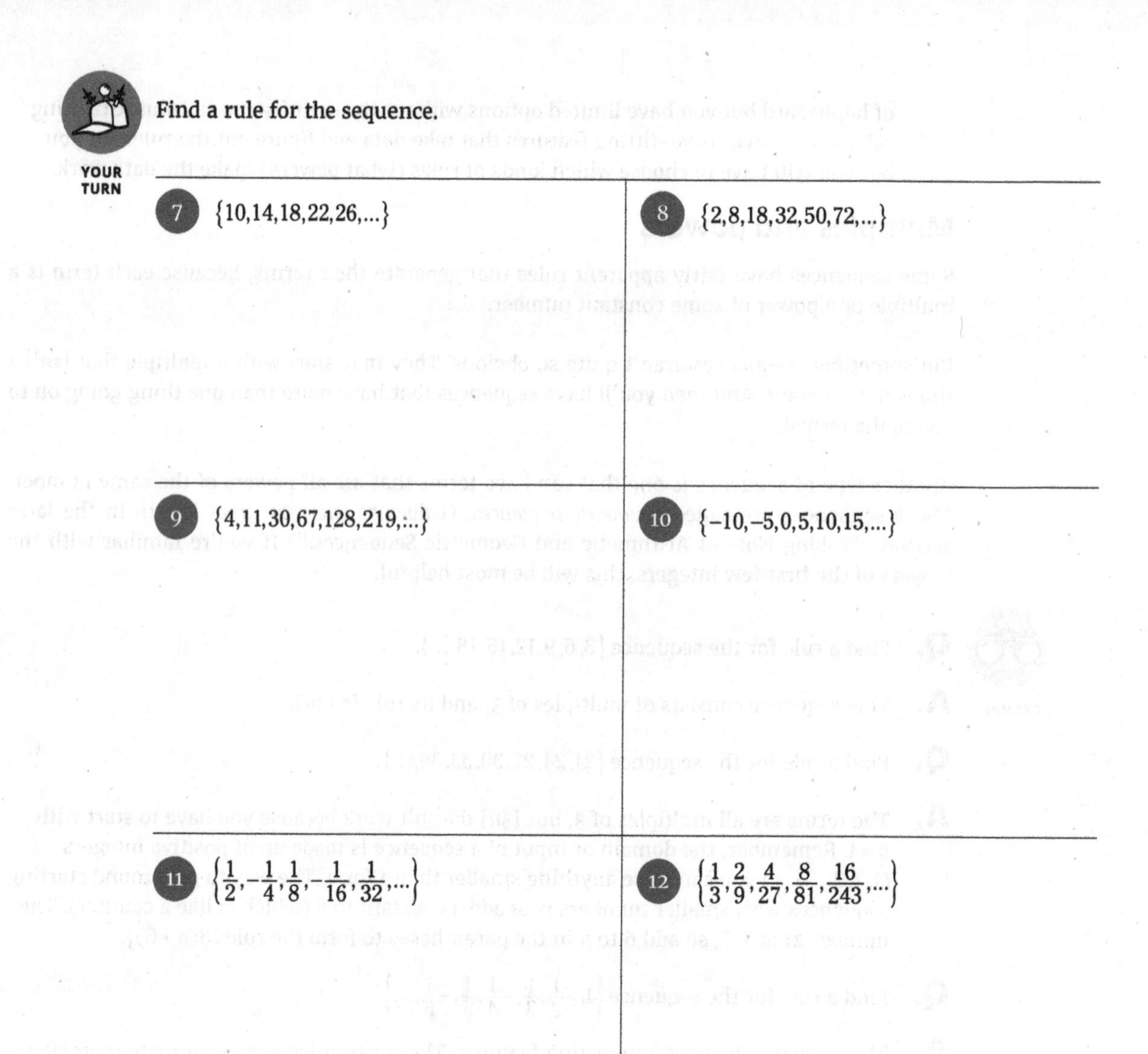

Find a rule for the sequence.

YOUR TURN

7. $\{10, 14, 18, 22, 26, \ldots\}$

8. $\{2, 8, 18, 32, 50, 72, \ldots\}$

9. $\{4, 11, 30, 67, 128, 219, \ldots\}$

10. $\{-10, -5, 0, 5, 10, 15, \ldots\}$

11. $\left\{\frac{1}{2}, -\frac{1}{4}, \frac{1}{8}, -\frac{1}{16}, \frac{1}{32}, \ldots\right\}$

12. $\left\{\frac{1}{3}, \frac{2}{9}, \frac{4}{27}, \frac{8}{81}, \frac{16}{243}, \ldots\right\}$

Taking Note of Arithmetic and Geometric Sequences

Arithmetic and geometric sequences are special types of sequences that have many applications in mathematics. Because you can usually recognize arithmetic or geometric sequences and write their general rules with ease, these sequences have become a mathematician's best friends. Arithmetic and geometric sequences also have very nice formulas for the sums of their terms, which opens up a whole new branch of mathematical activity.

Finding common ground: Arithmetic sequences

Arithmetic sequences (pronounced ah-*rith*-ma-tick, with the emphasis on mat) are sequences whose terms have the same differences between them, no matter how far down the lists you go (in other words, how many terms the lists include).

The general formula for arithmetic sequences is $a_n = a_{n-1} + d$. This formula says that the *n*th term of the sequence is equal to the term directly before it (the $n-1$st term) plus the common difference, *d*.

Another equation to use with arithmetic sequences is $a_n = a_1 + (n-1)d$. This formula says that the *n*th term of the sequence is equal to the first term, a_1, plus $n-1$ times the common difference, *d*.

REMEMBER

The equation you use depends on what you want to accomplish. You use the first formula if you single out a term in the sequence and want to find the next one in line. You use the second formula if you want to find a specific term in a sequence. And you can solve for one of these rules for a sequence if you have just the right information.

EXAMPLE

Q. Find the next term after the number 201 in the sequence $a_n = a_{n-1} + 3$.

A. You just add 3 to the 201 and get that the next term is 204.

Q. Find the 50th term in the sequence where $a_n = 5 + (n-1)7$.

A. Replace the *n* with 50, subtract the 1, multiply by 7, add 5, and you get 348. All that may sound like a lot of work but it's easier than listing all 50 terms.

Q. Find the rule for the sequence that has the common difference between terms of 4 where the sixth term is 37.

A. Substitute this information into the equation $a_n = a_1 + (n-1)d$, letting $a_n = 37$, $n = 6$, and $d = 4$. You get the following: $a_6 = a_1 + (6-1)d$, which becomes $37 = a_1 + (6-1)4 = a_1 + 20$. Subtracting 20 from each side of the equation, you get that $a_1 = 17$. Now, armed with the first term and the difference, you can write the general term using $a_n = a_1 + (n-1)d$. Replacing a_1 with 17 and *d* with 4, you have $a_n = 17 + (n-1)4$. Distributing the 4 and simplifying, you have $a_n = 17 + 4n - 4 = 13 + 4n$. An arithmetic sequence that has a common difference of 4 and whose sixth term is 37 has the general rule $\{13 + 4n\}$.

Q. You and a group of friends have been hired to be ushers at a local theater performance, and your payment includes free tickets to the show. The theater allocates the whole last row in the middle section for your group. The first row in the middle section of the theater has 26 seats, and the number of seats in each row increases by one seat per row as you move backward for a total of 25 rows. How many seats are in the last row?

A. You can solve this problem quickly with an arithmetic sequence. Using the formula $a_n = a_1 + (n-1)d$, you replace a_1 with 26, *n* with 25, and *d* with 1:

$$a_n = a_1 + (n-1)d \rightarrow a_{25} = 26 + (25-1)\cdot 1 \rightarrow a_{25} = 26 + (24)\cdot 1 = 26 + 24 \rightarrow a_{25} = 50$$

The last row has 50 seats. It looks like your friends can bring their friends, too!

YOUR TURN

13 The number 124 is which term in the sequence $\{4+(n-1)3\}$?

14 Find the rule for the sequence with a common difference of 5 if the tenth term is 48.

Taking the multiplicative approach: geometric sequences

A geometric sequence is a sequence in which each term is different from the one that follows it by a common ratio. In other words, the sequence has a constant number that multiplies each term to create the next one. With arithmetic sequences, you add a constant; with geometric sequences, you multiply the constant.

A general formula or rule for a geometric sequence is $g_n = rg_{n-1}$. In this equation, r is the constant ratio that multiplies each term. The rule says that, to get the *n*th term, you multiply the term before it — the $(n-1)$st term — by the ratio, r.

Another way you can write the general rule for a geometric sequence is $g_n = g_1 r^{n-1}$. The second form of the rule involves the first term, g_1, and applies the ratio as many times as needed. The *n*th term is equal to the first term multiplied by the ratio $n-1$ times.

You use the first rule, $g_n = rg_{n-1}$, when you're given the ratio, r, and a particular term in the sequence, and you want the next term. You use the second rule, $g_n = g_1 r^{n-1}$, when you're given the first term in the sequence and you want a particular term.

EXAMPLE

Q. If the ninth term in a sequence whose ratio is 3 is 65,610, what is the tenth term?

A. Multiply 65,610 times 3 to get 196,830.

Q. You know that the first term of a sequence is 3 and the ratio is 2. What is the tenth term?

A. You find that term by multiplying 3 times 2^9, which is 1,536.

Q. You have a geometric sequence where the sixth term is 1,288,408 and the fifth term is 117,128. What is the ratio?

A. You can always find the ratio or multiplier, r, if you have two consecutive terms in a geometric sequence. Just divide the second term by the term immediately preceding it — the quotient is the ratio. In this case, you can find r by dividing 1,288,408 by 117,128 to get 11. The ratio, r, is 11.

Q. The rule for a geometric sequence is $\left\{360\left(\frac{1}{3}\right)^{n-1}\right\}$. What are the first six terms?

A. When $n=1$, the power on the fraction is zero, and you have 360 multiplying the number 1. So, $g_1 = 360$. When $n=2$, the exponent is equal to 1, so the fraction multiplies the 360, and you get 120. Here are the first six terms in this sequence:

$$\left\{360, 120, 40, \frac{40}{3}, \frac{40}{9}, \frac{40}{27}, \ldots\right\}$$

You can find each term by multiplying the previous term by $\frac{1}{3}$.

YOUR TURN

15 The rule for a geometric sequence is $\left\{8\left(\frac{1}{2}\right)^{n-1}\right\}$. What are the first five terms?

16 An unwise gambler bets a dollar on the flip of a coin and loses. Instead of paying up, they say, "Double or nothing," meaning that they want to flip the coin again; they'll pay two bucks if they lose, and their opponent gets nothing if the gambler wins. Oops! The gambler loses again, and again they say, "Double or nothing!" If they repeat this doubling-and-losing process 20 times, how much will they owe on the 21st try?

Recursively Defining Functions

An alternate way to describe the terms of a sequence in place of giving the general rule for the sequence, is to define the sequence recursively. To do so, you identify the first term, or maybe a few of the initial terms, and describe how to find the rest of the terms by using two or more of the terms that come before them.

The recursive rule for arithmetic sequences is $a_n = a_{n-1} + d$, and the recursive rule for geometric sequences is $g_n = rg_{n-1}$. You can also define sequences recursively by referring to more than one previous term.

EXAMPLE

Q. Given the recursively defined sequence $a_n = 2a_{n-1} + 3$, find the first five terms if you know that $a_1 = 6$.

A. The formula says that, to find a term in the sequence, you look at the previous term (a_{n-1}), double it $(2a_{n-1})$, and add 3. The first term is 6, so the second term is 3 more than the double of 6, or 15. The next term is 3 more than the double of 15, or 33. Here are some of the terms of this sequence listed in order: $\{6, 15, 33, 69, 141, \ldots\}$.

Q. Given the recursively defined sequence $a_n = 3a_{n-2} + a_{n-1}$, find the first six terms. This rule says that to find the *n*th term in the sequence, you have to look at the two previous terms [the $(n-2)$th and the $(n-1)$th terms], multiply the term two positions back by $3(a_{n-2})$, and then add the term one position back, a_{n-1}, to that product. To start writing the terms of this sequence, you need to identify two consecutive terms. To create terms in a particular sequence (let's name it *B*) with this rule, you decide to let $b_1 = 4$ and $b_2 = -1$. Just pick random numbers, but, after they're chosen, they determine what's going to happen to the rest of the numbers in the sequence — using the given rule.) Here's how the terms go (note that if you're looking for the sixth term, you need the $[n-1]$st term, which is the fifth term, and the $[n-2]$nd term, which is the fourth term):

$b_1 = 4,\ b_2 = -1$

$b_n = 3b_{n-2} + b_{n-1}$

$b_3 = 3b_1 + b_2 = 3(4) + (-1) = 12 - 1 = 11$

$b_4 = 3b_2 + b_3 = 3(-1) + 11 = -3 + 11 = 8$

$b_5 = 3b_3 + b_4 = 3(11) + 8 = 33 + 8 = 41$

$b_6 = 3b_4 + b_5 = 3(8) + 41 = 24 + 41 = 65$

So the terms of this sequence are: $\{4, -1, 11, 8, 41, 65,\}$.

YOUR TURN

17 The Fibonacci sequence is recursively defined by $a_n = a_{n-2} + a_{n-1}$. If the first two terms are {1, 1, }, then find the next six terms.

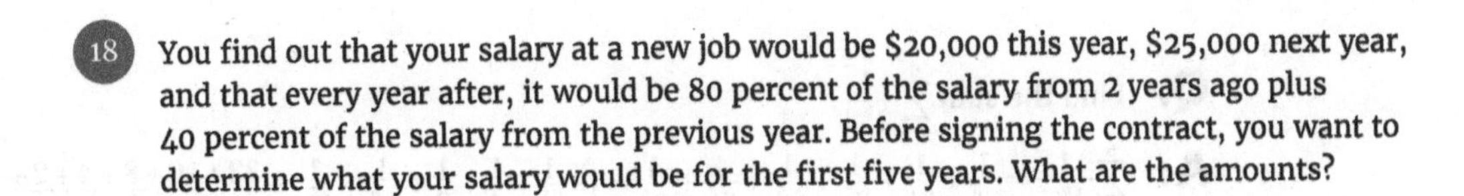

18 You find out that your salary at a new job would be $20,000 this year, $25,000 next year, and that every year after, it would be 80 percent of the salary from 2 years ago plus 40 percent of the salary from the previous year. Before signing the contract, you want to determine what your salary would be for the first five years. What are the amounts?

Making a Series of Moves

A series is the sum of a certain number of terms of a sequence. How many terms? That's part of the problem — it's either given to you or what you determine in order to answer some question.

Being able to list all the terms in a sequence is a handy tool to have in your algebra tool belt but you can do much more with sequences. Finding the sum of the sequence means adding as many terms as you need to figure out the desired total. This process doesn't sound like a chore, especially with hand-held calculators, but if the numbers get really big and you want the sum of many terms, the task can be daunting.

For this reason, many sequences used in business and financial applications have formulas for the sum of their terms. These formulas are a big help. In fact, for some geometric series, you can add all the terms — forever — and be able to predict the sum of all those terms.

Introducing summation notation

Mathematicians like to keep formulas and rules neat and concise, so they created a special symbol to indicate that you're adding the terms of a sequence. The special notation is sigma, Σ, or summation notation.

The notation $\sum_{k=1}^{n} a_k$ indicates that you want to add all the terms in the sequence that have the general rule a_k, all the way from $k = 1$ through $k = n$: $\sum_{k=1}^{n} a_k = a_1 + a_2 + a_3 + a_4 + \cdots + a_{n-1} + a_n$. If you want to add all the terms in a sequence, forever and ever, you use the symbol ∞ in the sigma notation, which looks like $\sum_{k=1}^{\infty} a_k$.

EXAMPLE

Q. Find the sum $\sum_{k=1}^{5}(k^2 - 2)$.

A. Find the first five terms, letting $k = 1, 2$, and so on; then finish by adding all those terms: $\sum_{k=1}^{5}(k^2 - 2) = (1^2 - 2) + (2^2 - 2) + (3^2 - 2) + (4^2 - 2) + (5^2 - 2) = -1 + 2 + 7 + 14 + 23 = 45$. You find that the sum is 45.

Q. Find the sum $\sum_{k=1}^{6} \frac{1}{2^{k-1}}$.

A. $\sum_{k=1}^{6} \frac{1}{2^{k-1}} = \frac{1}{2^0} + \frac{1}{2^1} + \frac{1}{2^2} + \frac{1}{2^3} + \frac{1}{2^4} + \frac{1}{2^5} = 1 + \frac{1}{2} + \frac{1}{4} + \frac{1}{8} + \frac{1}{16} + \frac{1}{32} = \frac{32+16+8+4+2+1}{32} = \frac{63}{32}$.

You will see in the later section "Summing geometrically" that there's a really nice formula for computing the sum of geometric series — and it will save you lots of work.

Summing arithmetically

An arithmetic sequence has a general rule (see the earlier section, "Finding common ground: Arithmetic sequences") that involves the first term and the common difference between consecutive terms: $a_n = a_1 + (n-1)d$. An arithmetic series is the sum of the terms that come from an arithmetic sequence. The sum of the first n terms of an arithmetic sequence, S_n, is

$$S_n = \frac{n}{2}[2a_1 + (n-1)d] = \frac{n}{2}(a_1 + a_n)$$

Here, a_1 and d are the first term and difference, respectively, of the arithmetic sequence $a_n = a_1 + (n-1)d$. The n indicates which term in the sequence you get when you put the value of n in the formula.

EXAMPLE

Q. Find the sum of the ten numbers 4, 9, 14, 19, 24, 29, 34, 39, 44, and 49.

A. This is the arithmetic sequence with the formula $a_n = 4 + (n-1)5 = 5n - 1$. You are given the first term, 4, and the tenth term, 49, so you can use the sum of an arithmetic sequence, $S_n = \frac{n}{2}(a_1 + a_n)$, giving you $S_{10} = \frac{10}{2}(a_1 + a_{10}) = 5(4 + 49) = 5(53) = 265$.

Q. Add the first 100 numbers in a sequence that starts with 13 and features a common difference of 2 between each of the terms: $13 + 15 + 17 + 19 + \ldots$, all the way to the 100th number.

A. You find the sum of these 100 numbers by using the first part of the sum formula:

$$S_n = \frac{n}{2}[2a_1 + (n-1)d]$$

$$S_{100} = \frac{100}{2}[2(13) + (100-1)2] = 50(26 + 198) = 50(224) = 11{,}200$$

Summing geometrically

A geometric sequence consists of terms that differ from one another by a common ratio. You multiply a term in the sequence by a constant number or ratio to find the next term. You can use two different formulas to find the sum of the terms in a geometric sequence. You use the first formula to find the sum of a certain, finite number of terms of a geometric sequence — any geometric sequence at all. The second formula applies only to geometric sequences that have a ratio whose absolute value lies between zero and 1 (a proper fraction); you use it when you want to add all the terms in the sequence — forever and ever (for more on geometric sequences, see the earlier section, "Taking the multiplicative approach: Geometric sequences").

Adding the first *n* terms

The formula you use to add a specific, finite number of terms from a geometric sequence involves a fraction where you subtract the ratio — or a power of the ratio — from 1. You can't reduce the formula, so don't try. Just use it as it is.

You find the sum of the first *n* terms of the geometric sequence $g_n = g_1 r^{n-1}$ with the following formula:

$$S_n = \frac{g_1(1-r^n)}{1-r}$$

The term g_1 is the first term of the sequence, and *r* represents the common ratio.

EXAMPLE

Q. Add the first ten terms of the geometric sequence $\{1, 3, 9, 27, 81, \ldots\}$.

A. You identify the first term, the 1, and then the ratio by which you multiply, 3. Substitute this information into the formula:

$$S_n = \frac{g_1\left(1-r^n\right)}{1-r} \rightarrow S_{10} = \frac{1\left(1-3^{10}\right)}{1-3} = \frac{1-59{,}049}{-2} = \frac{-59{,}048}{-2} = 29{,}524$$

Quite a big number! Isn't using the formula easier than adding $1+3+9+27+81+243+729+2{,}187+6{,}561+19{,}683$?

Q. Find the sum of $\frac{1}{3}, \frac{1}{9}, \frac{1}{27}, \ldots, \frac{1}{3^{10}}$.

A. This is the first eight terms of the geometric sequence with the rule $\left\{\frac{1}{3^n}\right\}$. The ratio is $\frac{1}{3}$ and the first term is $\frac{1}{3}$. Using the formula,

$$S_8 = \frac{\frac{1}{3}\left(1-\left(\frac{1}{3}\right)^8\right)}{1-\frac{1}{3}} = \frac{\frac{1}{3}\left(1-\frac{1}{3^8}\right)}{\frac{2}{3}} = \frac{\left(1-\frac{1}{3^8}\right)}{2} = \frac{\left(1-\frac{1}{6561}\right)}{2} = \frac{\frac{6{,}560}{6{,}561}}{2} = \frac{3{,}280}{6{,}561}.$$

Adding all the terms to infinity

Geometric sequences have a ratio, or multiplier, that changes one term into the next one in line. If you multiply a number by 4 and the result by 4 and keep going, you create huge numbers in a short amount of time. So, it may sound impossible to add numbers that seem to get infinitely large.

But algebra has a really wonderful property for geometric sequences with ratios between −1 and 1. The absolute values of the numbers in these sequences get smaller and smaller, and the sums of the terms in these sequences never exceed set, constant values.

If the ratio is bigger than 1, the sum just grows and grows, and you get no final answer. If the ratio is negative and between 0 and −1, the sum approaches a single, constant value. For ratios smaller than −1, you have chaos again.

In Figure 19-1, you see the terms in a sequence that starts with 1 and has a ratio of $\frac{1}{2}$. You also see the sum of all the terms up to and including that term.

n	g_n	S_n	Decimal
1	1	1	1
2	$\frac{1}{2}$	$1+\frac{1}{2}=\frac{3}{2}$	1.5
3	$\frac{1}{4}$	$\frac{3}{2}+\frac{1}{4}=\frac{7}{4}$	2.75
4	$\frac{1}{8}$	$\frac{7}{4}+\frac{1}{8}=\frac{15}{8}$	1.875
5	$\frac{1}{16}$	$\frac{15}{8}+\frac{1}{16}=\frac{31}{16}$	1.9375
6	$\frac{1}{32}$	$\frac{31}{16}+\frac{1}{32}=\frac{63}{32}$	1.96875
⋮	⋮	⋮	
12	$\frac{1}{2{,}048}$	$=\frac{4{,}095}{2{,}048}$	1.999511719
⋮	⋮	⋮	
n	$\left(\frac{1}{2}\right)^{n-1}$	$=\frac{2(2^{n-1})-1}{2^{n-1}}$	1.999999999......

FIGURE 19-1: Adding terms in a geometric sequence.

The general rule for this particular sequence is as follows:

$$g_n = g_1 r^{n-1} = 1\cdot\left(\frac{1}{2}\right)^{n-1} = \left(\frac{1}{2}\right)^{n-1}$$

So, you're really finding powers of $\frac{1}{2}$. Here are the first few terms:

$$\left\{1, \frac{1}{2}, \frac{1}{4}, \frac{1}{8}, \frac{1}{16}, \frac{1}{32}, \frac{1}{64}, \frac{1}{128}, \frac{1}{256}, \frac{1}{512}, \ldots\right\}$$

As you see in Figure 19-1, the sum of the terms is equal to almost 2 as the number of terms increases. The number in the numerator of the fraction of the sum is always 1 less than twice the denominator. The sum in Figure 19-1 approaches 2 but never exactly hits it. The sum gets so very, very close, though, that you can round up to 2. This business of approaching a particular value is true of any geometric sequence with a proper fraction (between zero and 1) for a ratio.

Algebra also offers a formula for finding the sum of all the terms in a geometric sequence with a ratio between zero and 1. You may think this formula should be much more complicated than the formula for finding only a few terms, but that isn't the case. This formula is actually much simpler.

The sum of all the terms from a geometric sequence whose ratio, r, has an absolute value between zero and 1 $\left(0<|r|<1\right)$ is $S_n \to \frac{g_1}{1-r}$, where g_1 is the first term in the sequence.

EXAMPLE

Q. Find the sum of the infinite sequence whose first term is 1 and whose common ratio is $\frac{1}{2}$.

A. Using the formula, $S_n \to \frac{g_1}{1-r} = \frac{1}{1-\frac{1}{2}} = \frac{1}{\frac{1}{2}} = 2$.

Q. Find the sum of all the terms of the geometric sequence whose first term is 1 and ratio is $-\frac{1}{2}$.

A. $S_n \to \frac{g_1}{1-r} = \frac{1}{1-\left(-\frac{1}{2}\right)} = \frac{1}{\frac{3}{2}} = \frac{2}{3}$. The sum is smaller than the sequence with a positive ratio, but the positives seem to outweigh the negatives!

YOUR TURN

Find the sum of the series.

19 $\sum_{k=1}^{6}\{k^2 - 2k\}$

20 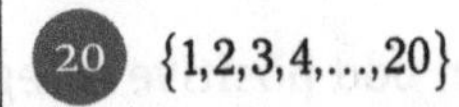 $\{1,2,3,4,\ldots,20\}$

21 $\{5,10,15,20,\ldots,200\}$

22 $\{3,6,12,24,\ldots,192\}$

23 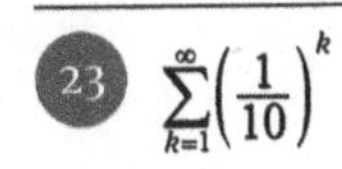 $\sum_{k=1}^{\infty}\left(\frac{1}{10}\right)^k$

Highlighting Special Formulas

Algebra offers several special types of sequences and series that you may use frequently in higher mathematics, such as calculus, and in financial and physics applications. For these applications, you have formulas for the sums of the terms in the sequences. Adding consecutive integers is a task made easier because you have formulas at your disposal. Counting tiles to be used in a floor or mosaic, computing the total amount of money in an annuity, and other such applications also use sums of sequences of numbers.

REMEMBER

Here are some of the special summation formulas you may see:

First n positive integers: $1+2+3+\cdots+n=\frac{n(n+1)}{2}$

First n squares of the positive integers: $1^2+2^2+3^2+\cdots+n^2=\frac{n(n+1)(2n+1)}{6}$

First n cubes of the positive integers: $1^3+2^3+3^3+\cdots+n^3=\frac{n^2(n+1)^2}{4}$

First n odd positive integers: $1+3+5+7+\cdots+(2n-1)=n^2$

EXAMPLE

Q. Find the sum of the first 10 squares of positive integers.

A. Use the formula $1^2+2^2+3^2+\cdots+n^2=\frac{n(n+1)(2n+1)}{6}$.

$$1+4+9+\cdots+100=\frac{10(10+1)(20+1)}{6}=\frac{10(11)(21)}{6}=385$$

The n is the number of the term, not the term itself.

Q. Add all the positive odd numbers from 1 to 49.

A. To use the formula for the sum of odd-numbered positive integers, you need to determine the number of the term — which odd number signifies the end of the sequence. To determine the 49th term, use the equation $2n-1$, set it equal to 49, and you see that $n=25$. Now use the formula for the sum of the first 25 odd integers: $1+3+5+7+\cdots+(2n-1)=n^2 \rightarrow 1+3+5+\cdots+49=25^2=625$.

YOUR TURN

24 Find the sum of the first 10 cubes.

Practice Questions Answers and Explanations

1. **{4, 10, 18, 28, 40, 54, ...}.** Replacing n with the first six positive integers, you have: $a_1 = 1^2 + 3(1) = 4, a_2 = 2^2 + 3(2) = 10$, $a_3 = 3^2 + 3(3) = 18$, $a_4 = 4^2 + 3(4) = 28$, $a_5 = 5^2 + 3(5) = 40$, $a_6 = 6^2 + 3(6) = 54$.

2. **{1, 2, 5, 12, 27, 58, ...}.** Replacing n with the first six positive integers, you have: $b_1 = 2^1 - 1 = 1$, $b_2 = 2^2 - 2 = 2$, $b_3 = 2^3 - 3 = 5$, $b_4 = 2^4 - 4 = 12$, $b_5 = 2^5 - 5 = 27$, $b_6 = 2^6 - 6 = 58$.

3. **{1, 3, 12, 60, 360, 2520, ...}.** Replacing n with the first six positive integers, you have: $c_1 = \frac{(1+1)!}{2} = \frac{2!}{2} = 1$, $c_2 = \frac{(1+2)!}{2} = \frac{3!}{2} = 3$, $c_3 = \frac{(1+3)!}{2} = \frac{4!}{2} = 12$, $c_4 = \frac{(1+4)!}{2} = \frac{5!}{2} = 60$, $c_5 = \frac{(1+5)!}{2} = \frac{6!}{2} = 360$, $c_6 = \frac{(1+6)!}{2} = \frac{7!}{2} = 2520$.

4. **{8, 11, 14, 17, 20, 23, ...}.** Replacing n with the first six positive integers, you have: $d_1 = 2 + 3(1+1) = 8$, $d_2 = 2 + 3(2+1) = 11$, $d_3 = 2 + 3(3+1) = 14$, $d_4 = 2 + 3(4+1) = 17$, $d_5 = 2 + 3(5+1) = 20$, $d_6 = 2 + 3(6+1) = 23$.

5. $\left\{-\frac{1}{2}, \frac{2}{3}, -\frac{3}{2}, \frac{24}{5}, \ldots\right\}$**.** Replacing n with the first four positive integers, you have: $a_1 = (-1)^1 \frac{1!}{1+1} = -\frac{1}{2}$, $a_2 = \frac{2}{3}$, $a_3 = -\frac{3}{2}$, $a_4 = \frac{24}{5}$.

6. **{1, -1, 2, -6}.** Replacing n with the first four positive integers, you have: $a_1 = (-1)^{1+1}(1-1)! = 1$, $a_2 = (-1)^{2+1}(2-1)! = -1$, $a_3 = (-1)^{3+1}(3-1)! = 2$, $a_4 = (-1)^{4+1}(4-1)! = -6$.

7. **{4n + 6}.** There is a common difference of 4 between the terms. Because you need the first term to be 10 when $n = 1$, write the rule as $\{10 + 4(n-1)\}$ so the 4 doesn't get added the first time. Then just simplify the terms in the braces and get $\{10 + 4n - 4\} = \{4n + 6\}$.

8. $\{2n^2\}$**.** There is a second difference of 4 between the terms:

```
2     8     18     32     50     72
   v     v      v      v      v
   6     10     14     18     22
      v      v      v      v
      4      4      4      4
```

This suggests a quadratic pattern. The terms are all twice the square of n: $\{2n^2\}$.

9. $\{n^3 + 3\}$**.** There is a third difference of 6 between the terms:

```
4     11     30     67     128     219
   v      v      v      v       v
   7      19     37     61      91
      v      v      v       v
      12     18     24      30
          v      v       v
          6      6       6
```

This suggests a cubic pattern. The terms are all 3 more than the cube of n: $\{n^3 + 3\}$.

(10) $\{5(n-3)\}$. The terms are all multiples of 5. The first term is –2 times 5, so subtract 3 from n to get $\{5(n-3)\}$.

(11) $\left\{\frac{(-1)^{n+1}}{2^n}\right\}$. The terms are alternating powers of $\frac{1}{2}$. Deal with the powers of –1 and the powers of the fraction separately: $\left\{(-1)^{n+1}\left(\frac{1}{2}\right)^n\right\}=\left\{\frac{(-1)^{n+1}}{2^n}\right\}$.

(12) $\left\{\frac{2^{n-1}}{3^n}\right\}$. The numerator is powers of 2 and the denominator powers of 3. The first power of 2 has to be 1, so the exponent has to be zero: $\left\{\frac{2^{n-1}}{3^n}\right\}$.

(13) **41.** Using $a_n=a_1+(n-1)d$, write the equation $124=4+(n-1)3$ and solve for n: $124=4+(n-1)3 \rightarrow 124=4+3n-3 \rightarrow 123=3n \rightarrow 41=n$.

(14) $\{3+(n-1)5\}$. Using $a_n=a_1+(n-1)d$, write the equation $48=a_1+(10-1)5$. Solving for the first term, $48=a_1+(10-1)5 \rightarrow 48=a_1+45 \rightarrow 3=a_1$. Use this value and the 5 in the formula.

(15) $\left\{8,4,2,1,\frac{1}{2}\right\}$. Starting with n = 1, you have

$$\left\{8\left(\frac{1}{2}\right)^{1-1},8\left(\frac{1}{2}\right)^{2-1},8\left(\frac{1}{2}\right)^{3-1},8\left(\frac{1}{2}\right)^{4-1},8\left(\frac{1}{2}\right)^{5-1},\ldots\right\}=\left\{8\left(\frac{1}{2}\right)^0,8\left(\frac{1}{2}\right)^1,8\left(\frac{1}{2}\right)^2,8\left(\frac{1}{2}\right)^3,8\left(\frac{1}{2}\right)^4,\ldots\right\}$$

$$=\left\{8(1)^0,8\left(\frac{1}{2}\right),8\left(\frac{1}{4}\right),8\left(\frac{1}{8}\right),8\left(\frac{1}{16}\right),\ldots\right\}=\left\{8,4,2,1,\frac{1}{2}\right\}$$

(16) **\$1,048,576.** Using the formula $g_n=g_1r^{n-1}$ (you know the first term — the one dollar — and the multiplier is 2), you replace the first term, g_1, with the number 1, r with 2 for the doubling, and n with 21: $g_{21}=1(2)^{21-1}=1(2)^{20}=1{,}048{,}576$. The gambler will owe over one million dollars if they keep going to 21. If they're unwilling to part with their first dollar, how are they going to deal with this number?

(17) **1, 1, 2, 3, 5, 8, 13, 21,** The rule says to add the two previous terms to get the next term. So 1 + 1 = 2, 1 + 2 = 3, 2 + 3 = 5, 3 + 5 = 8, 5 + 8 = 13, and 8 + 13 = 21.

(18) **{\$20,000, \$25,000, \$26,000, \$30,400, \$32,960,…}.** The rule involved reads: $b_n=0.8b_{n-2}+0.4b_{n-1}$. Using the first two terms and this rule, your salary for the first five years would be $b_1=20{,}000$, $b_2=25{,}000$, $b_3=0.8(20{,}000)+0.4(25{,}000)=26{,}000$, $b_4=0.8(25{,}000)+0.4(26{,}000)=30{,}400$, $b_5=0.8(26{,}000)+0.4(30{,}400)=32{,}960$.

(19) **49.** The terms of this sequence are: $\{1^2-2,2^2-4,3^2-6,4^2-8,5^2-10,6^2-12\}=\{-1,0,3,8,15,24\}$. The sum of the terms is 49.

(20) **210.** This is an arithmetic sequence starting with 1, a difference of 1, and containing 20 terms. Using $S_n=\frac{n}{2}(a_1+a_n)$, $S_{20}=\frac{20}{2}(1+20)=10(21)=210$.

(21) **4,100.** This is an arithmetic sequence starting with 5, and a difference of 5. To determine the number of terms, use $a_n=a_1+(n-1)d$ and solve for n: $200=5+(n-1)5 \rightarrow 200=5+5n-5 \rightarrow 200=5n \rightarrow 40=n$. Now, to find the sum using $S_n=\frac{n}{2}(a_1+a_n)$, $S_{40}=\frac{40}{2}(5+200)=20(205)=4{,}100$.

(22) **381.** This is a geometric sequence with first term 3 and ratio 2. To determine the number of terms, use $g_n=g_1r^{n-1}$ and solve for n: $192=3(2)^{n-1} \rightarrow 64=2^{n-1} \rightarrow 2^6=2^{n-1} \rightarrow 6=n-1 \rightarrow n=7$.

Now, using $S_n=\frac{g_1(1-r^n)}{1-r}$, $S_7=\frac{3(1-2^7)}{1-2}=\frac{3(1-128)}{-1}=\frac{3(-127)}{-1}=381$.

(23) $\frac{1}{9}$. This is an infinite geometric sequence with a ratio of $\frac{1}{10}$. Using $S_n \to \frac{g_1}{1-r}$,

$$S_n \to \frac{\frac{1}{10}}{1-\frac{1}{10}} = \frac{\frac{1}{10}}{\frac{9}{10}} = \frac{1}{10} \cdot \frac{10}{9} = \frac{1}{9}$$

(24) **3,025.** Using the formula $1^3 + 2^3 + 3^3 + \cdots + n^3 = \frac{n^2(n+1)^2}{4}$,

$$1^3 + 2^3 + 10^3 = \frac{10^2(10+1)^2}{4} = \frac{100(121)}{4} = 3{,}025.$$

Whaddya Know? Chapter 19 Quiz

Quiz time! Complete each problem to test your knowledge on the various topics covered in this chapter. You can then find the solutions and explanations in the next section.

1. Write the first 5 terms of the sequence $\{n^2 - n + 1\}$ where n is a positive integer.
2. Find the sum: $\sum_{k=1}^{\infty}\left(-\frac{1}{4}\right)^k$.
3. Write a rule describing the sequence $\{3, -6, 12, -24, 48, \ldots\}$
4. Find the sum of the series $\sum_{k=3}^{8}\frac{2}{k}$.
5. Find the 100th term in the sequence $\left\{1, \frac{3}{2}, 2, \frac{5}{2}, 3, \frac{7}{2}, \ldots\right\}$.
6. Find the sum of the first 30 cubes.
7. Write the first six terms of the sequence where $a_1 = 1$, $a_2 = 4$, and $a_n = 2a_{n-1} + a_{n-2}$.
8. Find the sum of the series $\sum_{k=1}^{6} 2k - 1$.
9. Find the sum of the first 80 positive integers.
10. Write the first 5 terms of the sequence $\left\{\frac{(-1)^{n+1}}{2^{n-1}}\right\}$ where n is a positive integer.
11. Find the sum of the first 40 squares.
12. Find the sum of the first 16 terms of the geometric series where $a_n = 2 \cdot 3^{n-1}$.
13. Write a rule describing the sequence $\{1, 4, 7, 10, 13, \ldots\}$.

Answers to Chapter 19 Quiz

1. **{1, 3, 7, 13, 21}.** Inserting the numbers 1 through 5 into the formula, $\left\{1^2-1+1, 2^2-2+1, 3^2-3+1, 4^2-4+1, 5^2-5+1\right\}$={1, 3, 7, 13, 21}.

2. $-\frac{1}{5}$. Using the formula, $S_n \to \frac{-\frac{1}{4}}{1-\left(-\frac{1}{4}\right)} = \frac{-\frac{1}{4}}{\frac{5}{4}} = -\frac{1}{5}$.

3. $\left\{3(-2)^{n-1}\right\}$. Each term after the first term is multiplied by –2: $\left\{3(-2)^{n-1}\right\}$.

4. $\frac{341}{140}$. $\sum_{k=3}^{8} \frac{2}{k} = \frac{2}{3}+\frac{2}{4}+\frac{2}{5}+\frac{2}{6}+\frac{2}{7}+\frac{2}{8} = \frac{280}{420}+\frac{210}{420}+\frac{168}{420}+\frac{140}{420}+\frac{120}{420}+\frac{105}{420} = \frac{1023}{420} = \frac{341}{140}$

5. $\frac{101}{2}$. The rule is $\left\{1+(n-1)\left(\frac{1}{2}\right)\right\} = \left\{\frac{1}{2}+\frac{n}{2}\right\}$, so the 100th term is $\frac{1}{2}+\frac{100}{2} = \frac{101}{2}$.

6. **216,225.** Using the special rule: $\frac{n^2(n+1)^2}{4} = \frac{30^2(31^2)}{4} = \frac{900(961)}{4} = 225(961) = 216{,}225$.

7. **1, 4, 9, 22, 53, 128.** This is a recursive formula, using the previous two terms.

8. **40.** $\sum_{k=1}^{6} 2k-1 = 1+3+5+7+11+13 = 40$

9. **3,240.** Using the special formula, $\frac{n(n+1)}{2} = \frac{80(81)}{2} = 40(81) = 3{,}240$.

10. $1, -\frac{1}{2}, \frac{1}{4}, -\frac{1}{8}, \frac{1}{16}$. $n=1,\ \frac{(-1)^2}{2^0} = \frac{1}{1} = 1$; $n=2,\ \frac{(-1)^3}{2^1} = \frac{-1}{2} = -\frac{1}{2}$; $n=3,\ \frac{(-1)^4}{2^2} = \frac{1}{4} = \frac{1}{4}$;
$n=4,\ \frac{(-1)^5}{2^3} = \frac{-1}{8} = -\frac{1}{8}$; $n=5,\ \frac{(-1)^6}{2^4} = \frac{1}{16} = \frac{1}{16}$

11. **22,140.** Using the special formula, $\frac{n(n+1)(2n+1)}{6} = \frac{40(41)(81)}{6} = \frac{\cancel{40}^{20}(41)(\cancel{81}^{27})}{\cancel{6}} = 22{,}140$.

12. $3^{17}-1$. Using the formula for the sum of a geometric series,
$\frac{a\left(1-r^{n+1}\right)}{1-r} = \frac{2\left(1-3^{17}\right)}{1-3} = \frac{2\left(3^{17}-1\right)}{2} = 3^{17}-1$.

13. $\{n+3(n-1)\} = \{4n-3\}$. This is an arithmetic sequence where the common difference is 3, and the first number is 1.

IN THIS CHAPTER

» Conquering set notation

» Performing algebraic operations on sets

» Circling the wagons with Venn diagrams

» Addressing factorials in sets

» Using permutations and combinations to count set elements

» Making set life easy with tree diagrams

Chapter 20
Everything You Wanted to Know about Sets and Counting

Back in the 1970s, a book titled *Everything You Always Wanted to Know About Sex — But Were Afraid to Ask* (Bantam) hit the shelves of stores everywhere. The book caused quite a buzz. Nowadays, however, people don't seem to be afraid to ask about — or inform you about — anything. So, in the modern spirit, this chapter lets it all hang out. Here you find out all the deep, dirty secrets concerning sets. I cover the union and intersection of sets, complementary sets, counting on sets, and drawing some very revealing pictures. Can you handle it?

Revealing the Set Rules

A set is a collection of items. The items may be people, pairs of shoes, or numbers, for example, but they usually have something in common — even if the only characteristic that ties them together is the fact that they appear in the same set. The items in a set are called the set's elements.

Mathematicians have developed some specific notation and rules to help you maneuver through the world of sets. The symbols and vocabulary aren't difficult to master; you just have to remember what they mean. Familiarizing yourself with set notation is like learning a new language.

Listing elements with a roster

The name of a set appears as a capital letter to distinguish it from other sets. To introduce the elements of a set, you put them in *roster notation*, which just means to list the elements.

EXAMPLE

Q. Create set A, which contains the first five whole numbers.

A. $A = \{0, 1, 2, 3, 4\}$. You list the elements in the set inside the braces and separate them by commas. The order in which you list the elements doesn't matter. You could also say that $A = \{0, 2, 4, 1, 3\}$, for example.

Q. Create set P, which contains the first six prime numbers.

A. $P = \{2,3,5,7,11,13\}$

Building sets from scratch

A simple way to describe a set (other than set notation; see the previous section) is to use *rule* or *set-builder notation*. With this notation you name a variable, insert a vertical bar, and then describe what rules apply to that variable.

EXAMPLE

Q. Write the set $A = \{0, 1, 2, 3, 4\}$ using set-builder notation.

A. $A = \{x \mid x \in W, x < 5\}$. You read the set-builder notation as "A is the set containing all *x* such that *x* is an element of W, the whole numbers, and *x* is less than 5." The vertical bar, |, separates the variable from its rule, and the epsilon, ∈, means "is an element of." This is all wonderful math shorthand. And it's better than having to list all 1,000 elements in a set.

Q. Use set-builder notation to describe the set containing all the odd numbers between 0 and 100.

A. $B = \{x \mid x \text{ is odd}, 0 < x < 100\}$. When dealing with sets that have huge numbers of elements, using set-builder notation can save you time and busy work. And if the pattern of elements is obvious (always a tricky word in math), you can use an ellipsis. For this set, you could also use $B = \{1, 3, 5, 7, \ldots, 99\}$. Both methods are easier than listing all the elements in the set.

YOUR TURN

1 Create a set C that contains the multiples of 3 between 0 and 29. Use roster notation, set-builder notation, and roster with an ellipsis.	**2** Given $D = \{x \mid x = 4n, n \in Z, 0 \le x < 50\}$, describe the set using roster notation.

Going for all (universal set) or nothing (empty set)

Consider the sets F = {Iowa, Ohio, Utah} and I = {Idaho, Illinois, Indiana, Iowa}. Set F contains three elements and set I contains four elements. Within these sets, you can branch out to a set that you call the universal set for F and I. You can also distinguish a set called the *empty set*, or *null set*. This all-or-nothing business lays the foundation for performing set operations (see the section, "Operating on Sets," later in this chapter).

REMEMBER

The following list presents the characteristics of these all-or-nothing sets:

» **Universal set:** A universal set for one or more sets contains all the possible elements in a particular category. The writer of the situation must decide how many elements need to be considered in a particular problem. But one characteristic is pretty standard: The universal set is denoted U.

  For example, you could say that the universal set for F = {Iowa, Ohio, Utah} and I = {Idaho, Illinois, Indiana, Iowa} is U = {States in the United States}. But the universal set for F and I doesn't have to be a set containing all the states; it can just be all the states that start with a vowel or all the contiguous states. There are many choices; it just depends on the particular situation.

» **Empty (or null) set:** The opposite of the universal set is the empty set (or null set). The empty set doesn't contain anything (no kidding!). The two types of notation used to indicate the empty set are ∅ and { }. The first notation resembles a zero with a slash through it, and the second is empty braces. You must use one notation or the other, not both at the same time, to indicate that you have the empty set.

  For example, if you want to list all the elements in set G, where G is the set that contains all the state names starting with the letter Q, you write G = { }. You have the empty set because no state names in the United States start with the letter Q.

Subbing in with subsets

The real world provides many special titles for the little guy. Apartments can have sublets; movies can have subtitles; and ships can be submarines, so of course sets can have subsets. A subset is a set completely contained within another set — no element in a subset is absent from the set it's a subset of. Whew! That's a mouthful, and my sentence ends with a preposition. For the sake of good English, I'll start calling the sets subsets and supersets; a superset represents what the subset is a subset of (there I go again!).

Indicating subsets with notation

To indicate that one set is the subset of another, use the notation $\subset$, which indicates "subset of." If you write $R \subset T$, then you've said that set R is a subset of set T.

EXAMPLE

Q. Given sets B and C, use the subset notation to indicate how they are related: $B = \{2, 4, 8, 16, 32\}$ and $C = \{x \mid x = 2^n,\ n \in Z\}$.

A. Set B is a subset of set C, because B is completely contained in C. Set C consists of all the numbers that are powers of 2, where the powers are all elements of the set of integers (Z). The notation for "subset of" is $\subset$, and you write $B \subset C$ to say that B is a subset of C.

Q. Given sets D and E, use the subset notation to indicate how they are related: $D = \{Z: \text{ the set of integers}\}$ and $E = \{W: \text{ the set of whole numbers}\}$.

A. The set of integers contains all positive and negative whole numbers and zero. The set of whole numbers contains all positive integers and zero. So $W \subset Z$.

When one set is a subset of another, and the two sets aren't equal (meaning they don't contain the same elements), the subset is called a *proper* subset, indicating that the subset has fewer elements than the superset. Technically, any set is its own subset, so you can say that a set is an *improper* subset of itself. You write that statement with the subset notation and a line under it to indicate "subset and, also, equal to." To say that set B is its own subset, you write $B \subseteq B$. This may seem like a silly thing to do, but as with all mathematical rules, you have a good reason for doing so. One of the reasons has to do with the number of subsets of any given set and another has to do with results of operations on sets.

Counting the number of subdivisions

Have a look at the following listings of some selected sets and all their subsets. Notice that I include the empty set in each list of subsets. I do so because the empty set fulfills the definition that no element of the subset is absent from the superset.

EXAMPLE

Q. List all the subsets of $A = \{3, 8\}$.

A. **The set A has four subsets: two subsets with one element, one with no elements, and one with both elements from the original set. The subsets are $\{3\}, \{8\}, \varnothing, \{3, 8\}$.**

Q. List all the subsets of $B = \{\text{dog, cat, mouse}\}$.

A. **The set B has eight subsets: three subsets with one element, three with two elements, one with no elements, and one with all the elements from the original set. The subsets are {dog}, {cat}, {mouse}, {dog, cat}, {dog, mouse}, {cat, mouse}, ∅, {dog, cat, mouse}.**

Q. List all the subsets of $C = \{r, s, t, u\}$.

A. **The set C has 16 subsets: four subsets with one element, six with two elements, four with three elements, one with no elements, and one with all the elements. The subsets are {r}, {s}, {t}, {u}, {r, s}, {r, t}, {r, u}, {s, t}, {s, u}, {t, u}, {r, s, t}, {r, s, u}, {r, t, u}, {s, t, u}, ∅, {r, s, t, u}.**

Q. List all the subsets of D = {3}.

A. **The set has 2 subsets: one with no elements and one with only the element — ∅ and {3}.**

Have you determined a pattern yet? See if the following information helps:

A set with two elements has four subsets.

A set with three elements has eight subsets.

A set with four elements has sixteen subsets.

A set with one element has two subsets.

The number of subsets produced by a set is equal to a power of 2. If a set A has n elements, it has 2^n subsets.

You can apply this rule to the set $Q = \{1, 2, 3, 4, 5, 6\}$, for example. The set has six elements, so it has $2^6 = 64$ subsets. Knowing the number of subsets a set has doesn't necessarily help you list them all, but it lets you know if you've missed any. (Check out the later section, "Mixing up sets with combinations," to find out how many subsets of each type [number of elements] a set has.)

YOUR TURN

3 List all the subsets of the set $E = \{3, 5, 7\}$.

List all the subsets of the set $F = \{m, a, t, h\}$.

Operating on Sets

Algebra provides three basic operations that you can perform on sets: union, intersection, and complement (negation). The operations union and intersection take two sets at a time to perform them — much like addition and subtraction take two numbers. Finding the complement of a set (or negating it) is like finding the opposite of the set, so you perform it on only one set at a time.

An additional process, which isn't really an operation, is counting up the number of elements contained in a set. This process has its own special notation to tell you to do the counting, just like the union, intersection, and negation operations have their own special symbols.

Celebrating the union of two sets

Finding the union of two sets is like merging two companies. You put them together to make one big set (forming unions probably isn't as lucrative, however).

To find the union of sets A and B, denoted $A \cup B$, you combine all the elements of both sets by writing them into one set. You don't duplicate any elements that the sets have in common. And you can also say that each set is a subset of the union of the sets.

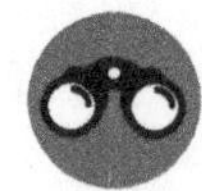

EXAMPLE

Q. Find the union of sets A and B: $A = \{10, 20, 30, 40, 50, 60\}$ and $B = \{15, 30, 45, 60\}$.

A. You only use the elements 10 and 60 once: $A \cup B = \{10, 15, 20, 30, 40, 45, 50, 60\}$.

Q. Find the union of the sets R, S, and T: $R = \{\text{rabbit, bunny, hare}\}$, $S = \{\text{bunny, egg, basket, spring}\}$, and $T = \{\text{summer, fall, winter, spring}\}$.

A. Again, you mention each element only once in the union of the sets: $R \cup S \cup T = \{\text{rabbit, bunny, hare, egg, basket, spring, summer, fall, winter}\}$.

Q. Find the union of the sets G and H: $G = \{1, 2, 3, 4, 5, 6\}$ and $H = \{2, 4, 6\}$.

A. This is a special case where one of the sets is a subset of the other. You can write the union as $G \cup H = G$ because the union of the sets is just the set G. H is a subset of G because every element in H is contained in G. Also, because H is a subset of G, H must also be a subset of the union of the two sets, $G \cup H$.

Q. Find the union of the set T and the empty set.

A. Because the empty set is a subset of every set, you can say $T \cup \varnothing = T$. This is true regardless of the size of the other set. Compare this to how adding zero to a number affects the number — it doesn't.

YOUR TURN

5 Find the union of the sets $G = \{2,3,5,7,11,13,17\}$ and $H = \{3,13,23,33\}$.

6 Find the union of the sets $J = \{2,4,8,16,32,64\}$ and $K = \{16,32,2,64,8\}$.

Looking both ways for set intersections

The intersection of two sets is like the intersection of two streets. If Main Street runs east and west and University Street runs north and south, they share the street where they intersect. The street department doesn't have to pave the intersection twice, because the two streets share that little part.

To find the intersection of sets A and B, denoted $A \cap B$, you list all the elements that the two sets have in common. If the sets have nothing in common, their intersection is the empty set.

EXAMPLE

Q. Find the intersection of set $A = \{a, e, i, o, u\}$ and set $B = \{v, o, w, e, l\}$.

A. They have the letters e and o in common: $A \cap B = \{e, o\}$.

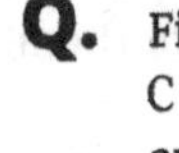

Q. Find the intersection of set $C = \{a, c, e, g, i, k, m, o, q, s, u, w, z\}$ and set $A = \{a, e, i, o, u\}$.

A. Because set A is a subset of set C, the intersection of A and its superset is just the set A. So $A \cap C = \{a, e, i, o, u\} = A$. The intersection of any set with the empty set is just the empty set.

YOUR TURN

7 Find the intersection of the sets $M = \{2,4,6,8,10\}$ and $N = \{1,2,3,5,6,7\}$.

8 Find the intersection of the sets $P = \{4,14,24,34,44\}$ and $Q = \{41,43,45,47\}$.

Feeling complementary about sets

The complement of set A, written A′ (or sometimes $\bar{A}$), contains every element that doesn't appear in A. Every item that doesn't appear in a set can be plenty of things, unless you limit your search.

To determine the complement of a set, you need to know the universal set. The complement of a set contains all the elements from the universal set that are not in the set in question.

Q. Find the complement of A if $A = \{p, q, r\}$, and the universal set, U, contains all the letters of the alphabet.

A. $A' = \{a, b, c, d, e, f, g, h, i, j, k, l, m, n, o, s, t, u, v, w, x, y, z\}$. You still deal with a lot of elements, but you limit your search to letters of the alphabet.

Q. Find the complement of B if $B = \{1,2,3,\ldots,99\}$ and the universal set, U, contains all the whole numbers less than 100.

A. Because all the whole numbers less than 100 start with zero and go up through 99, that just leaves zero: So $B' = \{0\}$.

The number of elements in a set plus the number of elements in its complement always equals the number of elements in the universal set. You write this rule as $n(A) + n(A') = n(U)$. This relationship is very useful when you have to deal with large numbers of elements and you want some easy ways to count them.

Counting the elements in sets

Situations often arise when you need to be able to tell how many elements a set contains. When sets appear in probability, logic, or other mathematical problems, you don't always care what the elements are — just how many of them sit in the sets. The notation indicating that you want to know the number of elements contained in a set is $n(A) = k$, meaning "The number of elements in set A is k." The number k will always be some whole number: 0, 1, 2, and so on. The number of elements in a set is sometimes referred to the set's *cardinal* number.

The union and intersection of sets have an interesting relationship concerning the number of elements:

$$n(A \cup B) = n(A) + n(B) - n(A \cap B)$$

You won't always be solving for the number in the union of two sets, but the equation is very useful when solving for one of the terms or another.

EXAMPLE

Q. If set A contains 20 elements and set B contains 16 elements, then how many elements are there in their union if their intersection contains 4 elements?

A. Using $n(A \cup B) = n(A) + n(B) - n(A \cap B)$, you have $n(A) = 20$, $n(B) = 16$, and $n(A \cap B) = 4$, so $n(A \cup B) = 20 + 16 - 4 = 32$.

Q. If set C contains 26 elements and set D contains 36 elements, then how many elements are there in their union if their union contains 50 elements?

A. Using $n(C \cup D) = n(C) + n(D) - n(C \cap D)$, $n(C) = 26$, $n(D) = 36$, and $n(C \cup D) = 50$, so $50 = 26 + 36 - n(C \cap D) \rightarrow 50 = 62 - n(C \cap D) \rightarrow n(C \cap D) = 62 - 50 = 12$.

YOUR TURN

9 Set E contains 50 elements and set F contains 40 elements. If their intersection contains 30 elements, then how many elements are in the union of the sets?

10 Set E contains 50 elements and the universal set of set E contains 200 elements. How many elements are in E′?

Drawing Venn You Feel Like It

Venn diagrams are pictures that show the relationships between two or more sets and the elements in those sets. The adage, "a picture is worth a thousand words," is never truer than with these diagrams. Venn diagrams can help you sort out a situation and come to a conclusion. Many problems that you solve by using Venn diagrams come in paragraphs — plenty of words and numbers and confusing relationships. Labeling the circles in the diagrams and filling in numbers helps you determine how everything works together and allows you to see if you've forgotten anything.

You usually draw Venn diagrams with intersecting circles. You label the circles with the names of the sets, and you encase the circles with the universal set (a rectangle around the circles). The elements shared by the sets are inserted in the overlapping parts of the respective circles.

EXAMPLE

Q. Create a Venn diagram consisting of set A, which contains the letters of the alphabet that spell *encyclopedia*, and set B, which contains the letters in the alphabet that rhyme with see. Both sets are encased in the universal set — all the letters in the alphabet.

A. The Venn diagram in Figure 20-1 shows set A, which contains the letters of the alphabet that spell *encyclopedia*, and set B, which contains the letters in the alphabet that rhyme with see. Both sets are encased in the universal set — all the letters in the alphabet. Notice that the letters c, d, e, and p appear in both sets. The Venn diagram makes it easy to see where the elements are and what characteristics they have — which elements are shared and which are special to each set.

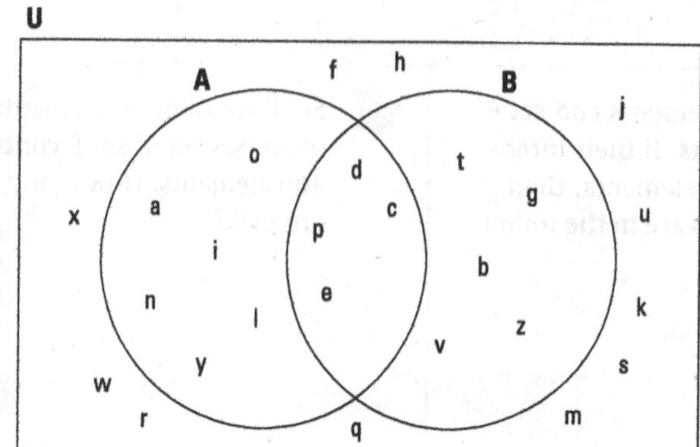

FIGURE 20-1: A Venn diagram with two sets enclosed by the universal set.

Applying the Venn diagram

The business of sorting letters based on their placement in a word or what they rhyme with, which you see in examples in the previous section, may not seem very fulfilling. The actual applications get a little more complicated but the examples here show you the basics. Some actual uses of Venn diagrams appear in the world of advertising (charting types of advertising and the results), of politics (figuring out who has what opinions on issues and how to make use of their votes), of genetics and medicine (looking at characteristics and reactions based on symptoms and results), and so on.

EXAMPLE

Q. A Chicago-area newspaper interviewed 40 people to determine if they were Chicago White Sox fans and/or if they cheered for the Chicago Bears. (If you couldn't care less about sports, the White Sox are a baseball team and the Bears are a football team.) Of the people interviewed, 25 are White Sox fans, 9 like both the Sox and the Bears, and 7 don't care for either team. How many people are Bears fans?

A. As you can see, 25 White Sox + 9 who like both + 7 who don't care for either is 41, which is more than the 40 people interviewed. The process has to have some overlap. You can sort out the overlap with a Venn diagram. You create one circle for White Sox fans, another for Bears fans, and a rectangle for all the people interviewed. Start by putting the seven fans who don't cheer for either team outside the circles (but inside the rectangle). You then put the 9 who like both teams in the intersection of the two circles (see Figure 20-2a).

Surveyed

Sox

Bears

9

7

a

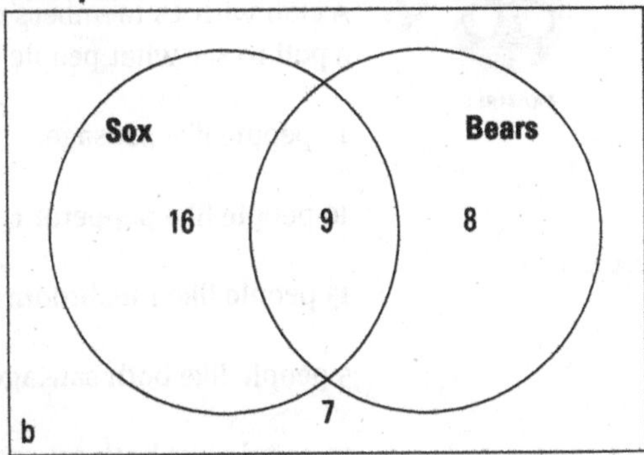

FIGURE 20-2: Watch for the overlap created by combining two groups; 17 Chicagoans root for Da Bears.

The number of White Sox fans is 25, so, if you put 16 in only the Sox's circle, the number in the entire circle (including the overlap) sums to 25. The only area missing is the circle where people are Bears fans but not Sox fans. So far, you have a total of $25 + 7 = 32$ people. Therefore, you can say that 8 people root for the Bears but not the Sox. You put an 8 in that area, and you see that the number of Bears fans totals 17 people (refer to Figure 20-2b).

Adding a set to a Venn diagram

Showing the relationship between the elements of two sets with a Venn diagram is pretty straightforward. But as with most mathematical processes, you can take the diagram one step further and illustrate the relationship between three sets. You can also handle four sets, but the diagram gets pretty hairy and isn't all that useful because of the way you have to draw the figure.

When two sets overlap, you divide the picture into four distinct areas: outside both circles, the overlap of the circles, and the two parts that have no overlap but appear in one circle or the other. When you overlap three circles in a Venn diagram, you create eight distinct areas. You can describe the eight different areas determined by the intersection of three circles as follows:

1. All the elements in A only
2. All the elements shared by A and B but not by C
3. All the elements in B only
4. All the elements shared by A and C but not by B
5. All the elements shared by A, B, and C
6. All the elements shared by B and C but not by A
7. All the elements in C only
8. All the elements not in A, B, or C

You may have to use a setup like this to sort out information and answer questions.

EXAMPLE

Q. A club with 25 members decides to order pizza for its next meeting; the secretary takes a poll to see what people like:

14 people like sausage.

10 people like pepperoni.

13 people like mushrooms.

5 people like both sausage and pepperoni.

10 people like both sausage and mushrooms.

7 people like both pepperoni and mushrooms.

4 people like all three toppings.

Here's the big question: How many people don't like any of the three toppings on their pizzas?

A. As you can see, the preferences sum to many more than 25, so you definitely have to account for some overlaps. You can answer the question by drawing three intersecting circles, labeling them Sausage, Pepperoni, and Mushrooms, and filling in numbers, starting with the last on the given list. Figure 20-3a shows the initial Venn diagram you can use.

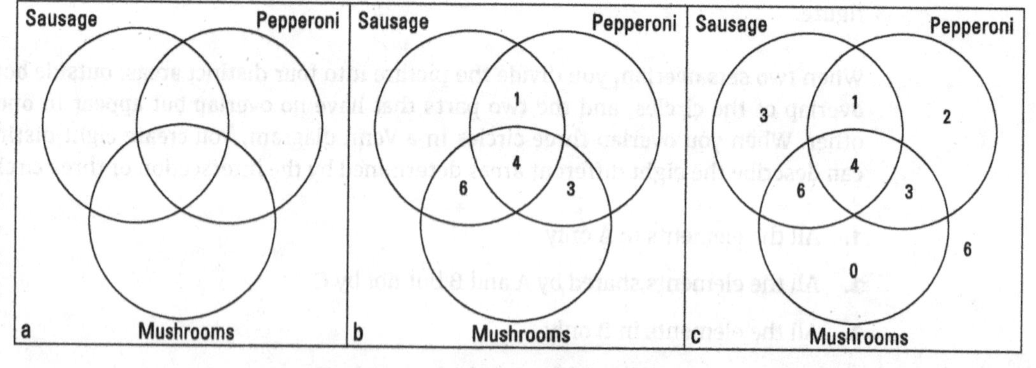

FIGURE 20-3: With a Venn diagram, you can tell how many people want a plain cheese pizza.

You put the 4, representing people who like all three toppings, in the middle section. Seven people like both pepperoni and mushrooms, but you already account for four of them in the middle, so put a 3 in the area for pepperoni and mushrooms but not sausage. Ten people like both sausage and mushrooms, but you already put four of those people in the middle area, so put a 6 in the area for sausage and mushrooms but not pepperoni. Five people like both sausage and pepperoni but you already account for four; put a 1 in the sausage and pepperoni section but not mushrooms. Figure 20-3b shows all these entries.

Now you can fill in the rest of the circles. Thirteen people like mushrooms, and you already have 13 people in that circle, so put zero for people who like mushrooms only. Ten people like pepperoni; you already have 8 in that circle, so put a 2 for pepperoni only. Fourteen people like sausage, so put a 3 in the section for sausage only to account for the difference. Look at all those filled-in numbers in Figure 20-3c!

You finish by adding all the numbers in Figure 20-3c. The numbers add to 19 people. The club has 25 members, so you conclude that 6 of them must not like sausage, pepperoni, or mushrooms. You can order a plain cheese pizza for these picky folk.

YOUR TURN

11 Create a Venn diagram of the months of the year where one set contains all the months with the letter *e* in their name and the other contains all the months with the letter *a* in their name. How many months have both letters? And how many months have neither letter?

12 Create a Venn diagram containing the first 20 positive integers and the sets "multiples of 2," "multiples of 3," and "primes." How many of the integers are in all three sets, and how many are in none of the sets?

Focusing on Factorials

When you perform a factorial operation, you multiply the number you're operating on times every positive integer smaller than that number. The factorial operation is denoted by an exclamation mark, !. I introduce the factorial operation in Chapter 1 and use it in Chapter 19, but it doesn't truly come into its own until you use it with permutations, combinations, and probability problems (as you can see in the section, "How Do I Love Thee? Let Me Count Up the Ways," later in this chapter).

One of the main reasons you use the factorial operation in conjunction with sets is to count the number of elements in the sets. When the numbers are nice, discrete, small values, you have no problem. But when the sets get very large — such as the number of handshakes that occur when everyone in a club of 40 people shakes hands with everyone else — you want a way to do a systematic count. Factorials are built into the formulas that allow you to do such counting.

Making factorials manageable

When you apply the factorial operation to a counting problem or algebra expansion of binomials, you run the risk of producing a very large number. Just look at how fast the factorial grows:

$1! = 1 \quad 2! = 2\cdot 1 = 2 \quad 3! = 3\cdot 2\cdot 1 = 6 \quad 4! = 4\cdot 3\cdot 2\cdot 1 = 24$

$5! = 5\cdot 4\cdot 3\cdot 2\cdot 1 = 120 \quad 6! = 6\cdot 5\cdot 4\cdot 3\cdot 2\cdot 1 = 720 \quad 7! = 7\cdot 6\cdot 5\cdot 4\cdot 3\cdot 2\cdot 1 = 5{,}040$

$8! = 8\cdot 7\cdot 6\cdot 5\cdot 4\cdot 3\cdot 2\cdot 1 = 40{,}320 \quad 9! = 9\cdot 8\cdot 7\cdot 6\cdot 5\cdot 4\cdot 3\cdot 2\cdot 1 = 362{,}880$

$10! = 10\cdot 9\cdot 8\cdot 7\cdot 6\cdot 5\cdot 4\cdot 3\cdot 2\cdot 1 = 3{,}628{,}800$

Check out the last two numbers in the list. You see that 10! has the same digits as 9! — only with an extra zero. This observation illustrates one of the properties of the factorial operation. Another property is the definition of 0! (zero factorial).

The two properties (or rules) for the factorial operation are as follows:

- $n! = n\cdot(n-1)!$: To find $n!$, you multiply the number n times the factorial that comes immediately before it.
- $0! = 1$: The value of zero factorial is 1. You just have to trust this one.

Simplifying factorials

The process of simplifying factorials in fractions, multiplication problems, or formulas is simple enough, as long as you keep in mind how the factorial operation works. For instance, if you want to simplify the fraction $\frac{8!}{4!}$, you can't just divide the 8 by the 4 to get 2!. The factorials just don't work that way. You'll find that reducing the fraction by using factorial properties is much easier and more accurate than trying to divide the huge numbers created by the factorials.

EXAMPLE

Q. Simplify $\frac{8!}{4!}$.

A. You have two options to simplify this factorial.

You can write out and cancel out all the factors of 4! and multiply what's left:

$$\frac{8!}{4!} = \frac{8\cdot 7\cdot 6\cdot 5\cdot \cancel{4}\cdot \cancel{3}\cdot \cancel{2}\cdot \cancel{1}}{\cancel{4}\cdot \cancel{3}\cdot \cancel{2}\cdot \cancel{1}} = \frac{8\cdot 7\cdot 6\cdot 5}{1} = 1{,}680$$

You can take advantage of the rule $n! = n(n-1)!$ (see the previous section) when writing the larger factorial:

$$\frac{8!}{4!} = \frac{8\cdot 7\cdot 6\cdot 5\cdot 4!}{4!} = \frac{8\cdot 7\cdot 6\cdot 5\cdot \cancel{4!}}{\cancel{4!}} = \frac{8\cdot 7\cdot 6\cdot 5}{1} = 1{,}680$$

Q. Compute $\frac{40!}{37!3!}$.

A. The value of 40! is huge, and so is the value of 37!. But the reduced form of the example fraction is quite nice. You can reduce the fraction by using the first property of factorials:

$$\frac{40!}{3!37!} = \frac{40 \cdot 39 \cdot 38 \cdot 37!}{3 \cdot 2 \cdot 1 \cdot 37!} = \frac{40 \cdot 39 \cdot 38 \cdot \cancel{37!}}{3 \cdot 2 \cdot 1 \cdot \cancel{37!}} = \frac{40 \cdot \cancel{39}^{13} \cdot \cancel{38}^{19}}{\cancel{3} \cdot \cancel{2} \cdot 1} = 40 \cdot 13 \cdot 19 = 9{,}880$$

The values 39 and 38 in the numerator are each divisible by one of the factors in the denominator. You multiply what you have left for your answer.

YOUR TURN

13 Compute $\frac{100!}{99!}$.

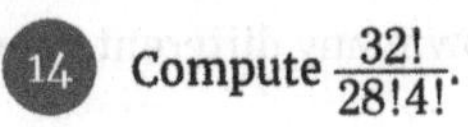

14 Compute $\frac{32!}{28!4!}$.

How Do I Love Thee? Let Me Count Up the Ways

You may think that you know everything there is to know about counting. After all, you've been counting since your parents asked you, a three-year-old, how many toes you have and how many cookies are on the plate. Counting toes and cookies is a breeze. Counting very, very large sets of elements is where the challenge comes in. Fortunately, algebra provides you with some techniques for counting large sets of elements more efficiently: the multiplication principle, permutations, and combinations. (For more in-depth information on these topics, check out *Probability For Dummies* [John Wiley & Sons, Inc.], by Deborah Rumsey, PhD.)

REMEMBER

You use each counting technique in a decidedly different situation, although deciding which technique to use is often the biggest challenge.

Applying the multiplication principle to sets

The multiplication principle is true to its name: It calls for you to multiply the number of elements in different sets to find the total number of ways that tasks or other things can be done. If you can do Task 1 in m_1 ways, Task 2 in m_2 ways, Task 3 in m_3 ways, and so on, you can perform all the tasks in a total of $m_1 \cdot m_2 \cdot m_3 \cdots$ ways.

EXAMPLE

Q. If you have six shirts, four pairs of pants, eight pairs of socks, and two pairs of shoes, then how many different outfits can you create using these items?

A. You have $6 \cdot 4 \cdot 8 \cdot 2 = 384$ different ways to get dressed. Of course, this doesn't take into account any color or style issues.

Q. How many different license plates are available in a state with the following rules:

- All the plates have three letters followed by two numbers.
- The first letter can't be O.
- The first number can't be zero or one.

A. Think of this problem as a prime candidate for the multiplication principle. Here's the first rule for the license plate:

$$(\text{\# ways letter1}) \cdot (\text{\# ways letter2}) \cdot (\text{\# ways letter3}) \cdot (\text{\# ways number1}) \cdot (\text{\# ways number2})$$

You can't use O for the first letter, so that leaves 25 ways to choose the first letter. The second and third letters have no restrictions, so you can choose any one of the 26 letters for them. The first number can't be zero or one, so that leaves you eight choices. The second number has no restrictions, so you have all ten choices. You simply multiply these choices together to get your answer: $25 \cdot 26 \cdot 26 \cdot 8 \cdot 10 = 1{,}352{,}000$. You can assume that this isn't a very big state if it has only a million or so license-plate possibilities.

YOUR TURN

15 The local pizza parlor is having a special on pizzas. You can order a pizza with four toppings where you can choose one of the 3 types of crust, one of the 4 choices of meat, one of the 3 choices of vegetable, either mushrooms or anchovies, and one of the 2 choices of cheese. How many different pizzas can be created?

16 You're trying to figure out that password that you've forgotten. You know that you put in either 2014 or 2015, and then you put in a month (using two digits), and then you put in a letter from the alphabet, and then you put in @ or # or * or &. How many different possibilities are there?

Arranging permutations of sets

The permutation of some set of elements is the rearrangement of the order of those elements. For example, you can rearrange the letters in the word "act" to spell six different words (well, not actual words; they could possibly be acronyms for something): act, cat, atc, cta, tac, tca. Therefore, you say that for the word "act," you have six permutations of three elements.

Counting the permutations

Knowing how many permutations a set of elements has doesn't tell you what those permutations are but you at least know how many permutations to look for. When you find the number of permutations for a set of elements, you select where you get the elements from and how many you need to rearrange. You assume that you don't replace any of the elements you select before you choose again.

You find the number of permutations, P, of n elements taken r at a time with the formula $_nP_r = \frac{n!}{(n-r)!}$ (see the section, "Focusing on Factorials," earlier in this chapter for an explanation of !).

EXAMPLE

Q. If you want to arrange four of your six vases on a shelf, how many different arrangements (orders) are possible?

A. If you label your vases A, B, C, D, E, and F, some of the arrangements are ABCD, ABCE, ABCF, BCFD, BFDC, and so on. Using the formula to count the number of arrangements possible, you let $n = 6$ and $r = 4$:

$$_6P_4 = \frac{6!}{(6-4)!} = \frac{6!}{2!} = \frac{6 \cdot 5 \cdot 4 \cdot 3 \cdot \cancel{2!}}{\cancel{2!}} = 6 \cdot 5 \cdot 4 \cdot 3 = 360$$

With that number, I hope you don't plan on taking a picture of each arrangement!

Q. Say that the seven dwarfs want to line up for a family photo. How many different ways can the dwarfs line up for the picture?

A. In algebraic terms, you want a permutation of seven things taken seven at a time. Using the previous formula, you find the following:

$$_7P_7 = \frac{7!}{(7-7)!} = \frac{7!}{0!} = \frac{7 \cdot 6 \cdot 5 \cdot 4 \cdot 3 \cdot 2 \cdot 1}{1} = 7 \cdot 6 \cdot 5 \cdot 4 \cdot 3 \cdot 2 \cdot 1 = 5{,}040$$

Remember that $0! = 1$ (which I explain in the section, "Making factorials manageable," earlier in the chapter). In the formula, you end up with a 0! in the denominator, because the number of items chosen is the same as the number of items available. Because $0! = 1$, the denominator becomes a 1, and the answer becomes the value in the numerator of the fraction.

When using the formula for permutations of *n* elements taken *r* at a time, there are some special situations that help save you time when computing:

» ${}_nP_n = n!,\ {}_nP_n = \frac{n!}{(n-n)!} = \frac{n!}{0!} = n!.$

» ${}_nP_1 = n!,\ {}_nP_1 = \frac{n!}{(n-1)!} = \frac{n\cdot\cancel{(n-1)}\cancel{(n-2)}\cancel{(n-3)}\cdots\cancel{3}\cdot\cancel{2}\cdot\cancel{1}}{\cancel{(n-1)}\cancel{(n-2)}\cancel{(n-3)}\cdots\cancel{3}\cdot\cancel{2}\cdot\cancel{1}} = n.$

» ${}_nP_0 = 1,\ {}_nP_0 = \frac{n!}{(n-0)!} = \frac{n!}{n!} = 1.$

Some problems call for a mixture of permutations and the multiplication principle.

EXAMPLE

Q. You have 10 people vying for the positions of president, vice president, and treasurer of your club, and two different places at which your group will meet. If the names of the officers are drawn at random, how many different results will there be?

A. First, choosing the officers uses a permutation of 10 things taken 3 at a time — the order makes a difference. Then, once the slate of officers is chosen, you multiply that times the two options for places to meet.

$$\left({}_{10}P_3\right)\cdot(2) = \frac{10!}{7!3!}\cdot(2) = \frac{10\cdot 9\cdot 8\cdot \cancel{7!}}{\cancel{7!}3!}\cdot(2) = \frac{10\cdot \cancel{9}^3\cdot \cancel{8}^4}{\cancel{3}\cdot\cancel{2}\cdot 1}\cdot(2) = 120(2) = 240$$

Q. You want to make up a new password for your computer account. The first rule for the system is that the first two entries must be letters with no repeats, and you can't use O or I. The next four entries must be digits with no restrictions. You can then enter two more letters with no restrictions. Finally, the last three entries must be three digits with no repeats. How many different passwords are possible?

A. Break down the problem into its four different parts, setting up the multiplication principle to combine the four parts:

(2 letters, no repeats, no O or I)·(4 digits)·(2 letters)·(3 digits, no repeats)

You can now set up the permutations and products:

- The arrangement of the first two letters is a permutation of 24 elements taken 2 at a time.
- The next four digits give you 10 choices each time — multiplied together.
- The following two letters give you 26 choices each time — multiplied together.
- The final three digits represent a permutation of 10 elements taken 3 at a time.

The computation goes as follows:

$$\left({}_{24}P_2\right)\cdot(10\cdot 10\cdot 10\cdot 10)\cdot(26\cdot 26)\cdot\left({}_{10}P_3\right) = \frac{24!}{22!}\cdot 10^4\cdot 26^2\cdot\frac{10!}{7!}$$
$$= 552\cdot 10{,}000\cdot 676\cdot 720 = 2{,}686{,}694{,}400{,}000$$

You calculate over two-and-a-half trillion different passwords. Hopefully, the rules will keep the hackers out. Now you just have to remember your password!

Distinguishing one permutation from another

If you're lining up the books on your bookshelf based on color — for eye appeal — rather than subject or author, you aren't distinguishing between the different red books and the different blue books. And if you want to know the total number of permutations of the letters in the word "cheese," you can't distinguish between $\text{cheese}_{\text{che1e2se}}$ and $\text{cheese}_{\text{che2e1se}}$, where you switch the e's around. You can't tell the permutations apart, so you don't count rearrangements of the same letter as different permutations. The way you counter this effect is to use the formula for *distinguishable permutations.*

If a set of n objects has k_1 elements that are alike, k_2 elements that are alike, and so on, you find the number of distinguishable permutations of those n objects with the following formula:

$$\frac{n!}{k_1!k_2!k_3!\cdots}$$

EXAMPLE

Q. Determine the number of distinguishable permutations of the word "cheese" (see the section, "Focusing on Factorials," for the division part of this formula).

A. Cheese has six letters altogether, and the three e's are all the same.

$$\frac{6!}{3!} = \frac{6 \cdot 5 \cdot 4 \cdot \cancel{3!}}{\cancel{3!}} = 120$$

Q. You have ten blue books, five red books, six black books, one green book, and one gray book. How many possible distinguishable arrangements of the books are there?

A. You can determine this number by using the formula for distinguishable permutations. You have 23 books total and can make 23! arrangements of all the books at once and then find the distinguishable number by dividing by the repeated colors:

$$\frac{23!}{10!5!6!} = 8.245512475 \times 10^{10}$$

Q. Referring to the previous problem, now consider another arrangement. Let's say you want to put all the books of the same color together, and you don't care about the order of the books in each color grouping. How many different arrangements are possible?

A. This problem is a permutation of the five colors (blue, red, black, green, and gray). A permutation of 5 elements taken 5 at a time is $5! = 120$ different arrangements (see the section, "Focusing on Factorials," for more on this calculation).

Okay. Back to the same book problem. Being the finicky decorator that you are, you now decide that the arrangements within each color matter, and you want the books arranged in groups of the same color. How many different arrangements are possible? This problem involves the permutations of the different colors first and then permutations of the books in the blue, red, and black groups. Doing the math, you find the following:

$$\begin{aligned}(\text{Colors}) \cdot (\text{Blues}) \cdot (\text{Reds}) \cdot (\text{Blacks}) &= ({}_5P_5)({}_{10}P_{10})({}_5P_5)({}_6P_6) \\ &= (5!)(10!)(5!)(6!) = 6.2705664 \times 10^{11}\end{aligned}$$

You have quite a few ways to arrange the books on the bookshelf. Maybe alphabetical order makes more sense.

17 How many permutations are there when you choose 3 of the letters from the word TEAR? (And how many form actual words?)

18 The first-grade teacher wants to line up the 16 students in a different order each day. If there are 180 school days each year, will there be enough arrangements to have a different line-up each day?

19 You have 4 dogs, 3 cats, and 2 rabbits. If you want to put their individual bowls in a line consisting of dogs, then cats, then rabbits, how many different arrangements are possible for Toby, Hank, Bucky, Eddie, Cassie, Buffy, Gabby, Robert, and Fred?

20 How many ways can you arrange the letters in the word *Pennsylvania*?

Mixing up sets with combinations

Combination is a precise mathematical term that refers to how you can choose a certain number of elements from a set. Unlike permutations, with combinations the order in which the elements appear doesn't matter. Combinations, for example, allow you to determine how many different ways you can choose three people to serve on a committee (the order in which you choose them doesn't matter, unless you want to make one of them the chairperson).

You find the number of possible ways to choose r elements from a set containing n elements, C, with the following formula (see the section, "Focusing on Factorials," for an explanation of !):

$$_nC_r = \frac{n!}{(n-r)!r!}$$

The bulk of this formula should look familiar; it's the formula for the number of permutations of n elements taken r at a time (see the previous section) — with an extra factor in the denominator. The formula is actually the permutations divided by $r!$.

You can also indicate combinations with two other notations: $C(n, r)$ and $\binom{n}{r}$.

The different notations all mean the same thing and have the same answer. For the most part, the notation you use is a matter of personal preference or what you can find on your calculator. You read all the notations as "n choose r."

EXAMPLE

Q. You have tickets to the theater, and you want to choose three people to join you. How many different ways can you choose those three people if you have five close friends to choose from? Your friends are Violet, Wally, Xanthia, Yvonne, and Zeke. You can take

{Violet, Wally, Xanthia}, {Violet, Wally, Yvonne}, {Violet, Wally, Zeke},

{Wally, Xanthia, Yvonne}, {Wally, Xanthia, Zeke}, {Xanthia, Yvonne, Zeke}

You find six different subsets. Is that all? Have you forgotten any?

A. You can check your work with the formula for the number of combinations of five elements taken three at a time:

$$_5C_3 = \frac{5!}{(5-3)!3!} = \frac{5\cdot \cancel{4}^2\cdot \cancel{3}\cdot \cancel{2}\cdot \cancel{1}}{\cancel{2}\cdot 1\cdot \cancel{3}\cdot \cancel{2}\cdot \cancel{1}} = 10$$

Oops! You must have forgotten some combinations. (See the later section, "Drawing a tree diagram for a combination," to see what you missed.)

Q. You have 10 cartons of ice cream in your freezer, each a different flavor. You're going to choose 4 cartons to serve at a party. How many different ways can you choose them?

A. Just use $_{10}C_4 = \frac{10!}{(10-4)!4!} = \frac{10!}{6!4!} = \frac{10\cdot \cancel{9}^3\cdot \cancel{8}^2\cdot 7\cdot \cancel{6!}}{\cancel{6!}\cdot \cancel{4}\cdot \cancel{3}\cdot \cancel{2}\cdot \cancel{1}} = 10\cdot 3\cdot 7 = 210$ different combinations.

YOUR TURN

21 You have six choices for extracurricular activities: art class, fencing, bowling, band, soccer, and chorus. You can choose only two. How many different ways can you choose the two activities?

22 You need to bring some fruit salad to a luncheon. You have: bananas, oranges, pears, peaches, blueberries, raspberries, strawberries, and cherries. If you want to combine 5 of the fruits, how many different kinds of salad can you make?

Branching Out with Tree Diagrams

Combinations and permutations (see the previous section) tell you how many subsets or groupings to expect within certain sets, but they don't shine a light on what elements are in those groupings or subsets. A nice, orderly way to write out all the combinations or permutations is to use a tree diagram. The name *tree diagram* comes from the fact that the drawing looks something like a family tree — branching out from the left to the right. The left-most entries are your first choices, the next column of entries shows what you can choose next, and so on.

You can't use tree diagrams in all situations dealing with permutations and combinations — the diagrams get too large and spread out all over the place — but you can use tree diagrams in many situations involving counting items. Tree diagrams also appear in the worlds of statistics, genetics, and other studies.

Picturing a tree diagram for a permutation

Tree diagrams are very satisfying to those of us who are very visual. You know what they should look like ahead of time and you just fill in the details. These diagrams are very useful to ensure that you haven't made an error or left anything out.

To create a tree diagram, go out on a limb and follow these simple steps:

1. **Start out with the original set of *n* items and list them vertically — one under the other — leaving some space for your tree to grow to the right.**
2. **Connect the next set of entries to the first set by drawing short line segments — the branches.**

   You'll need $n-1$ branches for each item.
3. **Keep connecting more entries until you complete the task as many times as needed.**

   Each time, you decrease the number of branches by 1.

EXAMPLE

Q. Make a tree diagram showing all the two-letter words possible from the letters in SEAT.

A. When you choose two different letters from a set of four letters to form different "words," you can find the number of different words possible with the permutation $_4P_2 = \frac{4!}{(4-2)!} = \frac{4\cdot 3\cdot \cancel{2!}}{\cancel{2!}} = 12$. You can form 12 different two-letter words. To make a list of the words, and not repeat yourself or leave any out, you create a tree diagram.

Using the steps given, in the first column, list the four letters from the word SEAT. Then move to your right, and after each of those four letters, connect them, one at a time, with each of the other three letters. Your connections give you all the possibilities for creating two-letter "words." Figure 20-4 shows the tree diagram that organizes these permutations. You see that you have 12 permutations, as expected.

First Choice | Second Choice | Two-Letter Words

S — E ---- SE
S — A ---- SA
S — T ---- ST

E — S ---- ES
E — A ---- EA
E — T ---- ET

A — S ---- AS
A — E ---- AE
A — T ---- AT

T — S ---- TS
T — E ---- TE
T — A ---- TA

FIGURE 20-4: Creating two-letter words (permutations) from SEAT.

Drawing a tree diagram for a combination

When you use a tree diagram to list all of the possibilities in a combination, this diagram is similar to that for a permutation. But there's a big difference. Because the order of the choices doesn't matter, your tree diagram won't be quite as symmetric. It will change as you move downward.

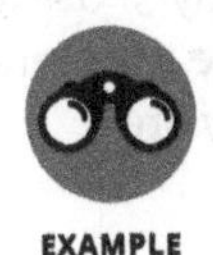

EXAMPLE

Q. You are given the job of choosing three friends from a set of five to accompany you to the theater, and you find that you have ten different ways to choose them. The order in which you choose them doesn't matter; they get to go or they don't. (Does this sound familiar? Well, yes, it should. You also find this situation in the earlier section, "Mixing up sets with combinations.") Use a tree diagram to determine the ten different groups of friends.

A. The tree diagram in Figure 20-5 shows a way to determine all the arrangements without repeating yourself.

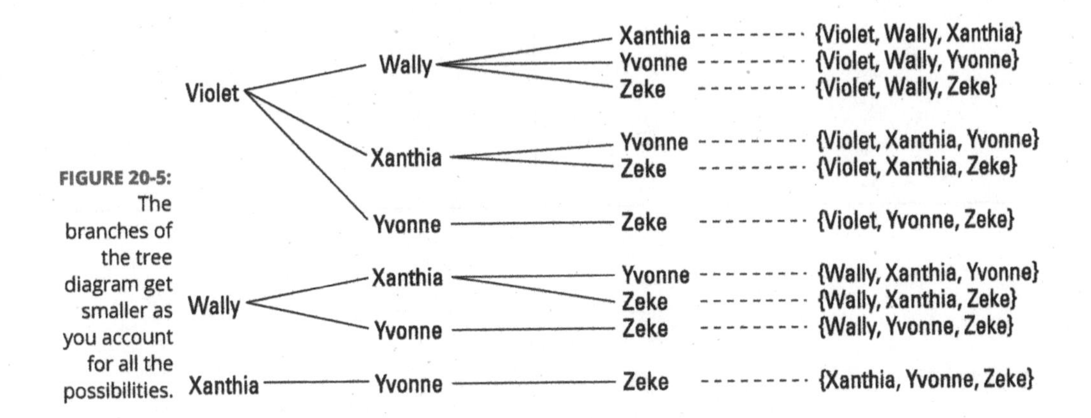

FIGURE 20-5: The branches of the tree diagram get smaller as you account for all the possibilities.

You figure out the different arrangements by starting out with any three people. It doesn't matter whom you start with — you get the same number of groupings with the same combinations of people in the groupings. (I didn't include poor Yvonne and Zeke in the first column of entries because their names come last in the alphabet.)

Because the order doesn't matter, you first figure out all the arrangements with Violet and Wally, and then Violet and Xanthia, and then Violet and Yvonne. The number of branches decreases because you can't repeat groupings. You then move to Wally and Xanthia and then Wally and Yvonne. You finally get to Xanthia and stop there, because you've accounted for all the groupings.

Do you see which sets I missed in the section, "Mixing up sets with combinations"? You can easily miss some arrangements if you don't have an orderly method for listing them. And you may be concerned that Violet has a much better chance for selection than her peers but that isn't the case. If you look through all ten sets of three people, you see that each name appears six times.

YOUR TURN

23 You have three tiles numbered 2, 4, and 5 in a jar and will draw two of them at random — keeping track of the order. You want to determine the chances of drawing two numbers that form a number divisible by 3. How many different numbers can be formed?

24 You have three tiles numbered 2, 4, and 5 in a jar and will draw two of them at random — keeping track of the order. You want to determine the chances of drawing two numbers that form a number divisible by 3. Use a tree diagram to list all the numbers — and determine how many are divisible by 3.

Practice Questions Answers and Explanations

1. **{3, 6, 9, 12, 15, 18, 21, 24, 27}**, $C = \{x \mid x = 3n, n \in Z, 0 < x < 29\}$, **C = {3, 6, 9, ... , 27}.** List all the numbers that apply, separating them with commas. For the set-builder notation, use $n \in Z$ to limit the values to being integers. You don't use the multiple zero, because that isn't included in the inequality. For the ellipsis notation, write enough terms to show the pattern.

2. $D = \{0, 4, 8, 12, 16, \ldots, 48\}$. Yes, I should have inserted the other multiples, but the pattern is clear, so I used the ellipsis.

3. **8.** You will find $8 = 2^3$ subsets: $\{3, 5, 7\}, \{3, 5\}, \{3, 7\}, \{5, 7\}, \{3\}, \{5\}, \{7\}, \{\ \}$

4. **16.** You will find $16 = 2^4$ subsets: $\{m, a, t, h\}, \{m, a, t\}, \{m, a, h\}, \{m, t, h\}, \{a, t, h\}, \{m, a\}, \{m, t\}, \{m, h\}, \{a, t\}, \{a, h\}, \{t, h\}, \{m\}, \{a\}, \{t\}, \{h\}, \{\ \}$

5. $\{2, 3, 5, 7, 11, 13, 17, 23, 33\}$. Do not repeat the elements 3 or 13: $G \cup H = \{2, 3, 5, 7, 11, 13, 17, 23, 33\}$.

6. $\{2, 4, 8, 16, 32, 64\}$. The set K is a subset of set J. So $J \cup K = J = \{2, 4, 8, 16, 32, 64\}$.

7. **{2, 6}.** The two sets have the elements 2 and 6 in common: $M \cap N = \{2, 6\}$.

8. **∅.** The two sets have nothing in common: $P \cap Q = \{\ \}$.

9. **60.** Using $n(A \cup B) = n(A) + n(B) - n(A \cap B)$, $n(E \cup F) = 50 + 40 - 30 = 60$.

10. **150.** Using $n(A) + n(A') = n(U)$, $50 + n(E') = 200 \rightarrow n(E') = 150$.

11. **February, July.** Draw a rectangle and insert two overlapping circles. Label the circles. Place the names of the months in their correct position.

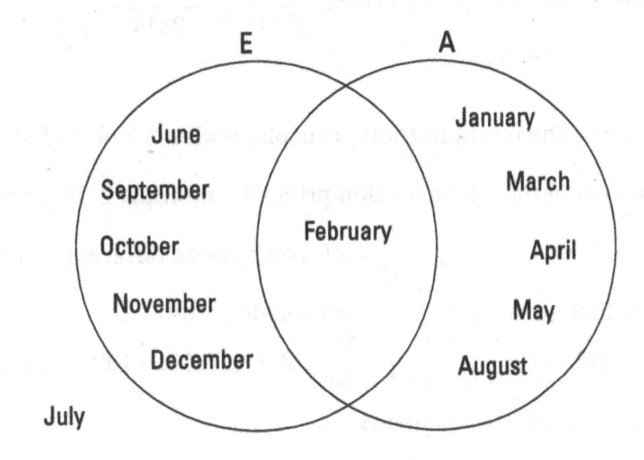

Only February contains both letters and only July has neither.

12. **None, 1.** Draw a rectangle and insert three overlapping circles. Label the circles. Place the "multiples of 2," "multiples of 3," and "primes."

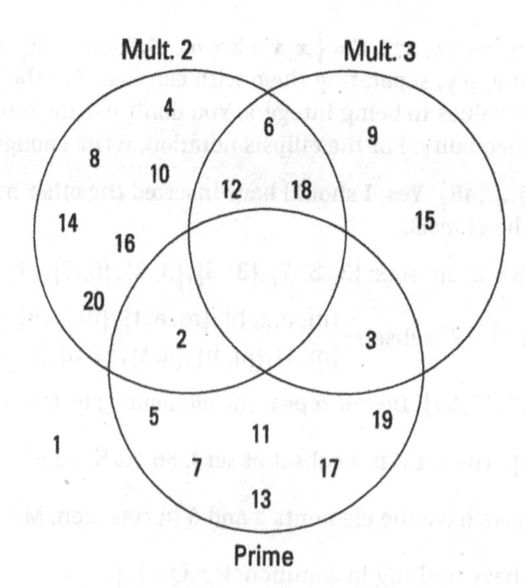

None of the integers has all three qualities, and only the number 1 doesn't qualify for any of them.

13. **100.** Using the factorial rules: $\frac{100!}{99!} = \frac{100 \cdot 99!}{99!} = \frac{100 \cdot \cancel{99!}}{\cancel{99!}} = 100.$

14. **35,960.** Using the factorial rules: $\frac{32!}{28!4!} = \frac{32 \cdot 31 \cdot 30 \cdot 29 \cdot 28!}{28!4 \cdot 3 \cdot 2 \cdot 1} = \frac{\cancel{32}^8 \cdot 31 \cdot \cancel{30}^5 \cdot 29 \cdot \cancel{28!}}{\cancel{28!}\cancel{4} \cdot \cancel{3} \cdot \cancel{2} \cdot 1} =$ $8 \cdot 31 \cdot 5 \cdot 29 = 35{,}960.$

15. **144.** Using the multiplication principle, multiply $3 \cdot 4 \cdot 3 \cdot 2 \cdot 2 = 144.$

16. **2,496.** Using the multiplication principle, multiply $2 \cdot 12 \cdot 26 \cdot 4 = 2{,}496.$ Lotsa luck!

17. **24.** Using ${}_4P_3 = \frac{4!}{(4-3)!} = \frac{4!}{1!} = 24$, you have 24 different "words" possible. I can only come up with 9 that are actual words. Can you do better?

18. **Yes.** Using ${}_{16}P_{16} = \frac{16!}{(16-16)!} = \frac{16!}{0!} = 2.092278989 \times 10^{13}$ — oh, yes, there will be no problem finding enough arrangements.

19. **288.** Find the product $({}_4P_4) \cdot ({}_3P_3) \cdot ({}_2P_2) = 4! \cdot 3! \cdot 2! = 24 \cdot 6 \cdot 2 = 288.$

20. **39,916,800.** There are 12 letters in the state's name. The *n* is used 3 times and the *a* is used 2 times. To solve, $\frac{12!}{3! \cdot 2!} = \frac{12 \cdot 11 \cdot 10 \cdot 9 \cdot 8 \cdot 7 \cdot 6 \cdot 5 \cdot \cancel{4}^2 \cdot \cancel{3!}}{\cancel{3!} \cdot \cancel{2} \cdot 1} = 39{,}916{,}800.$

21) **15.** To solve, $_6C_2 = \frac{6!}{(6-2)!2!} = \frac{6!}{4!2!} = \frac{\cancel{6}^3 \cdot 5 \cdot \cancel{4!}}{\cancel{4!} \cdot \cancel{2} \cdot 1} = 15.$

22) **56.** There are 8 fruit possibilities: $_8C_5 = \frac{8!}{(8-5)!5!} = \frac{8!}{3!5!} = \frac{8 \cdot 7 \cdot \cancel{6} \cdot \cancel{5!}}{\cancel{3 \cdot 2} \cdot 1 \cdot \cancel{5!}} = 56.$

23) **6.** This is a permutation of '3 choosing 2': $_3P_2 = \frac{3!}{(3-2)!} = \frac{3!}{1!} = 6$. There are six possible arrangements.

24) **4.** You have determined that there are six different possibilities. List the three numbers vertically and create a tree by connecting each possibility with one of the other numbers.

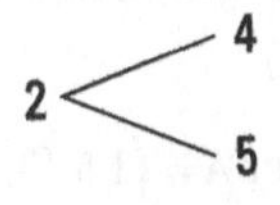

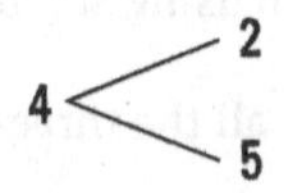

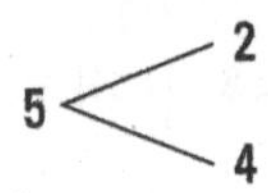

You see that 4 of the 6 two-digit numbers are divisible by 3: 24, 42, 45, and 54.

Whaddya Know? Chapter 20 Quiz

Quiz time! Complete each problem to test your knowledge on the various topics covered in this chapter. You can then find the solutions and explanations in the next section.

1. Simplify: $\frac{20!}{4!16!}$

2. Write the elements of set A using roster notation: $A = \{x \mid x = 2n+1,\ n \in W,\ 0 \le x \le 20\}$.

3. Use a Venn diagram to determine how many days of the week contain the letter *S* but not the letter *N*.

4. How many subsets does set $A = \{1,5,13,25,41\}$ have?

5. Write the elements of set B using set-builder notation: $B = \{1,11,21,31,41,51,61,71,81,91\}$.

6. Use a tree diagram to find all the three-letter "words" that can be formed from PEAR.

7. Given $A = \{1,3,5,7,9,11,13\}$ and $B = \{2,3,5,7,11,13,17\}$, find $A \cup B$.

8. Given $C = \{x \mid x = 5n,\ 0 < n < 8,\ n \in Z\}$ and $D = \{x \mid x = 4n,\ 0 < n < 10,\ n \in Z\}$, find $C \cap D$.

9. Given the universal set $U = \{n \mid n \in \text{even integers},\ 0 \le n < 50\}$, $C = \{0,4,8,12,16,\ldots,28,32\}$, and $D = \{0,6,12,18,\ldots,36,42\}$, find $(C \cup D)'$.

10. How many three-letter "words" can be created from the letters in the word STAND? (Some are: and, tan, ant, sta, nad,; not all are actually words.)

11. Rachel and Roger just got engaged and want to get married in the next year. They are happy with any month but would like to have the ceremony on a Friday, Saturday, or Sunday. They have the choice of four different venues and know of three different caterers they'd prefer. How many different arrangements are possible for this wedding?

12. Hank is going fishing and has room for 12 lures in his tackle box. He has 20 different lures to choose from. How many different combinations are possible?

13. Given $A = \{1,3,5,7,9,11,13\}$ and $B = \{2,3,5,7,11,13,17\}$, find $A \cap B$.

14. Given the universal set $U = \{\text{letters in the English alphabet}\}$ and $A = \{\text{letters in } \textit{United States of America}\}$, find A'.

Answers to Chapter 20 Quiz

1. **4,845.** $\frac{20!}{4!16!} = \frac{20 \cdot 19 \cdot 18 \cdot 17 \cdot 16!}{4 \cdot 3 \cdot 2 \cdot 1 \cdot 16!} = \frac{\cancel{20}^5 \cdot 19 \cdot \cancel{18}^3 \cdot 17 \cdot \cancel{16!}}{\cancel{4} \cdot \cancel{3} \cdot \cancel{2} \cdot 1 \cdot \cancel{16!}} = 5 \cdot 19 \cdot 3 \cdot 17 = 4{,}845$

2. **{1, 3, 5, 7, 9, 11, 13, 15, 17, 19}.** List the elements, separating them using commas. These are the odd whole numbers between zero and 20.

3. **Tuesday, Thursday, Saturday.** Create a Venn diagram showing the set of days with an S and days with an N. Tuesday, Thursday, and Saturday have an S but not an N.

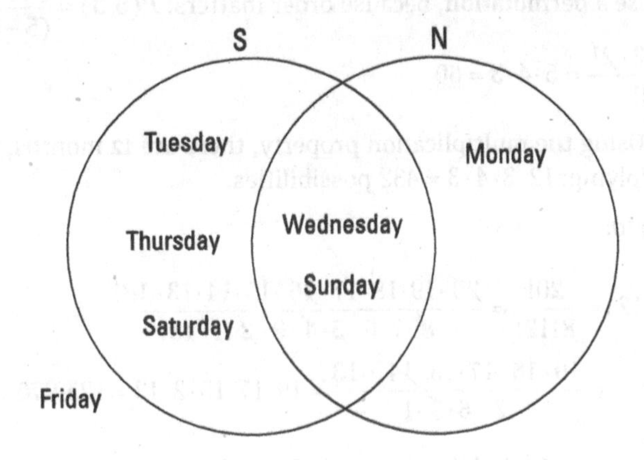

4. **32.** Because there are 5 elements in set A, there are $2^5 = 32$ subsets.

5. Write the elements of set B using set-builder notation (W: whole numbers):

$$B = \{x \mid x = 10n + 1,\ n \in W,\ 0 \le x \le 9\}$$

6. **See the figure.** Using $P(4,3) = \frac{4!}{(4-3)!} = \frac{4!}{1!} = 24$, you will find 24 different "words." List the four letters and connect each letter with one of the others and each of those with the remaining two. There are words like: PEA, PER, EAR, REP, and many more!

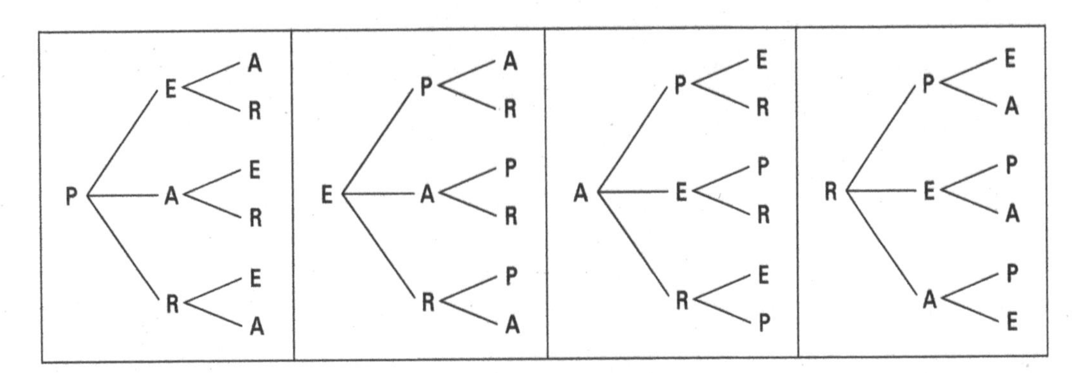

7. **{1, 2, 3, 5, 7, 9, 11, 13, 17}.** Don't repeat the 3, 5, 7, 11, or 13.

8. $\mathbf{C \cap D = \{20\}}$**.** First, create rosters of the two sets: $C = \{5,10,15,20,25,30,35\}$ and $D = \{4,8,12,16,20,24,28,32,36\}$. Thus, $C \cap D = \{20\}$.

9. $\mathbf{(C \cup D)' = \{2,10,14,22,26,34,38,40,44,46,48\}}$**.** First, find the union of C and D: $C \cup D = \{0,4,6,8,12,16,18,20,24,28,30,32,36,42\}$. Now find the complement of that union of sets: $(C \cup D)' = \{2,10,14,22,26,34,38,40,44,46,48\}$.

10. **60.** Use a permutation, because order matters: $P(5,3) = \frac{5!}{(5-3)!} = \frac{5!}{2!} = \frac{5 \cdot 4 \cdot 3 \cdot 2!}{2!} = \frac{5 \cdot 4 \cdot 3 \cdot \cancel{2!}}{\cancel{2!}} = 5 \cdot 4 \cdot 3 = 60$

11. **432.** Using the multiplication property, there are 12 months, 3 days, 4 venues, and 3 caterers. Multiplying: $12 \cdot 3 \cdot 4 \cdot 3 = 432$ possibilities.

12. **125,970.**

$$C(20,12) = \frac{20!}{8!12!} = \frac{\cancel{20} \cdot 19 \cdot 18 \cdot 17 \cdot \cancel{16} \cdot 15 \cdot 14 \cdot 13 \cdot \cancel{12!}}{\cancel{8} \cdot 7 \cdot 6 \cdot \cancel{5} \cancel{4} \cdot 3 \cdot \cancel{2} \cdot 1 \cdot \cancel{12!}}$$

$$= \frac{19 \cdot \cancel{18} \cdot 17 \cdot 15 \cdot \cancel{14}^{2} \cdot 13}{\cancel{7} \cdot \cancel{6} \cancel{3} \cdot 1} = 19 \cdot 17 \cdot 15 \cdot 2 \cdot 13 = 125,970$$

13. **{3, 5, 7, 11, 13}.** Find the elements shared by both sets.

14. $\mathbf{A' = \{b,g,h,j,k,l,p,q,v,w,x,y,z\}}$**.** Find the letters that do NOT appear in the spelling of *United States of America.*

5 Applying Known Formulas

In This Unit . . .

IN THIS CHAPTER

» Expanding binomials in two different ways

» Checking out formulas to use when graphing

» Taking aim at geometric formulas

» Solving a formula for a variable

Chapter 21
Manipulating Formulas

Formulas are the basis of so many processes in business, agriculture, finance, and, of course, mathematics. A formula is an equation that always works. It's something that has been proven to be reliable and helpful and usable. You have geometric formulas that tell you the area of a trapezoid or the volume of a cone. Financial formulas tell you how much your monthly mortgage payment will be. Temperature formulas help you know how to pack for that trip to a country that uses Centigrade measures. Formulas are useful, but you have to know how to use them not abuse them.

Expanding Binomials

A binomial is an expression with two terms. A common process that occurs with binomials is expanding them, or raising them to higher powers. You can raise the binomial $(a+b)$ to the second power, third power, or actually any power by just multiplying it over and over and over again. But, stop! You have other options: There's the *Binomial Theorem*, which is a formula for raising a binomial to any whole-number power; and there's *Pascal's Triangle*, which provides a quick and neat method for those expansions.

Using the Binomial Theorem

The *Binomial Theorem* says that you can take the binomial $(a+b)$ and raise it to some whole-number power using a formula based on summation, combinations, and exponents. (For a review of combinations, see Chapter 20.) Here's the theorem:

$$(a+b)^n = \sum_{k=0}^{n} \binom{n}{k} \cdot a^{n-k} b^k$$

This says to add up all the terms that are created by computing the combination of n things taken k at a time (that's what the $\binom{n}{k}$ notation means) and multiplying it by powers of a times powers of b. The combination result is the coefficient of the term.

EXAMPLE

Q. Use the Binomial Theorem to expand $(a+2)^4$.

A. First, note that there will be five terms. The first will contain a^5 and the last will contain a^0. Filling in the n's and k's and creating the terms, you have:

$$(a+2)^4 = \binom{4}{0} a^{4-0} 2^0 + \binom{4}{1} a^{4-1} 2^1 + \binom{4}{2} a^{4-2} 2^2 + \binom{4}{3} a^{4-3} 2^3 + \binom{4}{4} a^{4-4} 2^4$$

Note the patterns here: decreasing numbers in the combinations, increasing exponents on the second term in the binomial, and increasing terms in the exponential subtraction. (Noting patterns helps!) Simplify all the exponents and fill in all the combination results.

$$= \binom{4}{0} a^4 \cdot 1 + \binom{4}{1} a^3 \cdot 2 + \binom{4}{2} a^2 \cdot 4 + \binom{4}{3} a^1 \cdot 8 + \binom{4}{4} 1 \cdot 16$$
$$= 1 \cdot a^4 \cdot 1 + 4 \cdot a^3 \cdot 2 + 6 \cdot a^2 \cdot 4 + 4 \cdot a^1 \cdot 8 + 1 \cdot 1 \cdot 16$$

Simplifying the terms, you find that $(a+2)^4 = a^4 + 8a^3 + 24a^2 + 32a + 16$.

Q. Use the Binomial Theorem to expand $(x+y)^7$.

A. This expansion will have eight terms. Using the Binomial Theorem:

$$(x+y)^7$$
$$= \binom{7}{0} \cdot x^7 y^0 + \binom{7}{1} \cdot x^6 y^1 + \binom{7}{2} \cdot x^5 y^2 + \binom{7}{3} \cdot x^4 y^3 + \binom{7}{4} \cdot x^3 y^4 + \binom{7}{5} \cdot x^2 y^5 + \binom{7}{6} \cdot x^1 y^6 + \binom{7}{7} \cdot x^0 y^7$$

Notice how the powers of x decrease as you move from left to right, and the powers of y increase as you move from left to right. Now, inserting the coefficients from the combination formula,

$$(x+y)^7 = 1 \cdot x^7 y^0 + 7 \cdot x^6 y^1 + 21 \cdot x^5 y^2 + 35 \cdot x^4 y^3 + 35 \cdot x^3 y^4 + 21 \cdot x^2 y^5 + 7 \cdot x^1 y^6 + 1 \cdot x^0 y^7$$
$$= x^7 + 7x^6 y + 21x^5 y^2 + 35x^4 y^3 + 35x^3 y^4 + 21x^2 y^5 + 7xy^6 + y^7$$

YOUR TURN

1 Use the Binomial Theorem to expand $(z+1)^5$.	2 Use the Binomial Theorem to expand $(2x-3)^3$.

Applying Pascal's Triangle

Pascal's Triangle (named for Blaise Pascal, a French mathematician and philosopher) is a triangular array of the coefficients used when expanding binomials. You get the next row of numbers by putting 1s on the ends and adding the two numbers diagonally above to fill in the spaces. The bonus: The numbers in the triangle are all the answers to the combination problems! Here's the beginning of the triangle:

1
1 1
1 2 1
1 3 3 1
1 4 6 4 1
1 5 10 10 5 1
1 6 15 20 15 6 1
1 7 21 35 35 21 7 1

This is just the beginning because the triangle goes on down forever. And the rows of numbers are the coefficients of the powers of $(a+b)$. The first row is just 1, because $(a+b)^0=1$. The second row, 1 and 1, shows the two coefficients of $(a+b)^1=a+b$. The third row gives you the coefficients of $(a+b)^2=a^2+2ab+b^2$. The fourth row gives you the coefficients of $(a+b)^3=a^3+3a^2b+3ab^2+b^3$ and so on forever! This is extremely helpful when raising binomials to powers.

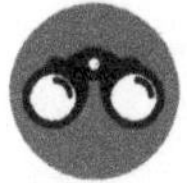

EXAMPLE

Q. Use Pascal's Triangle to find $(x-y)^6$.

A. First, go to the row that starts with "1 6". Write down all the coefficients, leaving space for the powers of the two terms.

1 6 15 20 15 6 1

Now write in decreasing powers of *x*.

$1x^6 \quad 6x^5 \quad 15x^4 \quad 20x^3 \quad 15x^2 \quad 6x^1 \quad 1x^0$

Next, add increasing powers of $-y$. Or, if you prefer, start on the right and move leftward, writing in decreasing powers of $-y$.

$1x^6(-y)^0 \quad 6x^5(-y)^1 \quad 15x^4(-y)^2 \quad 20x^3(-y)^3 \quad 15x^2(-y)^4 \quad 6x^1(-y)^5 \quad 1x^0(-y)^6$

Finally, simplify the terms.

$x^6 - 6x^5y + 15x^4y^2 - 20x^3y^3 + 15x^2y^4 - 6xy^5 + y^6$

Q. Use Pascal's Triangle to find $(2a + 3b)^4$.

A. First, go to the row that starts with "1 4". Write down all the coefficients, leaving space for the powers of the two terms.

1 4 6 4 1

Write in decreasing powers of $2a$ and increasing powers of $3b$.

$1(2a)^4(3b)^0 \quad 4(2a)^3(3b)^1 \quad 6(2a)^2(3b)^2 \quad 4(2a)^1(3b)^3 \quad 1(2a)^0(3b)^4$

Finally, simplify the terms.

$1(16a^4)(1) \quad 4(8a^3)(3b) \quad 6(4a^2)(9b^2) \quad 4(2a)(27b^3) \quad 1(1)(81b^4)$

$= 16a^4 + 96a^3b + 216a^2b^2 + 216ab^3 + 81b^4$

YOUR TURN

 Use Pascal's Triangle to find $(x - 1)^7$.

 Use Pascal's Triangle to find $(3x + y)^4$.

Taking On the Graphing Formulas

When you are graphing functions, you are actually graphing formulas. It's just that functions are special types of formulas. What you find in this section, though, are some handy, dandy formulas that help you work your way around graphs of many types of curves — some functions and some not.

Formulating properties between two points

When you choose two points on the coordinate plane, you can work with several relationships that are formed. You can find the distance between the points, you can find the slope of the segment between the points, and you can find the midpoint between the points.

Distancing yourself

The distance between two points on the plane can be found with a formula that's actually created from the Pythagorean Theorem. You're finding the length of the hypotenuse of a right triangle. That's why your answers will often involve a radical.

The distance between the points (x_1, y_1) and (x_2, y_2) is found with $d = \sqrt{(x_2 - x_1)^2 + (y_2 - y_1)^2}$.

EXAMPLE

Q. Find the distance between the points (4,–6) and (8,–9).

A. Using the formula, $d = \sqrt{(8-4)^2 + (-9-(-6))^2} = \sqrt{4^2 + (-3)^2} = \sqrt{16+9} = \sqrt{25} = 5$.

Q. Find the distance between the points (0,7) and (5,2).

A. Using the formula, $d = \sqrt{(5-0)^2 + (2-7)^2} = \sqrt{5^2 + (-5)^2} = \sqrt{25+25} = \sqrt{50} = 5\sqrt{2}$.

With the distance formula, it really doesn't matter in what order you do the subtractions, because the difference will be squared. This won't be the case with the other formulas, so try to keep things in order. And always be sure that you're subtracting *x*'s from *x*'s and *y*'s from *y*'s.

Getting a good slant on things

The slope of a segment or line is a positive or negative number that indicates how steep or flat it is and whether it moves upward or downward as you move from left to right. For all the details on the slope, see Chapter 6.

The slope of a segment between the points (x_1, y_1) and (x_2, y_2) is found with $m = \frac{y_2 - y_1}{x_2 - x_1}$.

REMEMBER

Unlike the situation with the distance formula, you have to be very careful about the placement of the numbers; the *y*'s go in the numerator. And be sure to subtract the coordinates in the same order, from the same points.

EXAMPLE

Q. Find the slope between the points (4,–6) and (8,–9).

A. Using the formula, $m = \frac{-9-(-6)}{8-4} = \frac{-3}{4} = -\frac{3}{4}$.

Q. Find the slope between the points (6,7) and (5,2).

A. Using the formula, $m = \frac{2-7}{5-6} = \frac{-5}{-1} = 5$.

Making your way to the middle

The midpoint of a segment is just what the word says: These are the coordinates of the point that is smack dab in the middle. It divides the segment into two equal parts.

The midpoint of a segment between the points (x_1, y_1) and (x_2, y_2) is found with $M = \left(\frac{x_2+x_1}{2}, \frac{y_2+y_1}{2}\right)$.

REMEMBER

Note that you're finding the sum rather than the difference between the coordinates.

EXAMPLE

Q. Find the midpoint of the points (–2,–7) and (8,–9).

A. Using the formula, $M = \left(\frac{-2+8}{2}, \frac{-7+(-9)}{2}\right) = \left(\frac{6}{2}, \frac{-16}{2}\right) = (3,-8)$.

Q. Find the midpoint of the points (5,0) and (–5,3).

A. Using the formula, $M = \left(\frac{-5+5}{2}, \frac{3+0}{2}\right) = \left(\frac{0}{2}, \frac{3}{2}\right) = \left(0, \frac{3}{2}\right)$.

YOUR TURN

5 Find the distance between the points (–1,3) and (4,–9).

6 Find the slope of the segment between the points (–1,3) and (4,–9).

7 Find the midpoint of the segment between the points (–1,3) and (4,–9).

Measuring the distance between a point and a line

Measuring the distance between a point and a line requires just one qualification: This has to be the shortest distance between the two objects. You can't consider a distance that is along a slanted segment to the line; the distance is a perpendicular distance, so that's why you see our friend Pythagoras and his theorem in this formula.

The distance between the point (x_1, y_1) and the line $Ax + By + C = 0$ is found with $d = \frac{|Ax_1 + By_1 + C|}{\sqrt{A^2 + B^2}}$. You use the absolute value of the sum in the numerator to ensure a positive number for the distance.

EXAMPLE

Q. Find the distance between the point $(-3,2)$ and the line $3x - 4y = 3$.

A. First, rewrite the equation of the line to fit the formula: $3x - 4y - 3 = 0$. Now, using the formula, $d = \frac{|3(-3) - 4(2) - 3|}{\sqrt{3^2 + (-4)^2}} = \frac{|-9 - 8 - 3|}{\sqrt{9 + 16}} = \frac{|-20|}{\sqrt{25}} = \frac{20}{5} = 4$

Q. Find the distance between the point $(4,-1)$ and the line $x = y$.

A. First, rewrite the equation of the line to fit the formula: $x - y + 0 = 0$. Now, using the formula, $d = \frac{|4(1) - 1(-1) + 0|}{\sqrt{1^2 + (-1)^2}} = \frac{|4 + 1 + 0|}{\sqrt{1 + 1}} = \frac{|5|}{\sqrt{2}} = \frac{5}{\sqrt{2}} = \frac{5\sqrt{2}}{2}$.

YOUR TURN

8 Find the distance between the point $(4,0)$ and the line $6x + 8y = 11$.

Find the distance between the point $(-1,-2)$ and the line $2x - 3y = 4$.

Getting All Geometric

Geometric formulas are found everywhere. To figure out how much a carton of Rubik's Cubes will cost, you first have to figure how many cubes will fit in the carton before multiplying by the price of each cube. To find out how many 10 oz. glasses can be filled from that keg, you need to know the volume of the keg — based on measurements. What you'll find in this section is not the usual array. You have done perimeter and area and volume problems before. These are some of the special formulas!

Telling the plain truth with plane geometry

When you're dealing with planes in geometry, you're talking about things that are flat. They have two dimensions. In plane geometry, you do things like find the distance around an object (perimeter) and the amount of flat surface (area).

Let's go fly a kite

A kite is a four-sided figure with two pairs of equal sides. Instead of the equal sides being opposite one another, they're next to one another. To find the area of a kite, you measure the two diagonals (call them d_1 and d_2). The formula for the area of a kite is:

$$A = \frac{d_1 \cdot d_2}{2}.$$

EXAMPLE

Q. You are going to hang a giant kite from a bridge to celebrate your best friend's birthday. It'll be decorated with the appropriate message for the world to see. The two metal supports of the kite (the diagonals) measure 20 feet and 50 feet. How many square feet of fabric will you have to purchase to create the kite?

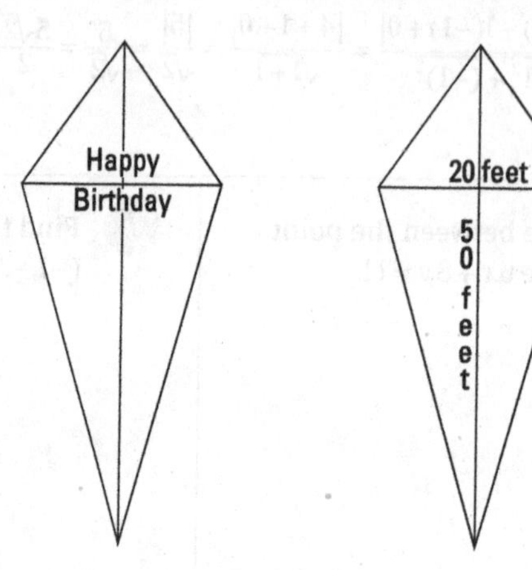

A. Multiply the two diagonal measures together and divide by 2: $A = \frac{20 \cdot 50}{2} = \frac{1000}{2} = 500$. You'll need 500 square feet of fabric. Start saving your money.

Tackling tricky triangles

The area of a triangle is easy to find if you know the measure of a side and the height of the triangle drawn from that side. But what if you can't measure the height? Why, you can use Heron's formula! All you need is the measure of each of the sides (call them *a*, *b*, and *c*). With those measures, figure the semi-perimeter, *s* (half the perimeter), do some subtracting and multiplying, and then find a square root. Here's what Heron's formula says: $A = \sqrt{s(s-a)(s-b)(s-c)}$.

EXAMPLE

Q. A math professor we all know well was sitting in her office one sunny day when the department secretary forwarded a phone message from a local businessman. He wanted to know how to find the area of a plot of land that he needed to fix up. He was thinking of purchasing sod to make it all a lovely, grassy area. The plot was triangular and had sides measuring 2000, 1500, and 1400 feet. What was the area of the plot of land?

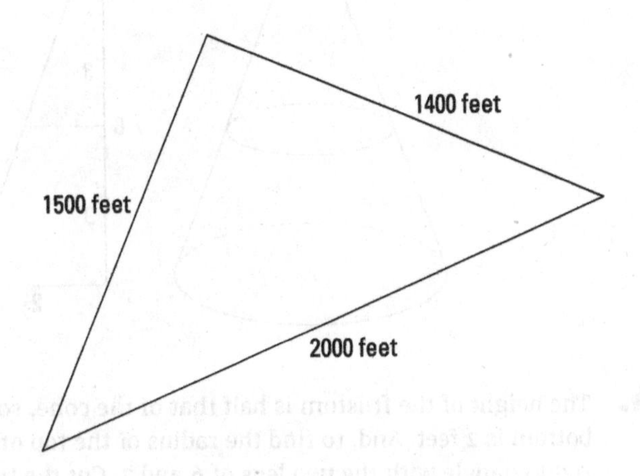

A. Aha! The math professor knew exactly what to do. And, of course, a calculator came in handy. The semi-perimeter is: $s = \frac{2000 + 1500 + 1400}{2} = \frac{4900}{2} = 2450$. Inserting all this into Heron's formula: $A = \sqrt{2450(2450 - 2000)(2450 - 1500)(2450 - 1400)} = \sqrt{2450(450)(950)(1050)} \approx 1{,}048{,}687$.

Yes, that's over a million square feet of area. He got his answer and the professor never heard from him again.

Keeping it solid

Just increase a plane object by one dimension and you have a solid. The geometric solids include: cubes, spheres, pyramids, and cylinders. Each has its own very interesting properties and methods for finding volume and surface area.

Fixing frustum frustration

A *frustum* is the bottom part of a cone. In Chapter 13 you see how slicing a cone in different ways creates the conic sections. In this chapter I slice the cone in half, parallel to the base, keeping the bottom part (the frustum). You may have seen these objects at the circus; the elephants like to perch on them!

To find the volume of a frustum, use $V = \frac{\pi h}{3}\left(R^2 + Rr + r^2\right)$, where h is the height of the frustum, R is the radius of the base, and r is the radius of the top.

EXAMPLE

Q. **I start with a cone measuring 6 feet high and with a base radius of 2 feet. Slicing the top cone off halfway up the cone, I am left with a frustum 3 feet high. What is its volume?**

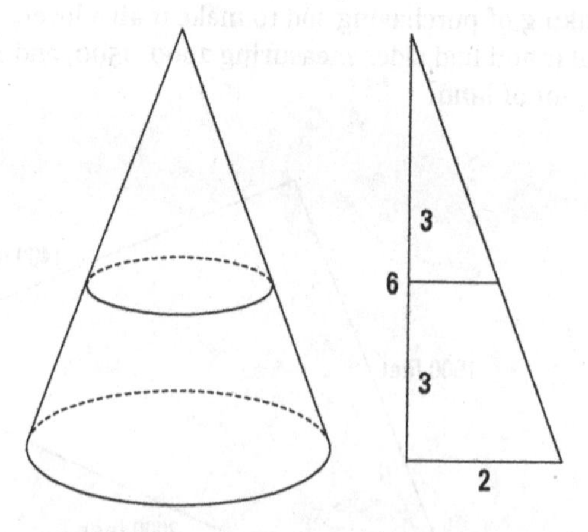

A. **The height of the frustum is half that of the cone, so $h = 3$. The radius of the bottom is 2 feet. And, to find the radius of the top of the frustum, just consider a right triangle with the two legs of 6 and 2. Cut the triangle in half, drawing from the side measuring 6. You have two similar triangles, so you can write a proportion: $\frac{\text{height big triangle}}{\text{base big triangle}} = \frac{\text{height small triangle}}{\text{base small triangle}}$, which becomes $\frac{6}{2} = \frac{3}{x} \rightarrow 6x = 6 \rightarrow x = 1$. The radius of the top of the frustum is 1.**

So now, with all the necessary information, $V = \frac{\pi h}{3}\left(R^2 + Rr + r^2\right) = \frac{\pi(3)}{3}\left(2^2 + (2)(1) + 1^2\right) = \pi(4 + 2 + 1) = 7\pi$. Let $\pi \approx 3.14$, and you have a volume of $7(3.14) = 21.98$ or almost 22 cubic feet.

Seeking a silly cylinder solution

This is just the area of the base times the height — a common situation in geometric shapes.

EXAMPLE

Q. **At a recent party, you provided a cylindrical tank/keg of apple juice. (I'm keeping this tuned to all ages.) The tank has a diameter of 2 feet and a height of 2 feet. You're providing cups that hold 16 ounces of fluid. How many servings are in the tank? (There are 32 fl. oz. in 57.75 cu. in.)**

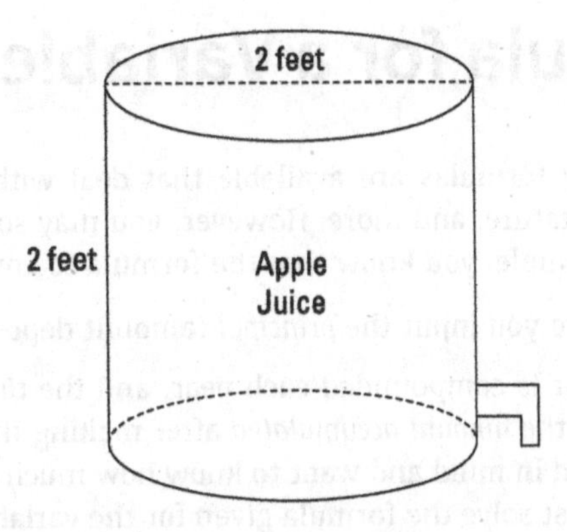

A. **You can change the feet to inches first or change the cubic feet to cubic inches later. Because you'll be using a calculator, anyway, I would say to change the feet to inches now and get it over with. The diameter is 2 feet, making the radius 1 foot, or 12 inches. The height is 2 feet, or 24 inches. To find the volume using the formula, $V = \pi(12)^2(24) = 3{,}456\pi \approx 10{,}851.84$ cubic inches. Divide that by 57.75, and you'll get the number of 32-ounce servings: $\frac{10{,}851.84}{57.75} \approx 187.91$. Multiply that by 2 to get the number of 16-ounce servings, and you have 375.82 servings. Depending on how many you've invited and how thirsty they are do you have enough?**

YOUR TURN

10 **A man buys a string 25,000 miles long and sets out to stretch it around the circumference of the earth. When he reaches his starting point, he discovers that the string is, in fact, 25,000 miles and one yard long. Rather than cut the string, he decides to tie the ends together and distribute the extra 36 inches evenly around the entire circumference. How far does the string stand out from the earth because of the extra yard? (Circumference $= \pi d$)**

---

11 **You are planning on building a new home with a dome-like roof. The roof will be constructed of 10 triangular pieces, each with sides measuring 10 feet by 24 feet by 26 feet. What will be the total area of the roof?**

Solving a Formula for a Variable

Many, many handy formulas are available that deal with geometric figures, money, algebraic issues, temperature, and more. However, you may sometimes want to change a formula's format. For example, you know that the formula to compute interest on an investment is $A=P\left(1+\frac{r}{n}\right)^{nt}$, where you input the *principal* (amount deposited), the interest *rate*, the *number* of times the interest is compounded each year, and the *time* you'll leave the money there in years. The result is the *amount accumulated* after making the deposit. But what if you have an amount accumulated in mind and want to know how much to deposit? Do you need a different formula? No! You just solve the formula given for the variable P.

EXAMPLE

Q. Solve the formula $A=P\left(1+\frac{r}{n}\right)^{nt}$ for P.

A. Divide each side of the equation by the multiplier of P.

$$A=P\left(1+\frac{r}{n}\right)^{nt}\rightarrow\frac{A}{\left(1+\frac{r}{n}\right)^{nt}}=\frac{P\cancel{\left(1+\frac{r}{n}\right)^{nt}}}{\cancel{\left(1+\frac{r}{n}\right)^{nt}}}\rightarrow\frac{A}{\left(1+\frac{r}{n}\right)^{nt}}=P$$

The result can be made to look a little nicer using a negative exponent but this works fine.

Q. Solve for $C°$ in the temperature formula $F°=\frac{9}{5}C°+32$.

A. First, subtract 32 from each side, and then multiply each side by $\frac{5}{9}$.

$$F°=\frac{9}{5}C°+32\rightarrow F°-32=\frac{9}{5}C°\rightarrow\frac{5}{9}(F°-32)=\frac{\cancel{5}}{\cancel{9}}\cdot\frac{\cancel{9}}{\cancel{5}}C°\rightarrow\frac{5}{9}(F°-32)=C°$$

Q. Solve for y in the equation of the circle: $(x-2)^2+(y+3)^2=25$.

A. First, subtract the term with x from each side. Then take the square root of each side.

$$(x-2)^2+(y+3)^2=25\rightarrow(y+3)^2=25-(x-2)^2\rightarrow\sqrt{(y+3)^2}=\pm\sqrt{25-(x-2)^2}\rightarrow$$
$$y+3=\pm\sqrt{25-(x-2)^2}$$

Now subtract 3 from each side and simplify the terms under the radical.

$$y+3=\pm\sqrt{25-(x-2)^2}\rightarrow y=-3\pm\sqrt{25-(x-2)^2}\rightarrow y=-3\pm\sqrt{25-(x^2-4x+4)}\rightarrow$$
$$y=-3\pm\sqrt{25-x^2+4x-4}\rightarrow y=-3\pm\sqrt{21+4x-x^2}$$

Q. Solve for b_1 in the equation for finding the areas of a trapezoid: $A=\frac{1}{2}h(b_1+b_2)$.

A. Multiply each side by 2 and divide each side by h. Then subtract b_2 from each side.

$$A=\frac{1}{2}h(b_1+b_2)\rightarrow2\cdot A=\cancel{2}\cdot\frac{1}{\cancel{2}}h(b_1+b_2)\rightarrow\frac{2A}{h}=\frac{\cancel{h}(b_1+b_2)}{\cancel{h}}\rightarrow$$
$$\frac{2A}{h}=b_1+b_2\rightarrow\frac{2A}{h}-b_2=b_1$$

12 **Solve for y in the equation $Ax + By + C = 0$.**

13 **Solve for y in the equation of the hyperbola $\frac{(x-4)^2}{4} - \frac{(y+3)^2}{9} = 1$.**

14 **Solve for h in the formula for the surface area of a cylinder: $A = 2\pi r(r+h)$.**

Practice Questions Answers and Explanations

(1) $\mathbf{z^5 + 5z^4 + 10z^3 + 10z^2 + 5z + 1}$**.** First, fill in the combinations rules and exponents.

$$(z+1)^5 = \binom{5}{0} z^{5-0}1^0 + \binom{5}{1} z^{5-1}1^1 + \binom{5}{2} z^{5-2}1^2 + \binom{5}{3} z^{5-3}1^3 + \binom{5}{4} z^{5-4}1^4 + \binom{5}{5} z^{5-5}1^5$$

Now fill in the combinations results, and simplify the exponents and then the terms.

$$\begin{aligned}(z+1)^5 &= 1z^5 1^0 + 5z^4 1^1 + 10z^3 1^2 + 10z^2 1^3 + 5z^1 1^4 + 1z^0 1^5 \\ &= z^5 + 5z^4 + 10z^3 + 10z^2 + 5z + 1\end{aligned}$$

(2) $\mathbf{8x^3 - 36x^2 + 54x - 27}$**.** First, fill in the combinations rules and exponents. Be sure to include the negative sign with the second term, the –3.

$$(2x-3)^3 = \binom{3}{0}(2x)^{3-0}(-3)^0 + \binom{3}{1}(2x)^{3-1}(-3)^1 + \binom{3}{2}(2x)^{3-2}(-3)^2 + \binom{3}{3}(2x)^{3-3}(-3)^3$$

Now fill in the combinations results, and simplify the exponents and then the terms.

$$\begin{aligned}(2x-3)^3 &= 1(2x)^3(-3)^0 + 3(2x)^2(-3)^1 + 3(2x)^1(-3)^2 + 1(2x)^0(-3)^3 \\ &= 1(8x^3)(1) + 3(4x^2)(-3) + 3(2x)(9) + 1(1)^0(-27) \\ &= 8x^3 - 36x^2 + 54x - 27\end{aligned}$$

Note that the signs of the terms alternate between positive and negative.

(3) $\mathbf{x^7 - 7x^6 + 21x^5 - 35x^4 + 35x^3 - 21x^2 + 7x - 1}$**.** First, write the coefficients, and then fill in the powers of the terms and simplify.

1 7 21 35 35 21 7 1

$1x^7(-1)^0 \quad 7x^6(-1)^1 \quad 21x^5(-1)^2 \quad 35x^4(-1)^3 \ \ldots$

$\ldots\ 35x^3(-1)^4 \quad 21x^2(-1)^5 \quad 7x^1(-1)^6 \quad 1x^0(-1)^7$ K?

$$= x^7 - 7x^6 + 21x^5 - 35x^4 + 35x^3 - 21x^2 + 7x - 1$$

(4) $\mathbf{81x^4 + 108x^3y + 54x^2y^2 + 12xy^3 + y^4}$**.** First, write the coefficients, and then fill in the powers of the terms and simplify.

1 4 6 4 1

$1(3x)^4(y)^0 \quad 4(3x)^3(y)^1 \quad 6(3x)^2(y)^2 \quad 4(3x)^1(y)^3 \quad 1(3x)^0(y)^4$

$1(81x^4)(1) \quad 4(27x^3)(y) \quad 6(9x^2)(y^2) \quad 4(3x)(y^3) \quad 1(1)(y^4)$

$81x^4 \quad 108x^3y \quad 54x^2y^2 \quad 12xy^3 \quad y^4$

$$(3x+y)^4 = 81x^4 + 108x^3y + 54x^2y^2 + 12xy^3 + y^4$$

5. **13.** Using the formula, $d = \sqrt{(-9-3)^2 + (4-(-1))^2} = \sqrt{(-12)^2 + (5)^2} = \sqrt{144+25} = \sqrt{169} = 13$.

6. $-\frac{\mathbf{12}}{\mathbf{5}}$. Using the formula, $m = \frac{-9-3}{4-(-1)} = \frac{-12}{5} = -\frac{12}{5}$.

7. $\left(\frac{\mathbf{3}}{\mathbf{2}}, \mathbf{-3}\right)$. Using the formula, $M = \left(\frac{4+(-1)}{2}, \frac{-9+3}{2}\right) = \left(\frac{3}{2}, \frac{-6}{2}\right) = \left(\frac{3}{2}, -3\right)$.

8. $\frac{\mathbf{7}}{\mathbf{2}}$. First, change the equation of the line to read $6x + 8y - 11 = 0$. Then, using the formula,

   $d = \frac{|6(4)+8(0)+11|}{\sqrt{6^2+8^2}} = \frac{|24+0+11|}{\sqrt{36+64}} = \frac{|35|}{\sqrt{100}} = \frac{35}{10} = \frac{7}{2}$.

9. **0.** First, change the equation of the line to read $2x - 3y - 4 = 0$. Then, using the formula,

   $d = \frac{|2(-1)-3(-2)-4|}{\sqrt{2^2+3^2}} = \frac{|-2+6-4|}{\sqrt{4+9}} = \frac{|0|}{\sqrt{13}} = \frac{0}{\sqrt{13}} = 0$. The distance is zero, meaning that the point lies on the line.

10. **6 inches.** The circumference of the earth is 25,000 miles. That's how long the string was supposed to be. Change that to yards by multiplying by 1,760: $25{,}000(1{,}760) = 44{,}000{,}000$ yards. But he had an extra yard, so he had 44,000,001 yards. Next, see what the additional yard does to the diameter of the globe it can encircle. Using $C = \pi d$ and dividing by π, the diameter is $d = \frac{C}{\pi}$. Compute the diameter using the two different measures: $d = \frac{44{,}000{,}000}{3.14} = 14{,}012{,}738.85$ and $d = \frac{44{,}000{,}001}{3.14} = 14{,}012{,}739.17$. The difference between the two diameters is 0.321975 yards. That's about 11.6 inches. Add half of that to each radius and it increases the radius almost 6 inches. The string will be 6 inches above the surface of the earth all the way around.

11. **1,200 sq. ft.** Find the area of one of the triangular pieces using Heron's formula. The semi-perimeter is half of $10 + 24 + 26$, which is 30. To find the area: $A = \sqrt{30(30-10)(30-24)(30-26)} = \sqrt{30(20)(6)(4)} = \sqrt{14{,}400} = 120$. There are 10 of these pieces, so the total area is 1,200 square feet.

12. $\boldsymbol{y = -\frac{A}{B}x - \frac{C}{B}}$. Subtract the Ax and C from each side. Then divide both sides by B:

    $Ax + By + C = 0 \rightarrow By = -Ax - C \rightarrow y = \frac{-Ax-C}{B} = -\frac{A}{B} - \frac{C}{B}$.

13. $\boldsymbol{y = -3 \pm \frac{1}{2}\sqrt{9(x-4)^2 - 36}}$. First, multiply all the terms by 36 to get rid of the fractions:

    $\frac{(x-4)^2}{4} - \frac{(y+3)^2}{9} = 1 \rightarrow \cancel{36}^9 \cdot \frac{(x-4)^2}{\cancel{4}} - \cancel{36}^4 \cdot \frac{(y+3)^2}{\cancel{9}} = 36 \cdot 1 \rightarrow 9(x-4)^2 - 4(y+3)^2 = 36$.

    Next, add the y-term to each side and subtract 36 from each side: $9(x-4)^2 - 4(y+3)^2 = 36 \rightarrow 9(x-4)^2 - 36 = 4(y+3)^2$.

Find the square root of each side, and then multiply each side by $\frac{1}{2}$:

$$\pm\sqrt{9(x-4)^2-36}=\sqrt{4(y+3)^2}\rightarrow\pm\sqrt{9(x-4)^2-36}=2(y+3)\rightarrow\pm\frac{1}{2}\sqrt{9(x-4)^2-36}=y+3.$$

Finally, subtract 3 from each side: $-3\pm\frac{1}{2}\sqrt{9(x-4)^2-36}=y.$

(14) $h=\frac{A}{2\pi r}-r$**. Divide each side by $2\pi r$. Then subtract r from each side.**

$$A=2\pi r(r+h)\rightarrow\frac{A}{2\pi r}=\frac{\cancel{2\pi r}(r+h)}{\cancel{2\pi r}}\rightarrow\frac{A}{2\pi r}=r+h\rightarrow h=\frac{A}{2\pi r}-r$$

Whaddya Know? Chapter 21 Quiz

Quiz time! Complete each problem to test your knowledge on the various topics covered in this chapter. You can then find the solutions and explanations in the next section.

1. Find the slope of the line that goes through the points $(-2,3)$ and $(-8,-3)$.
2. Use the Binomial Theorem to expand $\left(x^2+1\right)^4$.
3. Solve for x in the equation of a parabola $y-k=a\left(x-h\right)^2$.
4. Find the surface area of the cylinder with a diameter of 8 inches and a height of 12 inches, using $SA=2\pi r\left(r+h\right)$.
5. Find the midpoint of the segment whose endpoints are $(-3,8)$ and $(-11,-6)$.
6. Use Pascal's Triangle to expand $\left(3x+2\right)^5$.
7. The total number of degrees in a polygon is determined with $D=180\left(n-2\right)$, where n is the number of sides of the polygon. What polygon has a total of 1,440 degrees?
8. Find the distance between the points $(-9,-1)$ and $(-14,11)$.
9. Use the Binomial Theorem to expand $\left(x-y\right)^6$.
10. Find the distance between the line $3x-4y=6$ and the point $(5,-1)$.

Answers to Chapter 21 Quiz

1. **1.** Using the formula, $m = \frac{-3-3}{-8-(-2)} = \frac{-6}{-6} = 1$.

2. $\mathbf{x^8 + 4x^6 + 6x^4 + 4x^2 + 1}$. Applying the formula, $(x^2+1)^4 = \binom{4}{0}(x^2)^{4-0}1^0 + \binom{4}{1}(x^2)^{4-1}1^1 + \binom{4}{2}(x^2)^{4-2}1^2 + \binom{4}{3}(x^2)^{4-3}1^3 + \binom{4}{4}(x^2)^{4-4}1^4$. Finding the combinations and simplifying the terms:

$$= (1)(x^2)^4 1^0 + (4)(x^2)^3 1^1 + (6)(x^2)^2 1^2 + (4)(x^2)^1 1^3 + (1)(x^2)^0 1^4$$
$$= x^8 + 4x^6 + 6x^4 + 4x^2 + 1$$

3. $\mathbf{x = h \pm \sqrt{\frac{y-k}{a}}}$. Divide each side of the equation by a, then find the square root of each side, and, finally, add h to each side.

$$y-k = a(x-h)^2 \rightarrow \frac{y-k}{a} = \frac{\cancel{a}(x-h)^2}{\cancel{a}} \rightarrow \frac{y-k}{a} = (x-h)^2 \rightarrow \pm\sqrt{\frac{y-k}{a}} = \sqrt{(x-h)^2} \rightarrow$$
$$\pm\sqrt{\frac{y-k}{a}} = x-h \rightarrow h \pm \sqrt{\frac{y-k}{a}} = x$$

4. **401.92 sq. in.** A diameter of 8 means the radius is 4. Applying the formula, $SA = 2\pi \cdot 4(4+12) = 8\pi(16) = 128\pi \approx 401.92$.

5. **(−7, 1).** Using the formula, $M = \left(\frac{-11+(-3)}{2}, \frac{-6+8}{2}\right) = \left(\frac{-14}{2}, \frac{2}{2}\right) = (-7,1)$.

6. $\mathbf{243x^5 + 810x^4 + 1{,}080x^3 + 720x^2 + 240x + 32}$. First, list the numbers from the row beginning with 1 and 5; leave spaces between the numbers. Then put in decreasing powers of $3x$ and increasing powers of 2. Finally, simplify the terms.

1		5		10		10		5		1
$1(3x)^5 \cdot 2^0$		$5(3x)^4 \cdot 2^1$		$10(3x)^3 \cdot 2^2$		$10(3x)^2 \cdot 2^3$		$5(3x)^1 \cdot 2^4$		$1(3x)^0 \cdot 2^5$
$243x^5$	+	$810x^4$	+	$1{,}080x^3$	+	$720x^2$	+	$240x$	+	32

7. **A 10-sided polygon.** Replace the D in the equation with 1,440 and solve for n:

$1{,}440 = 180(n-2) \rightarrow \frac{1{,}440}{180} = n-2 \rightarrow 8 = n-2 \rightarrow 10 = n$. It's a decagon.

8. **13.** Using the formula,

$d = \sqrt{(-14-(-9))^2 + (11-(-1))^2} = \sqrt{(-5)^2 + (12)^2} = \sqrt{25+144} = \sqrt{169} = 13$.

9. $\mathbf{x^6 - 6x^5y + 15x^4y^2 - 20x^3y^3 + 15x^2y^4 - 6xy^5 + y^6}$. Expanding with the Binomial Theorem:

$$(x-y)^6 = \binom{6}{0}x^{6-0}(-y)^0 + \binom{6}{1}x^{6-1}(-y)^1 + \binom{6}{2}x^{6-2}(-y)^2 + \binom{6}{3}x^{6-3}(-y)^3 + \cdots$$

$$\cdots + \binom{6}{4}x^{6-4}(-y)^4 + \binom{6}{5}x^{6-5}(-y)^5 + \binom{6}{6}x^{6-6}(-y)^6$$

Simplifying the terms:

$$(x-y)^6 = (1)x^6(-y)^0 + (6)x^5(-y)^1 + (15)x^4(-y)^2 + (20)x^3(-y)^3 + \cdots$$
$$\cdots + (15)x^2(-y)^4 + (6)x^1(-y)^5 + (1)x^0(-y)^6$$

Which gives you:

$$(x-y)^6 = x^6 - 6x^5y + 15x^4y^2 - 20x^3y^3 + 15x^2y^4 - 6xy^5 + y^6$$

(10) $\mathbf{\frac{13}{5}}$**. First, change the line to the** ***Ax + By*** **+ C = 0 form, giving you** $3x - 4y - 6 = 0$**. Then, using the formula,** $d = \frac{|3(5) - 4(-1) - 6|}{\sqrt{3^2 + (-4)^2}} = \frac{|15 + 4 - 6|}{\sqrt{9 + 16}} = \frac{|13|}{\sqrt{25}} = \frac{13}{5}$.

IN THIS CHAPTER

» Mixing it up with money, fluids, and foods

» Keeping track of the ages

» Going the distance with distance problems

» Working it all out with work problems

Chapter 22
Taking on Applications

Applications are what mathematics is all about. Okay, so you get a lot of pleasure out of adding decimals, subtracting exponentials, multiplying logarithms, and dividing well, anything! But the reason mathematics is so important is because, using it correctly, you can live and work successfully. Mathematical concepts and theorems were all developed to help answer questions. And here, in this chapter, you are getting some of those questions answered!

Making Mixtures Magically Mathematical

Mixtures come in all shapes and sizes — or all ingredients and values. You can mix up milk and chocolate sauce, and have differing intensities. You can mix up all those coins in the jar and always have the same amount of money. Mixing it up is what this section is all about. A common thread with mixture problems is that you will have a number of different items, with each item worth or weighing a different amount. To come up with a total, you multiply the amount times the value.

Keeping money matters solvent

You open your piggy bank. Look at all those pennies! You check out your cash drawer. Look at all those singles! You empty the coin bin in the soda machine! Oh, the quarters! Here are some examples involving such situations.

EXAMPLE

Q. You finally fill the pickle jar with coins. Your brother dumps them on the counter and announces that you have two fifty-cent pieces, six more than three times as many dimes as quarters, and twice as many nickels as quarters. You have a total of \$21.10. How many of each coin do you have?

A. First things first: Let the number of quarters be represented by q. The number of dimes is $6+3q$, and the number of nickels is $2q$. Multiply each number of coins times what they are worth. Add them together and set them equal to \$21.10.

fifty-cent pieces	quarters	dimes	nickels
2	q	$6+3q$	$2q$
0.50	0.25	0.10	0.05

Your equation: $0.50(2)+0.25(q)+0.10(6+3q)+0.05(2q)=21.10$

Simplify the equation: $1+0.25q+0.60+0.30q+0.10q=21.10 \rightarrow 0.65q+1.60=21.10$

Now solve for q: $0.65q+1.60=21.10 \rightarrow 0.65q=19.50 \rightarrow q=30$

You have 30 quarters. The number of dimes is $6+3(30)=96$. And the number of nickels is $2(30)=60$. That's 2 fifty-cent pieces, 30 quarters, 96 dimes, and 60 nickels.

Q. The cashier is done for the day and checking their cash drawer. They have a total of 170 bills in singles, fives, tens, and twenties. There are 10 more singles than fives, 20 more tens than fives, and twice as many fives as twenties. How much money is there in the drawer?

A. This time, not all the numbers are related to one of the values. But, if you start by letting the twenties be represented by t and then the fives by $2t$, you can relate the singles and the tens to the fives.

singles	fives	tens	twenties
	$2t$		t
$10+2t$	$2t$	$20+2t$	t

The total number of bills is 170, so add up the numbers and solve for t: $10+2t+2t+20+2t+t=170 \rightarrow 7t+30=170 \rightarrow 7t=140 \rightarrow t=20$. That means there are $10+2(20)=50$ singles, $2(20)=40$ fives, $20+2(20)=60$ tens, and 20 twenties.

Now multiply each money amount times the number of bills to get the total: $\$1(50)+\$5(40)+\$10(60)+\$20(20)=\$1,250$.

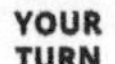

YOUR TURN

1. Casey has \$7.80 in dimes and quarters. If she has a total of 60 coins, then how many of each coin does she have?

2. Zeke has a total of \$3.60 in pennies, nickels, and dimes. He has twice as many dimes as pennies and 10 more nickels than dimes. How many of each coin does he have?

Flowing freely with fluids

Mixing fluids together just takes a big spoon or a good shake of the container. Some of them will keep your car running, and some of them will taste good! These mixture problems deal with how much of each ingredient is needed to create the required mixture.

EXAMPLE

Q. How many quarts of 70% apple juice do you need to add to 2 quarts of 40% apple juice to create a 50% mixture? (Figure 22-1a shows you the setup.)

A. You multiply the qualities and quantities and let x be the amount you're seeking.

$$70\%(x)+40\%(2)=50\%(x+2)\rightarrow 0.70x+0.80=0.50x+1\rightarrow$$
$$0.70x=0.50x+0.20\rightarrow 0.20x=0.20\rightarrow x=1$$

It will take just 1 quart of the 70% concentration.

Q. How many ounces of chocolate syrup do you need to add to 12 ounces of milk to make the mixture 25 percent chocolate syrup? (See Figure 22-1b.)

A. The concentration of syrup in the milk right now is 0%, and the concentration of the chocolate syrup is 100%. Let x be the number of ounces of chocolate syrup that you need to add.

$$100\%(x)+0\%(12)=25\%(x+12)\rightarrow 1x+0=0.25(x+12)\rightarrow$$
$$x=0.25x+4\rightarrow 0.75x=4\rightarrow x=\frac{16}{3}=5\frac{1}{3}$$

You need to add $5\frac{1}{3}$ ounces of syrup.

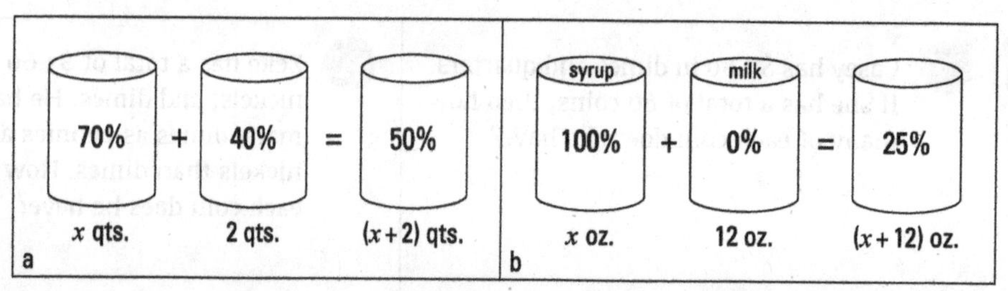

FIGURE 22-1: Mixing apple juice in Figure a and chocolate syrup in Figure b.

YOUR TURN

3 How many gallons of fertilizer that is 30% nutrient A do you need to mix with 200 gallons of fertilizer that is 18% nutrient A to bring the total mixture up to 25%?

4 To be efficient, the cleaning solution you use needs to have a 9% concentration. You have on hand some 5% concentration and some 15% concentration. How much of each of these do you need to make 4 quarts of the 9% concentration?

Mixing up many ingredients

When making a cake or some granola or fruit salad, you add many different ingredients. When the ingredients have to have a particular relationship to one another (so the cake will rise), you need to pay attention to the amounts.

EXAMPLE

Q. Your favorite pizza has twice as many ounces of sausage as onions and 4 more ounces of mushrooms than onions. The sausage costs 30 cents an ounce. The onions cost 5 cents an ounce. And the mushrooms cost 15 cents an ounce. If the total amount that you want to spend on these toppings can be up to $4.60, then how many ounces of each can you add to your pizza?

A. Let x represent the number of ounces of onions. Then 2x is the number of ounces of sausage and x + 4 is the number of ounces of mushrooms.

sausage	onions	mushrooms
$2x$	x	$x+4$
0.30	0.05	0.15

You are willing to spend as much as $4.60.

$$2x(0.30)+x(0.05)+(x+4)(0.15)=4.80 \rightarrow 0.60x+0.05x+0.15x+0.60=4.60 \rightarrow$$
$$0.80x+0.60=4.60 \rightarrow 0.80x=4.00 \rightarrow x=5$$

You can use 5 ounces of onions, 10 ounces of sausage, and 9 ounces of mushrooms.

Q. **You want to sell a nutty mixture of peanuts, cashews, and almonds. Peanuts cost $4.00 per pound, cashews cost $11.50 per pound, and almonds cost $7.50 per pound. You want to create a huge mixture that has 10 more pounds of almonds than peanuts and 10 fewer pounds of cashews than peanuts. You've got $420 available to spend on the mixture. How many pounds of each do you need?**

A. Let x represent the pounds of peanuts, $x+10$ the pounds of almonds, and $x-10$ the pounds of cashews.

peanuts	almonds	cashews
x	$x+10$	$x-10$
4.00	7.50	11.50

$4.00(x)+7.50(x+10)+11.50(x-10)=420 \rightarrow 4x+7.50x+75+11.50x-115=420 \rightarrow$
$23x-40=420 \rightarrow 23x=460 \rightarrow x=20$

The mixture will contain 20 pounds of peanuts, 30 pounds of almonds, and 10 pounds of cashews.

YOUR TURN

5 Stephanie has twice as many nickels as quarters and five fewer than three times as many dimes as quarters. If she has a total of $15.10, then how many of each coin does she have?

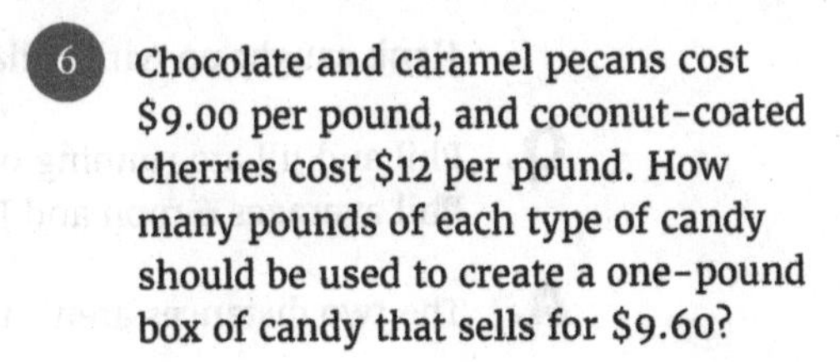

6 Chocolate and caramel pecans cost $9.00 per pound, and coconut-coated cherries cost $12 per pound. How many pounds of each type of candy should be used to create a one-pound box of candy that sells for $9.60?

Going the Distance

You can use a distance formula when graphing on the coordinate plane, but another distance formula is much more commonly used all the time. The formula says that the distance you travel depends on how fast you're going and how long you travel at that speed. So it's a really simply formula: $d=rt$, which says that the *distance* traveled is equal to the *rate* of speed times the *time* traveled.

REMEMBER

The rate and time have to be in the same units. If the rate is in miles per hour, then the time has to be in hours.

Equating distances

There are instances where one person is traveling at a certain speed and another is going to the same place but traveling slower or faster. You want to know when they get where they're going or when one of them passes the other.

EXAMPLE

Q. **Hillary left home at 6:00 a.m. traveling at 50 mph. At 7:00 a.m., Hank noticed that she had forgotten her wallet, jumped into his car, and took off after her, traveling at 70 mph. At what time did Hank catch up with Hillary?**

A. **Of course, you're assuming that they're taking the same route and that both speeds are permissible. What you need to do here is figure out when the two distances they've both traveled are the same: $d_{Hillary} = d_{Hank}$. Hillary's rate is 50 mph; let the time she traveled be represented by t. Hank's rate is 70 mph; the time Hank traveled is $t-1$. Setting these two distances together:**

$$d_{Hillary} = d_{Hank} \rightarrow 50t = 70(t-1) \rightarrow 50t = 70t - 70 \rightarrow$$

$$70 = 20t \rightarrow \frac{70}{20} = t \rightarrow 3.5 = t$$

Hank caught up with Hillary 3.5 hours after she left home. That was at 9:30 a.m.

Q. **Phil and Jill are running on the practice track. Phil gave Jill a quarter-mile head start. Phil averages 6 mph and Jill averages 5 mph. How long before Phil catches up with Jill?**

A. **The two distances aren't the same, so add $\frac{1}{4}$ mile to Phil's distance and solve the equation where the extra distance is added, $d_{Phil} = d_{Jill} + \frac{1}{4}$: $d_{Phil} = d_{Jill} + \frac{1}{4} \rightarrow 6t = 5t + \frac{1}{4} \rightarrow t = \frac{1}{4}$. The rates are in miles per hour, so it'll take $\frac{1}{4}$ hour, or 15 minutes, for Phil to catch up to Jill.**

YOUR TURN

7 **Harry left for Chicago at 9:00 a.m. in the morning, traveling at 45 mph. Cary left the same place at 9:30 a.m., heading for Chicago along the same route, traveling at 60 mph. How soon did Cary catch up to Harry?**

8 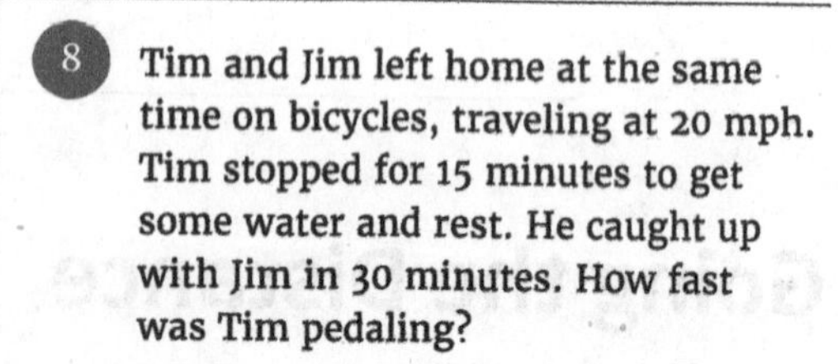
Tim and Jim left home at the same time on bicycles, traveling at 20 mph. Tim stopped for 15 minutes to get some water and rest. He caught up with Jim in 30 minutes. How fast was Tim pedaling?

Summing up the distances

Another type of distance problem is when you add up the two distances that people or objects have traveled to solve for the amount of time it took or how far one or the other traveled.

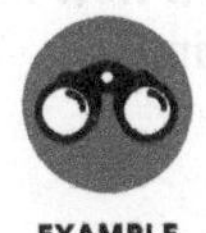

EXAMPLE

Q. **Larry and Terry left home at the same time, traveling in opposite directions. If Larry was traveling at 45 mph and Terry was traveling at 65 mph, then how long did it take for them to be 440 miles apart? (See Figure 22-2a.)**

A. **This is a matter of just adding up the distances they traveled. You want to solve $d_{Larry} + d_{Terry} = 440$. The time they traveled is the same, so the equation becomes: $d_{Larry} + d_{Terry} = 440 \rightarrow 45t + 65t = 440 \rightarrow 110t = 440 \rightarrow t = 4$. It took four hours.**

Q. **Esther and Norm both left home at 7:00 a.m. Esther headed east at 60 mph, and Norm headed north at 80 mph. At what time were they 100 miles apart? (Get the picture with Figure 22-2b.)**

A. **In this case, you don't add the two distances, because they are traveling at right angles to one another. (Again, you assume the roads are staying exactly eastward and exactly northward.) This time, you want the two legs (east and north) to represent the sides of the right triangle and the distance apart to be the hypotenuse. In comes Pythagoras! The equation you want is $(d_{Esther})^2 + (d_{Norm})^2 = 100^2$. Put in their rates of speed and solve for the time it took: $(d_{Esther})^2 + (d_{Norm})^2 = 100^2 \rightarrow (60t)^2 + (80t)^2 = 100^2 \rightarrow 3600t^2 + 6400t^2 = 10{,}000 \rightarrow 10{,}000t^2 = 10{,}000 \rightarrow t^2 = 1 \rightarrow t = 1$**

It took an hour for them to be 100 miles apart, so that was at 8:00 a.m.

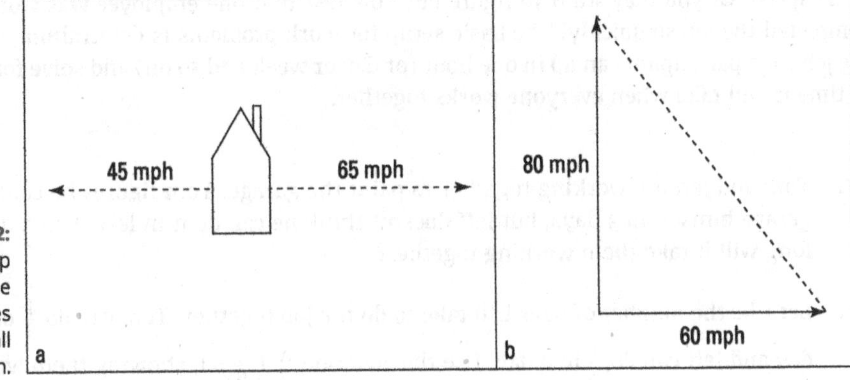

FIGURE 22-2: Adding up the distances and going all Pythagorean.

YOUR TURN

9 Two buses left the same station at the same time going in opposite directions. One bus was traveling at 40 mph, and the other bus was traveling at 55 mph. How long did it take for them to be 570 miles apart?

10 A plane and a helicopter left the airport at 4:00 p.m. The plane traveled south moving at 240 mph, and the helicopter headed west. In one hour, they were 250 miles apart. How fast was the helicopter traveling?

Working through the Challenges

Work problems involve work. Surprise, surprise! You may want to find out how long it'll take to complete a task with two or more people on the job when they don't perform the task at the same speed. Or you may want to figure out how fast that one employee was working to have completed the job so quickly! The basic setup for work problems is determining how much of the job each participant can do in one hour (or day or week and so on) and solve for the amount of time it will take when everyone works together.

EXAMPLE

Q. Tom and Jeff are working together to paint the garage. Tom figures he could paint the garage himself in 3 days, but Jeff doesn't think he can do it in less than 5 days. How long will it take them working together?

A. Let x be the number of days it'll take to do the job together. Tom can do $\frac{1}{3}$ of the job in a day and Jeff can do $\frac{1}{5}$ in a day. Use the equation $\frac{x}{3}+\frac{x}{5}=1$, showing them working together to do the whole job, and solve for x. The common denominator of the two fractions is 15, so multiply each term by 15: $\frac{x}{3}+\frac{x}{5}=1 \rightarrow 5x+3x=15 \rightarrow 8x=15 \rightarrow x=\frac{15}{8}$.

It will take them just short of 2 days to paint the garage, working together.

Q. The yard needs to be landscaped: new bushes, new flowers, repaired lawn, and so on. Ben, Jim, and Harry are taking it on. Working alone, it would take Ben 4 days, it would take Jim 5 days, and it would take Harry 6 days. How long will it take if they all work together?

A. Using the equation $\frac{x}{4}+\frac{x}{5}+\frac{x}{6}=1$, solve for x. The common denominator is 60, so multiply through by 60: $\frac{x}{4}+\frac{x}{5}+\frac{x}{6}=1 \rightarrow 15x+12x+10x=60 \rightarrow 37x=60 \rightarrow x=\frac{60}{37}=1\frac{23}{37}$. It will take less than 2 days to do the job!

YOUR TURN

11 Bart and Bev are moving and need to clear out the basement. If Bart does it alone, it'll take him 6 hours (because he loves to look through all the old boxes). Bev can do it alone in 2 hours, because she isn't interested in all the accumulated treasures. How long will it take if they do the job together?

12 The family swimming pool needs to be filled. With the two hoses running constantly, it'll take 12 hours to fill it. But no one noticed that there is a small leak. At the rate it's leaking, the pool could be emptied in 36 hours. With the filling and the leaking going on, how long will it take to fill the pool (and make them notice there's a leak)?

Acting Your Age

Age can be a touchy subject. Recently, I was asked how old I was. I replied, "My age is the product of two prime numbers." Well, it's definitely not 3 times 5!

The age problems here deal more with comparing one person's age with that of another, and then solving for all of the ages involved.

EXAMPLE

Q. Tom is twice as old as Harry. Seven years ago, Tom was three times as old as Harry. How old are they now?

A. Let t be Harry's age right now. That makes Tom $2t$ years old. Seven years ago, Tom was $2t-7$ and Harry was $t-7$. The equation that says that Tom's age seven years ago was three times Harry's age is: $2t-7=3(t-7)$. Solve the equation for t: $2t-7=3(t-7) \rightarrow 2t-7=3t-21 \rightarrow 14=t$. So Harry is 14 and Tom is 28. Seven years ago, they were 7 and 21, respectively. Suggestion: use a different variable than "t" for Harry's age. Try "h"?

Q. Molly's mother is 2 years more than 4 times as old as Molly. Six years ago, the sum of their ages was 50. How old are they now?

A. Let Molly's age be m. Then her mother is $4m+2$. Six years ago, Molly was $m-6$ and her mother was $4m+2-6=4m-4$. So the sum of their ages, which was 50, is written: $(m-6)+(4m-4)=50$. Solving for m, $(m-6)+(4m-4)=50 \rightarrow 5m-10=50 \rightarrow 5m=60 \rightarrow m=12$. Molly is 12 and her mother is $4(12)+2=50$.

YOUR TURN

13 Andrew is three times as old as Jack. Ten years ago, Andrew was five times as old as Jack. How old are they now?

14 Agnes, the mother of triplets, is 12 times as old as they are. In 4 years, the sum of the ages of all four of them will be 61. How old are the triplets?

Practice Questions Answers and Explanations

1. **48 dimes, 12 quarters.** Let d represent the number of dimes and $60-d$ the number of quarters. Solving for d in $0.10d+0.25(60-d)=7.80 \rightarrow 0.10d+15-0.25d=7.80 \rightarrow -0.15d=-7.20$. Divide each side by -0.15, and you have $d=48$. There are 48 dimes and $60-48=12$ quarters.

2. **10 pennies, 20 dimes, 30 nickels.** First, let p represent the number of pennies. Then the number of dimes is $2p$. The number of nickels is then $10+2p$. Multiply each number of coins times their value: $0.01p+0.10(2p)+0.05(10+2p)=3.60 \rightarrow 0.01p+0.20p+0.50+0.10p=3.60$. Combining like terms and solving for p: $0.31p+0.50=3.60 \rightarrow 0.31p=3.10 \rightarrow p=10$. There are $2p=20$ dimes and $10+2p=30$ nickels.

3. **280.** Multiplying the quality times the quantity: $30\%x+18\%(200)=25\%(x+200) \rightarrow 0.30x+0.18(200)=0.25(x+200)$. Solving for x: $0.30x+36=0.25x+50 \rightarrow 0.05x=14 \rightarrow x=280$. You need 280 gallons.

4. **2.4 qts. of 5% and 1.6 qts. of 15%.** Multiplying the quality times the quantity: $5\%x+15\%(4-x)=9\%(4) \rightarrow 0.05x+0.15\%(4-x)=0.09(4) \rightarrow 0.05x+0.6-0.15x=0.36$. Solving for x: $0.05x+0.6-0.15x=0.36 \rightarrow -0.10x=-0.24 \rightarrow x=2.4$. You need 2.4 quarts of 5% and 1.6 quarts of the 15%.

5. **24 quarters, 48 nickels, 67 dimes.** Let q represent the number of quarters. Then the nickels are $2q$ and the dimes are $3q-5$. Multiplying each number of coins times their worth and solving for q: $0.25q+0.05(2q)+0.10(3q-5)=15.10 \rightarrow 0.25q+0.10q+0.30q-0.50=15.10 \rightarrow 0.65q=15.60 \rightarrow q=24$. She has 24 quarters, 48 nickels, and 67 dimes.

6. **0.8 pound of pecan candies, 0.2 pound of cherry candies.** Let x represent the pounds of pecan candies and $1-x$ represent the pounds of cherry candies. Multiplying the costs times the amounts, you have: $9.00x+12.00(1-x)=9.60$. Solving for x: $9.00x+12.00-12.00x=9.60 \rightarrow -3.00x=-2.40 \rightarrow x=\frac{-2.40}{-3.00}=0.8$. You need 0.8 pound of the pecan candies and 0.2 pound of the cherry candies.

7. **11:00 a.m.** You want $d_{Harry}=d_{Cary}$. Their rates and times are different. Letting t represent the amount of time that Harry traveled, the equation representing these values is: $45t=60\left(t-\frac{1}{2}\right)$. Simplifying and solving for t: $45t=60t-30 \rightarrow -15t=-30 \rightarrow t=2$. So it was 2 hours after Harry left that Cary caught up to him; that's 11:00 a.m.

8. **30 mph.** You want $d_{Tim}=d_{Jim}$. Their rates and times are different. Let r represent the rate at which Tim traveled after the stop; and start counting the distance since the stop (and after Jim kept cycling). During the 15-minute stop, Jim kept going at 20 mph, so add that into the distance Jim traveled. The equation representing these values is: $r\left(\frac{1}{2}\right)=20\left(\frac{1}{2}\right)+20\left(\frac{1}{4}\right)$. Solving for r, $\frac{1}{2}r=20\left(\frac{1}{2}\right)+20\left(\frac{1}{4}\right)=10+5=15 \rightarrow \frac{1}{2}r=15 \rightarrow r=30$. Tim was traveling at 30 mph to catch up with Jim.

9. **6 hours.** Let t represent the number of hours the buses traveled. Add the two distances created by multiplying the rate times the time and solve for t: $40t+55t=570 \rightarrow 95t=570 \rightarrow t=\frac{570}{95}=6$. It took 6 hours.

(10) **70 mph.** The two distances form a right triangle. The distance they are apart in one hour is the hypotenuse of the triangle. Letting the rate of the helicopter be r, and creating the Pythagorean Theorem with the information: $[240]^2+[r]^2=[250]^2$. Now solve for r: $57{,}600+r^2=62{,}500 \rightarrow r^2=4900 \rightarrow r=70$. The helicopter is travelling at 70 mph.

(11) $1\frac{1}{2}$ **hours.** Letting x represent the amount of time it will take them to do the job working together, use the equation $\frac{x}{6}+\frac{x}{2}=1$. Multiply through by 6 and solve for x: $\frac{x}{6}+\frac{x}{2}=1 \rightarrow x+3x=6 \rightarrow 4x=6 \rightarrow x=\frac{6}{4}=1\frac{1}{2}$. It'll take them $1\frac{1}{2}$ hours working together.

(12) **18 hours.** Let x represent the amount of time it will take to fill the pool. The rate going in will be a positive value and the rate running out will be a negative value. When it's full, the job is done: $\frac{x}{12}-\frac{x}{36}=1$. Multiply through by 36 and solve for x:

$\frac{x}{12}-\frac{x}{36}=1 \rightarrow 3x-x=36 \rightarrow 2x=36 \rightarrow x=18$. It will take 18 hours to fill the pool.

(13) **Jack: 20, Andrew: 60.** Let k represent Jack's age. Then Andrew is $3k$. Their ages 10 years ago were $k-10$ and $3k-10$, respectively. Use the equation $3k-10=5(k-10)$ and solve for k: $3k-10=5(k-10) \rightarrow 3k-10=5k-50 \rightarrow 40=2k \rightarrow 20=k$. Jack is 20 and Andrew is 60.

(14) **3.** Let t represent the age of the triplets. Agnes is then $12t$. In four years, the triplets will each be $t+4$ and Agnes will be $12t+4$. Add up all the ages and set them equal to 61: $3(t+4)+(12t+4)=61 \rightarrow 3t+12+12t+4=61 \rightarrow 15t=45 \rightarrow t=3$. The triplets are 3 years old.

Whaddya Know? Chapter 22 Quiz

Quiz time! Complete each problem to test your knowledge on the various topics covered in this chapter. You can then find the solutions and explanations in the next section.

1. How many quarts of 80% antifreeze do you need to add to 8 quarts of 20% antifreeze to create a solution that's 60% antifreeze?

2. You have five times as many quarters as dimes, three more nickels than dimes, and two fewer than nine times as many pennies as dimes. How many of each coin do you have if they total $15.03?

3. At a local event, the cost of tickets was $15 for adults and $5 for children. The total receipts for the event were $27,500. If a total of 3,480 tickets were sold, then how many of the tickets were for children and how many for adults?

4. At a recent sale, the store owner sold twice as many $60 items as $70 items and five times as many $40 items as $60 items. If his total was $7,080, then how many of each item did he sell?

5. A train left the station at noon, traveling east. At 2:00 p.m., another train left a station traveling west on the same track as the first train and with a speed 20 mph greater than the first train. If the two stations were 1,100 miles apart, and the trains met at 9:00 p.m., how fast were they traveling? (They met, not crashed.)

6. Steve is three times as old as Sam. In 15 years, Steve will be only twice as old as Sam. How old are they now?

7. It will take Helen 6 hours to weed the garden. Jon can do the same job in just 4 hours. How long will it take them to weed the garden working together?

8. Two trains left the station at the same time — one traveling south and the other traveling east. If the southbound train was moving at 50 mph and the eastbound train at 120 mph, then how long did it take for them to be 520 miles apart?

9. Jake is 4 more years than twice Jack's age. Four years ago, Jake was 10 times Jack's age. How old is Jake right now?

10. Adam can paint the fence in 2 days. It'll take Ben 3 days to paint the fence. And Chuck needs 5 days to do the job. How long will it take them to paint the fence if they work together?

Answers to Chapter 22 Quiz

1. **16 qts.** Let x represent the quarts of 80% antifreeze. Create an equation where you multiply the qualities times the quantities: $80\%(x)+20\%(8)=60\%(x+8)$. Solve for x:

   $0.80(x)+0.20(8)=0.60(x+8) \rightarrow 0.80x+1.6=0.60x+4.8 \rightarrow 0.20x=3.2 \rightarrow x=\frac{3.2}{0.20}=16$.
   You need to add 16 quarts of the 80% antifreeze.

2. **10 dimes, 50 quarters, 13 nickels, 88 pennies.** Let d represent the number of dimes. Then $5d$ is the number of quarters, $3+d$ is the number of nickels, and $9d-2$ is the number of pennies. Multiply the monetary values times the number of each coin and create a sum equaling \$15.03: $0.10(d)+0.25(5d)+0.05(3+d)+0.01(9d-2)=15.03 \rightarrow$ $0.10d+1.25d+0.15+0.05d+0.09d-0.02=15.03 \rightarrow 1.49d+0.13=15.03 \rightarrow$ $1.49d=14.90 \rightarrow d=10$. There are 10 dimes, $5d$ = 50 quarters, $3+d=13$ nickels, and $9d-2=88$ pennies.

3. **2,470 children, 1,010 adults.** Let c represent the number of tickets for children. Then the number of tickets for adults is $3,480-c$. Multiply the ticket prices times the amount the tickets cost and find the sum: $5(c)+15(3480-c)=27,500 \rightarrow 5c+52,200-15c=27,500 \rightarrow$ $-10c=-24,700 \rightarrow c=2,470$. They sold 2,470 children's tickets and $3,480-2,470=1,010$ adult tickets.

4. **\$70: 12, \$60: 24, \$40: 120.** Let x represent the number of \$70 items. Then 2x is the number of \$60 items, and $5(2x)=10x$ is the number of \$40 items. Multiply the monetary values times the number of items and write the sum: $\$70(x)+\$60(2x)+\$40(10x)=\$7,080$. Simplify the terms and solve for x: $70x+120x+400x=7,080 \rightarrow 590x=7,080 \rightarrow x=\frac{7,080}{590}=12$. So there were 12 of the \$70 items, 24 of the \$60 items, and 120 of the \$40 items.

5. **60 and 80 mph.** Let the speed of the first train be r and the speed of the second train be r + 20. The first train traveled for 9 hours and the second train traveled for 7 hours. Multiply their rates times the time they traveled and sum the distances, making them equal to 1,100: $r(9)+(r+20)(7)=1,100 \rightarrow 9r+7r+140=1,100 \rightarrow 16r=960 \rightarrow r=60$. The first train was traveling at 60 mph and the second train at 80 mph.

6. **15 and 45.** Let Sam's age be represented by m. Then Steve is $3m$. Their ages in 15 years will be $m+15$ and $3m+15$, respectively. The equation needed is: $3m+15=2(m+15)$. Solving for m, $3m+15=2(m+15) \rightarrow 3m+15=2m+30 \rightarrow m=15$. So Sam is 15 and Steve is $3(15)=45$.

7. $2\frac{2}{5}$ **hours.** Write the equation $\frac{x}{6}+\frac{x}{4}=1$. Solve for x, the amount of time it will take them to do the job working together. Start by multiplying through by 12: $\frac{x}{6}+\frac{x}{4}=1 \rightarrow 2x+3x=$ $12 \rightarrow 5x=12 \rightarrow x=\frac{12}{5}$. It will take $2\frac{2}{5}$ hours.

8. **4 hours.** The two distances form the sides of a right triangle. The hypotenuse is the distance apart. Let t represent the amount of time it will take. Using the Pythagorean Theorem,

$(50t)^2+(120t)^2=520^2 \rightarrow 2{,}500t^2+14{,}400t^2=270{,}400 \rightarrow 16{,}900t^2=270{,}400 \rightarrow$

$t^2=\dfrac{270{,}400}{16{,}900}=16 \rightarrow t=4$. It took 4 hours for the trains to be 520 miles apart.

9. **14.** Let Jack's age be represented by k, which makes Jake's age $2k+4$. Their ages four years ago were $k-4$ and $2k+4-4=2k$, respectively. Using the equation $2k=10(k-4) \rightarrow 2k=10k-40 \rightarrow 40=8k \rightarrow 5=k$, you determine that Jack is 5 and Jake is $2(5)+4=14$.

10. $\mathbf{\dfrac{30}{31}}$ **of a day.** Using the equation $\dfrac{x}{2}+\dfrac{x}{3}+\dfrac{x}{5}=1$ and solving for x (the amount of time it will take with all of them working together), you first multiply each term by 30, the least common denominator: $\dfrac{x}{2}+\dfrac{x}{3}+\dfrac{x}{5}=1 \rightarrow 15x+10x+6x=30 \rightarrow 31x=30 \rightarrow x=\dfrac{30}{31}$.
It will take them less than a day to paint the fence!

Index

Symbols

A

B

C

D

E

F

G

L

M

Q

R

S

About the Author

Mary Jane Sterling is the author of many *For Dummies* products, including *Algebra I, Algebra II with Trigonometry, Math Word Problems, Business Math, Linear Algebra, Finite Math,* and *Pre-Calculus.* She is currently in a not-really-retired state, after forty-plus years of teaching mathematics. She loves sharing math-related topics through workshops and Zoom sessions to school-age through senior audiences. She and her husband Ted enjoy spending their leisure time with their children and grandchildren, traveling, and seeking out new adventures.

Dedication

I want to dedicate *Algebra II All-in-One* to the people of the Ukraine. They are and will be in our thoughts and prayers. Their strength and resilience are great models for any freedom-loving people.

Author's Acknowledgments

A big thank you to Chrissy Guthrie, my project editor, for pulling together all the elements of this project. This has been a big challenge, but she is, as always, up to the challenge! Thank you, also, to Michelle Hacker, my managing editor, for all her support and assistance. Another big thank you to Marylouise Wiack, my copyeditor, for her great catches and her way with words. And Amy Nicklin gets a big thanks as technical editor, making sure I have my numbers right! And, of course, I can't say thank you enough to Lindsay Lefevere, who made this project possible for me.

Publisher's Acknowledgments

Executive Editor: Lindsay Sandman Lefevere

Project Manager and Development Editor: Christina N. Guthrie

Managing Editor: Michelle Hacker

Copy Editor: Marylouise Wiack

Technical Editor: Amy Nicklin

Production Editor: Tamilmani Varadharaj

Cover Image: © Paul Aparicio/Shutterstock